# Lecture Notes in Nanoscale Science and Technology

Volume 7

Lecture Notes in Nanoscale Science and Technology (LNNST) aims to report latest developments in nanoscale science and technology research and teaching – quickly, informally and at a high level. Through publication, LNNST commits to serve the open communication of scientific and technological advances in the creation and use of objects at the nanometer scale, crossing the boundaries of physics, materials science, biology, chemistry, and engineering. Certainly, while historically the mysteries in each of the sciences have been very different, they have all required a relentless step-by-step pursuit to uncover the answer to a challenging scientific question, but recently many of the answers have brought questions that lie at the boundaries between the life sciences and the physical sciences and between what is fundamental and what is application. This is no accident since recent research in the physical and life sciences have each independently cut a path to the edge of their disciplines. As both paths intersect one may ask if transport of material in a cell is biology or is it physics? This intersection of curiosity makes us realize that nanoscience and technology crosses many if not all disciplines. It is this market that the proposed series of lecture notes targets.

Nirmal Mazumder • Rajib Biswas
Guan-Yu Zhuo
Editors

# Advanced Biophysical Techniques in Biosciences

Springer

*Editors*
Nirmal Mazumder
Department of Biophysics
Manipal School of Life Sciences
Manipal Academy of Higher Education
Manipal, Karnataka, India

Rajib Biswas
Department of Physics
Tezpur University
Tezpur, Assam, India

Guan-Yu Zhuo
Institute of Biophotonics
National Yang Ming Chiao Tung University
Taipei, Taiwan

ISSN 2195-2159 ISSN 2195-2167 (electronic)
Lecture Notes in Nanoscale Science and Technology
ISBN 978-3-031-94550-2 ISBN 978-3-031-94551-9 (eBook)
https://doi.org/10.1007/978-3-031-94551-9



This Springer imprint is published by the registered company Springer Nature Switzerland AG
The registered company address is: Gewerbestrasse 11, 6330 Cham, Switzerland

# Contents

# Advanced Biophysical Characterization for Biopolymer Based Drug Delivery Systems and Polymeric Scaffolds for Tissue Engineering

**Shakila Nishadani, E. G. Senali Kanchana, Hemanth Noothalapati, and Thennakoon M. Sampath Udeni Gunathilake**

**Abstract** Biopolymers are emerging as interesting materials in the context of controlled drug delivery systems and tissue engineering applications. In any case, their physicochemical properties should be known for ensuring their performance is maximized. Some of the methods, like FTIR spectroscopy, SEM, HPLC, compression tests, DMA, DSC, TGA, Raman spectroscopy, etc., are efficiently used for understanding the mechanical and chemical properties of biopolymer systems. Such techniques help assess the properties such as degradation, drug loading, drug release profile, porosity and tensile strength. An additional understanding of these properties facilitates the creation of appropriate and specific drug delivery systems as well as scaffolds for tissue repair. These synergistic methods improve the functional advancement of biopolymers in biomedical applications in terms of structure–property relationship.

S. Nishadani · E. G. S. Kanchana · T. M. S. U. Gunathilake (✉)
Department of Polymer Science, Faculty of Applied Science, University of Sri Jayewardenepura, Nugegoda, Sri Lanka
e-mail: sgunathilake@sjp.ac.lk

H. Noothalapati (✉)
Department of Biomedical Engineering, Chennai Institute of Technology, Chennai, Tamil Nadu, India

Department of Chemical Engineering, Indian Institute of Technology Hyderabad, Kandi, Sangareddy, Telangana, India

Faculty of Life and Environmental Sciences, Shimane University, Matsue, Japan
e-mail: nvhnag@life.shimane-u.ac.jp



N. Mazumder et al. (eds.), *Advanced Biophysical Techniques in Biosciences*, Lecture Notes in Nanoscale Science and Technology 7,
https://doi.org/10.1007/978-3-031-94551-9_1

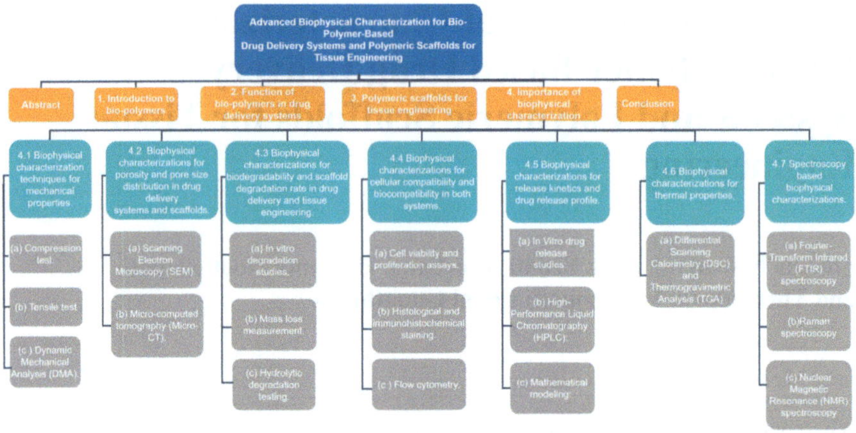


**Keywords** Polymer · Drug delivery · Tissue engineering · Biophysical characterization

## 1 Introduction to Biopolymers

Biopolymers are polymers composed of multiple monomers or units derived from living organisms, primarily plants and animals, rendering them both biodegradable and renewable for various applications. These materials are predominantly synthesized within the biosystems of organisms. Biopolymers serve numerous critical functions within the body, including: the formation of cellular tissues; support for cartilage and other types of connective tissue; and the provision of molecules that act as signaling factors in the human endocrine system [1]. The linkage of the monomers creates polymers, contributing to the structural integrity of these biopolymers. Common examples include nucleic acids such as DNA and RNA, polypeptides or proteins formed from amino acids encoded in nucleic acids, and polysaccharides, which are complex carbohydrates. These molecules play a crucial role in biological systems that maintain and regulate biological functions and systems. Furthermore, while biopolymers such as nucleic acids mainly store and transmit genetic information, proteins and polysaccharides drive cellular processes and maintain structural regularities [2].

Biopolymers not only fulfill various biological roles but also possess exceptional properties such as biodegradability, biocompatibility and the ability to replicate the natural biological environment, making them highly suitable for medical applications, including drug delivery systems (DDS) and tissue engineering. Their biodegradability allows for gradual breakdown over time, rendering them ideal for drug delivery systems and tissue engineering scaffolds that can be absorbed by the body, thus eliminating the necessity for surgical removal after their function. Their biocompatibility ensures safe interaction with biological tissues. Furthermore, biopolymers can undergo chemical modifications that enhance their affinity towards drugs or cells, facilitating controlled drug delivery or improved cellular proliferation. Additionally, their multifunctional structures enable them to be fabricated into films, fibers, hydrogels and nanoparticles, which are valuable for drug delivery and other medical applications such as tissue engineering.

## 2 Function of Biopolymers in Drug Delivery Systems

Recent research and advancements in biopolymers are facilitating the formulation of various biopolymer-based drug systems, which offer enhanced therapeutic benefits, increased patient acceptability, and reduced drug costs. In the case of drug delivery, a number of reinforcing biodegradable polymers have been used because these biopolymers undergo biological degradation and break into non toxic substances in the organism. Drug delivery systems are various methods for transporting drug candidates to predetermined locations in the body with controlled kinetics. A drug delivery system is defined as any system in which a drug entrapped within the system is released and exerts a suitable therapeutic effect [3]. To attribute controlled drug delivery and release, these drug delivery systems can be made from biodegradable and biocompatible materials, ensuring safe and efficient drug delivery. Drug delivery systems to be used include: microspheres, microcapsules, nanoparticles, polymeric micelles, dendrimers, beads, tyrosinase-containing composite hydrogels, patches, films, sponges, and scaffolds. These systems are designed in a way that reduces toxicological effects, improves drug stability and bioavailability [4].

The biopolymeric carriers are mainly used to transport drugs to the specific sites of action and at the same time afford drug protection due to their biocompatibility and biodegradability. These properties are essential for the biopolymers to exhibit nontoxicity for the therapeutic application. For instance, the poly(lactic-co-glycolic acid) (PLGA) forms the core of the majority of synthetic biopolymers incorporated in DDS because it breaks down to lactic acid and glycolic acid found naturally in the body. In another study by Zhang et al. (2017), PLGA microspheres were employed in the entrapping of paclitaxel, a chemotherapeutic agent which would only be released slowly over an extended period in the body, reducing the need for frequent dosing [5].

Another advantage of biopolymers when used in DDS is their ability to deliver a drug at the desired site and release it at a controlled rate over a certain period of time. Chitosan is a biopolymer that can produce hydrogels that experience swelling when in contact with water, hence altering the rate of drug release. Immobile anti-inflammatory drugs have been formulated from chitosan hydrogels to provide a controlled release character ideal for long-term diseases like rheumatoid arthritis. Venkatesan et al., 2018 also found that chitosan hydrogels were capable of releasing drugs such as diclofenac sodium in a sustained manner for a number of hours, thereby helping patients to feel the therapeutic impacts without requiring constant dosing [6].

Targeted drug delivery is one of the most significant benefits of biopolymers, as it allows for drugs to be delivered specifically to the disease site (e.g. tumors, infected tissue) while sparing healthy tissues. This reduces unwanted side effects and improves the therapeutic outcome. For targeted delivery, biopolymers are often functionalized with ligands or antibodies that recognize and bind to specific receptors overexpressed on target cells or tissues.

The encapsulation of drugs within biopolymeric carriers serves multiple functions: it shields sensitive drugs from degradation, enhances stability, and enables the drug to release at the site of action to be desired. This is especially accurate for proteins, peptides, and weakly soluble active agents. Cyclodextrins are cyclic oligosaccharides that have been extensively used for enhancing the solubility of water-insoluble drugs. Nishadani et al. (2024) developed curcumin-β-cyclodextrin complexes with carboxymethyl cellulose—fumaric acid hydrogel, significantly improving the hydrophobic drug's solubility and bioavailability, making it more effective in treating cancer [7]. These complexes can assist in the encapsulation of the hydrophobic drug molecules within their hydrophobic cavity, leading to improved dissolution in the aqueous phase.

Intelligent or stimuli-sensitive biopolymers in a smart system alter their characteristics in response to the stimuli, like temperature, pH, or light, and therefore allow precise targeting of drugs. These systems are especially useful when the drug must be delivered to certain organs or tissues, such as tumour or inflamed tissues, the environment of which is different from that of the healthy tissues.

## 3 Polymeric Scaffolds for Tissue Engineering

Polymeric scaffolds are critical in tissue engineering because they serve as a supporting structure for these cells while at the same time enhancing the growth and differentiation of tissues. These scaffolds simulate the shape and structure of the natural tissues' extracellular matrix (ECM), and promote the development of new tissues in vivo. Bio polymers, both natural and synthetic, have been preferred for scaffold fabrication due to their inherent biologic compatibility, degradation and possibility to be assessed for virtually all tissue remodelling requirements [8].

Polymeric scaffolds are fundamental in regenerative medicine, wound healing, and organ repair, to reinstate the function of the altered tissues and organs. Since the polymeric scaffolds are often exposed to an in vivo environment, they are also expected to have acceptable cell adhesiveness, support cell material interactions, demonstrate structural and mechanical stability, and have an appropriate microstructure that will allow tissue integration [9].

The design and development of polymeric scaffolds and their structure are a significant determinant of the result of tissue engineering. Evaluation of pore size, porosity, and interconnectiveness of the scaffolds is critical because the scaffold system needs to facilitate nutrient and oxygen diffusion as well as waste removal and cell migration. Pore size is commonly between 50 and 500 μm [10], smaller pore size may prevent cell infiltration and larger pore size may show poor mechanical stability. Scaffolds' fabrication methods involving electrospinning, three-dimensional printing, and freeze drying have also been considered to achieve better pore-size characteristics and mechanical requirements.

Electrospinning is one of the most predominant scaffolds' fabrication methods, which creates nanofibers that relatively mimic the functions and structures of the extracellular matrix (ECM). Among the fabrication techniques, electrospinning stands out because of its relatively easy technique of producing nanofibers that mimic components of the ECM. Due to the ability to replicate the properties of natural tissues, polymeric nanofibers have attracted significant attention in tissue engineering. Nanofibers from polymers like PLA, PCL and chitosan can produce porous structures of the nanoscale endowing the matrix with improved cell adhesion and proliferation [11]. A newer technology of the fabrication of tissue scaffolds is through 3D printing, otherwise called additive manufacturing, which provides the opportunity of constructing scaffolds that can fit into the specific tissue regeneration application. The use of biomaterials such as PLGA or GelMA enables scientists to fabricate scaffolds with appropriate mechanical characteristics and pore dimensions that resemble native tissues [12].

Polymer characterization is critical in identifying their properties in the areas of drug delivery and tissue engineering. It assists in ensuring that these materials are capable of releasing drugs effectively, compatible with the body tissue and capable of stimulating tissue regeneration.

## 4 Importance of Biophysical Characterization

The most important considerations in the development of various biopolymer-made biomedical applications are the choice of the biopolymer(s), which require knowledge on the surface and bulk properties of the biopolymers that can be functional in the construct of drug delivery systems and tissue engineering to achieve the best therapeutic result. As with most substances, the right properties are crucial for success, in this case, therapeutic success in the particular fields. These properties relate to the ability of materials to engage with biological systems, whether for the enhancement of drug delivery systems or tissue engineering applications. Therefore, through various characterization methods, the properties of biopolymers can be evaluated and altered depending on the needs of the drug delivery system and tissue engineering.

Mechanical, chemical and biophysical properties of the biopolymer matrices play a very crucial role in determining the release profile of the encapsulated drug in drug delivery systems. For instance, in controlled drug delivery systems, which can use PLGA or chitosan as a polymer that must degrade over time and release the drug slowly. The degradation of these polymers must be controlled to achieve a specific rate of drug release, not releasing the drug too early or too late. In the same way, tissue engineering requires scaffolds that should be able to support the regenerating tissue. They should replicate not only the mechanical characteristics of native tissue but also withstand physiological stresses, should facilitate cellular integration and tissue regeneration. For instance, scaffolds for bone tissue engineering technology require a high compressive modulus for the support of bone tissue formation, while scaffolds for soft tissue—like skin and cartilage—require more flexibility.

## *4.1 Biophysical Characterization Techniques for Mechanical Properties*

### 4.1.1 Compression Test

Compressive test is being of the basic methods that are applied in determining the mechanical characteristics of materials, especially biopolymer-drug delivery systems and polymeric scaffolds for tissue engineering. These tests determine the reaction of the material under force in mechanical testing to give information on strength, flexibility, or even toughness as well. These properties are important in the context of a material's ability to perform in biological settings; within these settings, a material can become subjected to various levels of stress and strain during drug delivery or tissue engineering [13].

In this test, a sample is mounted between two parallel plates, where a load is exerted along the vertical direction, reducing the thickness of the material. It is carried out to the point where the material has either been deformed or failed, and the various mechanical characteristics, including compressive strength, modulus and deformation, are determined.

In tissue engineering, and particularly in hard tissue applications such as bone or cartilage regeneration, the scaffolds must provide mechanical properties appropriate for the loading conditions experienced at the specific application site. For instance, in the case of bone scaffolds, they have to endure body loads and possess adequate mechanical strength during bone tissue formation. Compressive testing assists in realizing tissue-engineered polymeric scaffolds sufficiently strong to support cellular metabolic processes as well as the formation of tissue that should elicit a similar mechanical response to that of engineered tissues.

The method can also be applied to drug delivery systems that employ hydrogels or other polymer-based carriers. For instance, hydrogels intended to deliver a drug in a controlled fashion could be subjected to compressive loads in an application such as injection or implantation. The mechanical properties of those materials give them the capacity to maintain the physical integrity through such processes while preventing the early release of the drugs present.

Compressive strength is the extent of stress that a material may be subjected to without cracking or breaking. When selecting materials for structural applications, it is crucial to determine whether they can withstand crushing or squeezing forces. It indicates the material ability to maintain its integrity under stress [13]. The mechanical requirement for bone scaffold is high compressive strength is necessary because tissue regeneration is occurring in a structure that should not crumple down.

Elastic modulus or compressive modulus is the measure of the stress–strain relation up to the elastic limit of the material, showing how stiff it is. Material resistance to non-permanent deformations is demonstrated by the elastic modulus, and these parameters help to select suitable materials for specific mechanical requirements and to design structures. While a material with a high compressive modulus is rigid and withstands deformation when a load is passed on to it, on the other hand material with a considerably low modulus is considered a flexible material [14].

Strain behavior demonstrates the ability of the material to stretch before it feels permanent strain. For tissue engineering, the most suitable materials are required that have strain limits that allow the cells to grow on them without depolymerizing or collapsing.

Compressive testing is normally done with a universal testing machine, where the load is applied in compression to the material and its reaction is recorded. Load and displacement data are recorded and applied for the determination of the compressive strength and compressive modulus. For the softer materials, such as hydrogels, or when practicing thicker layers of scaffolds, indentation tests are more common. Then a very small needle-like penetrator is applied on the surface of the material with a focus on assessments of hardness and stiffness while emulating pressures in a compressive sense on a micro scale [15].

#### 4.1.2 Tensile Test

Tensile testing measures the response of a material to stretching or pulling force through to the respective mechanical property. Typically, this test is used to determine materials' tensile strength, strain hardening characteristics, Young's modulus and Poisson's ratio. It is crucial to test materials' tensile strength to determine materials quality, suitability and to select materials for various applications, ranging from plastics to metals. A sample is ordinarily made to extend under a uniaxial tensile load, and the stress–strain diagram that is obtained yields details on strength, elasticity, and ductility of the material [16].

Tensile testing is usually applied for the determination of flexural properties as well as mechanical performance of drug delivery matrices, which can be thin films or fibres. For instance, films or fibers in the transdermal drug delivery systems should possess a certain level of flexibility and mechanical strength to load the right dose of drugs for delivery while holding the structure intact. Tensile testing is especially useful in cases of soft tissue scaffolds such as skin, cartilage, or muscle tissue engineering. Muscle tissues such as skin and cartilage are viscoelastic, therefore their scaffolds must be able to withstand mechanical forces during cellular growth and tissue remodeling [17].

For instance, scaffolds employed in ligament or tendon tissue engineering require to mimic the native tissue mechanics features for instance, tensile strength sufficient to withstand stretching force and strain during certain activities.

Tensile Strength shows the maximum load or force that the material can handle without failure. Indeed, this is critical for scaffolds designed to support load-bearing tissues such as skin, ligaments, and tendons. Young's modulus or elastic modulus is defined as the ratio of stress experienced to the strain developed at the time when the material obeys Hooke's law of linear stress–strain relation [17]. A high tensile modulus is considered to make the material stiff, while a low value represents a more flexible material, which is important in developing soft tissues. Ductility measures the capacity of the material to be elongated before it breaks, as captured by the ultimate elongation (ductility). Where the scaffold is required for soft tissue regeneration (e.g., skin tissue), the materials should have higher levels of ductility to allow for tissue growth, as well as constricting in another context. Tensile stress–strain curve describes the stress–strain curve produced from the test, giving more information, like elastic zone, yield point and elastic limit point.

Standard tensile test can be carried out in a universal testing machine, the sample for the test is placed in a clamp and is pulled at a constant rate, force and elongation are measured. The sample sizes and shapes used in the tests and the pulling rate should be used according to the standards. The test produced a stress-strain curve of the material's mechanical properties, which can be extracted via analysis [18].

### 4.1.3 Dynamic Mechanical Analysis (DMA)

In tissue engineering, the stiffness of a construct could be altered depending on the biomechanical properties of the tissue it is intended to replace. The dynamic mechanical analysis (DMA) is one of the most effective thermal analysis techniques, which makes it possible to determine phase transformations and relaxation processes in diverse types of materials [19]. This technique presents the possibility to evaluate the dependence of the solid-state rheological properties of viscoelastic materials on temperature and frequency.

DMA utilizes an oscillatory (reciprocal) mechanical load for a sample as well as evaluating the sample's reaction to this load. This oscillatory deformation is usually applied in terms of stress or strain and the material's behavior is described in terms of the storage modulus, the real or elastic component and the loss modulus; the imaginary or viscous component.

Storage modulus (E′) represents the material's ability to store elastic energy. It is related to the stiffness of the material and indicates the solid-like behavior of the polymer. Loss modulus (E″), represents the material's ability to dissipate energy as heat and is associated with the material's viscosity or damping behavior, indicating how "rubbery" or "fluid-like" the material is. Tan delta (tan $\delta$) is the ratio of the loss modulus to the storage modulus (E″/E′). It provides insights into the material's damping characteristics and helps to identify the material's glass transition temperature ($T_g$), which is critical for understanding the thermal stability and flexibility of polymers [20].

## *4.2 Biophysical Characterizations for Porosity and Pore Size Distribution in Drug Delivery Systems and Scaffolds*

In both drug delivery and tissue engineering, the scaffold porosity plays a crucial role in material functions. In controlled release drug delivery systems, the permeability characteristics of the polymer matrix decide how the drug comes out of it over time. The highly porous scaffold could facilitate faster diffusion of a drug, while a less porous scaffold could slow down the rate at which the drug is released, depending on the intended therapeutic effect.

In tissue engineering, pore characteristics, specifically the pore size and distribution, are therefore very significant in terms of cell migration, nutrition exchange and revascularization. For example, in bone tissue engineering scaffold with a high level of porosity can allow the infiltration of osteoblasts for the deposition of the bone matrix and new tissue formation and remodeling. In addition to that, pore sizes and porosity of scaffolds affect to direct cellular responses and alter the mechanical properties of scaffolds [21].

### 4.2.1 Scanning Electron Microscopy (SEM)

Enables high-resolution of the scaffold surface and cross-sectional view, which is essential for the determination of the pore size required for the rate at which drugs diffuse out of drug delivery systems or the extent to which cells penetrate through the tissue engineering scaffolds. Through applying SEM, clear and detailed images of the sample are obtained and so other parameters, involving pore size, porosity, the surface roughness of the scaffold and other microstructure factors significant in the evaluation of drug delivery systems can be accurately determined.

In SEM, the secondary electron imaging technique uses a tightly focused beam of high-energy electrons that is passed over the sample surface. These electrons collide with the atoms present in the material during scanning and emit secondary electrons, which are used to form an image. The sample is usually made conductive if it is nonconductive by applying a thin layer of conducting material on the surface of the sample to avoid effects induced by charging during scan [21]. SEM is capable of functioning at different resolution levels, showing the surface of the samples at micrometer and in some cases even at the nanometer level.

### 4.2.2 Micro-Computed Tomography (Micro-CT)

Allows for visual imaging of the internal arrangement of the scaffold, as well as, the orientation and interconnectedness of pores. This is critical when it is necessary to assess how a given scaffold will facilitate cell movement in tissue engineering or how drug molecules will disperse in a drug delivery system. Micro-computed tomography (Micro-CT) is an excellent tool for the visualization of internal structures of material cross sections with sub-microscopic resolution without necessarily destroying the sample [22]. This technique offers a complete picture of the pore architecture and distribution, porosity and connectivity of scaffolds for tissue engineering and drug delivery, and allows for evaluation of how the scaffold will perform under biological conditions.

Micro-CT is based on taking a series of 2-D cross-sectional images (sections) when the X-ray beam is scanned through a sample at different angles. This "Micro-CT" provides image rotation, quantitative imaging parameters, and virtual "slicing" of specimens and these images could be rotated in all directions and cross sections of internal portions of specimens could be visualized from any angle. These images can then be fed into computational algorithms to perform a reconstruction of the sample [23]. Micro-CT can be as precise as micrometers, showing high-quality of the inner structure of the scaffold without destroying the sample.

## *4.3 Biophysical Characterizations for Biodegradability and Scaffold Degradation Rate in Drug Delivery and Tissue Engineering*

The rate of degradation of biopolymer matrices is important in both drug delivery services and tissue engineering. In drug delivery systems, such biopolymers as PLGA are adjusted to degrade at the desired rate, commonly corresponding to the rate of drug release. If degradation is too rapid, the burst release of the drug might occur, causing toxicity or reduced therapeutic effectiveness. If degradation is slow, on the other hand, it may mean that drug release is slow, therefore, having low therapeutic effectiveness.

Degradation is as critical in tissue engineering as the scaffold itself. As new tissue forms, the scaffold must break down in a manner that promotes the formation of tissue without creating a residue that is left behind. For instance, in cartilage regeneration, the scaffold, has been designed to degrade as the cartilage forms to minimize chronic foreign body reactions.

### 4.3.1 In Vitro Degradation Studies

When the drug-loaded scaffolds or tissue engineering scaffolds are placed in a physiologically relevant medium, the degradation rate can therefore be monitored quantitatively using weight loss, changes in pH and morphology.

### 4.3.2 Mass Loss Measurement

Weight measurements of scaffolds before and after degradation tests indicate useful information toward the degradation kinetics of the polymer, which is very critical in drug delivery as well as tissue engineering applications.

#### 4.3.3 Hydrolytic Degradation Testing

Specifically for synthetic polymers such as PLGA, assessing the degradation of the scaffold under water conditions can be highly useful in determining the rate of degradation of the scaffold, hence being useful information on controlling drug release and scaffold performance. This methodology enables the sequential screening of degradable polymers at physiological conditions and has the potential to be used for the material's preclinical evaluation [24].

### *4.4 Biophysical Characterizations for Cellular Compatibility and Biocompatibility in both Systems*

Specifically for tissue engineering scaffolds, the respective material has to promote cell adhesion, growth, differentiation, and formation of the desired tissues. A scaffold must also be biocompatible, that is, the scaffold should not trigger any immunity or toxicity responses as it is expected to create a suitable culture for cell tissue regeneration. For instance, collagen or gelatin-based bio polymers are very biocompatible in nature and encourage the cell to adhere and proliferate, making them suitable for application in cosmetic surgery, such as treatment of burns, formation of skin and bone grafts, etc.

In terms of drug delivery systems, this compatibility is also paramount since the polymer should not cause an immune or toxic response while encapsulating or delivering the drug. Chitosan and PLGA are preferred polymers for drug delivery purposes because of their low toxicity, and their degradation products are harmless to the system. However, stability, compatibility of the material with the drug, and potential effects on the efficacy of the drug must be analyzed.

#### 4.4.1 Cell Viability and Proliferation Assays

Evaluating the biocompatibility of the scaffolds and drug delivery systems is crucial in bioengineering to evaluate their suitability for continuous functioning and supporting the cell growth of biological systems. Cell adhesion, proliferation, and survival in these systems are assessed via numerous viability assay tests and cytotoxicity analysis. In drug delivery and tissue engineering, MTT, Alamar Blue, and Live Dead staining are some of the tests employed to show the ability of the cells to grow and survive on the scaffold or in the drug delivery system [25].

#### 4.4.2 Histological and Immunohistochemical Staining

These techniques are used to assess cellular behavior and ECM production in tissue engineering scaffolds, or to monitor any adverse effects in drug delivery systems. This histological staining involves using dyes in order to achieve microscopic visualization of cell morphology, ECM organization and tissue structure. Furthermore, analysis of cellular activity and disease markers is done by the immunohistochemical (IHC) staining technique, which uses specific antibodies. They can show whether the cells produce specific proteins or extracellular matrix components, indicating proper tissue formation [26].

### 4.4.3 Flow Cytometry

In drug delivery, flow cytometry can be used to analyze how drug-loaded scaffolds affect specific cell populations, which helps determine the therapeutic effectiveness and safety of the system. This technique facilitates the quantitative analysis of cell characteristics (Size, protein expression) by antibodies or dyes that are labeled with fluorescence. Therefore, it analyzes the cellular uptake of drug-loaded nanoparticles and immune responses in drug delivery systems [27].

## 4.5 *Biophysical Characterizations for Release Kinetics and Drug Release Profile*

To exert the desired therapeutic efficacy, in drug delivery systems, it is mandatory to manage the drug release kinetic profiles. Characterization of biophysical properties enables scientists to explore the release kinetics and to ensure that the drug is released systematically. For instance, the hydrophilic or hydrophobic character of the polymer can affect the release of the drug and knowledge of the relationship between the drug and the polymer matrix is crucial for controlling the rate of release. In vitro drug release profiles give information on how quickly and how much drug is released over time, using techniques such as sustained-release drug delivery systems, which minimize side effects and enhance therapeutic effects.

### 4.5.1 In Vitro Drug Release Studies

These studies are performed by immersing the drug-loaded scaffold in a buffer solution and monitoring the amount of drug released over time. By adjusting factors like polymer composition, pore size, and degradation rate, researchers can fine-tune the release profile.

### 4.5.2 High-Performance Liquid Chromatography (HPLC)

Because HPLC is a significant method of determining the concentration of drugs formed in biological and pharmaceutical matrices, it is an indispensable technique for drug release evaluations of biopolymer-based drug delivery systems. HPLC is used to quantify the time point drug concentrations thus offering a rich understanding of the kinetics of drug release from the polymeric delivery system.

HPLC is a chromatographic technique that separates compounds depending on their tendency to interact with a stationary phase (column packing material) and mobile phase (eluent or solvent or mixture of solvents). When the sample is being eluted through the column with high pressure, all the individual in the sample has different affinities with the stationary phase. The amount of time that each component takes to pass through the column is called the retention time and is the basis for identifying and measuring the components [28]. UV-Vis or fluorescence monitoring is employed as a detector for the components as they come out of the column, and quantifies the constituents.

Analysis of the concentration of the drug released during dissolution studies is highly dependent on HPLC. Since the portions of a solution give the concentration of the drug in it, it enables the creation of a time-concentration profile in research. It is required for the characterization of how the drug is liberated from the polymeric delivery system over time and is a critical requirement for assessing the efficacy of the drug release system [29].

Besides the determination of drug concentrations, HPLC can control the chemical stability of the drug during the course of release. Ideally, it could identify any degradation products or changes that occur in the drug's structure, which would help in determining the stability of the drug. The data obtained from HPLC can also be used to draw the profile of release kinetics of the drug by non-linear regression analysis and by entering the data into mathematical models like zero order, first order, Korsmeyer-Peppas, etc. [30]. Such models assist in determining the release mechanism, including diffusion-dependent, swelling-dependent, or degradation-dependent release, all of which are significant for designing and developing drug delivery systems.

This approach is considered to be very sensitive, together with a high level of accuracy, so it becomes very useful, especially when the concentration of the drug in a particular release medium happens to be very low. This factor makes it sensitive to detect small variations in drug concentrations with time. Also, for in vitro and vivo drug release studies, HPLC can be used for analyzing various types of samples such as aqueous release media, biological fluids and drug delivery systems. These features, when compounded with the high precision and reproducibility of HPLC, render this technique a very important tool in studying and manipulating drug release kinetics in biopolymer-based drug delivery systems.

### 4.5.3 Mathematical Modeling

By regression of experiment data to target release models such as zero order, first order or Higuchi models, investigators are able to extrapolate future release characteristics of drugs from polymeric matrices.

## 4.6 Biophysical Characterizations for Thermal Properties

The thermal behavior of biopolymers includes their melting point, glass transition temperature or $T_g$ and the thermal stability of the associated polymers and these are critical factors in predicting the response of a polymer under physiological conditions. For a drug delivery system, thermal stability guarantees that the polymer matrix will withstand the environment of the system until it reaches the target area. In tissue engineering, thermal properties are important in determining the applicability of the scaffold during electrospinning or in 3D printing.

Also referred to as the vitreous transition temperature, the glass transition temperature or $T_g$ is the temperature at which a polymer changes from a rigid, fracture-prone glassy state to an elastic one. This property is very important, especially in the application of the material in the delivery of drugs, since it defines the flexibility and thermal characteristics of the material. The melting temperature ($T_m$) designates the temperature of the change of state of a crystalline polymer from solid to the liquid state. The present property is most pertinent to biodegradable polymers; Tm has a direct connection with the rate of material degradation inside the body. In a drug delivery system if the $T_m$ is low, the polymer may release the drug rapidly, and if the $T_m$ is high, the drug may be released in a controlled way over a long period of time. Thermal stability of a material is the ability of that material to maintain its performance characteristics when exposed to heat. This is important in both drug delivery systems and tissue scaffolds since exposing the material to high temperatures in storage or during implantation could alter the biocompatibility of the material as well as its drug release rate. There is the thermal degradation of the polymer, which is whereby the polymer is exposed to high temperatures, so that its molecular structure changes. This may have an impact on the scaffold's mechanical characteristics and the whole release behavior of the drug.

### 4.6.1 Differential Scanning Calorimetry (DSC) and Thermogravimetric Analysis (TGA)

This technique determines the temperature range over which specific heat capacity occurs and gives information on the thermal stability of the polymer for drug delivery or for tissue engineering applications. In the case of drugs and drug delivery systems, DSC can tell how the drug interacts with the polymer to change its thermal characteristics of the system, including whether the drug loading or encapsulation process has impacted the material in any way. For scaffolds, DSC can understand how the incorporation of bioactive agents into the polymer material matrix alters the thermal behavior and degradation rates of the scaffold [31].

Thermal analysis in general and DSC in particular determines the heat flow into or out of the sample while heating or cooling, with reference to another sample that is usually a reference material, typically aluminum. Here, the heat flow information reveals changes in thermal events, including heat melting, glass transition, and crystallization temperatures of the material under investigation. If the sample undergoes a phase change, e.g., melting or undergoing a glass transition, then more or less heat will be absorbed/released during this transition and this is shown on the DSC curve by a peak; or if heat is released by the sample then it appears as a trough. DSC is used in analyzing the thermal properties of polymers and deciding the thermal stability, crystallinity and compatibility of polymers with drugs and other additives.

Thermogravimetric Analysis (TGA) is an analytical technique that measures the change in mass of the specimen when exposed to a temperature varying program. The mass loss is presented as a function of temperature, and the results obtained serve in the assessment of thermal stability, degradation mechanisms, and composition of the materials [32]. TGA presupposes that in the process of heating, components which are chemical solvents, plasticizers or sometimes decay products will evaporate or break down into elements at a specific temperature and the loss in mass is extremely valuable information on the material's behavior under heat treatment.

DSC and TGA are powerful tools for identifying the thermal behaviors of biopolymer-based materials employed in pharmaceutical carriers and tissue engineering scaffolds. Using DSC, investigators gain information about polymer thermal properties, crystallinity, and interactions with drugs, and can adjust material properties for therapeutic delivery and tissue restoration. On the other hand, TGA assists in determining the stability of the material at higher temperatures, the manner in which it degrades, and the quantity that is left behind so that the polymers remain functional and perform optimally under physiological conditions. Combined, these techniques offer a thorough insight into the thermal response and thermal stability of the material and these are critical components in the success of these biomedical uses [33].

## *4.7 Spectroscopy-Based Biophysical Characterizations*

Common analytical techniques of biomaterials include spectroscopy, with significant applicability for biopolymers employed in drug delivery systems and tissue engineering. These techniques offer important qualitative and quantitative data about the chemical makeup, functional groups, molecular configuration and compatibility of the materials under study. Some of the spectroscopic methods used in the characterization of polymeric material are described below.

### 4.7.1 Fourier-Transform Infrared (FTIR) Spectroscopy

Fourier-transform infrared (FTIR) spectroscopy is an analytical procedure employed for the determination of functional groups and chemical bonds originating in a selected sample through the evaluation of the degree of absorption of the infrared radiation within a specific range of wavelengths. The sample is exposed to infrared light, and the frequency with which the sample absorbs these wavelengths is equal to the vibrational frequency of certain chemical bonds (C–H, N–H, C=O etc.) [34]. The obtained FTIR spectrum is a graph that shows the relations between the absorption intensity and the wavelength (or wavenumber) with the details about the molecular structure of the material.

FTIR is employed in the characterization of drug delivery systems as well as functional groups of polymeric scaffold tissue. For instance, in natural polymers collagen or chitosan, FTIR can identify the amide, hydroxyl and amino groups within the sample. This is very important in order to know how the polymer shall react with the drugs, cells and other biological elements [35]. In drug delivery, FTIR is used to identify that the drugs are encapsulated in a polymer matrix and to evaluate interactions between drugs and polymers that may influence release characteristics.

FTIR can also be utilized to draw a correlation between the changes in the polymer's spectrum, thereby detecting molecular interaction upon the addition of a drug. For example, variation in order or alteration of the height of the peak can point to hydrogen bond, electrostatic interaction, or formation of a complex of the drug and polymer. This is especially so in the case of controlled-release systems, since they establish how the rate of release of the drug from the polymer matrix is affected.

FTIR spectroscopy enables quantitative assessment of chemical stability and deterioration of polymeric scaffolds. Because functional groups can be observed to change over time/with some conditions, researchers can assess how scaffold material changes chemically in order to support tissue stability/regeneration.

### 4.7.2 Raman Spectroscopy

Raman spectroscopy is a non-destructive vibrational spectroscopic technique involving analysis of reflection, transmission or scattering of light. When a sample is illuminated by monochromatic light, typically a laser, most of the light is scattered at the same wavelength; however, a portion of the light shifts to a different wavelength through interactions with the molecular vibrations of the sample [36]. This frequency shift gives an idea of how bonding and functional groups exist in the material.

Raman spectroscopy is closely associated with the chemical nature of materials and is applied for the investigation of polymer and drug structures. In polymeric scaffolds, it can help determine the extent of cross-linking, types of functional groups and chemical composition, therefore, give an indication of how the scaffold will degrade or react with cells. In the drug delivery systems, Raman can determine the drug loading into the polymer matrix and its release profile through molecular linkages [37].

Raman spectroscopy is applicable when examining the degradation of a polymeric scaffold, as the Raman spectra will be progressively different over the scaffold degradation time. For instance, Raman can track over time the loss of material properties in biodegradable polymers such as PLGA, where the polymer responds to hydrolytic degradation when exposed to moisture or other body fluids [37]. This is important in the development of scaffold systems that disintegrate in a pacing pattern with new tissue formation.

Raman spectroscopy is non-invasive, and samples do not need to be removed from the instruments for analysis, making this technique suitable for studies of live cell-scaffold interactions. Raman can engage in the examination of cell infiltration and behavior within scaffolds, as well as the formation of tissues and compatibility in real time.

### 4.7.3 Nuclear Magnetic Resonance (NMR) Spectroscopy

Nuclear magnetic resonance (NMR) spectroscopy is a technique that profiles the magnetic properties of atomic nuclei to produce information about the structures and dynamics of substances. A sample is placed in a strong magnetic field, and when the sample is irradiated with radiofrequency radiation, the nuclei emit energy at characteristic frequencies, which are then picked and used. The obtained spectra provide information on molecular structure, functional groups, as well as the mode of molecular motion [38]. NMR spectroscopy is particularly applied in the study of the structure of both synthetic and natural polymers. But it can furnish precise quantitative data regarding the monomer units, the status of polymerization and the functional groups that exist in the polymer chain. When it comes to the development of drug delivery systems, NMR may determine the atomic structure of polymers contained in delivery and confirm their purity.

In the drug delivery system, NMR is particularly good at helping to investigate the interaction between polymers and drugs. Chemical shifts and coupling constants information of nuclei in the NMR spectrum allows the researchers to study the process of the interaction of drugs with the polymer matrix. For instance, a change in the NMR spectrum due to hydrogen bonding or complexation of the drug with the polymer can be used to predict the drug release [38].

However, occasionally the technique can be used in vivo, particularly when coupled with MRI to track the distribution and release of DDS. The technique is especially important in understanding biodegradation and degradation rate of biodegradable polymers and polymer composites within the body, which is a critical factor when designing biomaterials to degrade in a specific manner.

### 4.7.4 Ultraviolet-Visible (UV-Vis) Spectroscopy

UV-Visible spectroscopy or electronic spectroscopy measures the per cent transmittance of light in the range of ultraviolet regions, which are between 200 and 400 nm and visible regions, which are between 400 and 700 nm. The concentration of analytes in a sample is analyzed using a UV-visible technique based on Beer–Lambert's law, which relates absorbance to concentration of the light by the analyte. Usually in bioengineering, such as drug delivery, this technique is employed to analyze interactions between drugs and polymers and study drug delivery profiles. In the case of samples, light at a certain wavelength excites the electrons within the molecules to higher energy levels [39]. The intensity of light absorbed at each wavelength contains details about the electronic structure of the molecular and conjugate systems, such as the aromatic rings.

UV-Vis spectroscopy is extensively used for the determination of drug-polymer systems released from polymeric matrices spectrum. By regularly monitoring the absorbance of the release media a choice of the amount of drug released can be closely monitored and hence optimized controlled release systems.

UV-Vis spectroscopy can be applied for the characterization of functionalized polymers in the drug delivery system. For instance, using the spectra of absorption, one can find out the characteristics of polymers containing chromophores or photosensitive groups. The information contained in this can be utilized to develop photoresponsive drug delivery systems or scaffolds that undergo degradation under certain wavelengths of light.

In another application, UV-Vis spectroscopy is also used to assess the interactions and uptake of cells in tissue engineering for drug delivery. Additionally, the study of protein adsorption on biomaterial surfaces, which crucial for the biocompatibility in tissue engineering, can also be analyzed via this technique. Fluorescent or UV-accessible custom can be employed in tracking how much drug or bioactive molecule is being released by the cells and tissues, and consequently can be effective in the assessment of the delivery systems. Monitoring the stability of drug formulations is done by UV-vis spectroscopy by detecting changes in absorbance over time [39].

# 5 Conclusion

Biophysical characterization will remain a fundamental part of the development process of biopolymer-based drugs as well as polymeric scaffolds in regenerative medicine. At both micro and macro levels, these techniques provide not only an insight but also a comprehensive understanding of the material properties like mechanical stability, porosity, biodegradability, cell interaction, drug release kinetics, and thermal behavior. The precise control and tailoring over these properties will play a significant role in optimizing drug delivery systems as well as advanced tissue scaffolds, rather instead of just harmless ones, to replicate the natural environment supporting cell proliferation, tissue formation and repair. Such continuous development in the understanding of material properties at several scales simultaneously enables the conception of novel drug delivery systems and regenerative therapies, thereby contributing to the ultimate aim of healthcare improvement. Biophysical characterization, combined more closely than today with computational modeling and in vivo studies, will indeed produce therapies that are increasingly predictive and effective in the near future, thus further accelerating the use of biopolymers in clinical practice. Last but not least, improvements in characterization techniques will further contribute to developing higher-order, sophisticated personalized therapeutic approaches within regenerative medicine for better patient outcomes.

## References


1. Singh, R., Gautam, S., Sharma, B., Jain, P., & Chauhan, K. D. (2021). Biopolymers and their classifications. In *Biopolymers and their industrial applications* (pp. 21–44). Elsevier. https://doi.org/10.1016/B978-0-12-819240-5.00002-X
2. Kaplan, D. L. (1998). Introduction to biopolymers from renewable resources. In *Biopolymers from renewable resources* (pp. 1–29). Springer. https://doi.org/10.1007/978-3-662-03680-8_1
3. Liechty, W. B., Kryscio, D. R., Slaughter, B. V., & Peppas, N. A. (2010). Polymers for drug delivery systems. *Annual Review of Chemical and Biomolecular Engineering, 1*(1), 149–173. https://doi.org/10.1146/annurev-chembioeng-073009-100847
4. Vilar, G., Tulla-Puche, J., & Albericio, F. (2012). Polymers and drug delivery systems. *Current Drug Delivery, 9*(4), 367–394. https://doi.org/10.2174/156720112801323053
5. Zhang, Z., Wang, X., Li, B., Hou, Y., Cai, Z., Yang, J., & Li, Y. (2018). Paclitaxel-loaded PLGA microspheres with a novel morphology to facilitate drug delivery and antitumor efficiency. *RSC Advances, 8*(6), 3274–3285. https://doi.org/10.1039/C7RA12683B
6. Ali, A., & Ahmed, S. (2018). A review on chitosan and its nanocomposites in drug delivery. *International Journal of Biological Macromolecules, 109*, 273–286. https://doi.org/10.1016/j.ijbiomac.2017.12.078
7. Nishadani, D. K. S., Gunathilake, T. M. S. U., Ching, Y. C., & Noothalapati, H. (2024). Carboxymethyl cellulose hydrogel for PH-responsive drug release of curcumin. *Iranian Polymer Journal, 33*(10), 1449–1467. https://doi.org/10.1007/s13726-024-01331-8
8. Gomes, M., Azevedo, H., Malafaya, P., Silva, S., Oliveira, J., Silva, G., Sousa, R., Mano, J., & Reis, R. (2008). Natural polymers in tissue engineering applications. In *Tissue engineering* (pp. 145–192). Elsevier. https://doi.org/10.1016/B978-0-12-370869-4.00006-9
9. Ozdil, D., & Aydin, H. M. (2014). Polymers for medical and tissue engineering applications. *Journal of Chemical Technology & Biotechnology, 89*(12), 1793–1810. https://doi.org/10.1002/jctb.4505
10. Annabi, N., Nichol, J. W., Zhong, X., Ji, C., Koshy, S., Khademhosseini, A., & Dehghani, F. (2010). Controlling the porosity and microarchitecture of hydrogels for tissue engineering. *Tissue Engineering. Part B, Reviews, 16*(4), 371–383. https://doi.org/10.1089/ten.teb.2009.0639
11. Shahverdi, M., Seifi, S., Akbari, A., Mohammadi, K., Shamloo, A., & Movahhedy, M. R. (2022). Melt electrowriting of PLA, PCL, and composite PLA/PCL scaffolds for tissue engineering application. *Scientific Reports, 12*(1), 19935. https://doi.org/10.1038/s41598-022-24275-6
12. Xiang, Y., Wang, W., Gao, Y., Zhang, J., Zhang, J., Bai, Z., Zhang, S., & Yang, Y. (2020). Production and characterization of an integrated multi-layer 3D printed PLGA/GelMA scaffold aimed for bile duct restoration and detection. *Frontiers in Bioengineering and Biotechnology, 8*, 971. https://doi.org/10.3389/fbioe.2020.00971
13. Dizlek, M. E., Guden, M., Turkan, U., & Tasdemirci, A. (2009). Processing and compression testing of Ti6Al4V foams for biomedical applications. *Journal of Materials Science, 44*(6), 1512–1519. https://doi.org/10.1007/s10853-008-3038-7

14. De Groot, J. H., Kuijper, H. W., & Pennings, A. J. (1997). A novel method for fabrication of biodegradable scaffolds with high compression moduli. *Journal of Materials Science. Materials in Medicine, 8*(11), 707–712. https://doi.org/10.1023/A:1018544124990
15. Wan, Y., Wu, H., Cao, X., & Dalai, S. (2008). Compressive mechanical properties and biodegradability of porous poly(caprolactone)/chitosan scaffolds. *Polymer Degradation and Stability, 93*(10), 1736–1741. https://doi.org/10.1016/j.polymdegradstab.2008.08.001
16. Longhitano, G. A., Larosa, M. A., Jardini, A. L., Zavaglia, C. A. C., & Ierardi, M. C. F. (2018). Correlation between microstructures and mechanical properties under tensile and compression tests of heat-treated Ti-6Al–4 V ELI alloy produced by additive manufacturing for biomedical applications. *Journal of Materials Processing Technology, 252*, 202–210. https://doi.org/10.1016/j.jmatprotec.2017.09.022
17. Griffin, M., Premakumar, Y., Seifalian, A., Butler, P. E., & Szarko, M. (2016). Biomechanical characterization of human soft tissues using indentation and tensile testing. *Journal of Visualized Experiments*, No. 118. https://doi.org/10.3791/54872
18. Elvira, C., Mano, J. F., San Román, J., & Reis, R. L. (2002). Starch-based biodegradable hydrogels with potential biomedical applications as drug delivery systems. *Biomaterials, 23*(9), 1955–1966. https://doi.org/10.1016/S0142-9612(01)00322-2
19. Pacelli, S., Paolicelli, P., Dreesen, I., Kobayashi, S., Vitalone, A., & Casadei, M. A. (2015). Injectable and photocross-linkable gels based on Gellan gum methacrylate: A new tool for biomedical application. *International Journal of Biological Macromolecules, 72*, 1335–1342. https://doi.org/10.1016/j.ijbiomac.2014.10.046
20. Bashaiwoldu, A. B., Podczeck, F., & Newton, J. M. (2004). Application of dynamic mechanical analysis (DMA) to determine the mechanical properties of pellets. *International Journal of Pharmaceutics, 269*(2), 329–342. https://doi.org/10.1016/j.ijpharm.2003.09.028
21. Loh, Q. L., & Choong, C. (2013). Three-dimensional scaffolds for tissue engineering applications: Role of porosity and pore size. *Tissue Engineering. Part B, Reviews, 19*(6), 485–502. https://doi.org/10.1089/ten.teb.2012.0437
22. Orhan, K. (Ed.). (2020). *Micro-computed tomography (micro-CT) in medicine and engineering*. Springer. https://doi.org/10.1007/978-3-030-16641-0
23. Tang, R., Buckley, J. M., Fernandez, L., Coopey, S., Aftreth, O., Michaelson, J., Saksena, M., Lei, L., Specht, M., Gadd, M., Yagi, Y., Rafferty, E., Brachtel, E., & Smith, B. L. (2013). Micro-computed tomography (micro-CT): A novel approach for intraoperative breast cancer specimen imaging. *Breast Cancer Research and Treatment, 139*(2), 311–316. https://doi.org/10.1007/s10549-013-2554-6
24. Fuoco, T., Cuartero, M., Parrilla, M., García-Guzmán, J. J., Crespo, G. A., & Finne-Wistrand, A. (2021). Capturing the real-time hydrolytic degradation of a library of biomedical polymers by combining traditional assessment and electrochemical sensors. *Biomacromolecules, 22*(2), 949–960. https://doi.org/10.1021/acs.biomac.0c01621
25. Ren, X., Tapias, L. F., Jank, B. J., Mathisen, D. J., Lanuti, M., & Ott, H. C. (2015). Ex vivo non-invasive assessment of cell viability and proliferation in bio-engineered whole organ constructs. *Biomaterials, 52*, 103–112. https://doi.org/10.1016/j.biomaterials.2015.01.061
26. Vallet, P. G., Guntern, R., Hof, P. R., Golaz, J., Delacourte, A., Robakis, N. K., & Bouras, C. (1992). A comparative study of histological and immunohistochemical methods for neurofibrillary tangles and senile plaques in Alzheimer's disease. *Acta Neuropathologica, 83*(2), 170–178. https://doi.org/10.1007/BF00308476
27. Ramanathan, M. (1997). Flow cytometry applications in pharmacodynamics and drug delivery. *Pharmaceutical Research, 14*(9), 1106–1114. https://doi.org/10.1023/A:1012178001348
28. Lozano-Sánchez, J., Borrás-Linares, I., Sass-Kiss, A., & Segura-Carretero, A. (2018). Chromatographic technique: High-performance liquid chromatography (HPLC). In *Modern techniques for food authentication* (pp. 459–526). Elsevier. https://doi.org/10.1016/B978-0-12-814264-6.00013-X
29. Abdu Hussen, A. (2022). High-performance liquid chromatography (HPLC): A review. *Annals of Advances in Chemistry, 6*(1), 010–020. https://doi.org/10.29328/journal.aac.1001026

30. Kumar, P., Mangla, B., Beg, S., Afzal, O., Saleh Alfawaz Altamimi, A., Almalki, W. H., Nasar Mir Najib Ullah, S., & Aggarwal, G. (2022). Optimization and validation of stability indicating RP-HPLC method for the quantification of gefitinib in bulk drug and nanoformulations: An application towards in vitro and ex vivo performance evaluation. *Arabian Journal of Chemistry, 15*(12), 104333. https://doi.org/10.1016/j.arabjc.2022.104333
31. Mano, J. F., Koniarova, D., & Reis, R. L. (2003). Thermal properties of thermoplastic starch/synthetic polymer blends with potential biomedical applicability. *Journal of Materials Science. Materials in Medicine, 14*(2), 127–135. https://doi.org/10.1023/A:1022015712170
32. Szlachta, M., Ordon, K., Nowicka, K., & Pielichowska, K. (2021). Thermal properties of polyurethane-based composites modified with chitosan for biomedical applications. *Journal of Thermal Analysis and Calorimetry, 143*(5), 3471–3478. https://doi.org/10.1007/s10973-020-09283-w
33. Filip, D., Macocinschi, D., & Vlad, S. (2011). Thermogravimetric study for polyurethane materials for biomedical applications. *Composites Part B: Engineering, 42*(6), 1474–1479. https://doi.org/10.1016/j.compositesb.2011.04.050
34. Kazarian, S. G., & Chan, K. L. A. (2006). Applications of ATR-FTIR spectroscopic imaging to biomedical samples. *Biochimica et Biophysica Acta (BBA)—Biomembranes, 1758*(7), 858–867. https://doi.org/10.1016/j.bbamem.2006.02.011
35. Srinivasan, B., Kumar, R., Shanmugam, K., Sivagnam, U. T., Reddy, N. P., & Sehgal, P. K. (2010). Porous keratin scaffold–promising biomaterial for tissue engineering and drug delivery. *Journal of Biomedical Materials Research. Part B, Applied Biomaterials, 92B*(1), 5–12. https://doi.org/10.1002/jbm.b.31483
36. Agrawal, G., & Samal, S. K. (2018). Raman spectroscopy for advanced polymeric biomaterials. *ACS Biomaterials Science & Engineering, 4*(4), 1285–1299. https://doi.org/10.1021/acsbiomaterials.8b00258
37. Fernández-Galiana, Á., Bibikova, O., Vilms Pedersen, S., & Stevens, M. M. (2024). Fundamentals and applications of Raman-based techniques for the design and development of active biomedical materials. *Advanced Materials, 36*(43), e2210807. https://doi.org/10.1002/adma.202210807
38. Rodin, V. V. (2018). Magnetic resonance in studying cells, biotechnology dispersions, fibers and collagen based tissues for biomedical engineering. In *Biological, physical and technical basics of cell engineering* (pp. 339–363). Springer. https://doi.org/10.1007/978-981-10-7904-7_15
39. An, J., Gou, Y., Yang, C., Hu, F., & Wang, C. (2013). Synthesis of a biocompatible gelatin functionalized graphene nanosheets and its application for drug delivery. *Materials Science and Engineering: C, 33*(5), 2827–2837. https://doi.org/10.1016/j.msec.2013.03.008

# Computational Approaches Employing Biophysical Principles for Drug Discovery and Biomolecular Interactions

**Nikhil Pathak, Vidhya Tangeda, Jhanvi Rachh, and Budheswar Dehury**

**Abstract** Computational biophysics has revolutionized drug discovery and biomolecular interaction studies by integrating theoretical models with experimental biophysical techniques. This chapter explores the synergy between computational approaches—such as molecular docking, molecular dynamics (MD) simulations, and free energy calculations—employ biophysical principles to investigate molecular interactions while simultaneously complementing techniques like X-ray crystallography, NMR spectroscopy, Cryo-EM, surface plasmon resonance (SPR), and isothermal titration calorimetry (ITC). These approaches provide atomic-level insights into molecular structures, binding affinities, and conformational dynamics, improving the efficiency of structure-based drug design (SBDD) and protein-ligand interaction studies. Furthermore, AI-driven methodologies, including machine learning-assisted docking and MD simulations, have enhanced predictive accuracy, accelerating drug screening and lead optimization. Case studies highlight real-world applications, demonstrating how computational strategies validated by experimental data contribute to drug repurposing, enzyme engineering, and biomolecular interaction research. The integration of computational and biophysical techniques continues to push the boundaries of molecular medicine, offering new avenues for precision drug discovery and therapeutic innovation.

**Keywords** Computational biophysics · Molecular docking · Molecular dynamics simulations · Structure-based drug design (SBDD) · Drug discovery

---

N. Pathak (✉)
Institute of Bioinformatics and Structural Biology, National Tsing Hua University (NTHU), Hsinchu, Taiwan

V. Tangeda
Department of Pharmacognosy, Centre for Molecular Cancer Research, Vishnu Institute of Pharmaceutical Education and Research (VIPER), Narsapur, Telangana, India

J. Rachh · B. Dehury (✉)
Department of Bioinformatics, Manipal School of Life Sciences, Manipal Academy of Higher Education, Manipal, Karnataka, India



N. Mazumder et al. (eds.), *Advanced Biophysical Techniques in Biosciences*, Lecture Notes in Nanoscale Science and Technology 7,
https://doi.org/10.1007/978-3-031-94551-9_2

# 1 Introduction: Computational and Biophysical Synergy in Biosciences

## *1.1 The Role of Computational Biophysics in Modern Drug Discovery*

Computational technologies built on biophysics principles have a transformative impact on modern drug discovery, enabling researchers to model and predict molecular structures, binding affinities, and conformational dynamics. Many of these computational methodologies, work in tandem with advanced biophysical experimental techniques such as X-ray crystallography, NMR, Cryo-EM, surface plasmon resonance (SPR), and isothermal titration calorimetry (ITC) to confirm the predicted observaations. These computational approaches by enabling robust theoretical predictions prior to experiments, improve our understanding of drug-target interactions thus accelerating drug discovery.

## *1.2 Synergy Between Biophysical and Computational Techniques and Their Applications*

While biophysical techniques provide high-resolution snapshots of biomolecular interactions, they often face limitations in terms of cost, accessibility, and the static nature of the data. Computational approaches complement biophysical techniques by addressing these challenges, offering a dynamic, reproducible, predictive framework for understanding molecular interactions. This is possible as many of the computational methodologies are themselves built by deeply integrating biophysical principles, thus sometimes refered to as computational biophysics. The synergism between various computational and experimental biophysical methods involved in solving complex biomedical challenge of drug discovery are summarized in Fig. 1.

## *1.3 Applications in Scientific Research*

A few major applications where computational biophysics has been proven crucial in drug discovery are discussed below. The comparison of the computational and biophysical methods pointing out to the reason for their synergism is summarized in Table 1.

### 1.3.1 Structure-Based Drug Design (SBDD)

For achieving successful drug discovery and optimization, researchers often employ a synergistic approach that combines experimental methods for biomolecular structure determination with computational tools for predictions and

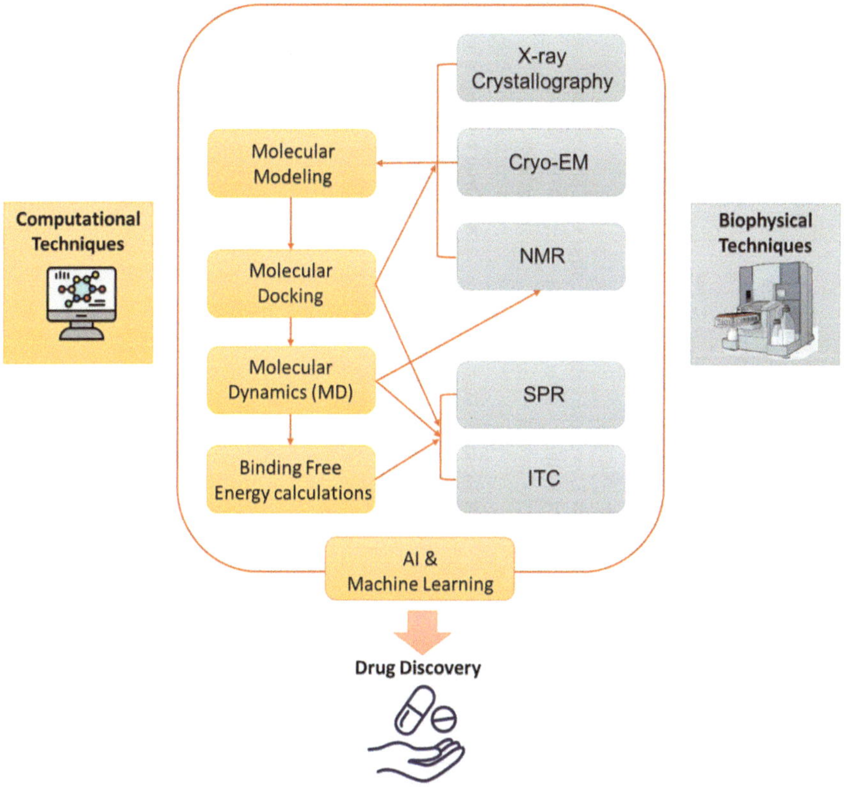


**Fig. 1** Synergy of computational and biophysical techniques in drug discovery. A schematic representation illustrating how computational methods (e.g., molecular modeling, docking, MD simulations, binding free energy calculations) complements experimental biophysical techniques (e.g., X-ray crystallography, Cryo-EM, NMR, SPR, ITC) in modern drug discovery. The brown arrows represent data flow between techniques

studying biomolecules. X-ray crystallography**,** Cryo-electron microscopy (EM), and SAXS provide high-resolution static structures of molecules of interest (such as proteins, nucleic acids, and ligands), while NMR spectroscopy reveals dynamic conformational states of these biomolecules as in physiology. These structures form the basis of SBDD and serve as a starting point for the implementation of computational techniques. Computational techniques that use biomolecular structures, such as molecular docking and molecular dynamics (MD) simulations, have revolutionized the process of drug discovery as they enable a systematic structure-driven strategy to discover, design, and optimize drug molecules, to achieve the desired pharmacological effects and mitigate side effects. Pipeline setups using many of these computational tools are affordable, scalable, and immensely accelerate the screening of large libraries of drugs virtually compared to experimental methods [1].

**Table 1** Comparison of biophysical and computational techniques in drug discovery

| Type | Technique | Brief description | Advantages | Limitations | Applications |
|---|---|---|---|---|---|
| Biophysical | X-ray crystallography | Determines 3D structures via X-ray diffraction on crystallized biomolecules | ✔ High-resolution structures<br>✔ Well-established method | ✖ Requires crystallization<br>✖ Captures static structures only | • Protein structure determination<br>• Drug design |
| | NMR spectroscopy | Uses magnetic fields to analyze biomolecules in solution | ✔ Studies molecules in near-native state<br>✔ Provides dynamic interaction data | ✖ Limited to small/medium-sized molecules<br>✖ Requires large sample amounts | • Protein folding & dynamics<br>• Drug screening |
| | Cryo-electron microscopy (Cryo-EM) | Captures flash-frozen biomolecules at cryogenic temperatures | ✔ No need for crystallization<br>✔ Suitable for large complexes | ✖ Lower resolution vs. X-ray<br>✖ Requires extensive data processing | • Large protein-protein/DNA/RNA/lipid complexes<br>• Structural virology |
| | Surface plasmon resonance (SPR) | Measures real-time binding interactions via refractive index changes | ✔ Kinetic data (on/off rates)<br>✔ Label-free detection | ✖ Requires surface immobilization<br>✖ Surface effects may interfere | • Drug binding affinity<br>• Protein-protein/DNA/RNA/lipid interactions |
| | Isothermal titration calorimetry (ITC) | Measures heat changes to determine binding thermodynamics | ✔ Provides complete thermodynamic profile<br>✔ Label-free and solution-based | ✖ Requires large sample amounts<br>✖ Less sensitive to very weak/strong interactions | • Binding affinity and thermodynamics<br>• Enzyme kinetics |

| | | | | | |
|---|---|---|---|---|---|
| Computational | Molecular docking | Predicts ligand-receptor binding modes and affinities | ✔ High-throughput virtual screening<br>✔ Guides experimental design | ✖ Limited accuracy if flexibility is ignored<br>✖ False positives possible | • Lead optimization<br>• Drug-target interaction studies |
| | Molecular dynamics (MD) simulations | Simulates atomic-level biomolecular motion over time | ✔ Captures conformational changes<br>✔ Models dynamic binding events | ✖ Computationally demanding<br>✖ Force-field dependent accuracy | • Protein-ligand stability<br>• Membrane protein studies<br>Protein-protein/DNA/RNA complex studies |
| | Free energy calculations (MM/GBSA, MM/PBSA, FEP) | Estimates ligand binding free energies and interaction strengths using molecular mechanics and statistical mechanics | ✔ Provide quantitative ranking of binding poses<br>✔ Useful for lead optimization<br>✔ Evaluate stability | ✖ Computationally expensive<br>✖ Limited to small QM regions | • Evaluate protein-ligand binding<br>• Study enzyme catalysis<br>• Enzyme engineering |
| | Machine learning and AI | Uses data-driven models to predict molecular properties and interactions | ✔ High predictive power<br>✔ Rapid analysis of large datasets | ✖ Requires high-quality datasets<br>✖ Risk of overfitting models | • AI-driven drug discovery<br>• Virtual screening |

A summary of experimental (e.g., X-ray crystallography, NMR, Cryo-EM, SPR) and computational (e.g., docking, MD, AI) approaches, comparing their advantages, limitations, and typical applications

### 1.3.2 Exploration of Biomolecular Interactions and Complexes

Computational biophysics plays a crucial role in studying diverse biomolecular complexes such as protein–protein, protein–DNA, protein–carbohydrate, and membrane-bound proteins and interactions at their interfaces. These complexes regulate essential biological functions like signal transduction, gene regulation, immune recognition, and cellular transport, which are central to normal cellular function and are often disrupted in diseases, making them key targets for therapeutic development [2].

For exploring protein–protein/DNA/RNA/lipid interactions, based on initial structures or predicted models of the biomolecules, computational docking tools such as HADDOCK [3] for protein-protein and protein-nucleic acid docking, can help identify potential interaction sites and binding modes which could be further evaluated by performing molecular dynamics (MD) simulations of the best complexes. Further biophysical experiments, such as Cryo-EM and SAXS, are deployed to help refine and validate computational binding predictions. These integrated approaches help identify the key interactions at interfaces and new druggable pockets, which could be exploited for protein–protein interaction inhibitor discovery. For instance, the cancer-associated protein complex MDM2-p53 interface interactions were targeted using computational approaches, yielding small-molecule inhibitors for cancer therapy [4]. Similarly, computational studies of protein-nucleic acid interactions aided in the development of anti-HIV therapies [5]. Membrane-bound proteins, like GPCRs and ion channels, require lipid–protein interaction modeling and MD simulations to study drug binding and to facilitate drug optimization [6].

### 1.3.3 Rational Enzyme Design

Enzymes, which are essentially proteins, are of great importance in industrial production, pharmaceutical development, and biomedical research. Enzyme engineering, which is tinkering with enzymes for improved function, stability, and efficiency, is an emerging field that relies majorly on synergistic approaches integrating computational modeling and experimental biophysics. By combining in-silico simulations with laboratory-based validation, researchers can iteratively refine enzyme modifications based on their target applications. These rational enzyme modifications employ computational techniques such as molecular modeling, docking, molecular dynamics (MD), and free energy calculations to design new proteins and predict enzyme-substrate binding energy and interactions, iteratively. Romero and Arnold highlighted how integrating iterative mutation modeling and computationally driven free energy calculations can improve enzyme catalytic efficiency and thermostability [7]. To validate these computational predictions, biophysical techniques such as calorimetry (e.g., ITC, DSC) provide real-time measurements of catalytic rates, thermodynamic stability, and conformational changes are employed. Thus, combining computational predictions with experimental biophysics reduces the trial-and-error accelerating field-specific engineering of enzymes.

# 2 Molecular Docking: Beyond Structural Predictions

## *2.1 Process of Molecular Docking*

Molecular docking is a computational technique that predicts the preferred binding orientation of a ligand in a receptor. Mostly, the receptor is a protein, while the ligand can be either a small molecule compound, peptide, another protein, or nucleic acids such as DNA/RNA, depending on the use case. The major steps involved in a typical molecular docking process are illustrated in Fig. 2. The first step is the selection of the receptor and ligand 3D structures for input, followed by conformational sampling of the ligand, scoring function for assigning scores, and evaluation of output docking poses. The execution of each of these steps will contribute to the accurate prediction of protein-ligand interactions. The choice of conformational sampling methods and the scoring functions have evolved and widely vary among different docking software are summarized in Table 2.

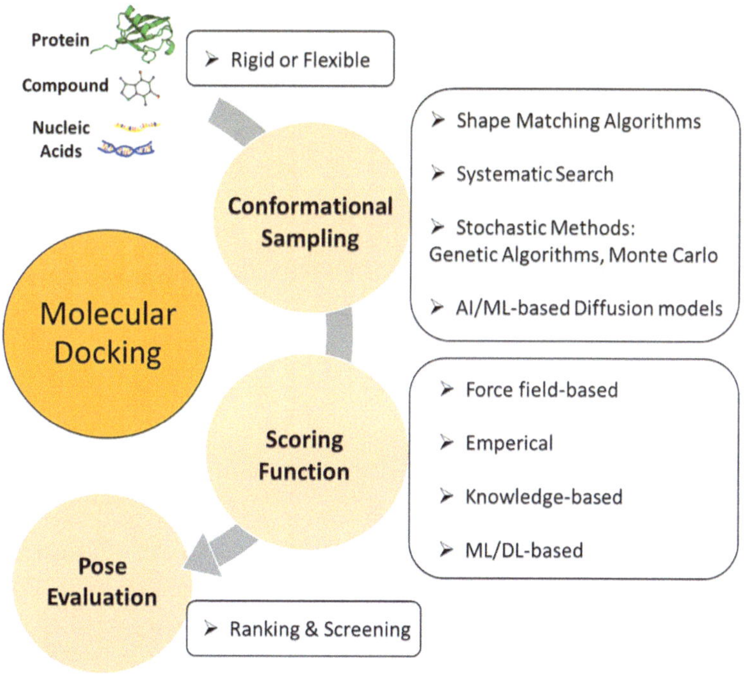


**Fig. 2** Schematic of the molecular docking process, showing various steps with underlying algorithms and biophysics principles. The prediction of receptor-ligand interactions is done by first performing ligand conformation sampling followed by the scoring of ligand poses and finally pose evaluation for ranking and screening the best interacting ligands

**Table 2** Summary of docking softwares with their scoring functions and biophysical considerations

| S. no | Software | Scoring function | Flexibility during docking | Biophysical considerations | Integration with experimental data | Reference |
|---|---|---|---|---|---|---|
| 1 | AutoDock | Lamarckian genetic algorithm | • Ligand flexible<br>• Receptor grid-based flexibility | Accounts for **electrostatics, solvation effects** | ✓ Can incorporate **NMR-derived inter-residue distances** and **X-ray crystallographic constraints** to refine docking poses<br>✓ Used in combination with **ITC-derived binding affinity data** for validating docking scores | [8] |
| 2 | Glide | GlideScore (empirical scoring function) | • Ligand fully flexible<br>• Receptor side-chains flexible | Incorporates **water-mediated interactions, hydrophobicity**, and **hydrogen bonding** to better model protein-ligand interactions | ✓ Can utilize **SPR-derived kinetic rate constants** and **Cryo-EM density maps** to refine receptor grids<br>✓ Works well for **structure-based drug design (SBDD)** when combined with **binding free energy calculations from ITC** | [11] |
| 3 | GOLD | Genetic algorithm with ChemScore or ASP | • Rigid receptor<br>• Some ligand flexibility | Utilizes **solvent-accessible surface calculations, electrostatics**, and **entropic contributions** for improved accuracy | ✓ Capable of **constraining docking poses with experimental NOE distance restraints from NMR**<br>✓ Often used with **SAXS-derived molecular envelopes** for validating protein conformations in docking studies | [12] |
| 4 | DOCK | Grid-based energy evaluation | • Ligand is flexible<br>• Receptor is rigid | Focuses on **geometric complementarity and steric fit** but does not model induced fit effects well | ✓ Can be used alongside **Cryo-EM and SAXS data** to ensure docked ligands conform to experimentally derived receptor conformations<br>✓ Often integrated with **thermodynamic data from ITC for ranking lead compounds** | [13] |
| 5 | FlexX | Incremental construction with empirical scoring | • Fully flexible protein-ligand docking | Considers **binding-site hydration, hydrogen bonding** patterns, and **metal coordination sites** in docking predictions | ✓ Can incorporate **metal-binding constraints from X-ray crystallography** for metalloproteins<br>✓ Used in combination with **SPR binding kinetics** to optimize **binding site occupancy models** | [14] |

| | | | | | | |
|---|---|---|---|---|---|---|
| 6 | Rosetta ligand | Physics-based energy function (Rosetta score) | • Ligand flexible<br>• Receptor side-chain movement | Models **ligand-induced receptor flexibility**, **loop dynamics**, and **side-chain rearrangements** | ✓ Often integrated with **X-ray/Cryo-EM datasets** for refining protein-ligand complexes<br>✓ It has been used with **ITC thermodynamics data** to refine binding free energy calculations | [15] |
| 7 | SwissDock | CHARMM-based free energy scoring | • Ligand flexible<br>• Receptor grid-based flexibility | Uses **molecular mechanics-based free energy** calculations for docking predictions | ✓ Can be used alongside **SPR and ITC experimental data** to validate docking predictions<br>✓ Often applied in **fragment-based drug discovery (FBDD)** workflows that require integration with **NMR-derived ligand binding hotspots** | [16] |
| 8 | HADDOCK | Physics-based (ambiguous interaction restraints) | • Ligand flexible<br>• Receptor flexible (side chains and loops) | Uses **experimental restraints** such as **NMR NOEs, chemical shifts, and SAXS** for docking accuracy | ✓ Integrates **multiple experimental data sources (NMR, SAXS, Cryo-EM)** for ensemble docking approaches<br>✓ Well-suited for protein-protein and protein-ligand docking | [17] |
| 9 | DiffDock | Deep learning-based energy function (transformer model) | • Ligand fully flexible<br>• Implicit receptor flexibility captured by AI model | Accounts for protein **backbone flexibility, ligand rotation, and solvent effects** using diffusion models | ✓ Uses **AI-trained protein-ligand interactions**, which can be refined using Cryo-EM and experimental constraints<br>✓ Shows **strong agreement** with experimental docking studies without requiring receptor rigidification | [18] |

A comparative table of traditional and newer molecular docking tools outlining their scoring functions, biophysical principles considered, and integration with experimental data

## 2.2 Biophysical Principles and Considerations in Docking

Docking employs scoring functions to estimate ligand binding affinity, making it a vital tool for virtual screening in drug discovery [8]. However, traditional docking approaches face limitations in accurately modeling protein flexibility, water-mediated interactions, and hydrogen bonding, which can significantly influence ligand binding and receptor conformations [9]. To address these challenges, docking algorithms have to incorporate biophysics-driven enhancements that improve accuracy and predictive power. Docking algorithms utilize advanced conformational sampling techniques, including shape-matching algorithms, systematic search methods, and stochastic approaches such as Genetic Algorithms and Monte Carlo simulations. More recently, AI/ML-based diffusion models have emerged, revolutionizing ligand-receptor binding predictions by capturing flexibility and induced-fit effects in a data-driven manner.

The receptor and ligand flexibility of various docking tools has been summarized in Table 2. For example, AutoDock/Vina and GOLD primarily use rigid receptor approximations, limiting their ability to model induced-fit effects [10]. In contrast, Glide, FlexX, and Rosetta Ligand implement induced fit modeling, allowing receptor flexibility to better predict ligand-induced conformational changes.

Additionally, water-mediated interactions play a crucial role in ligand binding. Conventional docking algorithms often neglect structured water molecules that mediate hydrogen bonding networks. However, newer approaches incorporate explicit and implicit solvent models to enhance docking predictions. Glide's WaterMap (Schrödinger) accounts for structured water molecules in binding pockets, improving docking accuracy by considering hydration effects and solvent displacement (see Table 2). FlexX and Rosetta Ligand also integrate binding-site hydration in docking predictions, refining hydrogen bonding patterns, while SwissDock employs CHARMM-based free energy scoring to incorporate implicit solvent modeling and hydration effects. Furthermore, scoring functions have advanced beyond traditional force field-based approaches to include empirical, knowledge-based, and ML/DL-based scoring models. These AI-driven techniques enhance docking accuracy by leveraging large-scale structural data and improving binding affinity predictions beyond conventional scoring functions.

To further improve the accuracy and reliability of virtual screening outcomes and to overcome the discrepancies between computational predictions and real-world biomolecular interactions modern docking software could incorporate constraints and validation from biophysical experiments such as X-ray crystallography, NMR spectroscopy, Cryo-EM, SAXS, surface plasmon resonance (SPR), and isothermal titration calorimetry (ITC). Several docking tools now support direct integration with experimental datasets (see Table 2). AutoDock and SwissDock incorporate ITC-derived binding affinity data, allowing for the ranking of docked ligand poses based on experimental thermodynamic parameters. Meanwhile, Glide and GOLD integrate SPR-derived kinetic binding rates and Cryo-EM density maps to optimize receptor flexibility and docking accuracy. HADDOCK uses NMR-derived NOE distance restraints and SAXS profiles to enhance docking accuracy, making it particularly effective for protein-protein and protein-ligand docking. Rosetta Ligand refines docking models with X-ray/Cryo-EM datasets to ensure that predicted ligand poses align with experimentally observed structures. These advancements highlight the synergy between computational docking and experimental biophysical techniques, leading to higher confidence in drug discovery and biomolecular modeling.

## 2.3 *The Era of AI-Driven Docking: DiffDock as an Example*

The advent of AI-driven docking tools has transformed ligand-receptor interaction predictions, surpassing traditional physics-based scoring methods. Unlike AutoDock [8], which relies on predefined energy minimization, DiffDock [18] learns from vast datasets to capture complex interaction patterns, enhancing binding pose prediction—especially in dynamic binding sites (see Table 3). By incorporating receptor flexibility and induced-fit effects, DiffDock overcomes the limitations of rigid docking models, making it more applicable to real-world biological systems.

**Table 3** Comparison of AI-driven and traditional docking technologies

| Aspect | Traditional docking (Ex: AutoDock) | AI-driven docking (Ex: DiffDock) |
|---|---|---|
| Approach | Uses physics-based scoring functions for ligand binding prediction | Uses machine learning models trained on large datasets to predict interactions |
| Data needs | Requires 3D structures of proteins and ligands | Rely on extensive training datasets; can predict interactions even with limited structural data |
| Flexibility | Assumes rigid proteins, with limited receptor flexibility | Better receptor flexibility and induced fit effects |
| Computational efficiency | Faster for single docking runs but slows with increasing system complexity | Involves a high training cost but offers rapid inference once the model is trained |
| Accuracy | Can struggle with flexible ligands or dynamic binding sites | Generally, it improves binding pose and affinity predictions through learned patterns from diverse datasets |
| Interpretability | Based on well-understood physical principles, making results easier to rationalize | Uses complex models that may reduce interpretability despite improved predictive performance |
| Experimental integration | Uses known binding sites, mutagenesis, and other experimental data to refine docking results | Integrates diverse experimental data (e.g., binding affinities, structural data) for enhanced predictions |

A table comparing various aspects of the traditional docking, using AutoDock as an example, with that of the AI-driven docking, using DiffDock as an example. Popular docking highlighting their accuracy, efficiency, and predictive power

DiffDock exemplifies efficiency, scalability, and predictive power in drug discovery. While traditional docking requires extensive 3D structural input and ligand-specific parameterization, DiffDock generalizes across diverse ligand classes, significantly accelerating high-throughput virtual screening. Though computationally demanding during training, its inference phase is rapid, enabling large-scale compound screening. Additionally, AI-driven docking integrates experimental data from Cryo-EM, SAXS, and binding affinity assays, refining predictions beyond static models. As AI-driven methodologies evolve, DiffDock marks a new era in structure-based drug discovery, offering a powerful alternative to traditional docking workflows and advancing computational biophysics.

## 2.4 Case Studies: Docking-Based Drug Discoveries

To demonstrate the impact of computational techniques in drug discovery, here we discuss some exemplified research studies that primarily employed molecular docking. Pathak and team conducted docking and virtual screening against protease enzymes from multiple viruses of the families *Flaviviridae* [19, 20] and *Coronaviridae* [21, 22] to identify drug hits that were further validated through experimental assays for anti-viral activities. A drug repurposing study by Pathak and team targeting SARS-CoV-2 main protease (Mpro) active site used molecular docking of approved drugs followed by virtual screening to discover Boceprevir, Telaprevir, and Nelfinavir whose antiviral efficacies were measured in experimental validation [22]. This study also deployed computational pharmacophore modeling to leverage the dynamic conformation of the Mpro active site, which enhanced the drug screening.

Dehury's work focuses on *Mycobacterium leprae* dihydropteroate synthase (MlDHPS), a key enzyme in the folic acid biosynthesis pathway to tackle dapsone resistance in the context of multidrug therapy through employing molecular docking and molecular dynamics (MD) simulations, demonstrates the application of in silico tools in drug discovery and optimization against drug-resistant diseases [23]. He assessed novel dapsone derivatives (DDs) for their binding affinity and stability against wild-type and mutant MlDHPS. He was able to identify vital ligand-binding interactions, highlighting conserved residues such as Phe19, Val107, Tr132, and Tyr141 that played important roles in stabilizing drug binding. The study also comprised free energy calculations, such as MM/PBSA analysis, to analyze the thermodynamic stability of the protein-ligand complexes. DD3 (2-naphthol-dapsone) demonstrated increased binding affinity compared to dapsone, indicating its utility as a more potent anti-leprosy drug [23].

# 3 Molecular Dynamics (MD) Simulations: From Structure to Dynamics to Function

## 3.1 MD Simulations for Research Applications

Molecular dynamics (MD) simulations provide a time-resolved view of biomolecular interactions, enabling researchers to study the motion, stability, and function of biomolecules at atomic resolution. By numerically solving Newton's equations of motion, MD generates detailed molecular trajectories, offering insights into protein-ligand stability, allosteric regulation, conformational dynamics, and membrane-protein interactions [24]. Unlike static structural models, MD simulations capture dynamic transitions, making them indispensable in modern drug discovery [25].

In Table 4, we summarize the major research problems in drug discovery that are addressed using MD Simulations. One of the key applications of MD is the identification of druggable sites, including allosteric pockets and cryptic binding sites that are often hidden in static structures, but are revealed by simulating the protein in its physiological environments. By simulating protein dynamics, MD enhances the target protein's druggability by enabling the design of allosteric modulators [26, 27]. MD is also crucial in identifying the equilibrated structures of a protein of interest,

**Table 4** Key applications of MD simulations in drug discovery

| Research application | Use of MD | Benefits | References |
|---|---|---|---|
| Identification of druggable sites | • For detecting<br>• Potential drug-binding pockets<br>• Allosteric sites<br>• Transient or cryptic sites (not evident in static structures) | – Enhances druggability assessment<br>– Aids in allosteric modulator design | [26, 27] |
| Investigation of protein conformational changes | • Captures dynamic transitions in protein structures<br>• Identifies functionally relevant conformational states involved in function or ligand binding | – Provides insights into molecular mechanisms<br>– Guides structure-based drug design | [25] |
| Validation and refinement of docking results | • Simulates protein-ligand interactions post-docking<br>• Evaluate stability and docking pose accuracy | – Improves ligand ranking reliability<br>– Enhances accuracy of docking predictions | [28] |
| Assessment of the impact of mutations | • Evaluates how genetic mutations affect structure and function<br>• Predicts drug resistance mechanisms | – Helps design drugs targeting mutant proteins<br>– Supports precision medicine approaches | [26] |
| Prediction of binding free energies | • Estimates interaction strength between drugs and targets using MD-based free energy calculations | – Facilitates lead optimization<br>– Ranks compounds based on affinity | [29] |

in their physiological environments and also in investigating protein conformational changes. Additionally, MD aids in the critical understanding of a biomolecule's functionally relevant states involved in molecular function (catalysis/transport), ligand binding, and cell signaling [25], which are the key insights to study the fundamental biological processes and mechanisms of disease.

In a research study, Tangeda and team employed MD simulations to study the physiologically and functionally relevant states of the mitochondrial membrane-bound protein NCLX (unpublished data), which is a calcium ion channel involved in cancer cell signaling pathways. Currently, there are no experimentally resolved crystal structures of NCLX, mostly due to the high difficulty of successful crystallization for membrane proteins. Tangeda and team employed systematic MD simulation approaches (as summarized in Fig. 3) on modeled structures of NCLX. The first step is embedding protein into a lipid bilayer membrane, and construction of a simulation box with water, and with physiological pH and salt concentrations. In the next step, we run molecular dynamics where the system is first subjected to energy minimization, equilibration at constant temperature and pressure (using restraints), and a production run where systems freely move. By analyzing the MD trajectories, Dr. Vidhya and the team were able to obtain the functional forms of the NCLX ion channels in the presence and absence of $Ca^{+2}$, which are key to studying the $Ca^{+2}$ ion transport mechanisms.

## 3.2 *Case Studies: MD Simulations in Drug Discovery*

One of the critical applications of MD simulations in drug discovery is in validating and refining the results of receptor-ligand docking (see Table 4). MD of receptor-ligand complexes helps us assess complex stability, binding pose accuracy,

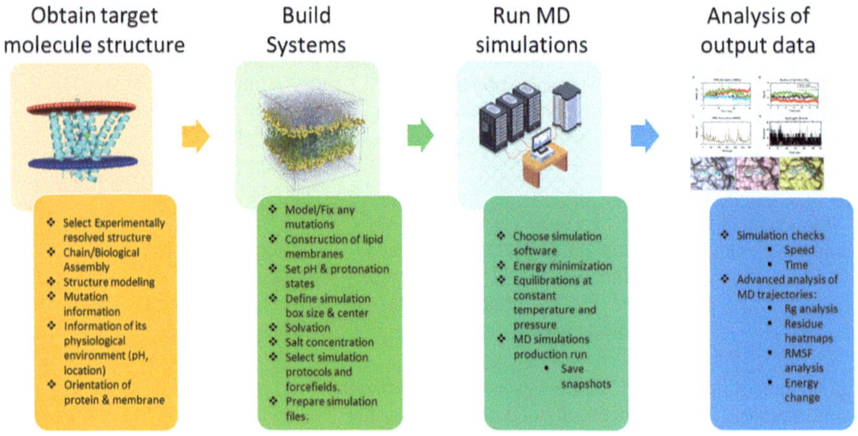


**Fig. 3** Schematic representation of MD simulation workflow. For the membrane protein NCLX, the steps for performing MD simulations, including the key considerations, are shown

ligand-induced conformational changes, and key interactions for binding. For instance, during simulations, the longer the ligand stays in the receptor with lower fluctuations, the more stable it is considered to be, and the higher potential it has to be an inhibitor. In the research study by Pathak and team on drug screening for influenza neuraminidases, they used MD simulations to evaluate the binding stability of the anti-viral inhibitors [30].

Dehury and team used MD of protein–drug complexes to analyze ligand stability and interaction dynamics in flaviviral inhibitors and SARS-CoV-2 receptor-binding domain interactions [31]. The investigation involved MM/GBSA and MM/PBSA calculations in the prediction of binding free energy, and PCA and RMSF analyses that checked for conformational flexibility as well as active site stability. Outcomes indicated that ligand binding minimized structural fluctuations and maximized compactness, signifying enhanced interactions and greater inhibitory capacity [32, 33]. Such findings strengthen the application of MD simulations in drug candidate optimization by offering deeper insights into protein-ligand binding mechanisms.

Another essential application of MD is for predicting binding free energies, where for the MD-equilibrated structures and complexes, the Gibbs free energy calculations are performed to estimate ligand-target interaction strength, facilitating lead optimization, or to study the stability of proteins for enzyme engineering [29]. This integration of MD into drug discovery pipelines ensures a more precise and mechanistic understanding of biomolecular interactions, ultimately accelerating drug development and optimization. Furthermore, MD is widely used to assess the impact of genetic mutations, predict mutation-induced structural disruptions, and aid in drug resistance studies to support precision medicine approaches [26].

For instance, Dehury et al. [32] employed MD simulations to investigate how SARS-CoV-2 RBD mutations affect ACE2 binding, identifying key residues such as Y449, Y489, and Y505 that are important for viral infectivity. Similarly, Kar et al. [33] screened the natural phytocompounds against SARS-CoV-2 main protease (Mpro) and spike protein, identifying Oleanderolide from *Nerium oleander*, Proceragenin A from *Calotropis procera*, and Balsaminone A from *Impatiens balsamina* as potential antiviral drug candidates. These compounds showed higher binding affinity compared to standard antiviral drugs, highlighting the potential of computational methods in accelerating drug discovery.

### 3.3 Integrating MD with Biophysical Techniques

To further enhance the predictive power of MD simulations, experimental biophysical techniques such as Cryo-EM, SAXS, SPR, ITC, immunoprecipitation, and microscopy have been integrated into computational workflows, ensuring that MD predictions align with experimentally validated biomolecular structures and interactions. This synergy between computational and experimental approaches refines MD-derived models, improves docking accuracy, and enhances the reliability of drug discovery pipelines.

Cryo-EM provides high-resolution static structures of biomolecular complexes but often lacks information about protein flexibility and transient conformational states. MD simulations complement Cryo-EM by capturing dynamic transitions, filling in missing disordered or flexible regions, and validating protein-ligand binding interactions [34]. For example, Vant et al. demonstrated how MD simulations helped refine Cryo-EM-derived structures, improving accuracy in binding site identification and structural flexibility modeling [35]. Similarly, small-angle X-ray scattering (SAXS) has been combined with MD to resolve low-resolution protein structures, particularly for disordered proteins and large biomolecular assemblies that are challenging to crystallize [36]. Beyond structural refinement, MD simulations have been integrated with experimental binding studies to validate ligand interactions. Surface plasmon resonance (SPR) provides real-time kinetic binding data, which has been incorporated into MD workflows to optimize ligand-binding pose predictions. Capelli et al. demonstrated that using SPR-derived kinetic parameters in MD simulations improved the accuracy of binding free energy calculations, ensuring that computationally predicted interactions align with experimentally measured affinities for the GPCR protein GPR17 [37]. Additionally, isothermal titration calorimetry (ITC), which directly measures binding thermodynamics, has been used alongside MD to refine ligand-protein interaction models, particularly in lead optimization [38].

Experimental assays such as immunoprecipitation (IP) and pull-down assays are widely used to validate computationally predicted biomolecular interactions, such as protein–protein interactions. IP uses cell lysates, while Pull-down assays use purified proteins, both employing antibodies to detect interaction partners of the target protein, providing direct evidence of the protein-protein interactions. For example, studies by Sung et al. and Tangeda et al. in cancer cells validated the interaction partners of the Lon protein using IP and pull-down assays, which were predicted from computational models [39, 40]. Furthermore, microscopy-based techniques that uncover protein colocalizations within specific cellular compartments add another layer of experimental validation for the predicted interactions [39, 40]. Thus, by integrating biophysical experimental techniques and MD simulations, researchers have been able to achieve better and significantly accelerate rational drug discovery and discover biomolecular interactions.

## 3.4 *Recent Advancements in Leveraging AI for MD Simulations*

There has been also a recent trend of integrating artificial intelligence (AI) with molecular dynamics (MD) simulations, as AI models are now becoming good at analyzing protein/compound structures, and predicting molecular interactions, leading to more efficient and accurate predictions of drug efficacy and safety. AI-driven molecular dynamics (MD) simulations have significantly improved drug discovery by enhancing binding site prediction, docking validation, and free energy calculations.

AI models now analyze vast datasets of protein structures and dynamics, identifying drug-binding pockets, allosteric sites, and transient or cryptic sites that may be hidden in static structures. This approach enhances druggability assessments and facilitates the design of allosteric modulators, as demonstrated in [41], where AI-driven feature extraction from MD trajectories enabled the identification of previously undetected functional binding sites. AI has also been leveraged to validate and refine docking results by integrating MD-based ligand-binding simulations. Traditional docking often suffers from inaccuracies due to rigid receptor assumptions, but AI-assisted MD simulations overcome this limitation by predicting ligand-induced conformational changes and interaction stability. As reported in [41], deep learning frameworks applied to MD trajectories have improved ligand ranking reliability and enhanced binding pose accuracy, leading to more precise drug candidate selection. Furthermore, AI-driven free energy calculations within MD simulations have optimized binding affinity predictions, accelerating lead optimization and drug screening. By training on MD-derived thermodynamic datasets, AI models can rank compounds based on predicted interaction strengths, reducing the need for extensive experimental validation [41]. These AI-enhanced methodologies improve accuracy and efficiency and lower computational costs, making MD simulations more accessible and practical for large-scale structure-based drug discovery.

# 4 Free Energy Calculations for Quantifying Biomolecular Interactions

## *4.1 Binding Free Energy Methods*

Free energy calculations provide a quantitative framework for ranking protein-ligand binding affinities, offering a more refined assessment than docking scores by incorporating entropy, solvation effects, and molecular flexibility. Several computational approaches have been developed, each with distinct strengths and limitations in predicting biomolecular interactions (see Table 5).

Among these, molecular mechanics/generalized Born surface area (MM/GBSA) and Poisson-Boltzmann surface area (MM/PBSA) methods are the two most widely used ones to estimate binding free energies by combining molecular mechanics force fields with solvation models. Genheden & Ryde reviewed these approaches and demonstrated that MM/GBSA/MM/PBSA methods provide rapid estimations of binding affinity [42, 43]. MM/GBSA is computationally efficient and suitable for post-processing MD trajectories, but it may lack accuracy in capturing entropic contributions. MM/PBSA, by contrast, provides a more detailed solvation model but is computationally more demanding [43, 44]. A more advanced method, Free Energy Perturbation (FEP), model's ligand modifications by simulating gradual free energy differences between two states, making it particularly useful for lead optimization. While FEP offers high accuracy in relative binding free energy predictions [45], it requires extensive sampling and is computationally intensive. These computational techniques play a pivotal role in drug discovery by enabling researchers to prioritize promising compounds for further experimental validation.

Parida et al. [46] illustrated the utility of binding free energy approaches to improve flaviviral inhibitors. In particular, molecular dynamics (MD) simulations and MM/GBSA calculations validated the stability of NS3 protease-ligand complexes, further enhancing their drug potential. Research on the dengue virus NS3 protease found PKP10, a Panduratin A derivative, to be a potential inhibitor. Binding free energy analysis confirmed its high affinity, with decreased flexibility and increased stability in the NS3-ligand complex. Principal component analysis (PCA) and RMSF calculations also emphasized significant conformational changes upon ligand binding. The results emphasize the significance of computational methods in structure-based drug design, showing that they are useful in improving antiviral compounds.

## 4.2 Biophysical Validation Using ITC and SPR

To ensure the reliability of computational free energy calculations, experimental validation is essential. Techniques such as isothermal titration calorimetry (ITC) and surface plasmon resonance (SPR) provide direct thermodynamic and kinetic measurements, serving as benchmarks for computational predictions (see Table 5). ITC measures the heat changes associated with ligand binding, determining binding constants, enthalpy, and entropy [38]. It provides a complete thermodynamic profile in a label-free format, making it applicable across a wide range of biomolecular interactions. However, ITC requires relatively large sample amounts and is less suitable for very weak or very strong interactions (Free Energy Methods in Drug Discovery—Introduction). SPR, on the other hand, detects binding interactions in real-time by monitoring refractive index changes on a sensor surface. It provides kinetic (on/off rates) and affinity data, enabling the refinement of computational binding models and ensuring that predicted ligand-target affinities align with experimental results [47]. However, SPR requires the immobilization of one binding partner, which may introduce surface effects that influence measurements. By integrating computational free energy calculations with biophysical techniques, researchers can validate and refine ligand binding predictions, ensuring that computational models accurately reflect real-world biomolecular interactions. This synergy significantly enhances the efficiency of drug discovery, reducing the need for exhaustive experimental screening while improving the accuracy of ligand ranking and lead optimization.

**Table 5** Summary of popular free energy calculations and experimental validation methods

| Method/technique | Type | Description | Advantages | Limitations | References |
|---|---|---|---|---|---|
| Molecular mechanics/ generalized born surface area (MM/ GBSA) | Computational | Estimates binding free energies by combining molecular mechanics energies with solvation terms from the generalized born model and surface area contributions | • Computationally efficient<br>• Suitable for post-processing of MD trajectories | • May lack accuracy in capturing entropic contributions<br>• Sensitive to the choice of dielectric constants | [42, 43] |
| Molecular mechanics/ Poisson-Boltzmann surface area (MM/ PBSA) | Computational | Similar to MM/GBSA but uses the Poisson-Boltzmann equation for solvation energy calculations | • Provides a more detailed solvation model<br>• Applicable to various biomolecular systems | • Computationally more demanding than MM/ GBSA<br>• May still struggle with entropic calculations | [43, 44] |
| Free energy perturbation (FEP) | Computational | Calculates free energy differences by simulating a gradual transformation between two states, often used for ligand modifications | • High accuracy in relative binding free energy predictions<br>• Useful for lead optimization | • Requires extensive sampling; computationally intensive<br>• Sensitive to the choice of intermediate states | [45] |
| Isothermal titration calorimetry (ITC) | Experimental | Measures the heat change associated with a binding event to determine binding constants, enthalpy, and entropy | • Provides a complete thermodynamic profile<br>• Label-free<br>• Applicable to a wide range of interactions | • Requires relatively large amounts of sample<br>• Less suitable for very weak or very strong interactions | [38] |
| Surface plasmon resonance (SPR) | Experimental | Monitors the binding of molecules in real-time by detecting changes in refractive index near a sensor surface | • Provides kinetic and affinity data<br>• Label-free<br>• Real-time analysis | • Requires immobilization of one binding partner<br>• Surface effects can influence results | [47] |

A comparison of MM/GBSA, MM/PBSA, and FEP free energy methods alongside biophysical validation techniques such as ITC and SPR

## 5 Future Perspectives and Challenges

The future of computational biophysics and drug discovery lies in deeper integration between computational and experimental techniques, the advancement of AI-driven simulations, and the development of more efficient biophysical validation methods. As these technologies evolve, new challenges and opportunities will emerge in drug discovery, biomolecular interaction studies, and enzyme engineering.

A key challenge in computational biophysics is ensuring that predicted biomolecular interactions align with real-world experimental data. Cryo-EM, SAXS, and MD simulations will continue to be integrated, enabling more accurate modeling of protein flexibility, binding dynamics, and biomolecular conformations. Additionally, surface plasmon resonance (SPR) and isothermal titration calorimetry (ITC) will play a central role in validating docking and free energy calculations, reducing the occurrence of false positives in virtual screening. As computational biophysics advances, refining these hybrid experimental-computational approaches will be crucial for improving drug-target interaction predictions.

AI-driven molecular docking and MD simulations have already transformed virtual screening and ligand-binding predictions, but the field is now moving toward AI-driven automation in drug discovery pipelines. AI-powered robotic systems capable of conducting autonomous wet-lab experiments are being developed to validate computationally predicted drug candidates, streamlining the hit-to-lead process. Furthermore, deep learning-based generative models will also facilitate the design of novel drug-like molecules, incorporating binding affinity, specificity, and pharmacokinetic properties to optimize drug candidates before experimental testing.

Advancements in structure-based drug discovery (SBDD), protein–protein interaction (PPI) modeling, enzyme engineering, and free energy calculations will continue to push the boundaries of biophysical and computational integration. Future developments will focus on improving force fields in MD simulations, optimizing docking accuracy with experimental restraints, and enhancing AI-driven free energy calculations. By leveraging data-driven models and physics-based simulations, researchers will further refine molecular interaction predictions, reducing drug development costs and improving therapeutic success rates.

## 6 Conclusion

This chapter has highlighted the transformative impact of computational biophysics, integrating computational and biophysical techniques in drug discovery and biomolecular interaction studies. By combining advanced approaches in molecular docking, molecular dynamics simulations, free energy calculations with experimental biophysical, methods, researchers can now achieve unprecedented accuracy in predicting drug-target interactions and biomolecular dynamics.

The synergy between computational and experimental approaches ensures better drug design strategies, with applications ranging from structure-based drug discovery, to biomolecular interaction investigation and inhibition, to enzyme engineering. As computational biophysics continues to evolve with advancements of techniques in AI, automation, and experimental validation, it will play a central role in precision medicine and the development of next-generation therapeutics.

The real-world applications of computational biophysics in drug discovery and bioscience research are exemplified by the key research studies from many research groups, including that of Pathak, Tangeda, and Dehury, further strengthening our understanding of how to leverage computational biophysics tools. The future holds great potential for hybrid experimental-computational approaches, which, integrated with machine learning models, will continue to push the boundaries of computational biosciences and molecular medicine.

## References


1. Shoichet, B. K. (2004). Virtual screening of chemical libraries. *Nature, 432*(7019), 862–865.
2. Kastritis, P. L., & Bonvin, A. M. (2013). On the binding affinity of macromolecular interactions: Daring to ask why proteins interact. *The Journal of the Royal Society Interface, 10*(79), 20120835.
3. Honorato, R. V., Trellet, M. E., Jimenez-Garcia, B., Schaarschmidt, J. J., Giulini, M., Reys, V., et al. (2024). The HADDOCK2.4 web server for integrative modeling of biomolecular complexes. *Nature Protocols, 19*(11), 3219–3241.
4. Shangary, S., & Wang, S. (2008). Targeting the MDM2-p53 interaction for cancer therapy. *Clinical Cancer Research, 14*(17), 5318–5324.
5. Mori, M., Manetti, F., & Botta, M. (2011). Targeting protein-protein and protein-nucleic acid interactions for anti-HIV therapy. *Current Pharmaceutical Design, 17*(33), 3713–3728.
6. Dror, R. O., Pan, A. C., Arlow, D. H., Borhani, D. W., Maragakis, P., Shan, Y., et al. (2011). Pathway and mechanism of drug binding to G-protein-coupled receptors. *Proceedings of the National Academy of Sciences of the United States of America, 108*(32), 13118–13123.
7. Romero, P. A., & Arnold, F. H. (2009). Exploring protein fitness landscapes by directed evolution. *Nature Reviews. Molecular Cell Biology, 10*(12), 866–876.
8. Morris, G. M., Huey, R., Lindstrom, W., Sanner, M. F., Belew, R. K., Goodsell, D. S., et al. (2009). AutoDock4 and AutoDockTools4: Automated docking with selective receptor flexibility. *Journal of Computational Chemistry, 30*(16), 2785–2791.
9. Meng, X. Y., Zhang, H. X., Mezei, M., & Cui, M. (2011). Molecular docking: A powerful approach for structure-based drug discovery. *Current Computer-Aided Drug Design, 7*(2), 146–157.
10. Trott, O., & Olson, A. J. (2010). AutoDock Vina: Improving the speed and accuracy of docking with a new scoring function, efficient optimization, and multithreading. *Journal of Computational Chemistry, 31*(2), 455–461.
11. Halgren, T. A., Murphy, R. B., Friesner, R. A., Beard, H. S., Frye, L. L., Pollard, W. T., et al. (2004). Glide: A new approach for rapid, accurate docking and scoring. 2. Enrichment factors in database screening. *Journal of Medicinal Chemistry, 47*(7), 1750–1759.
12. Jones, G., Willett, P., Glen, R. C., Leach, A. R., & Taylor, R. (1997). Development and validation of a genetic algorithm for flexible docking. *Journal of Molecular Biology, 267*(3), 727–748.

13. Oshiro, C. M., Kuntz, I. D., & Dixon, J. S. (1995). Flexible ligand docking using a genetic algorithm. *Journal of Computer-Aided Molecular Design, 9*(2), 113–130.
14. Rarey, M., Kramer, B., Lengauer, T., & Klebe, G. (1996). A fast flexible docking method using an incremental construction algorithm. *Journal of Molecular Biology, 261*(3), 470–489.
15. Meiler, J., & Baker, D. (2006). ROSETTALIGAND: Protein-small molecule docking with full side-chain flexibility. *Proteins, 65*(3), 538–548.
16. Grosdidier, A., Zoete, V., & Michielin, O. (2011). SwissDock, a protein-small molecule docking web service based on EADock DSS. *Nucleic Acids Research, 39*(Web Server issue), W270–W277.
17. Dominguez, C., Boelens, R., & Bonvin, A. M. (2003). HADDOCK: A protein-protein docking approach based on biochemical or biophysical information. *Journal of the American Chemical Society, 125*(7), 1731–1737.
18. Corso, G., Jing, B., Barzilay, R., & Jaakkola, T. (2023). DiffDock: Diffusion steps, twists, and turns for molecular docking. *ARXIV.*
19. Pathak, N., Kuo, Y. P., Chang, T. Y., Huang, C. T., Hung, H. C., Hsu, J. T., et al. (2020). Zika virus NS3 protease pharmacophore anchor model and drug discovery. *Scientific Reports, 10*(1), 8929.
20. Pathak, N., Lai, M. L., Chen, W. Y., Hsieh, B. W., Yu, G. Y., & Yang, J. M. (2017). Pharmacophore anchor models of flaviviral NS3 proteases lead to drug repurposing for DENV infection. *BMC Bioinformatics, 18*(Suppl 16), 548.
21. Chen, Y. T., Chang, Y. H., Pathak, N., Tzou, S. C., Luo, Y. C., Hsu, Y. C., et al. (2022). Methotrexate inhibition of SARS-CoV-2 entry, infection and inflammation revealed by bioinformatics approach and a hamster model. *Frontiers in Immunology, 13*, 1080897.
22. Pathak, N., Chen, Y. T., Hsu, Y. C., Hsu, N. Y., Kuo, C. J., Tsai, H. P., et al. (2021). Uncovering flexible active site conformations of SARS-CoV-2 3CL proteases through protease pharmacophore clusters and COVID-19 drug repurposing. *ACS Nano, 15*(1), 857–872.
23. Swain, S. S., Paidesetty, S. K., Dehury, B., Sahoo, J., Vedithi, S. C., Mahapatra, N., et al. (2018). Molecular docking and simulation study for synthesis of alternative dapsone derivative as a newer antileprosy drug in multidrug therapy. *Journal of Cellular Biochemistry, 119*(12), 9838–9852.
24. Karplus, M., & McCammon, J. A. (2002). Molecular dynamics simulations of biomolecules. *Nature Structural Biology, 9*(9), 646–652.
25. Dos Santos Nascimento, I. J., Gomes, J. N. S., de Oliveira Viana, J., de Medeiros, E. S. Y. M. S., Barbosa, E. G., & de Moura, R. O. (2024). The power of molecular dynamics simulations and their applications to discover cysteine protease inhibitors. *Mini Reviews in Medicinal Chemistry, 24*(11), 1125–1146.
26. AlRawashdeh, S., & Barakat, K. H. (2024). Applications of molecular dynamics simulations in drug discovery. *Methods in Molecular Biology, 2714*, 127–141.
27. Durrant, J. D., & McCammon, J. A. (2011). Molecular dynamics simulations and drug discovery. *BMC Biology, 9*, 71.
28. De Vivo, M., Masetti, M., Bottegoni, G., & Cavalli, A. (2016). Role of molecular dynamics and related methods in drug discovery. *Journal of Medicinal Chemistry, 59*(9), 4035–4061.
29. Ahmed, M., Maldonado, A. M., & Durrant, J. D. (2023). From byte to bench to bedside: Molecular dynamics simulations and drug discovery. *BMC Biology, 21*(1), 299.
30. Hsu, K. C., Hung, H. C., HuangFu, W. C., Sung, T. Y., Eight Lin, T., Fang, M. Y., et al. (2017). Identification of neuraminidase inhibitors against dual H274Y/I222R mutant strains. *Scientific Reports, 7*(1), 12336.
31. Dikhit, M. R., Ansari, M. Y., Vijaymahantesh, Kalyani, Mansuri, R., Sahoo, B. R., et al. (2016). Computational prediction and analysis of potential antigenic CTL epitopes in Zika virus: A first step towards vaccine development. *Infection, Genetics and Evolution, 45*, 187–197.
32. Dehury, B., Raina, V., Misra, N., & Suar, M. (2021). Effect of mutation on structure, function and dynamics of receptor binding domain of human SARS-CoV-2 with host cell receptor ACE2: A molecular dynamics simulations study. *Journal of Biomolecular Structure & Dynamics, 39*(18), 7231–7245.

33. Kar, B., Dehury, B., Singh, M. K., Pati, S., & Bhattacharya, D. (2022, July). Identification of phytocompounds as newer antiviral drugs against COVID-19 through molecular docking and simulation based study. Journal of Molecular Graphics and Modelling, 114, 108192. https://doi.org/10.1016/j.jmgm.2022.108192. Epub 2022 Apr 14. https://pubmed.ncbi.nlm.nih.gov/35468453/
34. Son, A., Kim, W., Park, J., Lee, W., Lee, Y., Choi, S., et al. (2024). Utilizing molecular dynamics simulations, machine learning, cryo-EM, and NMR spectroscopy to predict and validate protein dynamics. *International Journal of Molecular Sciences, 25*(17).
35. Vant, J. W., Sarkar, D., Nguyen, J., Baker, A. T., Vermaas, J. V., & Singharoy, A. (2022). Exploring cryo-electron microscopy with molecular dynamics. *Biochemical Society Transactions, 50*(1), 569–581.
36. Schneidman-Duhovny, D., Kim, S. J., & Sali, A. (2012). Integrative structural modeling with small angle X-ray scattering profiles. *BMC Structural Biology, 12*, 17.
37. Capelli, D., Parravicini, C., Pochetti, G., Montanari, R., Temporini, C., Rabuffetti, M., et al. (2019). Surface plasmon resonance as a tool for ligand binding investigation of engineered GPR17 receptor, a G protein coupled receptor involved in myelination. *Frontiers in Chemistry, 7*, 910.
38. Velazquez-Campoy, A., & Freire, E. (2006). Isothermal titration calorimetry to determine association constants for high-affinity ligands. *Nature Protocols, 1*(1), 186–191.
39. Tangeda, V., Lo, Y. K., Babuharisankar, A. P., Chou, H. Y., Kuo, C. L., Kao, Y. H., et al. (2022). Lon upregulation contributes to cisplatin resistance by triggering NCLX-mediated mitochondrial Ca(2+) release in cancer cells. *Cell Death & Disease, 13*(3), 241.
40. Sung, Y. J., Kao, T. Y., Kuo, C. L., Fan, C. C., Cheng, A. N., Fang, W. C., et al. (2018). Mitochondrial Lon sequesters and stabilizes p53 in the matrix to restrain apoptosis under oxidative stress via its chaperone activity. *Cell Death & Disease, 9*(6), 697.
41. Iwata, H. (2024). Transforming drug discovery: The impact of AI and molecular simulation on R&D efficiency. *Bioanalysis, 16*(23–24), 1211–1217.
42. Cournia, Z., & Chipot, C. (2024). Applications of free-energy calculations to biomolecular processes. A collection. *The Journal of Chemical Information, 64*(7), 2129–2131.
43. Genheden, S., & Ryde, U. (2015). The MM/PBSA and MM/GBSA methods to estimate ligand-binding affinities. *Expert Opinion on Drug Discovery, 10*(5), 449–461.
44. Chipot, C. (2023). Free energy methods for the description of molecular processes. *Annual Review of Biophysics, 52*, 113–138.
45. York, D. M. (2023). Modern alchemical free energy methods for drug discovery explained. *ACS Physical Chemistry Au, 3*(6), 478–491.
46. Parida, P., Yadav, R. N. S., Dehury, B., Ghosh, D., Mahapatra, N., Mitra, A., et al. (2017). Novel insights into the molecular interaction of a Panduratin A derivative with the non structural protein (NS3) of dengue serotypes: A molecular dynamics study. Current Pharmaceutical Biotechnology, 18(9), 769–782.(1), 54–61.
47. Rich, R. L., & Myszka, D. G. (2000). Advances in surface plasmon resonance biosensor analysis. *Current Opinion in Biotechnology, 11*(1), 54–61.

# Quantum Dots Mediated Drug Delivery Systems

**Kunal Biswas, Kameshwaran R, Inbathamizh L, Subashini R, and Prabhakar Singh**

**Abstract** In the field of targeted drug delivery, quantum dots (QDs) have emerged as the most promising tool with their peculiar optical and electronic properties. As a versatile platform for the advancement of nanoparticle-mediated drug delivery (NMDD) systems, the efficacy and precision of therapeutic interventions are enhanced by QDs, which are nanoscale semiconductor crystals. The sophisticated role of QDs in drug delivery is delved into in this chapter, especially with a focus on their application in targeting specific disease sites, such as cancerous tissues. The controlled release of therapeutic agents is not only permitted by the integration of QDs into drug delivery vehicles but synchronous diagnostic imaging is also enabled, thereby transversing the gap between therapy and diagnostics (theranostics). The principles of QD design and biofunctionalization that are critical for minimizing toxicity and enhancing biocompatibility are explored. Additionally, the potential of QDs in overcoming the challenges of traditional drug delivery systems, such as limited bioavailability and off-target effects, is discussed. The chapter concludes with an aspect on the future prospects of QDs in clinical applications, spotlighting their potential to revolutionize targeted drug delivery and personalized medicine.

**Keywords** Quantum dots · Targeted drug delivery · Nanomedicine · Biofunctionalization · Theranostics · Nanoparticles · Cancer therapy

K. Biswas (✉)
Centre for Nanoscience and Nanotechnology, International Research Centre, Sathyabama Institute of Science and Technology, Chennai, Tamil Nadu, India

K. R. · I. L. · S. R. · P. Singh
Department of Biotechnology, Sathyabama Institute of Science and Technology, Chennai, Tamil Nadu, India



N. Mazumder et al. (eds.), *Advanced Biophysical Techniques in Biosciences*, Lecture Notes in Nanoscale Science and Technology 7,
https://doi.org/10.1007/978-3-031-94551-9_3

# 1 Introduction

Drug delivery system (DDS) refers to the processes, formulations, technologies, and techniques used to transport drug substances through the body to deliver the desired therapeutic effect [1]. It involves medication delivery techniques for humans and animals for therapeutic purposes. The recent generation of DDSs introduced smart DDSs, designed to deliver drugs at the right time, dosage, and location to achieve maximum safety and effectiveness. NDDS, currently receiving much attention due to its advancements, is discussed in this chapter. These systems meet the need for appropriate drug demand and enhance the efficacy of both novel and existing drugs through targeted, controlled, and extended release [1].

DD is an emerging area in pharmaceutical science. Targeted delivery is one of the features of the four generations of DDSs, which are further divided into five generations [2]. Controlled or sustained DDSs aim to reduce the number of doses administered, regulate or prolong drug delivery, and improve drug effectiveness compared to conventional methods. One NDDS that integrates modifications of traditional preparations and delivery methods is the bilayered tablet. These tablets consist of two different or identical drugs combined into a single dose. This may involve either the sustained and immediate release of the same drug or sequential release of combined drugs, with one serving as the loading dose and the other as the maintenance dose [3].

While these developments in conventional DD represent promising advancements, certain DDSs still require major improvements. These include protein delivery, self-controlled insulin delivery, delivery of poorly soluble drug formulations, and targeted drug delivery systems (TDDS). Another significant achievement in DDS, leveraging nanotechnology, is targeted delivery to tumours [4]. NMDD enables the controlled release of drugs, allowing sufficient time for drugs to react with specific stimuli such as pH, light, heat, or enzymes to enhance therapeutic action [5]. TDDS combines molecular biology, polymer science, pharmacology, and bioconjugate chemistry. It involves delivering drugs to a specific site in the body rather than distributing them throughout the body or an organ. TDD aims to regulate and manage pharmacokinetics, pharmacodynamics, aspecific toxicity, immunogenicity, and biorecognition of therapeutic agents. Drug delivery includes methods, formulations, and technologies for transporting pharmaceutical substances across the body to achieve therapeutic effects [1]. It enables medication delivery in all forms for both humans and animals to achieve therapeutic efficacy. Recent advancements in DDSs have focused on smart DDSs that deliver drugs at the right dose, time, and location for maximum safety and efficiency [6]. NMDDSs have recently gained much attention. These systems fulfil the demand for proper drug delivery and enhance the therapeutic efficiency of both new and existing drugs through targeted, controlled, and sustained delivery [1]. Targeted delivery is a feature of the fourth generation of DDSs, divided into five generations [2]. Controlled or sustained DDSs have been developed to reduce dosing frequency, control or prolong drug release, and improve drug effectiveness compared to traditional methods. One example of NMDDS is the bilayer tablet, which combines two drugs (either different or identical) into a single dose. This allows for either sustained and immediate release of the same drug or sequential release, with one as the loading dose and the other as the maintenance dose [3]. While these innovations in classic DDSs are promising, certain classes still demand further development. Examples include protein delivery, auto-regulated insulin delivery, poorly soluble drug formulations, and TDDSs. Another advancement using nanotechnology is targeted delivery to tumours [4]. NMDD ensures controlled drug release, allowing drugs to respond to specific stimuli such as pH, light, heat, or enzymes for enhanced therapeutic action. TDDS incorporates disciplines like molecular biology, polymer science, pharmacology, and bioconjugate chemistry. Medication is administered to a specific site rather than the entire body or organ. This approach aims to control pharmacokinetics, pharmacodynamics, aspecific toxicity, immunogenicity, and biorecognition of therapeutic agents. These advancements are enabled by improved synthesis materials, better nanoparticle size control, and a deeper understanding of biological-material interactions and their in vivo implications. A schematic of the targeted drug delivery system using Quantum Dots is shown in Fig. 1.

Nanoparticles or nanocarriers are increasingly favoured in biomedicine due to their advantages over traditional drug delivery methods. These include enhanced solubility of hydrophobic drugs, improved physicochemical stability, and increased bioavailability at the targeted site. Nanoparticle-based drug delivery systems represent a transformative development in biomedicine. Cellular aspects of nanoparticles, such as targeting, absorption, and intracellular fate, ensure proper deployment of nanomedicine delivery systems. Quantum dots (QDs) vary significantly in molecular weights and have been conjugated with numerous compounds, enabling targeted intracellular delivery. For biomedical applications, QDs must be functionalized with biomolecules such as proteins, peptides, or antibodies. Over the years, significant efforts have been made to modify the surface of QDs to achieve stable bioactive labels that bind specifically to targets. Although functionalized QDs offer

**Fig. 1** Targeted drug—delivery using quantum dots

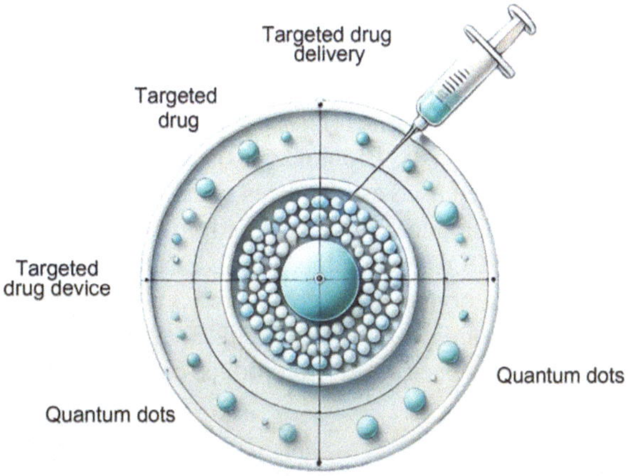

excellent photophysical properties, challenges such as complex surface chemistry and relatively large sizes remain. The biggest barrier preventing the clinical application of QDs is their toxicity and non-biodegradable nature. Considerable efforts are being made to develop biocompatible, non-toxic QD-based probes, with some progress reported in this area [7, 8]. Additionally, QDs could benefit from or contribute to NMDD fields such as biosensing, photodynamic therapy, tracking biologically active cargos, multimodal imaging with quantum dots, and nanoparticle-enhanced cellular responses [9].

# 2 Synthesis and Functionalization of Quantum Dots

## *2.1 Methods of Synthesis*

All "top-down" and "bottom-up" methods proposed for the synthesis of QDs fall into one of three categories: physical (involving no chemical transformation of matter), chemical (involving a chemical reaction that forms a substance), or biological (involving living organisms). The diameter of QDs can range from a few nanometers to a few micrometres, depending on the manufacturing process. Precise growth techniques allow for control of particle size distribution within a 2% range. Depending on the type of QDs—cadmium-based or cadmium-free—preparation techniques range from traditional methods to cutting-edge techniques [10], as illustrated in Fig. 2.

One of the earliest methods for producing QDs was organometallic synthesis. Approximately 20 years ago, the "bottom-up" organometallic route, which produces QDs with superior optical properties, was developed [11]. The process involves three stages—nucleation, growth, and termination—carried out in organic solvents under an inert atmosphere at high temperatures. By adjusting reaction conditions, the size of QDs and their emission wavelengths can be fine-tuned. For example, emitted light shifts in colour from red to blue as crystal size decreases. Due to their strong mono-dispersity, high quantum yields, and narrow emission peaks, QDs synthesized using this method are ideal for bioimaging applications [12]. Alternative methods for QD synthesis include the biomolecule-templated technique. This approach, based on bio-templating, enables the assembly of inorganic nanomaterials. Biomolecules such as peptides, polysaccharides, and nucleic acids have been used as templates to create QDs. Examples of this innovative technology include genetically modified viruses, protein cages made from heat shock proteins, and QDs assembled using DNA templates [10]. The synthesis of QDs within living organisms is considered a novel approach. Several bacterial species, such as *Gluconacetobacter xylinus* [13], *Bacillus megaterium* [14], and *Escherichia coli* [15], have been shown to produce cadmium selenide QDs.

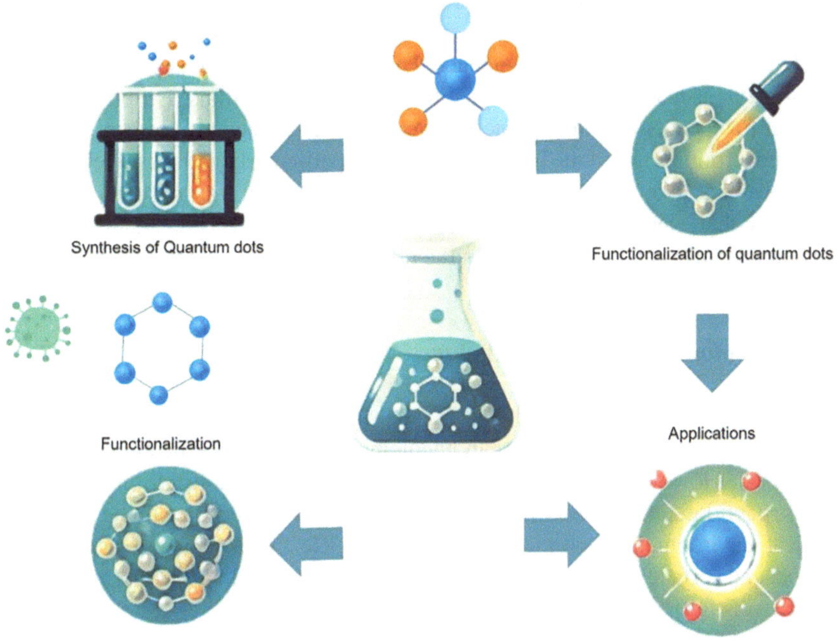


**Fig. 2** Synthesis and functionalization of quantum dots

## 2.2 *Surface Functionalization*

Despite their strong photostability and fluorescence, quantum dots (QDs) show significant promise for use in biology, according to recent advancements. However, biological applications cannot be directly implemented using naked, uncoated QDs. Surface passivation is essential to achieve targeted distribution, optimize surface characteristics, and ensure solubility in aqueous and physiologically relevant environments. Strategies for surface passivation include amphiphilic combination, surface silanization, and ligand exchange [16].

### 2.2.1 Functional Groups and Ligands

The process in which the natural hydrophobic ligands on the surface of quantum dots (TOP, TOPO, HDA) are replaced using mass action with hydrophilic ligands is known as cap exchange or ligand exchange. Bifunctional groups that bind to the ZnS shell on the QD surface include thiols (–SH) for binding to nanoparticles, carboxyls (–COOH), amines (–$NH_2$), and hydroxyls (–OH) for enhancing water solubility. Secondary biomolecules, such as proteins, drugs, and antibodies, can also be attached to the surface through these substituting ligands [17].

The large surface area of nanoparticles enables them to adsorb large molecules, including polymers, long-chain chemical compounds, proteins, enzymes, and nucleic acids, in a non-specific manner. This technique does not involve functional binding but instead relies on non-specific molecular interactions. The primary determinant for the adsorption approach is electrostatic interaction between oppositely charged terminal ends of the ligand-coated QDs, particularly with positively charged biomolecules like proteins. Through genetic engineering, positively charged domains that electrostatically self-assemble with negatively charged, carboxyl-terminated QDs can be expressed on the relevant protein molecules [18].

### 2.2.2 Biofunctionalization Techniques

For the treatment of diseases, delivering the right dosage to the correct sites is crucial. With the targeting ability of nanomedicine, drug delivery efficiency can now be improved. However, therapeutic reagents such as small-molecule drugs, nucleic acids, peptides, and proteins are prone to degradation or clearance in the pharmacological environment. Optimization of nanoparticle characteristics, such as size, shape, and surface chemistry, can aid in drug delivery at cellular, organ, and subcellular levels [19].

QD/drug nanoparticle compositions are being investigated for targeted drug delivery and therapy. Targeting biomolecules can be combined with drug formulations to achieve traceable medication delivery and therapy both in vitro and in vivo. Studies have shown that the inclusion of QDs in drug formulations does not diminish the effectiveness of the drug. However, due to toxicity concerns, cadmium-based QDs are unsuitable for in vivo drug delivery. Researchers are synthesizing cadmium-free QDs, such as silicon, indium phosphide, and copper indium sulfide QDs, for in vivo applications, including targeted tumour imaging. Another concern is the total volume of QDs, prompting ongoing research in areas such as passivation and customized surface treatments to reduce toxicity. In the future, applications of QDs in medicine are expected to focus on their integration with drug nanoparticles for early cancer detection and targeted medication therapy [20].

### 2.2.3 Tailoring Quantum Dots for Targeted Drug Delivery

Delivering the right dosage to the correct sites is essential for the effective treatment of diseases. Drug delivery efficiency can now be enhanced due to the targeting ability of nanotechnology. However, therapeutic reagents, such as small-molecule drugs, nucleic acids, peptides, and proteins, may degrade or be cleared in the pharmacological environment. Optimization of nanoparticle properties, such as size, shape, and surface chemistry, may further enhance drug delivery at cellular, organ, and subcellular levels.

QD/drug nanoparticle compositions are being studied for the development of targeted drug delivery and therapy. Targeted biomolecule formulations combined with drugs can enable traceable medication delivery and therapy both in vitro and in vivo. Studies have indicated that incorporating QDs into drug formulations does not compromise drug efficacy. However, due to toxicity concerns, cadmium-based QDs are unsuitable for in vivo drug delivery. To address this, researchers are synthesizing cadmium-free QDs, such as silicon, indium phosphide, and copper indium sulfide QDs, for in vivo applications, including targeted tumor imaging. Another concern is the total volume of QDs, which is being addressed through research in areas like passivation and tailored surface treatments to reduce toxicity. In the future, the medical applications of QDs are expected to focus on their integration with drug nanoparticles for early cancer detection and targeted medication therapies [20].

## 3 Mechanisms of Targeted Drug Delivery Using Quantum Dots

### *3.1 Principles of Targeted Drug Delivery*

The process of delivering a medication at a high concentration to the desired location while minimizing its presence in nontargeted areas is known as drug targeting. This approach enhances the therapeutic effects of the drug while reducing side effects caused by nontarget exposure, higher dosages, and multitarget interactions.

An ideal drug-targeting system should be non-toxic, non-immunogenic, biochemically inert, biodegradable, and stable. It must also be simple, reproducible, cost-effective, and capable of ensuring a predictable drug release pattern, while being rapidly and efficiently eliminated from the body. The preparation of targeted medicinal products must consider the unique properties of target cells and transport carriers.

Targeted drug delivery (TDD) has proven particularly valuable in the treatment of malignant tumors, as it enhances drug concentration at the site of infection and improves macrophage penetration. This is especially significant in novel nanomedicine and therapeutic applications [21].

### *3.2 Quantum Dot Interaction with Biological Systems*

Given the potential applications of QDs in biological applications and disease therapy, it is critical to understand their interaction with living systems, particularly mammalian cells. Quantum dots (QDs) are semiconductor nanocrystals that exhibit unique optical and electronic properties due to quantum confinement effects. When combined with bioactive agents as shown in Fig. 3, these nanocrystals can effectively interact with biological systems for applications such as targeted drug delivery and bioimaging. QDs enter mammalian cells through various routes, including micropinocytosis, caveolae-mediated endocytosis, and clathrin-mediated endocytosis. Different cells employ distinct routes, such as SK-Hep1 cells and alveolar macrophages. The pathway of QD endocytosis is also influenced by the serum content in the medium.

The subcellular distribution of QDs within cells depends on factors such as the type of QD and its conjugated ligands. Negatively charged carboxylic ligands of cadmium-based QDs exhibit a higher affinity for the endosome and lysosome compared to indium-based QDs. Cadmium-based QDs are highly cytotoxic, primarily causing apoptosis and growth inhibition in mouse liver and HeLa cells. Green QDs have been found to be more toxic than yellow QDs, indicating a size-dependent mechanism of QD toxicity. CdTe QDs disturb mitochondrial membrane potential, increase $Ca^{2+}$ levels, and lead to mitochondrial swelling, resulting in cytochrome C leakage. Similarly, CdSe QDs cause cell membrane shrinkage and elevated intracellular $Ca^{2+}$, which are lethal to hippocampal neuronal cells. In order to mitigate toxicity, InP QDs have been proposed as alternatives to Cd-based QDs, though studies on HeLa cells suggest similar damaging effects. In contrast, carbon dots exhibit much lower toxicity than their metal-based QD counterparts and demonstrate promising properties in NIH/3T3 and L929 cell lines. Two human mammary cell lines, MCF-7 and MCF-10A, have been used with various surface coatings to study the kinetics and mechanisms of QD cellular uptake. Fluorescence microscopy and laser scanning cytometry (LSC) revealed QD655-COOH accumulation in nearly all cells, with MCF-7 cells showing slightly higher endocytosis rates than MCF-10A. Live-cell fluorescence imaging demonstrated an increase in QD uptake over time, reaching significant levels after 40 h of incubation. Organellar staining confirmed QD localization in lysosomes. Transmission electron microscopy (TEM) identified a

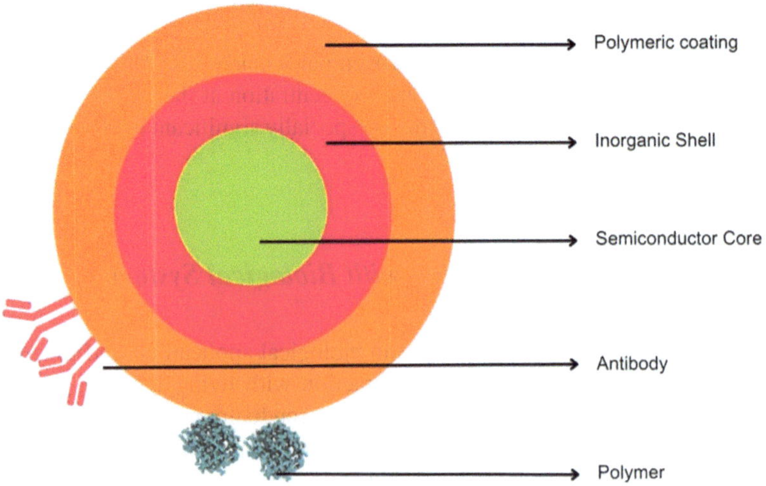


**Fig. 3** Structure of quantum dots using bioactive agents

clear pathway of QD uptake: endocytosis, confinement in early endosomes, and subsequent transport to lysosomes or late endosomes. After incubation with 0.8 nm QDs for 72 h, no cytotoxic effects were observed [22]. For these studies, cells were seeded at a density of 30,000 cells/cm$^2$ on tissue culture chamber slides and incubated with QD655-COOH, QD655-amine PEG, and QD655-PEG at final concentrations of 0.8, 0.5, and 0.8 nm, respectively, at 37 °C for 12 h. Cells were then fixed with 10% neutral-buffered zinc formalin for 45 min and subjected to laser scanning cytometry or fluorescence microscopy after staining with DAPI [22]. In nanotechnology, both active and passive targeting strategies are employed for effective targeting of cancer stem cells. While passive targeting is less effective for lung cancers, it remains essential for controlling nanoparticle size and surface properties. Active targeting enhances QD accumulation in tumor cells or tissues through attachment to targeting agents. In vivo, QD probes are directed to specific tumor sites using antibodies, enabling precise diagnosis.

## 4 Theranostics: Quantum Dots in Diagnostics and Therapeutics

Theranostics combines drug delivery with real-time evaluation, merging diagnostics and therapeutics. Quantum dots (QDs) offer synergistic advantages and properties as mentioned in the Fig. 4, over single-modal nanoparticles. QDs can carry diagnostic substances for real-time monitoring, target specific cell types, efficiently infiltrate cells, overcome intracellular barriers, and respond to local stimuli.

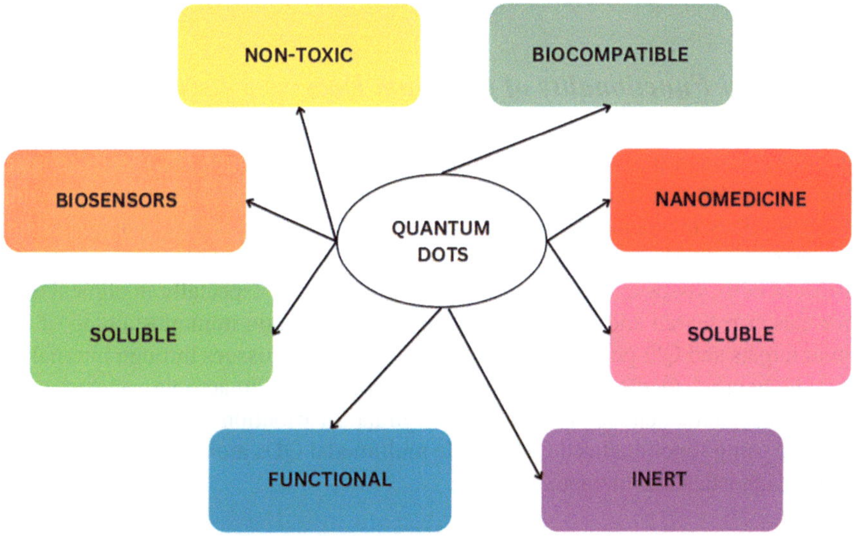


**Fig. 4** Properties of quantum dots

Theranostics represents a revolutionary approach in modern medicine that integrates diagnosis and therapy under the same umbrella, with this concept particularly becoming the heart of cancer treatment. Quantum dots, or semiconductor nanoparticles with exceptional optical and electronic properties, have significantly amplified this paradigm. These quantum dots possess the following unique features: one such prominent feature is the property called size-tunable fluorescence. These can emit light at several wavelengths, thereby aiding in multiplexing during imaging applications. In addition, quantum dots have high photostability, which means that the performance of QDs is consistent during extended imaging times, unlike conventional organic dyes [23, 24]. In diagnostics, QDs are increasingly used as fluorescent labels for bioimaging and biomolecular detection. Their surface can be easily functionalized with various biomolecules to target specific cells, which results in enhanced tumor imaging [25, 26]. Moreover, they are used in electrochemical sensors to detect cancer biomarkers with high sensitivity and selectivity, thus having the potential for early detection of cancer [27, 28]. Quantum dots have been promising in drug delivery and PDT due to their biocompatibility and the ability to encapsulate therapeutic agents, thus reducing systemic side effects [26, 29]. In addition, there is also the study of QDs for potential application in PDT, where they can destroy cancerous cells through the process of the generation of ROS upon photoactivation [30, 31]. However, despite such great potential of QDs in theranostics, challenges such as the toxicity of heavy metal cadmium are still posed [32]. However, advancements of carbon-based quantum dots that are non-toxic and research is being done into optimizing the surface chemistry to overcome regulatory hurdles to improve the biocompatibility. As continued work would overcome these limitations, an expectation is that QD technology shall be integrated with improved clinical practice to substantially change the patient outcomes [33].

## 4.1 Dual Functionality of Quantum Dots

The results show that the dual function in photoluminescence (PL) leads to spectrally invariant transitions and structural deformation in graphene quantum dots (GQDs). This technique is highly valuable for developing optical materials using graphene Wang et al. [34, 35]. Quantum dots (QDs) are used to identify specific types of cells, tissues, and organelles as imaging probes, especially in clinical trials for cancer research and therapy. They have enabled static immunostaining of cellular targets and QD-peptide conjugates targeting tumor tissues through internalization-inducing methods, such as phagocytosis, macropinocytosis, and receptor-mediated endocytosis. The intracellular distribution of plasmid DNA was tracked using functionalized QDs, while multimodal QDs are considered useful for both diagnostics and therapeutics [36].

### *4.2 Imaging Capabilities*

Recent research focuses on producing near-infrared (near-IR) luminous quantum dots (QDs) for biological imaging to collect real-time data during cellular activities. These QDs must be coupled with molecules that recognize specific targets, preventing aggregation and minimizing nonspecific binding. In vitro imaging has utilized QDs to label cells, while in vivo imaging uses them as nanoprobes and labels for tumor-targeted therapeutics. QDs have also been explored in fluorescence lymphangiography and quantum rods. QD-biomolecule conjugates have shown efficacy in both in vitro and in vivo imaging, though biocompatibility remains a significant challenge. Studies indicate that Cd-containing QDs are stable in the short term but degrade over time, raising concerns about chronic toxicity. Non-cadmium alternatives are in high demand, prompting research into novel fluorescent nanomaterials such as NaYF4 and NaGaF4. New methodologies and detection techniques demonstrate that QDs have applications in various fields, including DNA binding and surface charge imaging [37].

## 5 Integrating Quantum Dots with Other Nanocarriers

ZnO-Au-PLA-GPPS-FA nanocarriers, developed as targeted delivery systems for tumors, exhibited bright yellow fluorescence at 570 nm. These nanocarriers demonstrated promising anticancer drug release, likely due to strong hydrophobic interactions with PLA molecules. Both blank and CPT-loaded nanocarriers showed significant anticancer activity, possibly due to the presence of GPPS on the surface. . It is shown in Fig. 5 the results of NP encapsulation within a silica shell (A) and utilization of hydrophobic interactions for NP coating with amphiphilic polymers (B) and lipids (C) represent most common examples of this approach [38].

The CPT-loaded nanocarriers were selectively taken up by cancer cells, as the FA moiety enhanced their absorption. Additionally, scientists developed a magnetic-based multifunctional anticancer drug carrier for curcumin, a nontoxic hydrophobic

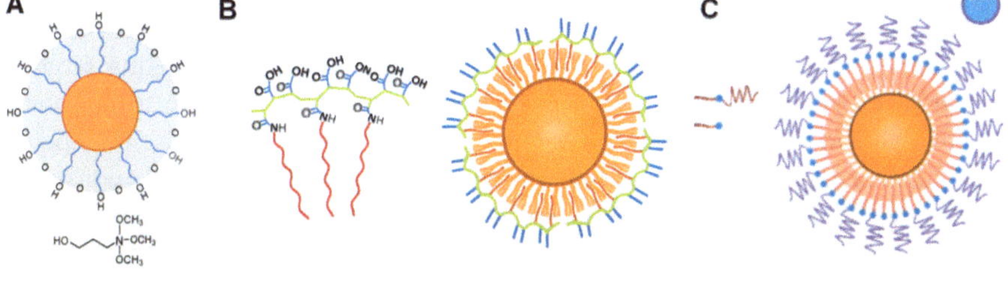


**Fig. 5** Encapsulation with "adaptor" coatings shields the NP core from the environment, providing a flexible interface for further functionalization and enabling application of the same surface coatings to variety of nanocarriers

drug. GQDs were prepared from graphene oxide (GO), and a magnetic nanocomposite was created using FA for cancer targeting. Characterization was performed using FTIR, VSM, XRD, and FE-SEM. The results indicated curcumin release at pH 5.5 and 7.5, with the nanocarrier showing a stronger response in MCF-7 and MG-63 cells Zadeh et al. [39, 40]. Researchers also utilized harmless $Fe_3O_4$ nanoparticles to prepare a nanocarrier for the targeted delivery of doxorubicin (DOX) into cancer cells. These nanocarriers, functionalized with PEG and graphene quantum dots, exhibited increased stability and drug-loading capability, with low toxicity and excellent superparamagnetic properties, achieving a 27% drug loading. The pH-dependent drug release showed comparable anticancer effects to free drugs in human breast cancer cells.

## 6 Clinical Applications of Quantum Dot-Based Drug Delivery

This work examines the clinical applications of quantum dots (QDs) in targeted therapy, biological imaging, and diagnostics. QDs offer enhanced photostability and brightness for in vivo imaging, allowing for long-term, detailed observation of cellular activities. Their stable fluorescence makes them ideal for tracking intracellular events and sensitively detecting chemical interactions, serving as cell labels and biosensors. QDs' ability to provide multiplexing in diagnostics enables the simultaneous detection of multiple biomarkers, which is essential for disease diagnosis. Their theranostic potential in cancer treatment paves the way for advancements in personalized medicine, enabling both therapeutic efficacy monitoring and targeted delivery [23].

## 7 Recent Advancements for Quantum Dots in Drug Delivery

Recent advancements in quantum dots (QDs) for drug delivery highlight their transformative potential in medicine. Quantum dots conjugated with mercapto groups exhibit enhanced aqueous solubility, increasing their biocompatibility. These nanostructures can encapsulate hydrophobic drugs between their inorganic core and polymer coating or bind hydrophilic drugs to their surface, facilitating targeted delivery. Innovations in drug release mechanisms, such as pH-sensitive, temperature-responsive, and enzyme-mediated systems, offer controlled and precise therapeutic applications. For instance, doxorubicin-graphene QD complexes have demonstrated improved cellular uptake and DNA cleavage. PEGylation of QDs enhances solubility and prolongs circulation, reducing nonspecific interactions. Furthermore, QDs are pivotal in cancer detection and tracking, particularly metastatic cells. Antibody-conjugated QDs provide real-time imaging, enabling the identification of circulating tumor cells even at low concentrations. FRET-based QDs further advance drug release studies by real-time monitoring of drug behavior within cells. These advancements emphasize QDs' versatility, offering targeted delivery, efficient drug release, and enhanced diagnostics for precision medicine [41].

## 8 Conclusion

This study concludes by emphasizing the transformative potential of quantum dots (QDs) in advancing diagnostic, bioimaging, and drug delivery technologies. In comparison to conventional drug carriers, QDs are particularly advantageous due to their unique optical properties, stability, and ability for precise, targeted delivery. Owing to the ongoing advancements in biocompatibility and reduced toxicity formulations, QDs are becoming increasingly viable for clinical applications, especially in theranostics, where diagnosis and treatment are integrated. In order to solidify their role in personalized medicine, future research should focus on further improving QD formulations for safety and efficacy, particularly for sensitive applications such as cancer treatment and neurological disease monitoring. Understanding the potentials of Quantum Dots and their different synthesis paradigms would aid in discovering the merits and significance of the zero-dimensional nano-scale systems in diverse drug delivery applications.

**Acknowledgments** The authors are thankful to Department of Biotechnology, Sathyabama Institute of Science & Technology, Chennai, India and Centre for Nanoscience and Nanotechnology, International Research Centre, Sathyabama Institute of Science & Technology, Chennai, India, for carrying out the work.

**Author Contributions** KR & SR drafted the manuscript; LI, PS and KB performed the revisions and necessary modifications for the submission of the book chapter.

**Conflicts of Interest** The authors declare no conflicts of interest among themselves.

## References

1. Mohammad, Z., Zeeshan, A., Faisal, S., Suhail, A., Sahar, I., Mohd, S., & Nazma, K. (2017). Vesicular drug delivery system used for liver diseases. *World Journal of Pharmaceutical Sciences*, 28–35.
2. Kumar, A., Nautiyal, U., Kaur, C., Goel, V., & Piarchand, N. (2017). Targeted drug delivery system: Current and novel approach. *International Journal of Pharmaceutical and Medical Research, 5*, 448–454.
3. Akhtar, M., Jamshaid, M., Zaman, M., & Mirza, A. Z. (2020). Bilayer tablets: A developing novel drug delivery system. *Journal of Drug Delivery Science and Technology, 60*, 102079.
4. Kwon, I. K., Lee, S. C., Han, B., & Park, K. (2012). Analysis on the current status of targeted drug delivery to tumors. *Journal of Controlled Release, 164*, 108–114.

5. Yoo, J., Park, C., Yi, G., Lee, D., & Koo, H. (2019). Active targeting strategies using biological ligands for nanoparticle drug delivery systems. *Cancers, 11*, 640.
6. Thakur, A., Roy, A., Chatterjee, S., Chakraborty, P., Bhattacharya, K., & Mahata, P. P. (2015). Recent trends in targeted drug delivery. *Philippine Journal of Science*.
7. Bilan, R., Nabiev, I., & Sukhanova, A. (2016). Quantum dot-based nanotools for bioimaging, diagnostics, and drug delivery. *ChemBioChem, 17*, 2103–2114.
8. Bilan, R., et al. (2016). Quantum dots for bioimaging and diagnostics. *Small*.
9. Delehanty, J. B., Boeneman, K., Bradburne, C. E., Robertson, K., & Medintz, I. L. (2009). Quantum dots: A powerful tool for understanding the intricacies of nanoparticle-mediated drug delivery. *Expert Opinion on Drug Delivery, 6*, 1091–1112.
10. Kargozar, S., Hoseini, S. J., Milan, P. B., Hooshmand, S., Kim, H. W., & Mozafari, M. (2020). Quantum dots: A review from concept to clinic. *Biotechnology Journal, 15*, 2000117.
11. Kim, J. Y., Voznyy, O., Zhitomirsky, D., & Sargent, E. H. (2013). 25th anniversary article: Colloidal quantum dot materials and devices: A quarter-century of advances. *Advanced Materials, 25*, 4986–5010.
12. Mozafari, M., Moztarzadeh, F., Seifalian, A. M., & Tayebi, L. (2013). Self-assembly of PbS hollow sphere quantum dots via gas–bubble technique for early cancer diagnosis. *Journal of Luminescence, 133*, 188–193.
13. Narayanan, K. B., & Sakthivel, N. (2010). Biological synthesis of metal nanoparticles by microbes. *Advances in Colloid and Interface Science, 156*, 1–13.
14. Prakash, A., Sharma, S., Ahmad, N., Ghosh, A., & Sinha, P. (2010). Bacteria mediated extracellular synthesis of metallic nanoparticles. *International Research Journal of Biotechnology, 1*, 071–079.
15. Sweeney, R. Y., Mao, C., Gao, X., Burt, J. L., Belcher, A. M., Georgiou, G., & Iverson, B. L. (2004). Bacterial biosynthesis of cadmium sulfide nanocrystals. *Chemistry & Biology, 11*, 1553–1559.
16. Karakoti, A. S., Shukla, R., Shanker, R., & Singh, S. (2015). Surface functionalization of quantum dots for biological applications. *Advances in Colloid and Interface Science, 215*, 28–45.
17. Thanh, N. T., & Green, L. A. (2010). Functionalisation of nanoparticles for biomedical applications. *Nano Today, 5*, 213–230.
18. Hitchcock, A. P., Leung, B. O., Brash, J. L., Scholl, A., & Doran, A. (2012). Soft X-ray spectromicroscopy of protein interactions with phase-segregated polymer surfaces. In Proteins at Interfaces III: State of the Art (pp. 731–760).
19. Algar, W. R., & Krull, U. J. (2010). Developing mixed films of immobilized oligonucleotides and quantum dots for the multiplexed detection of nucleic acid hybridization using a combination of fluorescence resonance energy transfer and direct excitation of fluorescence. *Langmuir, 26*, 6041–6047.
20. Yong, K.-T., Wang, Y., Roy, I., Rui, H., Swihart, M. T., Law, W.-C., Kwak, S. K., Ye, L., Liu, J., & Mahajan, S. D. (2012). Preparation of quantum dot/drug nanoparticle formulations for traceable targeted delivery and therapy. *Theranostics, 2*, 681.
21. Tewabe, A., Abate, A., Tamrie, M., Seyfu, A., & Abdela Siraj, E. (2021). Targeted drug delivery—From magic bullet to nanomedicine: Principles, challenges, and future perspectives. *Journal of Multidisciplinary Healthcare, 14*, 1711–1724.
22. Xiao, Y., Forry, S. P., Gao, X., Holbrook, R. D., Telford, W. G., & Tona, A. (2010). Dynamics and mechanisms of quantum dot nanoparticle cellular uptake. *Journal of Nanobiotechnology, 8*, 1–9.
23. Medintz, I. L., Uyeda, H. T., Goldman, E. R., & Mattoussi, H. (2005). Quantum dot bioconjugates for imaging, labelling, and sensing. *Nature Materials, 4*(6), 435–446.
24. Zhao, D., Sun, Q., & Li, J. (2021). Size-tunable fluorescence and photostability of quantum dots in cancer imaging. *ACS Nano, 15*(7), 12030–12044.
25. Cai, W., Chen, X., & Xie, J. (2007). Functionalized quantum dots for tumor imaging and targeted therapy. *Journal of Nanoscience and Nanotechnology, 7*(8), 1308–1318.
26. Liu, J., Zhang, Y., & Zhang, W. (2020). Surface functionalization of quantum dots for bioimaging and drug delivery applications. *Advanced Materials, 32*(18), 1906701.

27. Freitas, M., Pereira, C., & Matos, A. (2020). Electrochemical sensors based on quantum dots for cancer biomarker detection. *Analytical Chemistry, 92*(7), 4520–4530.
28. Zhang, L., Wang, H., & Zhao, X. (2019). Highly sensitive electrochemical detection of cancer biomarkers using quantum dots. *Biosensors and Bioelectronics, 133*, 110–116.
29. Shen, Y., Zhao, L., & Li, Z. (2022). Biocompatibility and therapeutic applications of quantum dots in drug delivery. *Nanomedicine, 17*(3), 151–162.
30. Kumar, A., Gupta, S., & Singh, R. (2021). Quantum dots for photodynamic therapy: Advances in ROS generation. *Photochemistry and Photobiology, 97*(5), 1235–1247.
31. Zhang, Y., Hu, W., & Wang, C. (2020). Quantum dots in photodynamic therapy: Recent developments and future trends. *Journal of Biomedical Nanotechnology, 16*(2), 329–341.
32. Branford, A. L., Kuo, W. H., & Chen, Y. (2018). Addressing cadmium toxicity in quantum dots: Progress and future perspectives. *Toxicological Sciences, 162*(4), 12–25.
33. Ali, X., Chen, Y., & Li, W. (2021). Advancements in carbon-based quantum dots for improved biocompatibility and clinical applications. *Journal of Nano Research, 14*(3), 87–99.
34. Wang, S., Cole, I. S., Zhao, D., & Li, Q. (2016). The dual roles of functional groups in the photoluminescence of graphene quantum dots. *Nanoscale, 8*, 7449–7458.
35. Wang, S., et al. (2016). Photoluminescence in quantum dot photocatalysis. *ACS Applied Materials & Interfaces, 8*, 7449–7458.
36. Ho, Y.-P., & Leong, K. W. (2010). Quantum dot-based theranostics. *Nanoscale, 2*, 60–68.
37. Jin, S., Hu, Y., Gu, Z., Liu, L., & Wu, H.-C. (2011). Application of quantum dots in biological imaging. *Journal of Nanomaterials, 2011*, 834139.
38. Probst, C. E., Zrazhevskiy, P., Bagalkot, V., & Gao, X. (2013). Quantum dots as a platform for nanoparticle drug delivery vehicle design. *Advanced Drug Delivery Reviews, 65*(5), 703–718. https://doi.org/10.1016/j.addr.2012.09.036. Epub 2012 Sep 20. PMID: 23000745; PMCID: PMC3541463.
39. Zadeh, E. S., Ghanbari, N., Salehi, Z., Derakhti, S., Amoabediny, G., Akbari, M., & Tokmedash, M. A. (2023). Smart pH-responsive magnetic graphene quantum dots nanocarriers for anticancer drug delivery of curcumin. *Materials Chemistry and Physics, 297*, 127336.
40. Zadeh, E. S., et al. (2023). Magnetic graphene quantum dots for cancer therapy. *Chemical Science, 297*, 127336.
41. Gidwani, B., et al. (2021). Quantum dots: Prospectives, toxicity, advances and applications. *Journal of Drug Delivery Science and Technology, 61*, 102308.

# Quantum Dot Facilitated Cancer Diagnosis and Treatment

**Debasmita Paul, Brajadulal Chattopadhyay, Kunal Biswas, and Abiral Tamang**

**Abstract** Quantum dots are semiconductor nanomaterials with exceptional optoelectronic properties showing significant potential in cancer diagnostics and treatments. Known for their fluorescence intensity, stability, and highly tunable emission, QDs are of considerable importance for their ability to image and target cancer cells at both in vitro and in vivo levels. This chapter explores the application of QDs in bioimaging, biosensing, and theranostics, with a focus on photodynamic therapy and the development of biocompatible QDs for real-time imaging in cancer research and treatments. QDs are critical in facilitating visualization of cellular uptake, cancer cell dynamics, and receptor profiling, which are crucial for early detection of the disease. QDs have also been employed for targeted drug delivery, where therapeutic molecules have been transported to the cancerous cells by surface-functionalized QDs with minimal impact on healthy tissues. The chapter also focuses on the challenges regarding QD toxicity and hence emphasizes the green surface-modified synthesis of QDs for reduced toxicity, enabling safe and biocompatible applications in cancer research and theranostics.

**Keywords** Quantum dots · Biosensing · Bioimaging · Theranostics · Cancer diagnostics

The original version of the chapter has been revised. A correction to this chapter can be found at https://doi.org/10.1007/978-3-031-94551-9_17

D. Paul · B. Chattopadhyay · A. Tamang (✉)
Department of Physics, Jadavpur University, Kolkata, India
e-mail: debasmitap.physics.rs@jadavpuruniversity.in; abiralt.physics@jadavpuruniversity.in

K. Biswas (✉)
Centre for Nanoscience and Nanotechnology, International Research Centre, Sathyabama Institute of Science and Technology, Chennai, India



N. Mazumder et al. (eds.), *Advanced Biophysical Techniques in Biosciences*, Lecture Notes in Nanoscale Science and Technology 7,
https://doi.org/10.1007/978-3-031-94551-9_4

# 1 Introduction

Cancer remains a prominent cause of mortality worldwide [1], presenting a complex challenge in terms of diagnosis and treatment due to its unique nature in each individual [2]. Recent advancements in nanomedicine have facilitated the detection and comprehensive analysis of tumour cells at a cellular level [3], marking a significant departure from traditional diagnostic methods. In recent years, there has been a considerable increase in the research and application of nanomaterials [4], particularly in the context of cancer treatment and diagnostics [5]. This field has demonstrated promising potential in addressing challenges related to early tumour detection, precise diagnoses [6], and personalized therapy. The primary advantage of nanomaterials in cancer treatment lies in the minuscule size of nanoparticles, which enables them to operate at the molecular level, thereby enhancing diagnostic capabilities and facilitating innovative targeting strategies at the molecular level. For instance, certain nanoparticles function by binding to cancer biomarkers such as circulating tumour cells, circulating tumour DNA, exosomes, and specific cancer-associated proteins [7]. Quantum dots (QDs) are special nanomaterials, often likened to "artificial atoms," and have emerged as a prominent subject in cancer nanomedicine. First described in 1981 by Alexey Ekimov [8], QDs are composed of a relatively small number of atoms (ranging from 100 to 10,000 atoms) of semiconductor materials from groups II-VI, III-IV, and IV-VI elements in the periodic table. Their minute dimensions are reflected in the term "dots," while "quantum" pertains to their properties and behaviour, which are extensively described by quantum mechanics. QDs, which are functional nanoparticles, have emerged as valuable tools for cancer diagnosis. These semiconducting nanocrystals, typically 2–10 nm in diameter, possess exceptional optoelectronic properties that can be finely tuned to provide sensitive environmental feedback. Materials at this scale exhibit unique properties due to their small size. These distinct properties are often attributed to quantum confinement and surface effects resulting from their diminutive dimensions. The phenomenon of quantum confinement confines the movement of electrons within a small volume, giving rise to distinctive optical and electronic effects. Additionally, the chemical reactivity of the surface typically increases as the size decreases, while the melting point usually decreases, leading to novel optical and thermal properties that can be leveraged for in vivo [9] and in vitro applications through active interactions with molecular components at the cellular level. Numerous nanoparticles have been explored for their potential in cancer diagnosis and therapy. This chapter focuses on the remarkable properties of semiconducting QDs and their application in cancer therapy and diagnostics by attaching reporting or binding molecules to the surface of the QD through conjugation [10]. Furthermore, the following chapter aims to provide a comprehensive summary of the latest progress in utilizing QDs for in vitro and in vivo imaging, as well as for targeted drug delivery systems in the context of research on cancer [5].

## 2 Characteristics of QDs

QDs exhibit versatile fluorescence properties that render them suitable for various biological applications. These properties include strong photoluminescence resulting from a combination of a high extinction coefficient and high quantum yield, broad absorption characteristics, and a narrow emission spectrum bandwidth [11]. In comparison to organic dyes, QDs have demonstrated emission that is independent of the excitation wavelength and offer emission maxima that are continuous and tunable owing to quantum size effects. Additionally, they possess an extended fluorescence lifetime that ranges from 10 to 40 ns [12], low susceptibility to photobleaching, and stability under relatively harsh environments. These distinctive features collectively position QDs as highly advantageous for biological applications.

A QD's size is proportioned inversely to its band gap energy level [13], which consequently modifies the frequency and colour of the emitted light (Fig. 1). Light of higher energy is emitted by smaller QDs, which decreases with increasing size of the particle.

Three Major Processes Influence the Fluorescence Properties of QDs:

1. Surface trapping of the excited electrons in the conduction band by the trap states incorporated on the surface of the QDs.
2. Recombination of the excited conduction band electrons in the valence band with the holes leading to photoluminescence.
3. Passivation of the surface traps with a duration of shell growth.

Since optical properties displayed by QDs are significantly influenced by their size [14], this characteristic can be modified during manufacturing to produce QDs

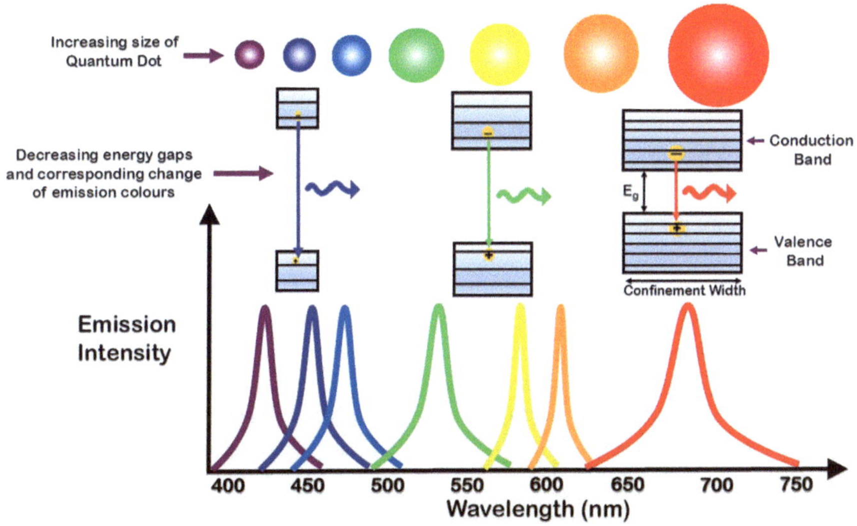


Fig. 1 Representation of quantum size effect; the principle behind quantum dots

suitable for imaging [15]. The size manipulation [16] is achieved through adjustments in the synthesis process, such as varying the extraction time of samples and the duration of shell growth [17]. It is crucial to carefully consider the shape and structure of QDs, as these factors directly impact the optical properties, including fluorescence colour, and the materials used in the synthesis process. Besides the importance of size on the energy of the emitted photoluminescence, the shape of QDs also influences the energy band level, thereby affecting the frequency of fluorescent light emitted or absorbed.

QDs are nanoparticles that are semiconductors [18] composed of periodic groups II-VI or III-V elements, with diameters typically ranging from 2 to 10 nm. This size range is comparable to or smaller than the exciton Bohr radius $a_B$, which refers to the separation between the negative electron and the positive hole within an exciton. The relationship between the QD‘s size and $a_B$ has defined three distinct regimes.

(a) Strong confinement regime—Here, the QD‘s radius is quite smaller than $a_B$, signifying that confinement energy towers above the Coulomb interaction.
(b) Weak confinement regime—Here the QD's radius is larger than $a_B$, suggesting that the confinement energy is lesser than the Coulomb interactions between electron and hole.
(c) Intermediate confinement regime—Here $a_B$ and confinement potential are comparable to one another.

The exciton's behaviour can be described by employing the particle in the box model, where the hole and the electron can be analogized to the hydrogen atom in the Bohr model, replacing the hydrogen nucleus with a positively charged hole and a negatively charged electron mass. The exciton's energy levels can be depicted as the solution to the particle in a box at the lowest energy level, i.e. $n = 1$, with the mass being substituted by the reduced mass. Hence, by altering the dimensions of the QD, it is possible to regulate the confinement energy of the exciton.

The quantum confinement effect occurs when the size of the particle is so small that it cannot be compared to the wavelength of the electron. The behaviour of an electron and a hole confined in nanocrystals is largely affected by the material properties, especially $a_B$. In III–V systems, $a_B$ is greater compared to II–VI systems, leading to more pronounced size quantization effects in QDs within III–V systems. The Bohr radius is a fundamental physical constant that denotes the most likely distance between the nucleus and the electron of a hydrogen atom in its lowest energy state. The impact of these effects is greater in larger nanocrystals and relies on material characteristics like the Bohr radius, $a_B = 2.34$ nm and $a_B$ of approximately 10 nm, which are specific to Cd-based compounds such as CdTe, CdZnTe, and CdTeSe. Some other types of QDs namely, $CuInS_2$ [19–26], GaAs [27], Carbon-QDs [6, 28–31], and Graphene-QDs [32–36] are also applied in cancer theronastics and bioimaging [33]. Here, the focus lies with the Cd-based QDs for its ease of fabrication and wider use across various studies. Semiconductor materials are commonly used in the fabrication of QDs and are usually composed of numerous atoms

(~thousands) of group II and VI elements (such as CdSe and CdTe) or group III and V elements (such as InP and InAs). The photoemission properties of colloidal II–VI group metal chalcogenide QDs, including CdS, CdSe, and CdTe, have been the subject of extensive research due to their size-dependent characteristics. This has resulted in substantial advancements in creating high-quality QD ensembles with a narrow size range.

The shape manipulation of colloidal nanostructures has been accomplished through alterations in the colloidal synthesis process to produce rod-shaped particles, known as quantum rods (QRs) [17]. These QRs exhibit a transition from zero-dimensional QDs to one-dimensional quantum wires [37]. In contrast to spherical dots, QRs demonstrate linearly polarized emission, as evidenced by fluorescence measurements on individual QRs and theoretical calculations. This unique property of linearly polarized emission, combined with the potential for wide spectral coverage and chemical accessibility, makes QRs highly promising as potential laser materials [38]. Additionally, QDs and QRs are anticipated to have lower lasing thresholds compared to two-dimensionally confined quantum wells, which form the basis of semiconductor diode laser devices widely used in information and telecommunication technologies [38].

## 3 Types of QDs

Metal chalcogenide nanocrystals [39], a type of colloidal semiconductor nanocrystals [40], have garnered significant attention due to their quantum confinement effects and size-dependent photoemission characteristics. Among these nanocrystals, cadmium-based chalcogenide nanocrystals, which encompass elements such as sulfur (S), selenium (Se), and tellurium (Te) [41] from the periodic table, are considered promising candidates for biological imaging applications. Additionally, these nanoparticles demonstrate potential utility in various optoelectronic technologies, including solar cells and laser technology [42].

Surface modification plays a crucial role in enhancing the photostability, solubility, and biocompatibility of QDs. This is achieved through the application of semiconductor shells on the cores [43], such as ZnS [44], ZnSe [45, 46], CdS [11, 47–55], and CdSe [56, 57], or by embedding QDs in biologically inert materials like polymers [58] or silica [59] (Fig. 2).

Over the years, a variety of QDs have been created to advance the field of biotechnology. These QDs can generally be classified into two categories based on their interaction with water: hydrophobic (water-hating) [60] and hydrophilic (water-loving) QDs [23]. The hydrophilic QDs are the topmost, feasible, and relatively safer contenders when it comes to the domain of cancer diagnostic studies or bio-based studies, owing to their biocompatibility with cells, which have a 70% water content.

## 4 Applications of QDs

Nanoparticles such as QDs [61] and QRs hold great promise in the field of nanotechnology (Fig. 3) [62]. Their potential applications extend beyond non-biological domains [63, 64], with ongoing research focused on mitigating their toxicity for biological use [65]. Efforts are being made to modify these nanoparticles to reduce their toxicity [66], allowing for their extensive use in biological science [67], including tagging biological molecules and drug delivery [35].

Research on doping QDs into polymers for sensor applications and the development of bio-inspired QDs for optoelectronic purposes are areas of emerging interest. Synthesis of temperature-dependent QDs for use as temperature sensors, as well as the development of efficient hydrophilic QDs [68] for early detection of cancer cells, are active areas of investigation [69]. Furthermore, QRs are under extensive

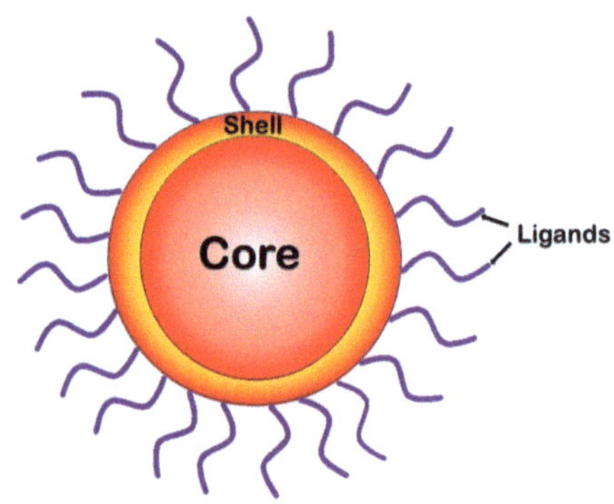


**Fig. 2** Structure of quantum dot

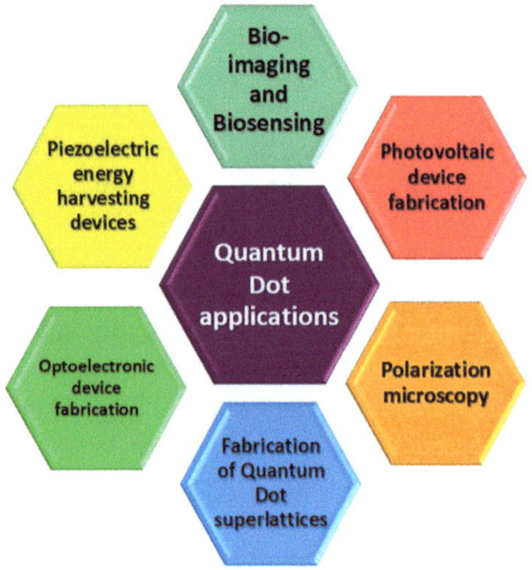


**Fig. 3** Applications of quantum dots

study for their potential applications in optoelectronics, such as in white LEDs, as well as polarization microscopy [38].

# 5 QDs in Cancer Detection and Treatment

## *5.1 Challenges of QD Utilization as Biological Probes*

QDs have exhibited exceptional optical properties for biomedical imaging applications, including high fluorescence intensity, broad absorption spectrum, and photostability compared to traditional organic dyes [70]. However, the conventional synthesis of QDs has been hindered by reaction conditions and toxicity concerns, impeding their advancement in the medical field for several decades. It has been reported that the cytotoxicity and stability of QDs are closely linked to surface modifications [71], affecting their solubility in different media and the duration of pathogen exposure to QDs [72]. QDs synthesized using organic methods are incompatible with numerous biological applications, resulting in inherent toxicity in various cells and bacterial pathogens. Furthermore, the increased toxicity of QDs has been associated with their exposure to the atmosphere, causing surface deterioration and reduced lifespan [69, 73]. Improper passivation leading to instability has been observed to cause QDs to accumulate as clogs on the surface of cells rather than penetrating them [21, 74]. Moreover, the use of heavy metals such as Hg, Pb, and Cd-based QDs, along with their synthetic methods, contributes to cytotoxicity and severe environmental effects.

To address these challenges, various synthetic methods for producing non-toxic heavy metal-free QDs with high fluorescence intensity and prolonged stability have emerged for biomedical applications [75]. QDs with an acceptable level of toxicity, such as CuInS/ZnS [24, 76], AgInS/ZnS, and ZnInS/ZnS, have shown promise in accelerating the use of these materials for diagnostics and therapy [21, 77]. The traditional synthesis methods are also believed to have resulted in a greater number of harmful effects on the environment and public health in our communities. Consequently, researchers are continually seeking more environmentally friendly approaches to counteract harmful production practices by proposing greener synthesis methods. Thus, eco-friendly methods involving the use of less toxic and biodegradable stabilizing agents for ternary nanomaterial production are being employed, alongside biologically friendlier systems. The concept of green synthesis is described as an effective method for sustainably creating nanomaterials using natural resources and methods that require minimal energy [78]. Cu and Ag-based multicomponent QDs have demonstrated even more remarkable absorption and emission tunability up to the near-infrared (NIR) range [76].

Capping ligands with high affinity for intracellular internalization have proven to be ideal candidates for the production of ternary QDs with high fluorescence intensity and improved cell internalization within bacteria and cell membranes [79].

Additionally, it has been emphasized that surface functionalization plays a crucial role in stabilizing and reducing the cytotoxicity of QDs while enhancing their utility as drug carriers and imaging tools [80, 81]. The impact of using a combination of thiol and protein-based ligands (TGA, sodium citrate, and gelatine MPA) as a dual capping group in the synthesis of NIR biocompatible QDs was investigated [43, 82, 83].

## 5.2 QDs as Fluorescent Probes

Methods like the large-scale synthesis of QDs were carried out using the commercial kitchen pressure cooker method, which has proven to be economically and environmentally viable. Gelatine, a micronutrient with a high protein content commonly used in food and medicinal packaging, was utilized along with thiols known to provide high internalization efficacy [84, 85]. The biocompatible NIR-emitting CuInSe/ZnS QDs synthesized in this manner demonstrated significant cellular uptake in both cancerous cells and bacteria, thereby contributing to the development of imaging probes for guided therapy and medicinal packaging materials. The feasibility of using TGA-citrate stabilized CuInSe/ZnS QDs as fluorescent probes was assessed by employing a confocal laser-scanning microscope to observe their uptake in both normal and cancer cells. The uptake of QDs was examined in pancreatic cancer cells (PC3), normal pancreatic cells (PNT), and human prostate adenocarcinoma cells (LNCaP). The results revealed a higher uptake of QDs in cancer cells (LNCaP and PC3) compared to PNT cells. The QDs were observed to be distributed in the cytoplasm of the cells, with minimal signals emitted from the nucleus. This selective uptake of QDs is attributed to the lack of lipid symmetry in cancer cells [86]. Quantification of fluorescence intensity confirmed the increased uptake of QDs in prostate cancer cells compared to normal prostate cells (Fig. 4A). The preferential affinity of CuInSe/ZnS QDs for cancer cells, in contrast to normal cells, was evident from the results with normal cells exhibiting less than 13- and 8-times fluorescence compared to prostate cancer cells (LNCaP) and pancreatic cancer cells (PC3), respectively (Fig. 4B) [24].

## 5.3 QRs for Multiplex Imaging

The utilization of QRs in bioimaging has recently gained attention due to advancements in controlled synthesis and surface-functionalization methods. Notably, Prasad's and Alivisatos' [87–94]groups have independently showcased the application of CdSe/CdS/ZnS QRs as targeting optical probes for live cancer cell imaging, highlighting their long-term biocompatibility and reduced photo-damage in two-photon fluorescence imaging. While CdSe/ZnS and CdTeSe alloy QDs have been previously used for multiplex labelling of cancer cells, there is limited information on the use of CdSe QRs for this purpose. Multiplex imaging [95], a valuable

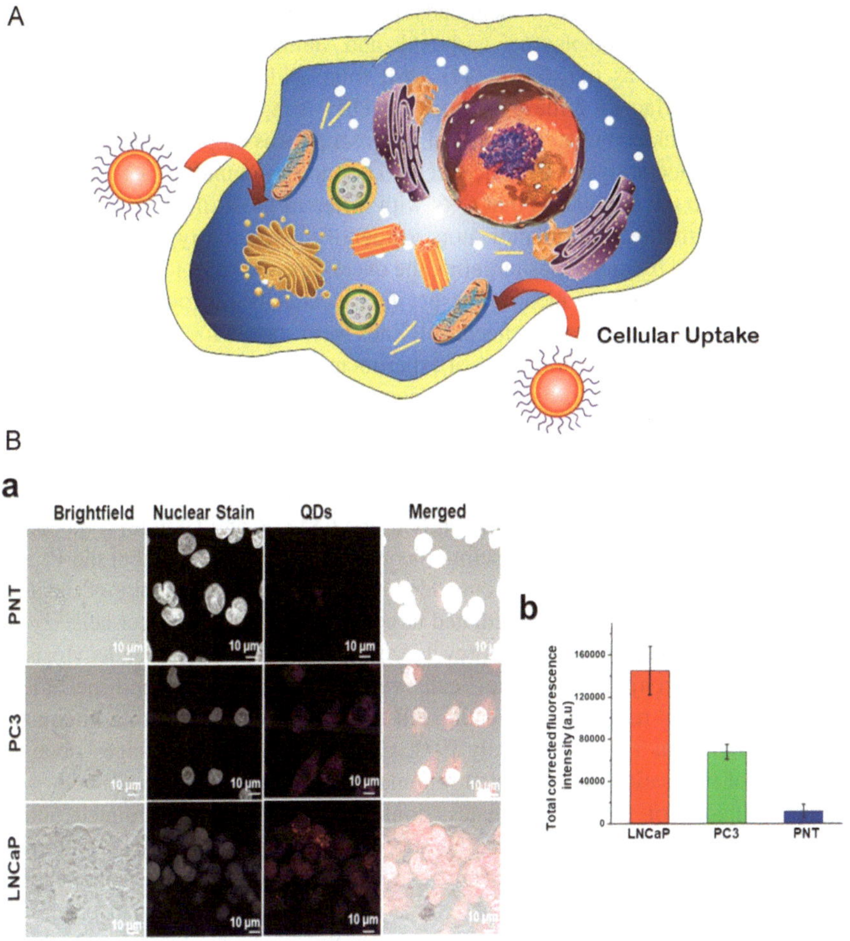


**Fig. 4** (**A**) Cellular uptake of quantum dots. (**B**) (**a**) Confocal images demonstrating the uptake of CuInSe/ZnS QDs in prostate cancer cells (PC3), human prostate adenocarcinoma cells (LNCaP) and prostate normal cells (PNT). (**b**) Image quantification of the fluorescence intensity

technique in in vitro microscopic imaging, typically involves the use of exogenous fluorescent probes or endogenous fluorescent proteins. However, compared to traditional fluorescent dyes, QR probes offer greater brightness, higher photostability, and larger absorption coefficients, making them promising candidates for multiplex labelling of overexpressed receptors on specific cancer cell lines. In this context, the application of CdSe/CdS/ZnS QRs has been reported, conjugated with cancer-specific biomolecules, for the multiplex labelling of live human cancer cells. The study utilized two different emission colours of QRs to demonstrate the multiplex labelling of cancer cells. The CdSe/CdS/ZnS QRs were synthesized by growing a CdS/ZnS graded shell on CdSe rods in a binary surfactant solution. The size of the rods was adjusted by varying the reaction time. Mercaptoundecanoic acid (MUA) displaced the hydrophobic surfactants from the QR surface, and lysine was then used to crosslink the carboxylic acid ends of the MUA on the QR surface, creating a hydrophilic shell displaying carboxyl and amine groups on its surface. Biomolecules such as transferrin (Tf) [96], folic acid (FA), or antibodies were linked to the surfaces of these hydrophilic QRs using carbodiimide chemistry, resulting in QR bioconjugates. Confocal microscopy was employed to analyze the uptake of QR-bioconjugates by cancer cells. The successful multiplex imaging with KB cells led to the hypothesis that QRs with different peak emission wavelengths can serve as effective optical indicators to identify a specific type of cancer cell, using conjugated biomolecules targeting the overexpressed receptors present on the cell type. The approach for identifying specific cancer cells was found to be straightforward and simple. Pancreatic cancer cell lines were chosen for these experiments. To validate the hypothesis, three sets of experiments were conducted. In the first two sets, 657 QRs were conjugated with a monoclonal antibody for specific targeting of human pancreatic cancer cells (Panc-1), using anti-Claudin 4 and anti-mesothelin antibodies, known to be overexpressed in both primary and metastatic pancreatic cancers. Additionally, 610 QRs were conjugated with the nonspecific antibody mouse-anti-IgG. In the last experiment, both the 657 QR-anti-Claudin 4 bioconjugates and the 610 QR-anti-mesothelin bioconjugates were used to label the Panc-1 cells. The results of these experiments demonstrated the specific nature of the anti-mesothelin and anti-Claudin 4-mediated targeting on pancreatic cancer cells.

The study aimed to verify the prevalence of receptor-mediated endocytosis of QR antibody bioconjugates in Panc-1 cells, which are known to lack anti-mesothelin and anti-claudin-4 receptors, by treating KB cells with 610 QR-anti-mesothelin and 657 QR-anti-Claudin-4 individually [97]. Confocal microscopy images obtained after 2 h of incubation demonstrated minimal detectable staining in KB cells with reduced uptake of the bioconjugates, confirming the receptor-mediated uptake observed in Panc-1 cells. This finding supported the hypothesis that a multiplex labelling technique can identify specific cancer cells through a simple set of experiments, potentially serving as a basis for a diagnostic tool for human pancreatic cancer. Considering the urgent need for robust imaging agents for the early detection of pancreatic cancer, this work may serve as a crucial milestone for future in vivo studies. The tunable emission spectra of QRs in the visible and NIR regions of the electromagnetic spectrum offer promising opportunities for bio-nanotechnology applications, such as simultaneous labelling of multiple antigens on normal and diseased cells, real-time tracking of intracellular processes, and single-copy level target detection. In summary, the study demonstrates the potential of QRs conjugated with targeting molecules as efficient optical tracking probes for multiplex labelling of cancer cells in vitro and presents novel optical probes, consisting of anti-Claudin 4 and anti-mesothelin-conjugated QRs, for in vitro detection of pancreatic cancer [97].

Some key points from the study have been highlighted:

- **Study Overview**:
  - KB cells were labelled with 610 and 657 QR–Tf bioconjugates.
  - QRs accumulated in vesicles via endocytotic uptake, confirmed to be receptor-mediated.

- **Control Experiments**:
  - Saturating cells with free Tf or using unconjugated QRs resulted in minimal uptake, validating specificity.
- **Selective Targeting**:
  - KB cells incubated with QR–Tf and QR–FA bioconjugates showed uptake driven by transferrin and folate receptors.
  - Hypothesis: QRs with distinct emission wavelengths can identify cancer cells by targeting overexpressed receptors.
- **Pancreatic Cancer Cell Line Experiments**:
  - **Experiment 1 & 2**: 657 QRs conjugated with anti-Claudin 4 and anti-mesothelin antibodies targeted Panc-1 cells, while 610 QRs conjugated with non-specific antibodies showed minimal labelling.
  - **Experiment 3**: Simultaneous use of 657 QR–anti-Claudin 4 and 610 QR–anti-mesothelin bioconjugates resulted in comparable staining, confirming specificity.
- **Receptor-Specific Uptake**:
  - Two-colour imaging demonstrated higher fluorescence intensity for receptor-specific QR-antibodies.
  - Minimal staining occurred in KB cells lacking Claudin 4 and mesothelin receptors.
- **Conclusions**:
  - Multiplex labelling using QRs can effectively identify specific cancer cells.
  - Technique shows promise as a diagnostic tool for pancreatic cancer, a condition requiring improved early detection methods.
- **Implications**:
  - Anti-Claudin 4- and anti-mesothelin-conjugated QRs show potential for pancreatic cancer detection.
  - The study highlights the potential of QRs in optical tracking and multiplex cancer cell labelling.

## *5.4 Bioconjugated QD-Based Theranostics for Cancer Cells*

Fluorescence imaging and photodynamic therapy (PDT) have demonstrated potential in clinical studies for the efficient identification and management of cancer [98]. Cancer fluorescence imaging includes the use of fluorescent dyes linked with anti-cancer antibodies to target tumour environments. The dyes are then excited using visible or NIR light sources. In PDT, cancer is treated by administering a photosensitizing (PS) drug followed by exposure to light. The basis of PDT is that when a PS drug is activated by light, it transfers energy or electrons to oxygen or other molecules, producing reactive oxygen intermediates (ROI) that harm essential biomolecules within cell organelles, ultimately leading to cell death. Fluorescence imaging offers a key benefit compared to other biomedical imaging methods like X-rays, CT, and PET. This advantage lies in the non-ionizing and less hazardous nature of visible and NIR excitation in fluorescence imaging [99]. PDT has a key advantage over chemotherapy and radiation therapy because it allows for targeted PS drug activation using visible or NIR light, providing selective therapy while preserving the immune system and normal cells. Fluorescence imaging and PDT for cancer encounter difficulties because of the restricted supply of photostable and NIR dyes as PS drugs. The key to fluorescence imaging and PDT for cancer is the targeted transportation of fluorescent dyes and PS drugs to tumour areas. The fundamental concept behind PDT involves the activation of a PS drug, which results in the generation of ROIs like singlet oxygen ($^1O_2$), hydroxyl radical (•HO), superoxide anion (−•$O_2$), and hydrogen peroxide ($H_2O_2$) through a sequence of energy and electron transfer processes. When a PS drug is photoactivated, it enters the excited singlet (S1) and triplet (T1) states. The T1 states for the majority of PS drugs typically last from several hundred nanoseconds to milliseconds, which is considerably longer than the S1 lifetime. In the T1 state, a PS drug can either release excess energy to molecular oxygen, causing it to relax to the ground state (S0), or transfer an electron (also in the S1 state) to oxygen, water, or a nearby molecule, thereby starting a sequence of photochemical reactions [100].

When a PS drug undergoes electron transfer, it generates ROI directly. However, this process also forms the cation radical of a PS, which then undergoes irreversible reactions with other molecules, leading to the chemical alteration of the PS. On the other hand, the generation of ROI through energy transfer in photosensitization is a sustainable process. For the longevity of PS drugs, energy transfer is favoured instead of electron transfer. Cell death in both processes is triggered by the photochemical reactions of ROI with biomolecules and cell organelles, including amino acids, endoplasmic reticulum, mitochondria, lysosomes, and the Golgi apparatus [100].

Further investigations have been undertaken to explore the targeted imaging of cancer cells utilizing QD-antibody conjugates. Receptors that are overexpressed in various cancers represent optimal targets for both the imaging and therapeutic intervention of malignancies. QDs conjugated with primary or secondary antibodies specific to over-expressed receptors have been extensively employed for the targeted visualization and sustained fluorescence imaging of cancer cells. A notable instance of cancer detection utilizing QDs was presented by Gao's group [101], who employed a conjugate of QDs and an antibody targeting prostate-specific membrane antigen (PSMA) to label human prostate cancer cells (C4-2). C4-2 cells exhibiting PSMA positivity were effectively targeted and labeled using the QD-antibody conjugate. Conversely, both the conjugate and QDs lacking an antibody failed to identify PSMA-negative PC-3 cells, as well as C4-2 and PC-3 cells. The Her2 receptor, a prominent biomarker frequently upregulated in various breast cancers, has been extensively utilized for the diagnosis and treatment of this malignancy. QD conjugated with Trastuzumab (Herceptin), a monoclonal antibody targeting the Her2 receptor, or immunoglobulin G (IgG) was utilized for the selective labelling and imaging of breast cancer cells. In a study conducted by Wu et al. [102], QDs were utilized to label human breast cancer cells (SK-BR-3) as well as mouse mammary tumour sections. This labelling approach involved initially targeting the cells and tissues with Herceptin, followed by the application of QD-IgG conjugates for enhanced visualization. SK-BR-3 cells were labeled utilizing a QD-streptavidin conjugate. This process involved initially targeting the cells with a humanized anti-Her2 antibody, followed by an application of biotinylated goat anti-human IgG. Yezhelyev et al. [103] advanced this methodology by selectively labelling MCF-7 and BT-474 breast cancer cell lines with visible and NIR QDs conjugated to antibodies targeting the human epidermal growth factor receptor 2 (Her2), epidermal growth factor receptor (EGFR), estrogen receptor (ER), progesterone receptor (PR), and the mammalian target of rapamycin (mTOR). Kawashima et al. [104] focused on the targeting of epidermal growth factor receptor (EGFR) single molecules in human ovarian epidermoid carcinoma cells (A431). This was achieved by employing cadmium selenide/zinc sulfide QDs conjugated with epidermal growth factor (EGF) and anti-EGFR antibodies. In a similar vein, Tada et al. [105] demonstrated that KPL-4 breast cancer cells can be selectively targeted and labelled utilizing NIR QDs that are conjugated with Herceptin. The work of Weng et al. [106] advanced the labelling of cancer cells with QDs by incorporating a multimodal strategy. This approach involved specifically targeting cancer cells through antibodies, facilitating drug delivery via immunoliposomes, and employing QDs for imaging purposes. Carboxylic acid-functionalized CdSe/ZnS QDs were covalently linked to a primary amine group within a liposome through the application of carbodiimide chemistry. Her2-overexpressing SK-BR-3 and MCF-7/HER2 breast cancer cell lines were specifically identified and quantified through the application of anti-Her2 antibodies conjugated to the surface of liposomes (Fig. 5A). The multimodal immunoliposome offers several advantages, including targeted labelling of cancer cells, enhanced contrast in fluorescence imaging, the ability to encapsulate chemotherapeutic agents such as doxorubicin, and efficient delivery of drugs at the intracellular level. Zhang et al. [107] recently identified QDs conjugated with anti-type 1 insulin-like growth factor receptor (IGF1R) as a promising approach for the targeted imaging of breast cancer cells. The critical aspect of this targeting approach was the identification of elevated levels of IGF1R expression in MCF-7 breast cancer cells through the use of QD-anti-IGFR1 conjugates. Yong et al. [108] employed QDs conjugated with anti-Claudin-4 antibodies and anti-prostate stem cell antigen (anti-PSCA) for the selective detection of human pancreatic cancer cells. The conjugates were identified by the membrane proteins Claudin-4 and PSCA, which exhibit elevated expression levels in both primary and metastatic pancreatic cancer cells. Figure 5B presents fluorescence images of human pancreatic cancer cells (MiaPaCa) that have been treated with InP/ZnS QD-conjugated anti-Claudin-4 antibodies.

Various strategies have concentrated on the targeted detection of cancer cells via QDs conjugated with anticancer antibodies. In a study by Kaul et al. [109], it was

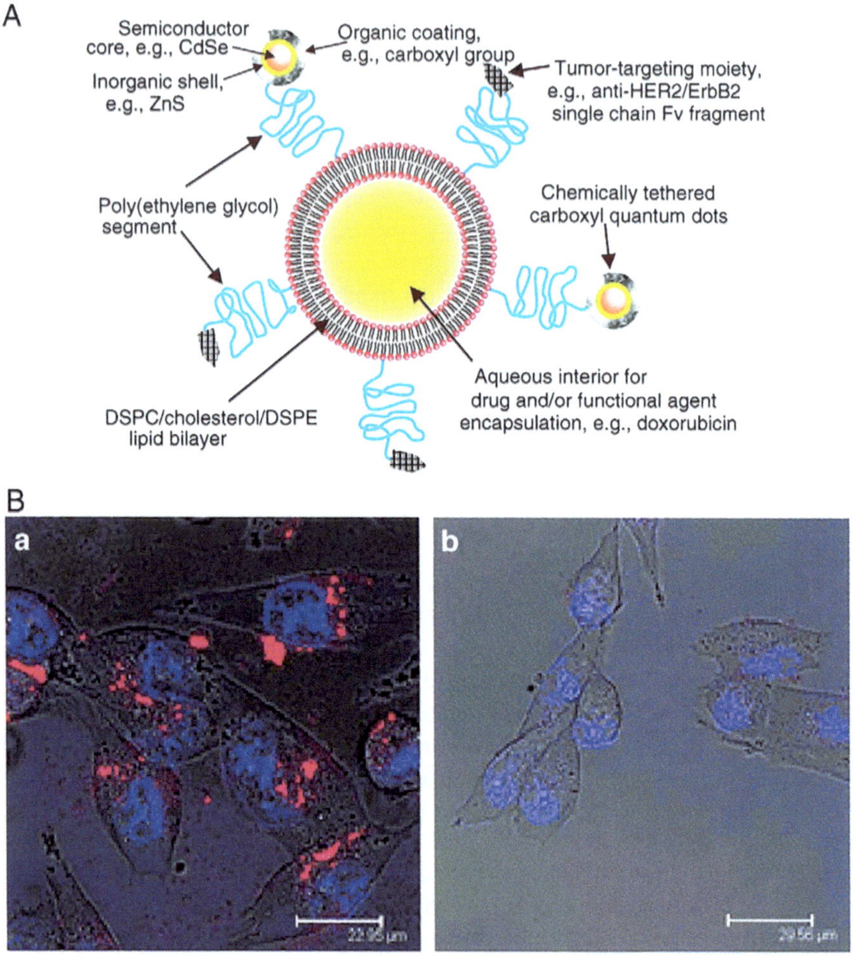


**Fig. 5** (**A**) Schematic presentation of an immunoliposome internalized with doxorubicin and conjugated with QDs and anti-Her2 antibody. (**B**) Fluorescence images of human pancreatic cancer cells incubated with (**a**) InP QD-anti-Claudin-4 antibody conjugate and (**b**) InP QD without antibody

demonstrated that cancer cells can be differentiated from normal cells utilizing QD-antibody conjugates. Normal human fetal fibroblast (WI-38) and osteosarcoma (U2OS) cells were subjected to immunostaining using an anti-mortalin antibody, followed by a biotinylated secondary antibody for detection. The biotinylated secondary antibody was subsequently conjugated with QD-streptavidin complexes. The study revealed that the intracellular localization of QDs varies between normal and cancerous cells, attributed to differences in mortalin distribution. In normal human cells, mortalin is consistently distributed throughout the cytoplasm, whereas in transformed cells, it predominantly localizes to the perinuclear region.

Additional investigations were conducted to examine in vivo targeted imaging of neoplastic cells utilizing QD-antibody conjugates.

The fundamental principles guiding in vitro targeting of cancer cells can also be effectively applied in vivo. The primary obstacles in the in vivo targeting and imaging of cancers utilizing QDs include the biodistribution of QD-bioconjugates, the penetration depth of excitation light and subsequent photoluminescence, tissue autofluorescence, as well as issues related to toxicity and pharmacokinetics. Bioconjugated QDs were utilized in vivo, administered either systemically to target deep-seated tumours or subcutaneously for the treatment of peripheral malignancies. In contrast to local administration, systemic administration requires greater consideration due to the potential interactions of QD-conjugates with components in the bloodstream and the subsequent activation of immune responses. While investigations have demonstrated that QDs conjugated with diverse anticancer antibodies exhibit selective and uniform distribution within the tumour microenvironment, there is limited evidence to suggest that QDs possess the capability to extravasate and access tumour cells in vivo. The biodistribution of QDs and their non-specific accumulation in the reticuloendothelial system, which encompasses the liver, spleen, and lymphatic system, represents a significant challenge for the in vivo applications of QDs.

The initial in vivo assessment of QDs was conducted by Akerman et al. [110]. In their study, they administered CdSe/ZnS QDs conjugated with peptides via tail vein injection in mice. The researchers observed a preferential accumulation of the injected QDs in the endothelial cells of pulmonary blood vessels. Through ex vivo fluorescence microscopic imaging of tissue sections, it was observed that the QD-peptide conjugates exhibited a preferential binding affinity for tumour tissues. Following this, QDs conjugated with different cancer biomarkers, including the PSMA antibody (Gao et al., [101]), RGD peptide (Cai et al., [111]), alpha-fetoprotein (Yu et al., [112]), and anti-Her2 antibody (Weng et al., [106]), were evaluated in vivo using murine models. Gao et al. [101] were among the first to utilize QD-antibody conjugates for in vivo examinations, which enabled detailed imaging of cancer in whole animal models. They systematically delivered QD-PSMA antibody conjugates to mice with subcutaneously implanted human prostate cancer. The conjugate of QDs and antibodies exhibited efficient and uniform distribution within the prostate tumour, attributable to the specific interaction between the prostate-specific membrane antigen (PSMA) present on prostate cancer cells and the PSMA antibody conjugated to the QDs (Fig. 6A). Cai et al. [111] explored in vivo targeting and imaging of tumours utilizing RGD peptide-conjugated NIR QDs. The research focused on targeting glioblastoma using a NIR QD conjugated with RGD peptides. The study assessed selective targeting capabilities through in vivo whole-animal imaging and ex vivo imaging of tumour tissues. The fundamental mechanism driving this targeting involves the specific interaction of the RGD peptide with the heightened expression of the $\alpha v\beta 3$ integrin found in U87MG glioblastoma cells and MDA-MB-435 human breast cancer cells. Figure 6B illustrates the signal-to-background ratio of NIR QD-RGD peptide conjugates in the context of cancer.

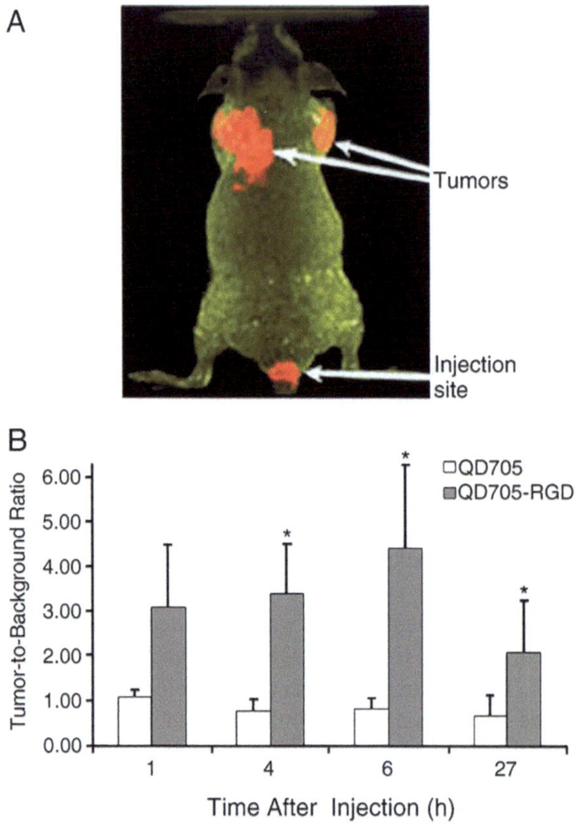


**Fig. 6** (**A**) Fluorescence image of human prostate cancer implanted in a mouse. The tumour is targeted with anti-PSMA antigen conjugated CdSe/ZnS QDs. (**B**) Histogram of fluorescence signal from U87MG tumour-bearing mice injected with an NIR QD-RGD peptide conjugate

In a study by Yu et al. [112], it was demonstrated that QDs conjugated with an antibody specific to alpha-fetoprotein (anti-AFP) serve as a promising agent for in vivo targeted imaging of the HCCLM6 human hepatocellular carcinoma cell line. Subcutaneously, HCCLM6 cancer cells were implanted in murine models, followed by the intravenous administration of QD-anti-AFP antibody conjugates. Alpha-fetoprotein (AFP), a key component of mammalian serum, serves as a significant biomarker for hepatic carcinoma. Consequently, the systemically delivered QD-anti-AFP conjugate demonstrated efficient accumulation within human hepatocellular carcinoma cells. Weng et al. [106] engineered multifunctional immunoliposomes aimed at in vivo targeted imaging of tumours, as well as for drug delivery and chemotherapy applications. In the preceding section, it was noted that NIR QDs were conjugated with anti-Her2 antibodies on the liposomal surface, while doxorubicin, an anti-cancer agent, was encapsulated within the liposome. The immunoliposome was utilized for targeted delivery in MCF-7/Her2 xenografts implanted in nude mice. This multimodal strategy for targeted imaging of tumours and therapeutic agent delivery holds significant promise for cancer imaging and photodynamic therapy (PDT).

Research has been undertaken to investigate the non-specific imaging of tumour vasculature and lymph nodes employing QDs. The pioneering research conducted by Larson et al. [113] on multi-photon in vivo fluorescence imaging utilizing QDs has garnered significant interest in cancer research. Their work demonstrated the effective targeting and two-photon imaging of tumour vasculature and lymph nodes through the use of bioconjugated QDs. Larson et al. [113] conducted a systematic study involving the administration of water-soluble cadmium selenide/zinc sulfide QDs in vivo in murine models. They employed NIR excited two-photon fluorescence microscopy to visualize capillary structures within adipose tissue and dermal layers. The significant two-photon absorption cross-section of QDs is essential for NIR excitation of visible QDs. Shortly following this report, the use of bioconjugated QDs for non-specific in vivo imaging of tumour vasculature, lymph nodes, and lymphatic drainage became a prominent area of research. Stroh et al. [114] focused on the identification and imaging of tumour vasculature related to MCaIV isogenic mouse adenocarcinoma tumour implants in C3H mice. They utilized PEG-phosphatidylethanolamine-conjugated core/shell QDs composed of CdS/ZnS and CdSe/ZnCdS, employing two-photon excitation for imaging. Kim et al. [115] utilized QDs for in vivo imaging and mapping of lymph nodes. The researchers administered subcutaneous injections of oligo-phosphine-coated NIR CdTe/CdSe QDs in both murine and porcine models. They observed that the QDs were rapidly transported away within a few minutes post-injection, particularly into the sentinel lymph node (SLN) located 1 cm beneath the skin. The NIR photoluminescence characteristics of QDs facilitated the visualization of their migration towards the sublingual nodes (SLN), as well as enabling image-guided surgical resection of the samples. Ballou et al. [116] demonstrated the successful imaging of lymph nodes in a mouse model utilizing QDs that lacked any specific surface functional groups. Intravenous and subcutaneous injections of CdSe/ZnS and CdSe/CdTe–ZnS QDs coated with negatively, positively, and neutrally terminated polyethylene glycol (PEG) molecules were administered to nude mice implanted with M21 melanoma. The lymph nodes were subsequently visualized using whole-animal fluorescence imaging (Fig. 7).

The study revealed that all three types of QDs were directed towards the inguinal lymph node, which was visualized through the skin using visible light for CdSe/ZnS QDs and NIR light for CdSe/CdTe–ZnS QDs. Interestingly, the study revealed that carboxylic acid-conjugated QDs exhibited a selective drainage toward the inguinal and axillary lymph nodes on the side of the tumour. The studies conducted by Kim et al. (2004) and Ballou et al. (2007) investigating the drainage of QDs devoid of specific biomolecules to sentinel lymph nodes (SLNs), inguinal lymph nodes, and lymph nodes adjacent to tumours highlight the potential of QDs for enhancing the visualization of tumour-associated lymph nodes as well as facilitating image-guided surgical applications. The retention of QDs in the initial draining lymph nodes, attributed to their substantial hydrodynamic size, limits their efficacy for imaging

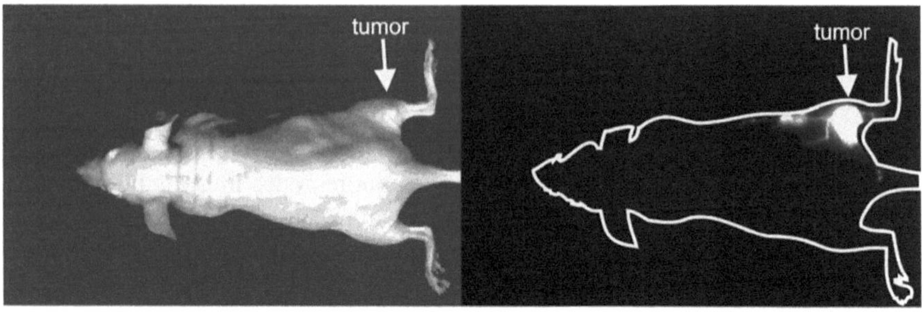


**Fig. 7** Nude mouse bearing M21 melanoma, dorsal view 3 min after injection into the tumour using 655 nm PEG 5 k-COOH quantum dots. Left, visible light; right, fluorescence at 655 nm

metastasis in distant lymph nodes and for delineating boundaries relevant to biopsy, surgical intervention, and photodynamic therapy (PDT).

## 5.5 *QD-Based Bimodal Imaging*

The fields of biomedical research heavily depend on magnetic resonance imaging (MRI), radiography, and fluorescence imaging due to their unique capabilities [117]. However, each modality possesses inherent limitations and cannot independently provide comprehensive imaging. Capturing top-notch images requires attributes like high spatial and temporal resolutions, 3D tomography, excellent signal-to-noise ratio, and non-invasiveness, which may not be present in individual imaging techniques. Consequently, the integration of multiple imaging modalities, known as multimodality imaging, has garnered considerable attention in both basic research and biomedical applications [118].

The use of independent imaging probes for different modalities is not always optimal due to differences in biodistribution and other pharmacodynamic properties. Hence, there is a growing interest in developing multimodal imaging probes that consolidate the properties required for different imaging modalities within a single chemical entity. These multimodal imaging probes feature components that synergistically enhance each other's functionalities. QDs show particular promise in this regard, as they offer the possibility of combining multiple probe characteristics [7].

QDs can be used for fluorescence imaging and can be integrated with MRI and radiography imaging by attaching them to molecules or materials that have paramagnetism and radioactivity on the surface of QDs. Bimodal imaging with QD probes can be seen in MRI-fluorescence imaging and scintigraphy-fluorescence imaging. The superior durability of QDs for multimodal imaging offers a notable edge over alternative probes, like organic dyes such as FITC and rhodamine, which are less favourable due to their susceptibility to photobleaching.

Researchers have explored various approaches for QD-based bimodal imaging, demonstrating successful targeting of specific cells and detection using fluorescence and MRI imaging. These approaches have been expanded to include silica nanoparticles to improve biocompatibility. Additionally, QDs have been employed for the detection of apoptosis and in vivo lymphatic imaging. The design and synthesis of suitable multimodal probes based on QDs offer a promising new dimension for biomedical imaging of cancer and other diseases, leveraging the rapid advancements in QD research and the wealth of information on the molecular mechanisms of diseases.

In PCR-based methods, the detection of mutated linear DNA molecules like circulating tumour DNA (ctDNA) [119] typically necessitates the presence of extensive nucleotide sequences on either side of the mutation for effective primer annealing. DNA fragments containing a mutation located nearer to the ends are less likely to be identified. Hence, there is a requirement for sensors that can detect linear DNA with specific mutations near the ends, offering advantages over existing methods. A biosensor [120] using electrochemiluminescence-resonance energy transfer (ECL-RET) has been explored. It involved capped CdS QDs and hairpin DNA probes with Au nanoparticle labels. This biosensor was designed for detecting epidermal growth factor receptor (EGFR) ctDNA containing a specific lung cancer mutation referenced as T790M. The ECL-RET system effectively identified different DNA molecules, such as 18-nucleotide single-stranded and 40-nucleotide single-stranded molecules, as well as 100-nucleotide double-stranded molecules. These molecules contained the single-nucleotide polymorphism (SNP) associated with T790M, positioned either in the middle or just 7 nucleotides from one end. The sensor showed excellent selectivity and achieved limits of detection (LODs) in the aM range for all target DNA. The 100-nt target linear ctDNA fragments were successfully identified with detection limits of 8.1 and 3.4 aM when the EGFR T790M SNP was positioned in the middle or at the end, respectively. These results show that ECL-RET systems are able to identify mutations in DNA fragments that might not be detected by conventional methods [7]. A comprehensive discussion of the mentioned study follows.

### 5.5.1 Identification of Small Single-Stranded Target DNA

Using T18-T790M as the target DNA dissolved in a potassium phosphate buffer, the ECL-RET sensor demonstrated remarkable linearity over a concentration range of 5 aM to 100 fM, achieving a limit of detection (LOD) of 3.5 aM ($3\sigma$, $n = 10$) as illustrated in Fig. 8A, B. This sensor exhibits a notably higher sensitivity for DNA molecules compared to the majority of electrochemical sensors documented in the current literature. The ECL-RET sensor demonstrated high specificity, with no detectable ECL signal observed beyond the anticipated background for the target DNA T18-wt, which corresponds to the wild-type allele of EGFR and contains only a single nucleotide polymorphism when compared to T18-T790M (Fig. 8A–C). The observed specificity of the sensor is influenced by

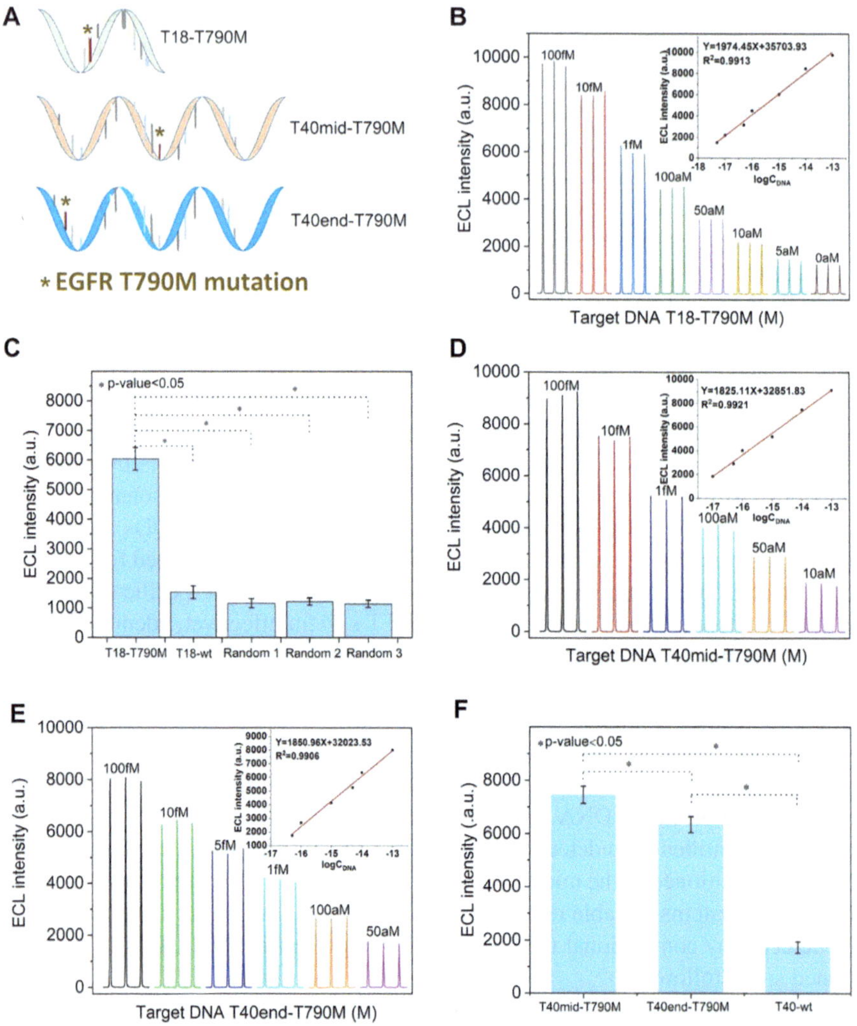


**Fig. 8** Detection of single-stranded DNA analytes with the CdS quantum dot ECL-RET sensor. (**A**) Evaluation of different single-stranded EGFR T790 M target DNA molecules. (**B**) ECL intensities and logarithmic curve (inset) with target DNA T18-T790M. (**C**) ECL intensities with 1 fM target DNA T18-T790M, T18-wt, or random 18-nt DNA molecules. ECL intensities and logarithmic curves (insets) with target DNA (D) T40mid-T790M and (**E**) T40end-T790M. (**F**) ECL intensities with 10 fM target DNA T40mid-T790M, T40end-T790M, and T40-wt. Each bar in (**C**) and (**F**) is the mean of at least three replicates with standard deviation; * indicates that the *p*-value is below 0.05

various operating conditions, including DNA concentration, buffer composition, and temperature. This specificity, which is reliant on the melting temperatures of the shared sequences between the haiDNA probe and the target analyte, is consistent with findings from numerous studies focusing on DNA detection systems that utilize similar FRET principles. Three distinct 18-nucleotide target DNA sequences, exhibiting various dissimilarities compared to T18-T790M, failed to produce an electrochemiluminescence (ECL) signal.

Furthermore, the ECL-RET sensor demonstrated consistent and reproducible outcomes, as illustrated in Fig. 8B. The measurement of 1 fM T18-T790M was further validated using six single-use sensing devices. Based on the logarithmic standard curve, the anticipated electrochemiluminescence (ECL) intensity for this concentration of the target analyte is 6093 (Fig. 8B, inset). The average relative error observed across the six replicates was merely 2.6%.

The performance of the EGFR T790M ECL-RET sensor was assessed using two distinct single-stranded target DNA molecules, each comprising 40 nucleotides (Fig. 8A, D–F). The first target, designated as T40mid-T790M, contained the C → T single-nucleotide polymorphism (SNP) associated with the T790M amino acid substitution in the central region of the linear DNA sequence. In contrast, the second target, T40end-T790M, featured the SNP located just 7 base pairs from the 5′ terminus. For the T40mid-T790M, the sensor demonstrated a linear detection range spanning from 10 aM to 100 fM, with a LOD recorded at 9.5 aM (Fig. 8D, F). In comparison, T40end-T790M exhibited a slightly narrower linear range of 50 aM to 100 fM, while demonstrating a lower LOD of 3.9 aM (Fig. 8E, F). The observed decrease in linear range for T40end-T790M as the analyte may be attributed to diminished electrochemiluminescence (ECL) intensities at equivalent concentrations when contrasted with T40mid-T790M (Fig. 8D, E). Specifically, T40end-T790M produced an ECL signal of 6347 ± 98.9 A.U. ($n = 3$) at a concentration of 10 fM, whereas T40mid-T790M yielded a higher ECL signal of 7460.7 ± 89.2 A.U. (Fig. 8F). One potential explanation for this observation is that the extended non-complementary region of T40end-T790M relative to the haiDNA probe may create steric hindrance, which could result in a slight reduction in the number of analyte molecules able to hybridize. Furthermore, the ECL-RET sensor demonstrated specificity even with longer single-stranded DNA, as no detectable ECL signal above background levels was observed for a 40-nucleotide target DNA containing the wild-type sequence (T40-wt). The sensing system presented herein is designed to detect single-stranded DNA of varying lengths and demonstrates a high specificity, enabling it to differentiate between targets that differ by a single-nucleotide polymorphism (SNP). Additionally, this sensor is effective in identifying the SNP, regardless of its position, whether it is situated at the center or at one terminus of the linear DNA target.

### 5.5.2 Identification of Double-Stranded Circulating Tumour DNA (ctDNA)-Like Entities

Subsequently, the ECL-RET sensor was assessed for its capacity to quantify double-stranded DNA targets measuring 100 nucleotides in length, which corresponds to the typical size of circulating tumour DNA (ctDNA) molecules (Fig. 9). In this experiment, analytes in a potassium phosphate buffer were subjected to a preheating process at 95 °C for 5 min to denature the double-stranded DNA, prior to their immediate application onto the ECL-RET device. The target sequence

T100mid-T790M, containing the SNP responsible for T790M positioned centrally, exhibited a linear detection range from 10 aM to 100 fM, with a LOD established at 8.1 aM (Fig. 9A, B). When the single-nucleotide polymorphism (SNP) was strategically placed 7 base pairs away from the 5′ terminus of the target DNA (T100end-T790M), the linear detection range was maintained, resulting in a lower LOD of 3.4 attomolar (aM) as illustrated in Fig. 9A, C. The ECL-RET sensor demonstrated a high level of specificity when assessing the shorter single-stranded target DNA, exhibiting a markedly diminished ECL signal for T100-wt, which contains the wild-type sequence of EGFR with just a single nucleotide polymorphism (SNP) variation, in comparison to T790M ctDNA-like molecules.

An electrochemical impedance spectroscopy (EIS) experiment was performed to verify the proper functioning of the ECL-RET sensor with extended DNA targets containing the single-nucleotide polymorphism (SNP) associated with the T790M mutation at various positions. In this analysis, the charge transfer resistance (Rct) derived from the Nyquist plots exhibits the following hierarchy: bare glassy carbon electrode (GCE) < AuNP-haiDNA/CdS QD/GCE < AuNP-haiDNA/CdS QD/GCE with T100-wild type < AuNP-haiDNA/CdS QD/GCE with

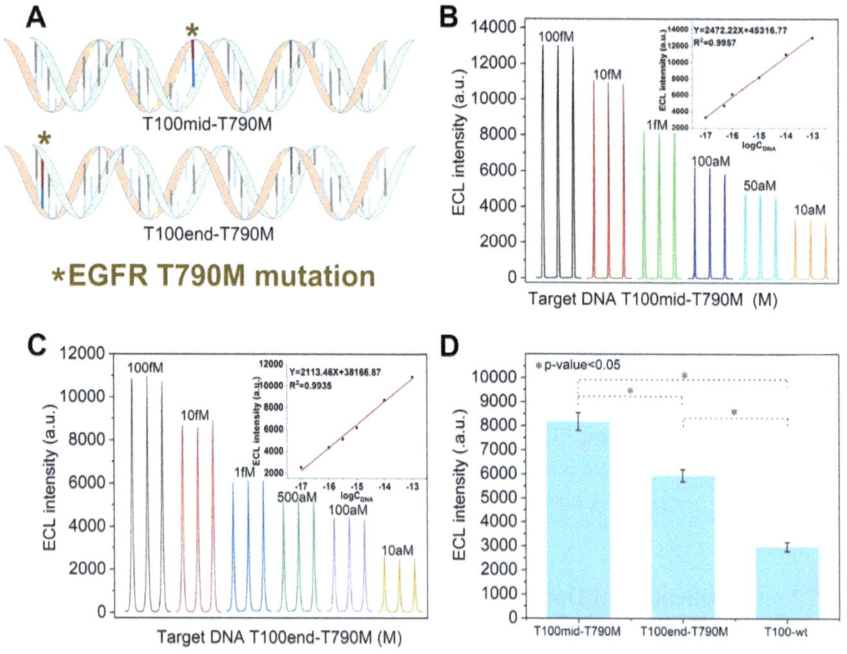


**Fig. 9** Detection of double-stranded EGFR T790M ctDNA-like molecules. (**A**) Double-stranded EGFR T790M target DNA molecules tested with the ECL-RET sensor. ECL intensities and logarithmic curves (insets) with target DNA (**B**) T100mid-T790M and (**C**) T100end-T790M. (**D**) ECL intensities with 1 fM target DNA T100mid-T790M, T100end-T790M, and T100-wt. Each bar in (**D**) is the mean of at least three replicates with standard deviation; * indicates that the $p$-value is below 0.05

T100end-T790M < AuNP-haiDNA/CdS QD/GCE with T100mid-T790M < CdS QD/GCE. The reduced charge transfer resistance (Rct) observed for the AuNP-haiDNA/CdS QD/GCE in the presence of T100-wt, compared to the sensors exposed to T100mid-T790M or T100end-T790M, indicates that the gold nanoparticles (AuNPs) are positioned closer to the glassy carbon electrode (GCE) surface. This proximity is attributed to the inadequate annealing between the wild-type target and the haiDNA probe, which subsequently enhances the electroconductivity of the system. The findings were consistent with the observed ECL signals, as evidenced by a comparison of the Nyquist plots of AuNP-haiDNA/CdS QD/GCE for the T100mid-T790M and T100end-T790M samples. The Rct exhibited a modest increase when the mutation was situated in the central region compared to instances where the mutation was confined to the 5′ end. The ECL-RET system may show a minor reduction in ECL intensity when comparing T100end-T790M to T100mid-T790M, potentially due to T100end-T790M having a slightly lower affinity for annealing to the haiDNA probe.

These findings demonstrate that the ECL-RET sensor is capable of detecting subtle sequence variations located near the terminals of elongated DNA fragments with high sensitivity. In contrast, conventional PCR methods aimed at detecting circulating tumour DNA (ctDNA) in biological fluids require a minimum of 50 nucleotides surrounding the mutation site for effective primer annealing (refer to Table S2). PCR is limited in its ability to detect DNA fragments that possess mutations situated too near either end of the fragment. Circulating tumour DNA (ctDNA) found in the bloodstream originates from the fragmentation of genomic DNA released by apoptotic or necrotic tumour cells. The process of DNA fragmentation in tumour cells remains inadequately explored; however, it is probable that circulating tumour DNA (ctDNA) molecules associated with a specific cancer mutation, such as T790M, are not uniform. Instead, they likely comprise a heterogeneous assortment of molecules varying in length, with the pertinent mutation situated at distinct locations within these fragments. In this context, PCR methods may yield false-negative results when the T790M SNP is located near the terminal region of the target linear DNA, whereas the ECL-RET sensor remains resistant to such limitations.

## 5.6 *QDs for Phagokinetic Tracking*

The mobility and relocation of cancer cells are extremely harmful as they lead to the development of additional tumours and the dissemination of cancer throughout the body [121, 122]. The movement of cells is closely associated with their potential to spread to other parts of the body, providing a way to evaluate the likelihood of metastasis by measuring how cells move. Observing the movement of cells currently presents practical limitations with existing techniques. The suggestion is that the latest developments in semiconductor nanotechnology might enhance cell motility tests by using the phagokinetic absorption of colloidal QDs.

The direct observation of cell movement, such as through time-lapse recordings of cells in culture, is limited by the ability to measure only a small number of cells at a time. This restricts its usefulness for conducting statistically significant studies of cell populations. The "scratched wound method" enables better data collection by removing a section of the cell culture substrate and measuring the time it takes for the "hole" to fill [123]. However, it has drawbacks such as losing the history of cell migration paths and relying on subjective analysis of intricate and inconsistent motion patterns. The Boyden Chamber invasion assay is a significant advancement. However, it is labour-intensive as it requires cell fixation and does not allow real-time monitoring of external conditions, despite its ability to monitor the rate of cell appearance on one side of a membrane after seeding cells on the other side [123].

Albrecht-Buehler suggested a technique for investigating cell movement by observing "phagokinetic tracks" [124]. In this approach, a cell moves across a layer of "markers," takes them in, and creates a blank area equal to the covered area, thus offering a fast and automated way to track cell movement while retaining specific path details. However, the technique has not been widely utilized because of challenges with finding suitable markers. The rapid bleaching of organic dyes has hindered the use of luminescence for phagokinetic tracking. Submicrometer Au particles [125] have been the sole markers utilized, leading to various constraints and impeding the widespread adoption of the technique [126]. The gold particles must be relatively large (0.15 μm) in order to be visible to the naked eye. On the other hand, these sizable particles exhibit weak adhesion to the substrate and need to be grown directly on it, leading to a significantly uneven distribution of particles.

The process of growth includes pouring a hot (almost boiling) water-based solution onto the surface, which restricts the types of substrates that can be used. Moreover, the considerable size of the Au microcrystals presents difficulties. When a cell moves by a distance equal to its diameter, it engulfs a volume of Au that is roughly 1% of the total cell volume, which could affect cell motility. This phenomenon is especially notable in small cells, such as epithelial cells, which are essential in the examination of metastasis and neurons. The film's uniformity is affected, and there is a significant distance between particles, which limits the resolution. Additionally, this approach provides a single level of contrast, which restricts measurements to two dimensions and makes it more challenging to link information about the chemical signals related to cell motility. In this approach, the initial use of nanometer-sized colloidal QDs as indicators for phagokinetic tracks is presented. The initial results outlined in this method indicated that QDs could potentially overcome the limitations mentioned earlier.

The study included applying thin layers of colloidal semiconductor nanocrystal QDs onto collagen-coated tissue culture substrates, and then seeding cells. The study utilized two different sizes of colloidal nanocrystal QDs: CdSe (2.8 nm)/ZnS/$SiO_2$ and CdSe (4.1 nm)/ZnS/$SiO_2$. The entry and movement of these QDs into the cells were studied using different types of cell lines, such as human mammary epithelial tumour cells (MDA-MB-231) and non-tumour cells (MCF-10 A). The findings indicated that the cells easily took in the colloidal QDs. Additionally, the small crystals were seen to be taken inside the cells and enclosed in tiny sacs. The research also showed that the QDs did not have any harmful effects on the cells within 10 days of being exposed to them.

The amount of QDs absorbed reflects the cells' migratory activity. MDA-MB-231 cells, which are a type of aggressive and undifferentiated breast cancer cells, effectively take in QDs as they move through the cell. This leads to the formation of an area around the cells that is free of QDs, allowing us to observe their migratory paths. In contrast, the MCF-10 A cell line, which is a highly differentiated non-tumour mammary epithelial cell line, exhibits minimal uptake of QDs and shows little indication of motility or migration. The way these cell types interact with QD-coated extracellular matrix is consistent with their usual behaviour when QDs are not present. The absence of QDs around MDA-MB-231 cells 24 h later is due to the cells engulfing the QDs as they migrate, rather than the QDs being displaced or degraded. The use of confocal microscopy clearly reveals the presence of numerous large vesicles filled with QDs in the MDA-MB-231 tumour cells after 24 h. In contrast, only a few small foci containing QDs were observed in the MCF-10 A cells [94].

Several significant results have been summarized here:

1. **QD Uptake and Localization**: All tested cell lines ingested QDs, which clustered in cytoplasmic vesicles, particularly near the nucleus. Confocal and two-photon microscopy confirmed internalization. QDs remained excluded from the nucleus and showed no degradation over time.
2. **Stability and Imaging**: Unlike organic dyes (FM 4–64), QDs resisted photobleaching during prolonged imaging. Vesicles containing QDs moved toward the nucleus at 0.1 μm/s, consistent with microtubule motor-mediated transport.
3. **Long-Term Retention**: Cells maintained QD-filled vesicles over a week and transferred them to daughter cells during division. No toxicity was observed within 10 days of exposure.
4. **Correlation with Migration**: Highly invasive MDA-MB-231 cells engulfed QDs during migration, leaving QD-free paths on coated substrates within 24 h. Non-migratory MCF-10 A cells retained intact QD layers, reflecting their stationary nature.
5. **Mechanism of Uptake**: QD uptake likely involved non-specific mechanisms like pino-, endo-, or phagocytosis. Adherence to the cell surface was attributed to interactions between QDs and glycoproteins or glycolipids.
6. **Applications of QD Tracking**: QDs proved ideal for phagokinetic tracking due to their robustness, resistance to photobleaching, and compatibility with fluorescence microscopy. They allow simultaneous monitoring of cell migration and protein dynamics, and the technique can extend to 3D matrix studies using QDs of varying sizes for depth contrast.

This approach offers a promising method to quantify cell motility and evaluate metastatic potential, particularly in cancer research.

# 6 Conclusion

The application of QDs with their versatile properties in cancer diagnostics and therapeutics represents a promising frontier in bioimaging, theranostics, and targeted drug delivery. Due to their photostability, tunable fluorescence, and ease of surface modifications with various functional molecules, QDs have a significant advantage over traditional organic fluorophores. These properties enable applications of QDs in real-time tracking of cellular processes and targeting of cancer cells, which is vital for precise and early diagnosis of cancer. Significant efforts have been made to improve the biocompatibility of QDs with reduced toxicity through advanced surface modifications and eco-friendly synthesis, however, the potential impact on the environment, long-term stability, and biodistribution remains a challenge. The prospects of QD applications in cancer studies lie in the development of heavy-metal-free QDs with biocompatible novel surface modifications to minimize the effect on non-target cells. QDs are excellent candidates for multimodal imaging as they have the promise of both diagnostic and treatment capabilities in oncological studies. The integration of QDs in imaging techniques such as PET and MRI can create simultaneous high-resolution insights into tumour morphology and precise treatment response. As QD technology advances, it could significantly improve the early detection, diagnosis, and effective management and treatment of cancer.

**Acknowledgments** The authors are thankful to Jadavpur University, Kolkata, India, and Sathyabama Institute of Science & Technology, Chennai, India, for providing the research facilities.

**Author Contributions** D.P. drafted and revised the manuscript, B.D.C., K.B. and A.T. performed the revisions and necessary modifications for the submission of the book chapter.

**Conflicts of Interest** The authors declare no conflict of interest.

# References

1. Sung, H., Ferlay, J., Siegel, R. L., Laversanne, M., Soerjomataram, I., Jemal, A., & Bray, F. (2021). Global Cancer Statistics 2020: GLOBOCAN estimates of incidence and mortality worldwide for 36 cancers in 185 countries. *CA: A Cancer Journal for Clinicians, 71*(3), 209–249. https://doi.org/10.3322/caac.21660
2. Khadem, H., Mangini, M., Farazpour, S., & De Luca, A. C. (2024). Correlative Raman imaging: Development and cancer applications. *Biosensors, 14*(7), 324. https://doi.org/10.3390/bios14070324
3. Liu, X., Zhou, P., Liu, H., Zhan, H., Zhang, Q., Zhao, Y., & Chen, Y. (2018). Design of bright near-infrared-emitting quantum dots capped with different stabilizing ligands for tumour targeting. *RSC Advances, 8*(8), 4221–4229. https://doi.org/10.1039/c7ra10824a
4. Moradi, S., Khaledian, S., Abdoli, M., Shahlaei, M., & Kahrizi, D. (2018). Nano-biosensors in cellular and molecular biology. Cellular and Molecular Biology, 64(5), 85–90. https://doi.org/10.14715/cmb/2018.64.5.14

5. Hamidu, A., Pitt, W. G., & Husseini, G. A. (2023). Recent breakthroughs in using quantum dots for cancer imaging and drug delivery purposes. Nanomaterials, 13. https://doi.org/10.3390/nano13182566
6. Wu, F., Su, H., Wang, K., Wong, W. K., & Zhu, X. (2017, October). Facile synthesis of N-rich carbon quantum dots from porphyrins as efficient probes for bioimaging and biosensing in living cells. International Journal of Nanomedicine, 12, 7375–7391. https://doi.org/10.2147/IJN.S147165
7. Yang, F., Gong, J. B., Li, M., Jiang, X., Zhang, J., Liao, M., Zhang, H., Tremblay, P. L., & Zhang, T. (2023). Electrochemiluminescent CdS quantum dots biosensor for cancer mutation detection at different positions on linear DNA Analytes. Analytical Chemistry, 95(37), 14016–14024. https://doi.org/10.1021/acs.analchem.3c02649
8. Ekimov, A. I. L., Efros, A. A., & Onushchenko, A. (1985). Quantum size effect in semiconductor microcrystals. Solid State Communications, 56(11), 921–924.
9. Bentolila, L. A., Ebenstein, Y., & Weiss, S. (2009). Quantum dots for in vivo small-animal imaging. Journal of Nuclear Medicine, 50, 493. https://doi.org/10.2967/jnumed.108.053561
10. Li, C., Ji, Y., Wang, C., Liang, S., Pan, F., Zhang, C., Chen, F., Fu, H., Wang, K., & Cui, D. (2014). BRCAA1 antibody- and Her2 antibody-conjugated amphiphilic polymer engineered CdSe/ZnS quantum dots for targeted imaging of gastric cancer. *Nanoscale Research Letters, 9*(1), 1–13. https://doi.org/10.1186/1556-276X-9-244
11. Rong, X. L., Zhao, Q., & Tao, G. H. (2012). Aqueous synthesis of CdSe and CdSe/CdS quantum dots with controllable introduction of Se and S sources. *Chinese Chemical Letters, 23*(8), 961–964. https://doi.org/10.1016/j.cclet.2012.05.029
12. Díaz, S. A., Aragonés, G. L., Buckhout-White, S., Qiu, X., Eunkeu, O., Susumu, K., Melinger, J. S., Huston, A. L., Hildebrandt, N., & Medintz, I. L. (2017). Bridging lanthanide to quantum dot energy transfer with a short-lifetime organic dye. *Journal of Physical Chemistry Letters, 8*(10), 2182–2188. https://doi.org/10.1021/acs.jpclett.7b00584
13. Ornes, S. (2016). Quantum dots. *Proceedings of the National Academy of Sciences of the United States of America, 113*, 2796. https://doi.org/10.1073/pnas.1601852113
14. Heyne, B., Arlt, K., Geßner, A., Richter, A. F., Döblinger, M., Feldmann, J., Taubert, A., & Wedel, A. (2020). Mixed mercaptocarboxylic acid shells provide stable dispersions of InPZnS/ZnSe/ZnS multishell quantum dots in aqueous media. *Nanomaterials, 10*(9), 1–24. https://doi.org/10.3390/nano10091858
15. Kargozar, S., Hoseini, S. J., Milan, P. B., Hooshmand, S., Kim, H. W., & Mozafari, M. (2020). Quantum dots: A review from concept to clinic. *Biotechnology Journal, 15*, e2000117. https://doi.org/10.1002/biot.202000117
16. Segets, D. (2016). Analysis of particle size distributions of quantum dots: From theory to application. *KONA Powder and Particle Journal*. https://doi.org/10.14356/kona.2016012
17. Shieh, F., Saunders, A. E., & Korgel, B. A. (2005). General shape control of colloidal CdS, CdSe, CdTe quantum rods and quantum rod heterostructures. *Journal of Physical Chemistry B, 109*(18), 8538–8542. https://doi.org/10.1021/jp0509008
18. Yoffe, A. D. (2001). Semiconductor quantum dots and related systems: Electronic, optical, luminescence and related properties of low dimensional systems. *Advances in Physics, 50*(1), 1–208. https://doi.org/10.1080/00018730010006608
19. Bagnall, A. J., Eliasson, N., Hansson, S., Chavarot-Kerlidou, M., Artero, V., Tian, H., & Hammarström, L. (2024). Ultrafast electron transfer from CuInS2 quantum dots to a molecular catalyst for hydrogen production: Challenging diffusion limitations. *ACS Catalysis, 14*(6), 4186–4201. https://doi.org/10.1021/acscatal.3c06216
20. Chen, B., Zheng, W., Chun, F., Xu, X., Zhao, Q., & Wang, F. (2023). Synthesis and hybridization of CuInS2 nanocrystals for emerging applications. *Chemical Society Reviews*. https://doi.org/10.1039/d3cs00611e
21. Fu, M., Luan, W., Tu, S. T., & Mleczko, L. (2015). Green synthesis of CuInS2/ZnS nanocrystals with high photoluminescence and stability. *Journal of Nanomaterials*. https://doi.org/10.1155/2015/842365

22. Iida, K., & Kim, D. (2022). Temperature-dependent photoluminescence properties of water-soluble CuInS2and CuInS2/ZnS quantum dots. *Journal of Applied Physics, 132*(19). https://doi.org/10.1063/5.0105290
23. Kang, X., Yang, Y., Huang, L., Tao, Y., Wang, L., & Pan, D. (2015). Large-scale synthesis of water-soluble CuInSe2/ZnS and AgInSe2/ZnS core/shell quantum dots. *Green Chemistry, 17*(8), 4482–4488. https://doi.org/10.1039/c5gc00908a
24. Ncapayi, V., Ninan, N., Lebepe, T. C., Parani, S., Girija, A. R., Bright, R., Vasilev, K., et al. (2024, September). Facile green synthesis of dual stabilized near-infrared CuInSe/ZnS quantum dots as fluorescent probes for cancer-bacteria imaging. *Inorganic Chemistry Communications, 167*, 112795. https://doi.org/10.1016/j.inoche.2024.112795
25. Peng, Z., Gu, Y., Li, X., Lu, B., Zhang, X., Zhang, Y., Liao, K., Li, W., & Chen, K. (2024, August). High-performance CuInS2 quantum dot sensitized solar cells through I−/MPA dual-ligands passivation. *Chemistry—An Asian Journal, 19*, e202400836. https://doi.org/10.1002/asia.202400836
26. Pons, T., Pic, E., Lequeux, N., Cassette, E., Bezdetnaya, L., Guillemin, F., Marchal, F., & Dubertret, B. (2010). Cadmium-free CuInS2/ZnS quantum dots for sentinel lymph node imaging with reduced toxicity. *ACS Nano, 4*(5), 2531–2538. https://doi.org/10.1021/nn901421v
27. Zhang, H., Ross, I. M., & Walther, T. (2013). Study of site controlled quantum dot formation on focused ion beam patterned GaAs substrate. *Journal of Physics: Conference Series, 471*, 012047. https://doi.org/10.1088/1742-6596/471/1/012047
28. Aladesuyi, O. A., & Oluwafemi, O. S. (2023, July). Synthesis of N, S co-doped carbon quantum dots (N,S-CQDs) for sensitive and selective determination of mercury ($Hg^{2+}$) in Oreochromis niloctus (Tilapia fish). *Inorganic Chemistry Communications, 153*. https://doi.org/10.1016/j.inoche.2023.110843
29. Àmbade, R. B., Ali, M., Lee, K. H., Jeong, W., Jeong, S. H., & Han, T. H. (2023). Nitrogen and sulfur co-doped carbon quantum dot-engineered $TiO_2$ graphene on carbon fabric for photocatalysis applications. *ACS Applied Nano Materials, 6*(17), 15782–15794. https://doi.org/10.1021/acsanm.3c02661
30. Kaur, J., Bhattu, M., Rawat, M., Varma, R. S., Acevedo, R., Shaban, M., Al-Saeedi, S. I., & Singh, J. (2023, November). Facile synthesis of carbon quantum dot/silver nanocomposite and its antimicrobial, catalytic and sensing applications. *Environmental Research, 237*, 116919. https://doi.org/10.1016/j.envres.2023.116919
31. Mowery, J. L., & Chen, J. (2024). Recent biomedical applications of carbon quantum dots in cancer treatment. *Journal of Physical Chemistry C*. https://doi.org/10.1021/acs.jpcc.4c04847
32. Chen, W., Lv, G., Hu, W., Li, D., Chen, S., & Dai, Z. (2018). Synthesis and applications of graphene quantum dots: A review. *Nanotechnology Reviews*. https://doi.org/10.1515/ntrev-2017-0199
33. Dar, M. S., Rosaiah, P., Bhagyalakshmi, J., Ahirwar, S., Khan, A., Tamizhselvi, R., Reddy, V. R. M., Palaniappan, A., & Sahu, N. K. (2025). Graphene quantum dots as nanotherapeutic agents for triple-negative breast cancer: Insights from 3D tumour models. *Coordination Chemistry Reviews, 523*, 216247. https://doi.org/10.1016/j.ccr.2024.216247
34. Osorio, H. M., Castillo-Solís, F., Barragán, S. Y., Rodríguez-Pólit, C., & Gonzalez-Pastor, R. (2024). Graphene quantum dots from natural carbon sources for drug and gene delivery in cancer treatment. *International Journal of Molecular Sciences, 25*(19), 10539. https://doi.org/10.3390/ijms251910539
35. Prabhakar, A. K., Ajith, M. P., Ananthanarayanan, A., Routh, P., Mohan, B. C., & Thamizhchelvan, A. M. (2022, November). Ball-milled graphene quantum dots for enhanced anticancer drug delivery. *OpenNano, 8*, 100072. https://doi.org/10.1016/j.onano.2022.100072
36. Shih, C. Y., Wang, P. T., Chung, W. P., Wang, W. H., Ting Chiang, I., Su, W. C., Huang, W. L., & Teng, H. (2023). Concise nanotherapeutic modality for cancer involving graphene oxide dots in conjunction with ascorbic acid. *Nanoscale, 15*(24), 10232–10243. https://doi.org/10.1039/d3nr00431g

37. Dalvand, P., & Mohammadi, M. R. (2011). Controlling morphology and structure of nanocrystalline cadmium sulfide (CdS) by tailoring solvothermal processing parameters. *Journal of Nanoparticle Research, 13*(7), 3011–3018. https://doi.org/10.1007/s11051-010-0198-6
38. Kazes, M., Lewis, D. Y., Ebenstein, Y., Mokari, T., & Banin, U. (2002). Lasing from semiconductor quantum rods in a cylindrical microcavity. *Advanced Materials, 14*(4), 317–321. https://doi.org/10.1002/1521-4095(20020219)14:4<317::AID-ADMA317>3.0.CO;2-U
39. Ghasempour, A., Dehghan, H., Ataee, M., Chen, B., Zhao, Z., Sedighi, M., Guo, X., & Shahbazi, M. A. (2023). Cadmium sulfide nanoparticles: Preparation, characterization, and biomedical applications. *Molecules*. https://doi.org/10.3390/molecules28093857
40. Mohagheghpour, E., Rabiee, M., Moztarzadeh, F., Tahriri, M., Jafarbeglou, M., Bizari, D., & Eslami, H. (2009). Controllable synthesis, characterization and optical properties of ZnS:Mn nanoparticles as a novel biosensor. *Materials Science and Engineering C, 29*(6), 1842–1848. https://doi.org/10.1016/j.msec.2009.02.012
41. Farkhani, S. M., & Valizadeh, A. (2014). Review: Three synthesis methods of CdX (X = Se, S or Te) quantum dots. *IET Nanobiotechnology*. https://doi.org/10.1049/iet-nbt.2012.0028
42. Bera, D., Qian, L., Tseng, T. K., & Holloway, P. H. (2010). *Quantum dots and their multimodal applications: A review*. https://doi.org/10.3390/ma3042260
43. Lin, W., Niu, Y., Meng, R., Huang, L., Cao, H., Zhang, Z., Qin, H., & Peng, X. (2016). Shell-thickness dependent optical properties of CdSe/CdS core/shell nanocrystals coated with thiol ligands. *Nano Research, 9*(1), 260–271. https://doi.org/10.1007/s12274-016-1014-0
44. Lien, V. T. K., Ha, C. V., Ha, L. T., & Dat, N. N. (2009). Optical properties of CdS and CdS/ZnS quantum dots synthesized by reverse micelle method. *Journal of Physics: Conference Series, 187*. https://doi.org/10.1088/1742-6596/187/1/012028
45. Okamoto, A., Bai, H., Toda, S., Huang, M., Kajii, H., Kawai, K., & Murakami, H. (2023). Controlling thickness of ZnSe intermediate shell narrows FWHM of green-emitting spectra of InP/ZnSe/ZnS multi-shell quantum dots. *ChemNanoMat, 9*(5). https://doi.org/10.1002/cnma.202200534
46. Zhang, J., Li, C., Li, J., & Peng, X. (2023). Synthesis of CdSe/ZnSe core/shell and CdSe/ZnSe/ZnS core/shell/shell nanocrystals: Surface-ligand strain and CdSe-ZnSe lattice strain. *Chemistry of Materials, 35*(17), 7049–7059. https://doi.org/10.1021/acs.chemmater.3c01333
47. Ashokan, A., Hutchison, J. A., & Mulvaney, P. (2024). Spectroelectrochemistry of CdSe/CdS core-shell quantum dots. *Chemistry of Materials, 36*(4), 1810–1817. https://doi.org/10.1021/acs.chemmater.3c01843
48. Bao, Y., Li, J., Wang, Y., Yu, L., Wang, J., Du, W., Lou, L., Zhu, Z., Peng, H., & Zhu, J. (2012). Preparation of water soluble CdSe and CdSe/CdS quantum dots and their uses in imaging of cell and blood capillary. *Optical Materials, 34*(9), 1588–1592. https://doi.org/10.1016/j.optmat.2012.03.033
49. Chen, Y., Xing, W., Liu, Y., Zhang, X., Xie, Y., Shen, C., Liu, J. G., Geng, C., & Xu, S. (2020). Efficient and stable CdSe/CdS/ZnS quantum rods-in-matrix assembly for white LED application. *Nanomaterials, 10*(2). https://doi.org/10.3390/nano10020317
50. Cirillo, M., Aubert, T., Gomes, R., Van Deun, R., Emplit, P., Biermann, A., Lange, H., Thomsen, C., Brainis, E., & Hens, Z. (2014). Flash synthesis of CdSe/CdS core-shell quantum dots. *Chemistry of Materials, 26*(2), 1154–1160. https://doi.org/10.1021/cm403518a
51. Das, K., Bhatt, N., Parambil, A. M., Kumari, K., Kumar, R., Rawat, K., Rajamani, P., et al. (2023). Divergent responses of hydrophilic CdSe and CdSe@CdS core–shell nanocrystals in apoptosis and in vitro cancer cell imaging: A comparative analysis. *Journal of Functional Biomaterials, 14*(9). https://doi.org/10.3390/jfb14090448
52. Das, K., Priyadarshini, E., Prasad, T., Rawat, K., & Bohidar, H. B. (2024). CdSe@CdS core-shell quantum dots as antifungal agents: Significance of particle size and shell thickness. *Journal of Nanoparticle Research, 26*(1). https://doi.org/10.1007/s11051-024-05931-6
53. Lin, Y. W., Hsieh, M. M., Liu, C. P., & Chang, H. T. (2005). Photoassisted synthesis of CdSe and core-shell CdSe/CdS quantum dots. *Langmuir, 21*(2), 728–734. https://doi.org/10.1021/la049489q

54. Nguyen, H. A., Hammel, B. F., Sharp, D., Kline, J., Schwartz, G., Harvey, S. M., Nishiwaki, E., et al. (n.d.). Colossal core/shell CdSe/CdS quantum dot emitters. *ACS Nano*. https://doi.org/10.26434/chemrxiv-2024-md0hn
55. Sabah, A., Shafaqat, I., Naifar, A., Albalawi, H., Alqahtani, M. S., Ashiq, M. G. B., & Shabbir, S. A. (2023, August). Investigation of band parameters and electrochemical analysis of multi core-shell CdSe/CdS/ZnS quantum dots. *Optical Materials, 142*, 114065. https://doi.org/10.1016/j.optmat.2023.114065
56. Kim, S., Fisher, B., Eisler, H. J., & Bawendi, M. (2003). Type-II quantum dots: CdTe/CdSe(core/shell) and CdSe/ZnTe(core/shell) heterostructures. *Journal of the American Chemical Society, 125*(38), 11466–11467. https://doi.org/10.1021/ja0361749
57. Naifar, A., & Hasanirokh, K. (2024, June). Exploring the microscopic effects of surrounding matrices ($HfO_2$ and $SiO_2$) on CdSe/ZnS and ZnS/CdSe cylindrical core/shell quantum dot for microelectronic transport applications. *Materials Science and Engineering: B, 304*, 117325. https://doi.org/10.1016/j.mseb.2024.117325
58. Ko, J., Jeong, B. G., Chang, J. H., Joung, J. F., Yoon, S. Y., Lee, D. C., Park, S., et al. (2020). Chemically resistant and thermally stable quantum dots prepared by shell encapsulation with cross-linkable block copolymer ligands. *NPG Asia Materials, 12*(1). https://doi.org/10.1038/s41427-020-0200-4
59. Pham, X. H., Park, S. M., Ham, K. M., Kyeong, S., Son, B. S., Kim, J., Hahm, E., et al. (2021). Synthesis and application of silica-coated quantum dots in biomedicine. *International Journal of Molecular Sciences, 22*(18). https://doi.org/10.3390/ijms221810116
60. Sperling, R. A., & Parak, W. J. (2010). Surface modification, functionalization and bioconjugation of colloidal inorganic nanoparticles. *Philosophical Transactions of the Royal Society A: Mathematical, Physical and Engineering Sciences*. https://doi.org/10.1098/rsta.2009.0273
61. Dessale, M., Mengistu, G., & Mengist, H. M. (2022). Nanotechnology: A promising approach for cancer diagnosis, therapeutics and theragnosis. *International Journal of Nanomedicine, 17*, 3735. https://doi.org/10.2147/IJN.S378074
62. Mohamed, W. A. A., El-Gawad, H. A., Mekkey, S., Galal, H., Handal, H., Mousa, H., & Labib, A. (2021). Quantum dots synthetization and future prospect applications. *Nanotechnology Reviews*. https://doi.org/10.1515/ntrev-2021-0118
63. Abdolahi, G., Dargahi, M., & Ghasemzadeh, H. (2020). Synthesis of starch-g-poly (acrylic acid)/ZnSe quantum dot nanocomposite hydrogel, for effective dye adsorption and photocatalytic degradation: Thermodynamic and kinetic studies. *Cellulose, 27*(11), 6467–6483. https://doi.org/10.1007/s10570-020-03198-3
64. Khan, M. M. R., Mitra, T., & Sahoo, D. (2020). Metal oxide QD based ultrasensitive microsphere fluorescent sensor for copper, chromium and iron ions in water. *RSC Advances, 10*(16), 9512–9524. https://doi.org/10.1039/c9ra09985a
65. Wang, J., Han, S., Ke, D., & Wang, R. (2012). Semiconductor quantum dots surface modification for potential cancer diagnostic and therapeutic applications. *Journal of Nanomaterials*. https://doi.org/10.1155/2012/129041
66. Valizadeh, A., Mikaeili, H., Samiei, M., Farkhani, S. M., Zarghami, N., Kouhi, M., Akbarzadeh, A., & Davaran, S. (2012). Quantum dots: Synthesis, bioapplications, and toxicity. *Nanoscale Research Letters, 7*. https://doi.org/10.1186/1556-276X-7-480
67. Wang, Y., Yu, L., Kong, X., & Sun, L. (2017). Application of nanodiagnostics in point-of-care tests for infectious diseases. *International Journal of Nanomedicine, 12*, 4789. https://doi.org/10.2147/IJN.S137338
68. Ham, K. M., Kim, M., Bock, S., Kim, J., Kim, W., Jung, H. S., An, J., et al. (2022). Highly bright silica-coated InP/ZnS quantum dot-embedded silica nanoparticles as biocompatible nanoprobes. *International Journal of Molecular Sciences, 23*(18). https://doi.org/10.3390/ijms231810977
69. Gidwani, B., Sahu, V., Shukla, S. S., Pandey, R., Joshi, V., Jain, V. K., & Vyas, A. (2021). Quantum dots: Prospectives, toxicity, advances and applications. *Journal of Drug Delivery Science and Technology*. https://doi.org/10.1016/j.jddst.2020.102308

70. Tsolekile, N., Parani, S., Matoetoe, M. C., Songca, S. P., & Oluwafemi, O. S. (2017). Evolution of ternary I–III–VI QDs: Synthesis, characterization and application. *Nano-Structures and Nano-Objects*. https://doi.org/10.1016/j.nanoso.2017.08.012
71. Liang, Z., Khawar, M. B., Liang, J., & Sun, H. (2021). Bio-conjugated quantum dots for cancer research: Detection and imaging. *Frontiers in Oncology, 11*. https://doi.org/10.3389/fonc.2021.749970
72. Oh, E., Liu, R., Nel, A., Gemill, K. B., Bilal, M., Cohen, Y., & Medintz, I. L. (2016). Meta-analysis of cellular toxicity for cadmium-containing quantum dots. *Nature Nanotechnology, 11*(5), 479–486. https://doi.org/10.1038/nnano.2015.338
73. Liu, Q., Ding, X., Pang, Y., Cao, Y., Lei, J., Wu, J., & Zhang, T. (2022). New insights into the safety assessment of quantum dots: Potential release pathways, environmental transformations, and health risks. *Environmental Science: Nano*. https://doi.org/10.1039/d2en00252c
74. Ncapayi, V., Ninan, N., Lebepe, T. C., Parani, S., Girija, A. R., Bright, R., Vasilev, K., et al. (2021). Diagnosis of prostate cancer and prostatitis using near infra-red fluorescent AgInSe/Zns quantum dots. *International Journal of Molecular Sciences, 22*(22). https://doi.org/10.3390/ijms222212514
75. Jing, L., Kershaw, S. V., Li, Y., Huang, X., Li, Y., Rogach, A. L., & Gao, M. (2016). Aqueous based semiconductor nanocrystals. *Chemical Reviews*. https://doi.org/10.1021/acs.chemrev.6b00041
76. Tsolekile, N., Nahle, S., Zikalala, N., Parani, S., Sakho, E. H. M., Joubert, O., Matoetoe, M. C., Songca, S. P., & Oluwafemi, O. S. (2020). Cytotoxicity, fluorescence tagging and gene-expression study of CuInS/ZnS QDS—meso (hydroxyphenyl) porphyrin conjugate against human monocytic leukemia cells. *Scientific Reports, 10*(1), 4936. https://doi.org/10.1038/s41598-020-61881-8
77. Oluwafemi, O. S., May, B. M. M., Parani, S., & Tsolekile, N. (2020, January). Facile, large scale synthesis of water soluble AgInSe2/ZnSe quantum dots and its cell viability assessment on different cell lines. *Materials Science and Engineering C, 106*, 110181. https://doi.org/10.1016/j.msec.2019.110181
78. Huston, M., Debella, M., Dibella, M., & Gupta, A. (2021). Green synthesis of nanomaterials. *Nanomaterials, 11*. https://doi.org/10.3390/nano11082130
79. Hashemkhani, M., Loizidou, M., Macrobert, A. J., & Acar, H. Y. (2022). One-step aqueous synthesis of anionic and cationic $AgInS_2$ quantum dots and their utility in improving the efficacy of ALA-based photodynamic therapy. *Inorganic Chemistry, 61*(6), 2846–2863. https://doi.org/10.1021/acs.inorgchem.1c03298
80. Kumar, B. A., Kumar, P., Elangovan, T., Ramalingam, G., Ravi, G., Mohanapriya, P., & Natarajan, T. S. (2021, September). Surface functionalization of core-shell QDs for solar photovoltaic and anti-cancer applications. *Applied Surface Science Advances, 5*, 100122. https://doi.org/10.1016/j.apsadv.2021.100122
81. Zhen, N., Wang, X., Li, X., Xue, J., Zhao, Y., Wu, M., Zhou, D., Liu, J., Guo, J., & Zhang, H. (2022). Protein-based natural antibacterial materials and their applications in food preservation. *Microbial Biotechnology, 15*, 1324. https://doi.org/10.1111/1751-7915.13918
82. Jeong, S., Achermann, M., Nanda, J., Ivanov, S., Klimov, V. I., & Hollingsworth, J. A. (2005). Effect of the thiol-thiolate equilibrium on the photophysical properties of aqueous CdSe/ZnS nanocrystal quantum dots. *Journal of the American Chemical Society, 127*(29), 10126–10127. https://doi.org/10.1021/ja042591p
83. Thangadurai, P., Balaji, S., & Manoharan, P. T. (2008). Surface modification of CdS quantum dots using thiols—structural and photophysical studies. *Nanotechnology, 19*(43), 435708. https://doi.org/10.1088/0957-4484/19/43/435708
84. Pastorino, J., Hansen, C. L., & McMahon, D. J. (2003). Effect of sodium citrate on structure-function relationships of Cheddar cheese. *Journal of Dairy Science, 86*(10), 3113–3121. https://doi.org/10.3168/jds.S0022-0302(03)73912-5
85. Taylor, D. A., Teku, J. A., Cho, S., Chae, W. S., Jeong, S. J., & Lee, J. S. (2021). Importance of surface functionalization and purification for narrow FWHM and bright green-emitting

InP core-multishell quantum dots via a two-step growth process. *Chemistry of Materials, 33*(12), 4399–4407. https://doi.org/10.1021/acs.chemmater.1c00348

86. Aizik, G., Waiskopf, N., Agbaria, M., Ben-David-Naim, M., Levi-Kalisman, Y., Shahar, A., Banin, U., & Golomb, G. (2019). Liposomes of quantum dots configured for passive and active delivery to tumour tissue. *Nano Letters, 19*(9), 5844–5852. https://doi.org/10.1021/acs.nanolett.9b01027
87. Alivisatos, P. (2004). The use of nanocrystals in biological detection. *Nature Biotechnology, 22*, 47. https://doi.org/10.1038/nbt927
88. Brinn, R. M., Yan, C., Abbas, A. S., Ondry, J., & Paul Alivisatos, A. (2024, July). Quantitative comparison of recombination rates of core/shell quantum dots in colloidal solutions and self-assembled monolayer superlattices. *The Journal of Physical Chemistry C, 128*, 12131. https://doi.org/10.1021/acs.jpcc.4c01434
89. Gerion, D., Pinaud, F., Williams, S. C., Parak, W. J., Zanchet, D., Weiss, S., & Paul Alivisatos, A. (2001). Synthesis and properties of biocompatible water-soluble silica-coated CdSe/ZnS semiconductor quantum dots. *Journal of Physical Chemistry B, 105*(37), 8861–8871. https://doi.org/10.1021/jp0105488
90. Gleason, S. P., Dahl, J. C., Elzouka, M., Wang, X., Byrne, D. O., Gababa, M., Cho, H., et al. (2024, July). *AuNR-SMA: Automated gold nanorod spectral morphology analysis pipeline*. http://arxiv.org/abs/2407.08769
91. Liu, H., Owen, J. S., & Paul Alivisatos, A. (2007). Mechanistic study of precursor evolution in colloidal group II-VI semiconductor nanocrystal synthesis. *Journal of the American Chemical Society, 129*(2), 305–312. https://doi.org/10.1021/ja0656696
92. Manna, L., Scher, E. C., & Paul Alivisatos, A. (2000). Synthesis of soluble and processable rod-, arrow-, teardrop-, and tetrapod-shaped CdSe nanocrystals. *Journal of the American Chemical Society, 122*(51), 12700–12706. https://doi.org/10.1021/ja003055
93. Manna, L., Scher, E. C., Li, L. S., & Paul Alivisatos, A. (2002). Epitaxial growth and photochemical annealing of graded CdS/ZnS shells on colloidal CdSe nanorods. *Journal of the American Chemical Society, 124*(24), 7136–7145. https://doi.org/10.1021/ja025946i
94. Parak, W. J., Boudreau, R., Le Gros, M., Gerion, D., Zanchet, D., Micheel, C. M., Williams, S. C., Alivisatos, A. P., & Larabell, C. (2002). Cell motility and metastatic potential studies based on quantum dot imaging of phagokinetic tracks. *Advanced Materials, 14*(12), 882–885. https://doi.org/10.1002/1521-4095(20020618)14:12<882::AID-ADMA882>3.0.CO;2-Y
95. Han, M., Gao, X., Su, J. Z, & Nie, S. (2001). *Quantum-dot-tagged microbeads for multiplexed optical coding of biomolecules*. http://biotech.nature.com
96. He, X., Gao, L., & Ma, N. (2013). One-step instant synthesis of protein-conjugated quantum dots at room temperature. *Scientific Reports, 3*. https://doi.org/10.1038/srep02825
97. Yong, K. T., Roy, I., Pudavar, H. E., Bergey, E. J., Tramposch, K. M., Swihart, M. T., & Prasad, P. N. (2008). Multiplex imaging of pancreatic cancer cells by using functionalized quantum rods. *Advanced Materials, 20*(8), 1412–1417. https://doi.org/10.1002/adma.200702462
98. SalmanOgli, A. (2011). Nanobio applications of quantum dots in cancer: Imaging, sensing, and targeting. *Cancer Nanotechnology, 2*, 1. https://doi.org/10.1007/s12645-011-0015-7
99. Brunetti, J., Riolo, G., Gentile, M., Bernini, A., Paccagnini, E., Falciani, C., Lozzi, L., et al. (2018). Near-infrared quantum dots labelled with a tumour selective tetrabranched peptide for in vivo imaging. *Journal of Nanobiotechnology, 16*(1), 21. https://doi.org/10.1186/s12951-018-0346-1
100. Biju, V., Mundayoor, S., Omkumar, R. V., Anas, A., & Ishikawa, M. (2010). Bioconjugated quantum dots for cancer research: Present status, prospects and remaining issues. *Biotechnology Advances, 28*, 199. https://doi.org/10.1016/j.biotechadv.2009.11.007
101. Gao, X., Cui, Y., Levenson, R., et al. (2004). In vivo cancer targeting and imaging with semiconductor quantum dots. *Nature Biotechnology, 22*, 969–976. https://doi.org/10.1038/nbt994
102. Wu, X., Liu, H., Liu, J., et al. (2003). Immunofluorescent labeling of cancer marker Her2 and other cellular targets with semiconductor quantum dots. *Nature Biotechnology, 21*, 41–46. https://doi.org/10.1038/nbt764

103. Yezhelyev, M., Al-Hajj, A., Morris, C., Marcus, A., Liu, T., Lewis, M., Cohen, C., Zrazhevskiy, P., Simons, J., Rogatko, A., Nie, S., Gao, X., & O'Regan, R. (2007). In situ molecular profiling of breast cancer biomarkers with multicolor quantum dots†. *Advanced Materials, 19*, 3146–3151. https://doi.org/10.1002/adma.200701983
104. Kawashima, N., Nakayama, K., Itoh, K., Itoh, T., Ishikawa, M., & Biju, V. (2010). Reversible dimerization of EGFR revealed by single-molecule fluorescence imaging using quantum dots. *Chemistry—A European Journal, 16*, 1186–1192. https://doi.org/10.1002/chem.200902963
105. Tada, H., Higuchi, H., Wanatabe, T. M., & Ohuchi, N. (2007, February 1). In vivo real-time tracking of single quantum dots conjugated with monoclonal anti-HER2 antibody in tumors of mice. *Cancer Research, 67*(3), 1138–1144. https://doi.org/10.1158/0008-5472.CAN-06-1185
106. Weng, K. C., Noble, C. O., Papahadjopoulos-Sternberg, B., Chen, F. F., Drummond, D. C., Kirpotin, D. B., Wang, D., Hom, Y. K., Hann, B., & Park, J. W. (2008). Targeted tumor cell internalization and imaging of multifunctional quantum dotconjugated immunoliposomes in vitro and in vivo. *Nano Letters, 8*(9), 2851–2857. https://doi.org/10.1021/nl801488u. Epub 2008 Aug 20. PMID: 18712930.
107. Zhang, H., Sachdev, D., Wang, C., et al. (2009). Detection and downregulation of type I IGF receptor expression by antibody-conjugated quantum dots in breast cancer cells. *Breast Cancer Research and Treatment, 114*, 277–285. https://doi.org/10.1007/s10549-008-0014-5
108. Yong, K.-T., Ding, H., Roy, I., Law, W.-C., Bergey, E. J., Maitra, A., & Prasad, P. N. (2009). Imaging pancreatic cancer using bioconjugated InP quantum dots. *ACS Nano, 3*(3), 502–510. https://doi.org/10.1021/nn8008933
109. Kaul, Z., Yaguchi, T., Kaul, S., et al. (2003). Mortalin imaging in normal and cancer cells with quantum dot immuno-conjugates. *Cell Research, 13*, 503–507. https://doi.org/10.1038/sj.cr.7290194
110. Akerman, M. E., Chan, W. C., Laakkonen, P., Bhatia, S. N., & Ruoslahti, E. (2002). Nanocrystal targeting in vivo. *Proceedings of the National Academy of Sciences of the United States of America, 99*(20), 12617–12621. https://doi.org/10.1073/pnas.152463399. Epub 2002 Sep 16. PMID: 12235356; PMCID: PMC130509.
111. Cai, W., Shin, D.-W., Chen, K., Gheysens, O., Cao, Q., Wang, S. X., Gambhir, S. S., & Chen, X. (2006). Peptide-labeled near-infrared quantum dots for imaging tumor vasculature in living subjects. *Nano Letters, 6*(4), 669–676. https://doi.org/10.1021/nl052405t
112. Yu, X., Chen, L., Li, K., Li, Y., Xiao, S., Luo, X., Jia, L., Zhou, L., Deng, Y., Pang, D. W., & Wang, Q. (2007, January 1). Immunofluorescence detection with quantum dot bioconjugates for hepatoma in vivo. *Journal of Biomedical Optics, 12*(1), 014008. https://doi.org/10.1117/1.2437744
113. Larson, D. R., et al. (2003). Water-soluble quantum dots for multiphoton fluorescence imaging in vivo. *Science, 300*, 1434–1436. https://doi.org/10.1126/science.1083780
114. Stroh, M., Zimmer, J., Duda, D., et al. (2005). Quantum dots spectrally distinguish multiple species within the tumor milieu in vivo. *Nature Medicine, 11*, 678–682. https://doi.org/10.1038/nm1247
115. Kim, S., Lim, Y., Soltesz, E., et al. (2004). Near-infrared fluorescent type II quantum dots for sentinel lymph node mapping. *Nature Biotechnology, 22*, 93–97. https://doi.org/10.1038/nbt920
116. Ballou, B., Ernst, L. A., Andreko, S., Harper, T., Fitzpatrick, J. A. J., Waggoner, A. S., & Bruchez, M. P. (2007). Sentinel lymph node imaging using quantum dots in mouse tumor models. *Bioconjugate Chemistry, 18*(2), 389–396. https://doi.org/10.1021/bc060261j
117. Hong, R., Sun, H., Li, D., Yang, W., Fan, K., Liu, C., Dong, L., & Wang, G. (2022). A review of biosensors for detecting tumour markers in breast cancer. *Life, 12*. https://doi.org/10.3390/life12030342
118. Omidian, H., Wilson, R. L., & Cubeddu, L. X. (2024). Quantum dot research in breast cancer: Challenges and prospects. *Materials, 17*. https://doi.org/10.3390/ma17092152

119. Burz, C., Pop, V.-V., Buiga, R., Daniel, S., Samasca, G., Aldea, C., & Lupan, I. (2018). Circulating tumour cells in clinical research and monitoring patients with colorectal cancer. *Oncotarget, 9*, 24561. www.oncotarget.com
120. Singhal, J., Verma, S., Kumar, S., & Mehrotra, D. (2021). Recent advances in nano-biosensing fabrication technology for the detection of oral cancer. *Molecular Biotechnology*. https://doi.org/10.1007/s12033-021-00306-x
121. Nogalski, M. T., Chan, G. C. T., Stevenson, E. V., Collins-McMillen, D. K., & Yurochko, A. D. (2012). A quantitative evaluation of cell migration by the phagokinetic track motility assay. *Journal of Visualized Experiments: JoVE, 70*. https://doi.org/10.3791/4165
122. Perfézou, M., Turner, A., & Merkoçi, A. (2012). Cancer detection using nanoparticle-based sensors. *Chemical Society Reviews, 41*(7), 2606–2622. https://doi.org/10.1039/c1cs15134g
123. Guy, J. B., Espenel, S., Vallard, A., Battiston-Montagne, P., Wozny, A. S., Ardail, D., Alphonse, G., Rancoule, C., Rodriguez-Lafrasse, C., & Magne, N. (2017). Evaluation of the cell invasion and migration process: A comparison of the video microscope-based scratch wound assay and the Boyden chamber assay. *Journal of Visualized Experiments, 2017*(129). https://doi.org/10.3791/56337
124. Fokkelman, M., Roosmalen, W., Rogkoti, V.-M., Le Dévédec, S., Geiger, B., & Water, B. (2016). PhagoKinetic track assay: Imaging and analysis of single cell migration. *Bio-Protocol, 6*(1). https://doi.org/10.21769/bioprotoc.1699
125. Ali, A. A., Abuwatfa, W. H., Al-Sayah, M. H., & Husseini, G. A. (2022). Gold-nanoparticle hybrid nanostructures for multimodal cancer therapy. *Nanomaterials, 12*. https://doi.org/10.3390/nano12203706
126. Sperling, R. A., Gil, P. R., Zhang, F., Zanella, M., & Parak, W. J. (2008). Biological applications of gold nanoparticles. *Chemical Society Reviews, 37*(9), 1896–1908. https://doi.org/10.1039/b712170a

# Biopolymers Modification in Andean Grains: Sustainable Processes and Techniques for Characterization of Starch and Protein for Food Applications

**Julio Rueda, Ileana Gremasqui, Rita Miranda, Manuel Lobo, and Norma Sammán**

**Abstract** Andean crops like quinoa, amaranth, maize, kañiwa, and local legumes are known for being more nutritious than many conventional grains. They also contain natural compounds that support health, making them excellent ingredients for functional foods. These grains are especially rich in starch and protein, which give them unique properties that are attracting growing interest in the food industry. The chapter takes a closer look at how traditional and modern processing methods are being used to enhance the functionality of key Andean grains—most notably quinoa and amaranth. These pseudocereals have drawn growing interest for their nutritional value and their potential in sustainable food systems. Techniques such as extrusion, alkaline extrusion, nixtamalization, roasting, germination, enzymatic hydrolysis, and fermentation, which could potentially alter the structural properties of starches and proteins, are discussed in detail. Special attention is given to the use of biophysical tools—like X-ray diffraction (XRD), Fourier transform infrared spectroscopy (FTIR), Raman spectroscopy, and confocal laser scanning microscopy (CLSM)—to understand the molecular changes occurring during these processes. These techniques offer valuable insights, either on their own or in combination with chemical analyses, helping researchers decode how starches, proteins, and other macronutrients behave and interact.

**Keywords** Quinoa · Amaranth · Extrusion · Hydrolysis · Macromolecules modification · Confocal and spectroscopic analysis

---

J. Rueda · I. Gremasqui · R. Miranda · M. Lobo · N. Sammán (✉)
Facultad de Ingeniería, Centro Interdisciplinario de Investigaciones en Tecnologías y Desarrollo social para el NOA (CIITeD), Consejo Nacional de Investigaciones Científicas y Técnicas—Universidad Nacional de Jujuy, San Salvador de Jujuy, Jujuy, Argentina
e-mail: jrueda@fca.unju.edu.ar; nsamman@fi.unju.edu.ar



N. Mazumder et al. (eds.), *Advanced Biophysical Techniques in Biosciences*, Lecture Notes in Nanoscale Science and Technology 7,
https://doi.org/10.1007/978-3-031-94551-9_5

# 1 Introduction

Andean crops like quinoa (*Chenopodium quinoa Willd.*), kiwicha or amaranth (*Amaranthus caudatus*), maize (*Zea mays*), kañiwa (*Chenopodium pallidicaule Allen*), and some legumes are comparatively more nutritious than traditional crops and contain bioactive compounds that make them ideal for formulating functional foods [1]. These grains are mainly composed of starch (50–75%) and proteins (9–24%) [2], which have gained interest in the food-chain sector for their unique, distinctive physicochemical properties [3]. This chapter explores sustainable processes for improving the properties of starch and proteins isolated or contained in Andean grain and legume flours, with minimal negative changes in their nutritional properties, emphasizing environmentally friendly approaches. Biophysical techniques for evaluating structural and functional changes in modified starches and proteins are reviewed, providing insights for optimizing food formulations.

# 2 Biopolymers in the Food Industry

Native starches and proteins from Andean crops may lack the solubility, strength, thermal stability, or digestibility needed by industry, so they often require modification. For the development of various food products, it is important to evaluate hydration, techno-functional, and nutritional properties among other properties of the raw materials. The treatments applied could negatively affect the structure and functionality of its main components. For example, heat treatments (cooking, drying, and extrusion) modify the conformational organization of starch and protein of grain and legume flours, which would affect important technological functions, and simultaneously, improve the protein and starch digestibility. These effects vary depending on factors such as temperature, exposure time, and the presence of water, so the control of these factors is of importance [4]. The use of main plant biopolymers (among them starches, proteins) has grown in importance among people consuming vegetal products, as well as those concerned about environmental issues [5].

# 3 Sustainable Food Systems

Current paradigms in industrialization, Industry 5.0, promote a shift from industrial automation (Industry 4.0) to more sustainable, human-centred production systems that valorise natural resources [6] and address the need to reshape food systems to be more productive, inclusive and environmentally sustainable, providing healthy and nutritious food for all [7]. Of global greenhouse gas (GHG) emissions, 30% accounts for the food-industry sector [8] and over the past decade, food production has grown by around 45%, with increased GHG emissions by the same proportion

[9]. Cereals are the most widely cultivated crops for feeding the growing population, and cultivation and processing of traditional cereals have led to increased GHG emissions and the extensive use of water for irrigation and cultivable soils in a detrimental tendency due to increased temperatures and drought conditions. One of the possible strategies to contribute to this situation is crop diversification by utilizing more environmentally friendly crop species. For example, daily $CO_2$ emissions observed for quinoa production (27.4 kg $CO_2$-C/ha/day/kg seed) are nearly half of those of maize (40–60 kg $CO_2$-C/ha/day/kg seed) [10]. Pseudocereals fulfil many of these requirements for their adaptive ability to extreme drought, salinity, frost, and dry climate/soil conditions of arid regions [11].

## 4 Processes for Modification of Starch and Protein Biopolymers in Andean Crops

Sustainable food systems demand more energetically efficient processes for food transformation. Ancestral and low-input technologies such as cooking, germination, fermentation, nixtamalization, extrusion, and the use of enzymes (Table 1) offer advantages for their low water requirements, minimal effluent generation, affordable machinery, and low energy consumption, among others. These processes modify starches, proteins, and other components to achieve improved properties for different applications [12]. Germination, fermentation, roasting, and enzymatic hydrolysis result in improved protein digestibility and bioavailability, reduction of anti-nutritional components, and increased mineral solubility and bioavailability. Furthermore, these treatments may lead to protein hydrolysis, lipolysis, and significantly affect starch degradation, gelatinization, and crystallinity, thus impacting techno-functional properties [13–16].

Nixtamalization is an ancestral process in which corn grains are cooked in an alkaline solution. The grain is hydrated and absorbs calcium from lime which, in addition to softening the pericarp of kernels, produces the gelatinization and solubilization of starch and lipids; it increases the availability of some essential amino acids [17, 18]. Doughs from nixtamalized kernels are used to make tortillas, tamales, etc. Nixtamalization requires large amounts of water and long cooking times, resulting in nutrients and the generation of wastewater [17, 19]. Alternatively, alkaline extrusion, as a replacement of nixtamalization, retains most of the nutrients and drastically reduces the use of water and wastewater production [17]. Chemical changes associated are the formation of amylose–lipid complexes affecting structural properties in starch granules [19].

Processing by extrusion combines temperature, pressure, and shear forces. Physical and chemical changes in food matrices depend on moisture, granulometry, residence time, temperature, die diameter, among others. Extrusion of starch-based raw material results in starch gelatinization, with partial or complete destruction of the crystalline structure, molecular fragmentation of starch and repolymerization of dextrin, proteins denaturation, aggregation, cross-linking and partial hydrolysis,

**Table 1** Changes produced by different processes applied to Andean crops

| Process applied | Characteristics | Andean grain/food product | Properties modified | Technologic characteristic modified | Reference |
|---|---|---|---|---|---|
| Extrusion | Simple-screw extruder: Feeding velocity 50 rpm, extruding temperature 69–95 °C, and screw velocity 149 rpm<br>Twin-screw extruder: Feeding velocity 30 kg/h, extruding temperature 80–180 °C (heating zones and die zone), and screw velocity 300 rpm | Quinoa flour | Protein denaturation and crosslinking ↑<br>Starch gelatinization ↑<br>Starch crystallization ↓ | **Rheology**<br>Fluidity ↑, elasticity ↓<br>Strength ↓, fluid-like viscoelastic behavior<br>**Thermals**<br>Starch gelatinization ↑<br>**Pasting**<br>PV ↓, TV ↓,BD ↓, FV ↓, SB↓, PT ↓<br>**Hydration**<br>WAI ↑, WSI ↑, SP↑ | [12, 13] |
| | Twin-screw extruder: Feeding velocity 32 kg/h, extruding temperature 60–145 °C (zones 1–5), and screw velocity 250 rpm<br>Twin 40/20 screw extruder: Feeding velocity 20 rpm, extruding temperature 80–100 °C, and screw velocity 200 and 400 rpm | Corn flour | Starch gelatinization ↑<br>Starch crystallization ↓<br>V-type crystallinity<br>Interaction of starch molecules or intramolecular hydrogen bonds | **Hydration**<br>WSI ↑<br>WAI ↑<br>**Pasting**<br>PV ↓, FV ↓ | [14, 15] |
| | Single screw extruder, L/D ratio 15.5:1, 3.55:1 compression ratio screw, 3.6 mm die diameter, feed rate at 150 g/min (dry matter). The mild extrusion utilized a defatted flour with 15 g/100 g moisture, at 120 °C and 158 rpm, whereas the severe extrusion (SE) utilized a whole flour with 25 g/100 g moisture, at 180 °C and 237 rpm | Amaranth, BRS Alegria variety<br>Defatted and whole flour | Starch gelatinization ↑<br>Gelation of protein | **Hydration**<br>WSI ↑<br>WAI ↑<br>**Pasting**<br>PV ↓,BD ↓, FV ↓, SB↓<br>PT and Dynamic oscillatory temperature sweep test<br>Viscous behavior | [16] |
| | Twin-screw extruder, 18% moisture, temperature 135 °C, diameter die 2.66 mm, and screw speed 251–253 rpm | Quinoa flour (scarified) | Starch digestibility↑<br>Protein digestibility↑ | **Pasting**<br>PV ↓,BD ↓, FV ↓, SB↓<br>PT | [17] |
| | Co-rotating twin-screw extruder, two screws with a length of 600 mm, L/D = 20, 17.1–17.8 moisture, screw speed 250 rpm, temperature 150 °C | Quinoa flour<br>Amaranth flour | | **Hydration properties**<br>WBC ↑, WHC ↑, SV ↑↓,<br>WRI ↑↓, WSI ↑<br>**Functional properties**<br>OBC ↓, Protein solubility ↑, FC ↓, FS<br>**Pasting**<br>PV ↓,BD ↓, FV ↓, SB↓ | [18] |

| | | | | | |
|---|---|---|---|---|---|
| Alkaline extrusion | With 0.25 g of Ca (OH)2/100 g of flour and 40% moisture<br>Single screw 19/25D, nominal compression ratio 2:1, feed rate 20 rpm, screw speed 60 rpm. Temperature 70, 80, and 90 °C, and die diameter of 3 mm | Maize flour, Cuzco race | Starch gelatinization ↑ | **Pasting**<br>PV ↓, BD ↓, FV ↓, SB↓<br>Dynamic oscillatory<br>Fluid-like viscoelastic behavior | [19] |
| | With 0.3 g of lime/100 g of flour and 25 g water/100 g flour<br>Single screw 19/25 D, nominal compression ratio of 1:1, a 3.0 mm die, feed speed 45 rpm, screw speed 145 rpm, residence time of 13 min, and an operating pressure of 5.68 atm (83.6 psi). Extruder outlet temperature 90 °C | Commercial hybrids of white corn (*Zea mays* L.) H-430 and H-431 | Resistance starch ↑ | **Hydration**<br>WAC ↑, WAI ↑, WSI ↑<br>**Rheology**<br>Dynamic oscillatory<br>Fluid-like viscoelastic behavior | [20] |
| Germination | At 32 °C for 48 h | Quinoa flour<br>Amaranth flour | Starch degradation, hydrolysis of protein lipolysis | **Pasting**<br>PV ↓, TP↓, FV ↓, BD ↓↑ | [21] |
| | At 32 °C for 24 h | Amaranth flour Quitensis and<br>Caudatus variety | Starch degradation<br>Hydrolysis of protein<br>Lipolysis<br>Starch crystallization ↓↑ | **Hydration**<br>WHC ↑, WBC ↓, WAI ↓<br>WSI ↑<br>**Pasting**<br>PV ↓, SB↓, FV ↓, BD ↓↑<br>**Rheology**<br>Fluid-like viscoelastic behavior<br>Flow behavior<br>Pseudoplastic | [22] |
| | At 28 °C for 48 h | Amaranth flour, hypochondriacus variety | Starch crystallization ↑<br>Absorption ratios 1047/1022 $cm^{-1}$ and 1022/995 $cm^{-1}$ ↓ | **Rheology**<br>Solid elastic-like behavior<br>Instantaneous and maximum creep compliance ↓ | [23] |

(continued)

**Table 1** (continued)

| Process applied | Characteristics | Andean grain/food product | Properties modified | Technologic characteristic modified | Reference |
|---|---|---|---|---|---|
| Roasting | Heated on a hot plate 180 °C for 10 s | Quinoa flour<br>Amaranth flour | Starch gelatinization ↑<br>Lipolysis | **Rheology**<br>PV ↓, TP↓, FV ↓, BD ↓↑ | [21] |
| | Heated 177 °C for 10 min in conventional oven | Amaranth flour | Starch gelatinization = | **Hydration**<br>WHC ↑, WSS ↓, WAI ↓<br>SP ↓<br>**Pasting**<br>PV =, FV =<br>**Rheology**<br>Fluid-like viscoelastic behavior<br>Flow behavior<br>Pseudoplastic<br>**Thermal**<br>Starch gelatinization = | [24] |
| | Heated on a hot plate at 140 °C for 10 min in a convection oven | Amaranth flour | | **Hydration**<br>WHC ↑, WSS↓<br>**Pasting**<br>PV =, BD = | [25] |
| Nixtamalization | Thermo-alkaline cooking: Corn water 1 kg–2L, 1% Ca $(OH)_2$, moisture: 36%, temperature 92 °C, time: 40 min<br>Steeping: Moisture: 46–50%, time: 8 h | Crude corn flour<br>Nixtamalized corn flour<br>Dough<br>Tortilla | Starch gelatinization ↑<br>Collapsed starch granules ↑<br>Resistance starch ↑ | **Thermals**<br>Starch gelatinization ↑<br>Walls interaction between starch and $Ca^{++}$ ions<br>lipid-starch<br>interaction<br>resistance starch<br>hydro-gel formation<br>**Pasting**<br>PV ↓, apparent viscosity ↓ | [26] |
| | Corn kernel samples (3 kg) cooked with water (1:3) and 1 g lime/100 g, boiling at 96 °C for 20 min. Cooked corn steeped for 14h | Commercial hybrids of white corn, H-430 and H-431 | | **Hydration**<br>WAC ↑, WAI ↑, WSI ↑<br>**Rheology**<br>Dynamic oscillatory<br>Fluid-like viscoelastic behavior | [20] |

| | | | | | |
|---|---|---|---|---|---|
| Fermentation | Milled seeds with distilled water (1:3 w/v). Incubated at 37 °C for 16 h. Dried in hot air oven at 40 °C for 14 h, after which it was re-milled<br>Strain: Spontaneous fermentation | Amaranth flour<br>Hypochondriacus variety | Starch crystallization ↑<br>Absorption ratios 1047/1022 $cm^{-1}$ and 1022/995 $cm^{-1}$ ↓ | **Rheology**<br>Fluid-like viscoelastic behavior ↑<br>Instantaneous and maximum creep compliance ↓ | [23] |
| | 37 °C; 0, 4, 8, 12, and 24 h<br>Strains: *Lactobacillus plantarum* ATCC 8014 and spontaneous fermentation | Quinoa flour | | **Rheology**<br>Fluid-like viscoelastic behavior ↓ | [27] |
| | 28 °C, 14 days<br>Strain:<br>*P. ostreatus* strain was acquired from the Spanish Culture Type Collection (CECT20311) | Quinoa flour | | **Hydration**<br>WHC ↑, SC ↑,<br>**Functional properties**<br>OHC ↓↑<br>EC ↓<br>FC ↓↑ | [28] |
| Enzymatic hydrolysis | 55 °C, pH 8, 60 min, Alcalase<br>55 °C, pH 7, 120 min, Flavourzyme | Quinoa protein | Modification of the secondary structure of the protein: Sheets between 1620 and 1640 $cm^{-1}$:<br>Structural intensity ↑<br>β-sheet and β-turns structures ↑ | | [29] |
| | 40 °C, pH 2, 16 h<br>Endogenous aspartic protease | Amaranth protein | | Antithrombotic activity ↑<br>Antioxidant activity ↑ | [30] |
| | 50 °C, 3 h, pH 8<br>Alcalase and Flavourzyme | Quinoa flour | | Functional properties<br>Solubility ↑, EAI ↑, FC↑, FS↑, WHC ↓, OHC = | [31] |

**Properties: Pasting.** *PV* Peak viscosity, *TV* trough viscosity, *BD* breakdown, *FV* final viscosity, *SB* setback, *PT* pasting temperature. **Hydration.** *WAI* water absorption index, *WSI* water solubility index, *SP* swelling power, *WBC* water binding capacity, *WHC* water holding capacity, *SV* swelling volume, *WRI* water retention index, *WAC* water absorption capacity, *WSS* water soluble solids, *SC* swelling capacity. **Functional.** *OBC* oil binding capacity, *OHC* oil holding capacity, *FC* foaming capacity, *FS* foaming stability, *EC* emulsifying capacity, *EAI* emulsifying activity index
*nd* not detectable

inactivation of enzymes, toxins or microorganisms, formation of resistant starch and modification of fiber levels, formation of starch–lipids complexes [20, 21]. When starchy matrices are extruded with sodium or calcium salts (carbonates, phosphates) or hydroxides, the process is known as alkaline extrusion. Alkaline salts solubilize part of the pericarp of kernels, generating polymeric carbohydrate subunits that affect the rheology of the extruded products. The level of gelatinization of starch and the formation of complexes with lipids and calcium depend on the extrusion conditions [22]. Consequently, the nutritional value and composition of the food are affected [17, 18]. In the extrusion process, extruded quinoa and corn flour have shown an increase in hydration properties. Furthermore, extrusion and nixtamalization also reduce pasting parameters [23–27].

Enzymatic hydrolysis is a widely used bioprocess for the development of food ingredients with appropriate technological functionalities and nutritional properties [28–31]. However, traditional enzymatic hydrolysis faces challenges such as high operating costs, long hydrolysis times, and waste generation [32, 33] along with the notable water use in batch reactors. Recently, the combination of extrusion simultaneously with enzymatic hydrolysis has been applied in cereal processing as an attractive alternative due to its low cost and continuous processing capacity, less wastewater generation, and a notable reduction in water use [34–36]. Although reactive extrusion or bio-extrusion is a process still in development and expansion in the food industry, its application to Andean grains represents a valuable asset for modifying food biopolymers and considering sustainable technologies. Table 1 resumes technological modifications produced in flours from Andean grains by different processes.

The study of hydration properties after germination of *Amaranthus quitensis* and *Amaranthus caudatus* showed that in *A. caudatus*, the water binding capacity (WBC) decreased, but in *A. quitensis*, this property remained constant. The water solubility index was higher in *A. quitensis*. Germination decreased the WAI and increased the water holding capacity (WHC) [37, 38]. Singh and Liu [39] reported similar changes during the roasting of amaranth flour. In fermented quinoa, improvements in the functional properties of starch swelling capacity (SC) and WHC, deteriorations in protein properties, emulsifying and foaming capacity (EC and FC, respectively), and decrease in thickening and gelling power were observed compared to native flours [40]. Modifications in pasting properties and changing parameters such as peak viscosity (PV), pasting temperature (PT), final viscosity (FV), setback (SB) and breakdown (BD), depends on the degradation mechanism and the formation of new links with starch in each process and vegetable species [22, 37–39, 42–44]. The study of rheological properties shows that doughs prepared with germinated, roasted, or fermented amaranth flour show fluid-like viscoelastic behaviour, pseudoplastic behaviour, and creep-recovery curves [13, 37, 38]. Fluid-like viscoelastic behaviour was observed in quinoa flour fermented with *Lactobacillus plantarum* and by spontaneous fermentation [45] and in corn treated by alkaline extrusion and nixtamalization [22, 46]. However, the effect of alkaline treatment on rheological properties is different when comparing traditional nixtamalization and nixtamalization by alkaline extrusion. Further research on this treatment applied to pseudocereals would shed light on the subject [47–49]. Starch and proteins are the majority components in Andean grains and the different processes induce changes in macromolecules such as protein denaturation and crosslinking, protein hydrolysis, lipolysis, starch–protein interactions, starch degradation and gelatinization, and changes in crystallization of starch. The study or characterization of these changes is of importance since they can affect the swelling of starch granules, modifying starch retrogradation, and affect rheological and techno-functional properties. Biophysical techniques are useful for the study of these interactions [50, 51]. Table 2 shows the biophysical techniques used for evaluating modifications of proteins and starch produced by the processing of Andean grains that lead to distinctive technological properties.

## 5 Characterization of Biopolymers in Pseudocereals by Biophysical Techniques

Biophysical techniques present several advantages over traditional chemical-based methods for the study of biopolymers modifications of seeds, flours or formulated products from Andean crops. Biophysical methods provide spatial and structural information of microstructures, as well as the distribution of molecules and single components, allowing simultaneous characterization in their own natural state. Raman and Fourier Transformed-Near Infrared Radiation (FT-NIR) spectroscopy techniques, or imaging techniques like Confocal Laser Scanning Microscopy (CLSM), enable the study of micro-structures and the inner arrangement of single nutrients without altering structures or composition. CLSM and X-ray diffraction provide information on biopolymers distribution and crystallinity patterns in grains or food products, predicting techno-functional or nutritional properties. Molecules or functional groups and the interaction of starch, lipids and proteins or pigments can be assessed by FTIR and Raman spectroscopy [52]. Furthermore, changes induced by the processing of food can be monitored with these non-destructive techniques, which additionally minimize the use of resources, reagents, and require minimal sample amount and preparation. Pseudocereals hold unique chemical (lower starch content, gluten-free, rich in bioactive compounds) and structural properties (small starch granules and typical crystallinity pattern) which eventually impact techno-functional properties such as gelatinization temperatures, retrogradation rates, textural and sensory attributes, among others. The uniqueness of Andean grains demands accurate and reliable instrumental techniques for exploring nutritional or techno-functional properties of these neglected crops. This section summarizes updated information based on instrumental techniques of chemical, spatial and structural characteristics of raw materials and formulated food from common pseudocereals.

In *Raman spectroscopy*, a monochromatic laser beam directed to a sample interacts with molecules and emerges as scattered photons at different energy levels, offering a unique fingerprint spectrum of molecular vibrations and bands'

**Table 2** Biophysical techniques used for evaluating modifications of proteins and starch produced by the processing of Andean grains

| Andean grain/product | Process | Associated molecule | Properties modified | Biophysical technique | Reference |
|---|---|---|---|---|---|
| Quinoa, whole flour | Extrusion | Protein<br>Starch | Denaturation and crosslinking ↑<br>Gelatinization ↑<br>Crystallization ↓ | SEM<br>X-ray microtomography<br>SAXS<br>Wide-angle X-ray scattering | [12] |
| Corn flour | Extrusion | Starch | Gelatinization ↑<br>Crystallization ↓<br>V-type crystallinity<br>Interaction of starch molecules or intramolecular hydrogen bonds | XRD<br>FT-IR<br>SEM | [14] |
| Amaranth flour<br>Var Quitensis<br>Var Caudatus | Germination | Starch<br>Protein<br>Lipids | Degradation ↑<br>Crystallization ↓↑<br>Hydrolysis of protein Lipolysis | XRD | [22] |
| Amaranth flour<br>Var hypochondriacus | Germination and fermentation | Starch | Crystallization ↑<br>Absorption ratios 1047/1022 $cm^{-1}$ and 1022/995 $cm^{-1}$ ↓ | XRD<br>FT-IR | [23] |
| Corn | Nixtamalization | Starch | Gelatinization ↑<br>Collapsed granules ↑<br>Resistance ↑ | LV-SEM<br>TIRI<br>XRD | [26] |
| Quinoa | Enzymatic hydrolysis | Protein (isolate) | Modification of the secondary structure of the protein: Sheets between 1620 and 1640 $cm^{-1}$: Structural intensity ↑<br>β-sheet and β-turns structures ↑ | FT-IR | [29] |

*SEM* scanning electron microscopy, *XRD* X-ray diffraction, *FT-IR* Fourier transform-infrared spectroscopy, *TIRI* thermal infrared imaging, *LV-SEM* low-vacuum scanning electron microscope, *SAXS* small-angle X-ray scattering intensity is directly related to the analyte concentration. The identification of food constituents in complex matrices and the determination of geographical origin and traceability can be assessed by fingerprint spectra. Non-polar bonds like C=C, C–C, and S–S groups, can be easily detected by Raman spectra, so it is useful for the identification and quantification of protein's secondary structure. Some aromatic amino acids like tryptophan, tyrosine, and phenylalanine are highly fluorescent and should be especially considered as interference [53]. In a Raman spectrum of proteins, amide bands I, II, and III are the fingerprint bands. Particularly amide I region, α-helix (1649–1660 $cm^{-1}$), random coils (1660–1665 $cm^{-1}$), β-sheet structures (1620–1647 $cm^{-1}$, and 1665–1680 $cm^{-1}$), and β-turn structures (1680–1699 $cm^{-1}$), allows the quantification using the area under the curve after band-curve fitting by mathematical methods [53, 54]. Changes in the secondary structure of amaranth proteins have been identified in the amide I region, while changes in the tertiary structure were assessed at 760 $cm^{-1}$ and the I850/I830 ratio for hydrophobicity/polarity measurements due to aromatic amino acid residues [55]. In the same way, protein denaturation after extrusion could be monitored by the reduction in the intensity of the 1658 $cm^{-1}$ band in maize-quinoa flours extrudates [56]. Although polar functional groups in carbohydrates can be detected better with IR, Raman spectra allow the study of starch at 475, 865, 938, 1128, 1340, and 1460 $cm^{-1}$ bands. Starch scatters typically at 475 $cm^{-1}$ due to the skeletal vibration C—C—O linkage in the glucopyranose ring [57]. In pseudocereals, seeds and food products have been characterized by Raman spectroscopy. In quinoa seeds, starchy perisperm has shown predominant bands at 472–485 $cm^{-1}$ which are markers of starch presence while fingerprint bands appearing at 800–1200 $cm^{-1}$, (specifically 1114, 1076, 1042, 855 and 931 $cm^{-1}$) belong to the glycosidic linkages fingerprint [52] while 940 is mostly associated to amylose [57]. On the other hand, quinoa globulins display amide I (1651 $cm^{-1}$), II (1559 $cm^{-1}$), and amide III (1295 $cm^{-1}$) key regions due to stretching vibrations of C=O, N–H groups, N–O and C–N groups, respectively [52]. In food products, Raman spectroscopy has been able to differentiate single components in a complex mixture of wheat, quinoa and kiwicha flours through the intensity of typical bands in similar pattern spectra [58]. Furthermore, Raman imaging microscopy has proven to be useful in identifying botanical varieties of seeds based on the I480/I492 $cm^{-1}$ ratio of starches, indicative of crystallinity and amylose/amylopectin content. Rolandelli et al. [56] reported increased intensity at 476 and 940 $cm^{-1}$ (amylose) in extrudates of maize-quinoa, sorghum, millet, and canary seeds blends (possible amorphization, formation of dextrin or amylose-lipids complexes) and diminished intensity of bands at 1460 $cm^{-1}$ (loss in the crystalline starch phase) and at 1754 $cm^{-1}$ (ester C=O stretching of triglycerides) possibly due to the formation of amylose-lipid complexes. Piga et al. [59] demonstrated that the intensity of these bands decreased due to retrograded starch in amaranth gluten-free breads. The presence or changes of pigments like carotenoids can also be assessed at 1522 and 1157 $cm^{-1}$. Rolandelli et al. [56] found increased intensity bands after extrusion due to the release of these pigments from proteins.

*Fourier transform infrared (FTIR) Ssectroscopy* principle is based on infrared radiation (IR) absorption at specific frequencies by chemical bonds causing them to stretch, bend, or twist. Bands' spectra are specific for functional groups/linkages in molecules, creating a fingerprint of molecular vibrations [20]. In proteins, FTIR can detect changes in the protein's secondary structure during processing [53]. Proteins display nine (A, B, and I–VII) characteristic vibrational absorption bands (200–3300 $cm^{-1}$) which belong to different amino groups/peptide bonds in proteins. Secondary structure of proteins is commonly characterized in the amide I region due to stretching vibrations of C=O bonds in peptides [60, 61]. The study of starch by the FTIR technique enables the identification of functional groups: 3000–3700 $cm^{-1}$ for hydroxyl (OH) groups; 1000–1200 $cm^{-1}$ for glycosidic glucose linkages or 1047 and 1022 $cm^{-1}$ for organized and amorphous structures, respectively [62]. FTIR also allows the determination of amylose and amylopectin ratio, changes in functional groups after processing and water–starch interactions [60]. In pseudocereals, FTIR spectra of amaranth and buckwheat whole flours showed a similar pattern and sharp peaks at 1241–1244 $cm^{-1}$ (amylose and amylopectin), although different intensities could mean a distinctive amylose-amylopectin ratio [63]. Mathematical modelling tools such as deconvolution, multivariate analysis, and calibration standards of known samples allowed the authentication of flours from quinoa, amaranth, and kañihua by grouping samples with similar spectra in the fingerprint region of 1100–900 $cm^{-1}$ due to differences in starch structure, which showed flours adulteration [64]. Changes during lupine (*Lupinus albus*) germination could also be assessed by FTIR spectroscopy. Spectra of control and germinated lupine grains showed similar patterns of bands at 4000–800 $cm^{-1}$, while different intensities were observed at each day of germination, indicating a chemical modification. Variations in lupin bands were found at 3000–2800 $cm^{-1}$ (carbonyl groups of triglycerides (C-H)) and at 1700–1300 $cm^{-1}$ for protein [65]. FTIR coupled with the attenuated total reflection technique (ATR) allows the study of liquids and paste samples at surface or interface levels, particularly in the hydrated state [53]. Changes associated with extrusion can be assessed by comparing peaks or bands intensity. In extruded quinoa starch, the band at 1038 $cm^{-1}$ (crystalline phase) diminished while the band at 1010 $cm^{-1}$ (amorphous phase) increased due to quinoa starch gelatinization, formation of amylose–lipids complexes (2922, 2853, and 1745 $cm^{-1}$). Also, changes in protein structures were from α-helix before extrusion (1640 $cm^{-1}$) to random coil/unordered conformation after extrusion (increased intensity at 1627 $cm^{-1}$ [56].

*Confocal laser scanning microscopy* (CLSM) is an imaging non-destructive technique based on the detection of fluorescence emission and optical sectioning of microscopic structures at different depths, enabling the creation of a three-dimensional image of microstructures. In CLSM, a laser beam excites molecules, often coupled with fluorescent dyes, which emit fluorescence at specific wavelengths, making this technique valuable for the study of food microstructures and changes in food products. The distribution and interaction between components in a food system may explain other physicochemical parameters measured by other techniques. Applications of CLSM to pseudocereals are related to the study/

observation of microstructures in intact seeds as seen in Fig. 1, to the monitoring of structural changes of nutrients during in vitro simulated digestion [66, 67] or for detecting interactions of particular molecules after food processing. As an example, the degree of gelatinized starch could be easily evidenced by CLSM in pseudocereals. Alvarez-Jubete et al. [68] observed a continuous phase of gelatinized starch in breads formulated with amaranth, quinoa and buckwheat compared to control breads, and described intrinsic polar lipids forming complexes with starch acting as natural emulsifiers and contributing to cavity formation and gas stabilization prior to cooking. Imaging analysis allows not only the formation but also their distribution and interaction of proteins, lipids, and carbohydrates in a given sample. Rolandelli et al. [56] observed a homogeneous distribution of protein aggregates inside a continuous network of totally gelatinized starch and absent pores, which explained the hardness in extruded snacks formulated with maize, millet, quinoa, sorghum, and canary seed extrudates. Furthermore, the interaction of single components was also observed in gels after heat treatment which exposed hydrophobic sites in a quinoa protein isolate and formed an interconnected network with guar, locus beam and xanthan gums were confirmed/reinforced by observation of a protein network (gel) of heterogenous polysaccharides phase with large protein particles/aggregates forming a stable gel confirmed by rheological measurements [69]. One of the most common applications of imaging analysis is the study of emulsions. Droplet size measurement is not the only application but also distribution, interaction, and spatial interaction of single components and the monitoring phenomena such as flocculation, coalescence, and stability in quinoa protein isolates–oil mixtures [70]. The stability of a plant-based quinoa milk aided by oat glucans proved to be a consequence of glucans disposition around proteins, forming a three-dimensional network which prevented phase separation [71]. In ultrasound-assisted emulsions, CLSM was even able to distinguish the thickness of quinoa protein layers around oil droplets, which was affected by pH and also the changing

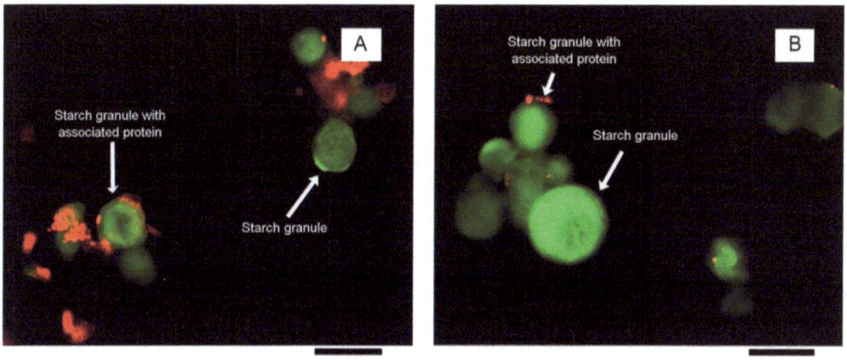


**Fig. 1** CLSM micrographs of maize starch stained with fluorescein isothiocyanate (FITC) (green-starch) and rhodamine B (red-protein), at excitation wavelengths of 488 and 543 nm, respectively. Andean maize races: (**A**) Bolita (higher protein content) than (**B**) Capia (low protein content). Scale bar 20 μm using a Zeiss LSM 980 (Carl Zeiss, Germany) CLSM

geometry of oil droplets at different oil concentrations [72]. Sumei et al. [73] evidenced that the adsorption of soybean proteins (4 and 8%w/v) around oil droplets is a key factor for stabilizing a plant-based quinoa milk; these data were reinforced by physical stability measurements. In Pickering emulsions [74], it was possible to observe that modified octenyl-succinate quinoa starch (OSQS) formed an outer layer at the interface of oil droplets, forming a strong film which strengthens the emulsion and the storage stability, explaining rheological behaviour in high oil proportion emulsions. Imaging analysis is useful in monitoring the effect of gastrointestinal stages on single components. Cao et al. [67] found that proteins and starch of whole quinoa flours interacted by forming binary complexes (protein wrapping starch and forming a barrier) that may slow the rate of protein and starch digestion due to proteins preventing the starch digestion. Similarly, Dong et al. [75] observed a low glycaemic index in thermally treated quinoa flours due to adhered proteins to starch agglomerates, which reduced accessibility to enzymes during in vitro digestion. With this data in consideration, CLSM has proven to be a powerful technique for the study of pseudocereals at the microstructure level as well as the interaction of single components and how processing affects them in food systems. From whole seeds to processed food products, passing through emulsions, the applications are diverse and with a meaningful impact in complementing other classical techniques.

When atoms, molecules, or particles arrange spatially in a symmetrical and repetitive pattern, they form a crystal, which can be detected by X-ray diffraction. When X-rays strike a crystal and interact with the periodic arrangement of atoms in the crystal lattice, they emerge as diffracted beams with intensities proportional to the distance between atoms in repetitive patterns, which are characteristics of unit cells in the crystal and can be used for the characterization of crystalline structures [54]. X-ray diffraction applied to pseudocereals provides information on crystalline structures (mostly amylopectin) in small starch granules. Crystalline structures A (orthorhombic), B (hexagonal), or C (hexagonal-orthorhombic structures) are associated with typical diffraction patterns in starch. Amylose–lipid complexes may form a type V crystalline structure, although they may form both a crystalline and an amorphous phase. At $2\theta$ values type A pattern has a peak at 15°, 23° and 17° and 18°(double peak). Type B occurs at 5.6°, 17°, and 20° with a single peak at 22°, 24° with a double. Type C is a type A and B combination. Finally, type V has peaks at 7.2°, 13.5°, and 19.5° [76]. X-Ray diffraction characterization has been used with pseudocereals in order to identify the crystalline type in starch and for the prediction of important properties such as thermal, water absorption, textural, and even digestibility properties. In X-ray amylopectin and amylose studies in starches, narrow and large peaks correspond to crystalline region (often amylopectin), alterations in these peaks by thermal treatments such as cooking or extrusion could mean a loss in the crystalline structure and a change to more disordered structures such as B, C, V types or a mixture of them [60] Similar peak patterns were observed for amaranth grains at 15.1°, 17.2°, 18.1°, 23.2° [77, 78] and quinoa at 15.1°, 17.2°, 18.1°, 23.2° [77, 79] showing a type A crystalline structure. Amaranth and buckwheat native starches also showed similar peaks at 15.05°, 16.90°, 17.50°, and 23° at $2\theta$, which demonstrated both possess an "A" type crystallinity, typical of cereals. Nevertheless, variations in crystallinity might indicate different origins, cultivation conditions, composition, and amylopectin content. For example, Durga amaranth variety showed higher crystallinity (lower amylose content) and buckwheat starch showed the lowest crystallinity (higher amylose content) [63]. Since crystallinity dictates the behaviour of other parameters during food processing and formulation, the effect of processing has been studied for some pseudocereals. Dong et al. [75] found that in quinoa whole flours (Type A pattern, peaks at $2\theta = 15.22°$, 17.12°, 18.04°, and 23.42°), the heat-moisture treatment (HMT) alone or combined with enzymes (P), microwave radiation (M) or citric acid (A) impacted differently on the starch crystallinity. The HMT + P samples showed a peak pattern of 15.22°, 17.12°, 20°, 22°, 24°, indicating a change from type A to type A + B + V-. Treatments HTM + A and HMT + M almost completely damaged the starch crystalline structure and the relative crystallinity was reduced from 25% in whole quinoa flour to 12%, with no initial peaks observed. The use of less intense treatments such as ultrasonication or 2-octenyl-1-succinic anhydride (OSA) modification appears to be a more controlled alternative for retaining a certain level of crystallinity. Wei et al. [80] found that after ultrasonication of quinoa starches (A-type peaks at 14.95°, 16.94°, 17.93°, and 22.99° at $2\theta$), there was no change in peaks observed in OSA samples which conserved initial orthorhombic crystalline structure exhibiting strong peaks at 15.1°, 17.2°, 18.1°, and 23.2° of $2\theta$ for quinoa and amaranth, aA-type crystalline pattern, which means the chemical modification occurred in amorphous region and not in the crystalline structure. OSA modified starches displayed better emulsion stability, swelling power, and digestibility after gelatinization, attributed to unique physical properties and new molecular structures in small granule starches such as quinoa and amaranth. X-ray diffraction has proven to be a powerful technique for assessing the crystalline structure mainly of starches, which correlates strongly with techno-functional properties, which can be modified and monitored by the information in X-ray peak patterns. By knowing these structural changes, it is possible to tailor the functionality of starch for desired food applications.

## 6 Conclusion

There is abundant scientific literature regarding the processing of Andean crops (especially quinoa and amaranth) to improve functional properties, in which the structural modifications of starches and proteins are studied in detail. Traditional affordable technologies and biotechnologies such as extrusion, alkaline extrusion, nixtamalization, roasting, germination, enzymatic hydrolysis, and fermentation have proven its suitability for achieving desired techno-functional properties. Explanations on structural modifications and the interaction of biopolymeric constituents, starch and protein, among other macronutrients, have been described with high precision by biophysical techniques. Lately, there has been a surge in the use of classical biophysical techniques such as X-ray diffraction (XRD), Fourier transform-infrared spectroscopy (FTIR), Raman spectroscopy and, more recently, confocal laser scanning microscopy (CLSM) for characterizing emerging crops and food products prepared from pseudocereals for their distinctive chemical and structural properties.

They are standing-alone effective tools or complementary to chemical studies for the elucidation of key principles in food systems. Modified biopolymers from Andean crops offer substantial benefits in food applications, aligning with the industry's push for sustainability. Further research and industry adoption can promote these crops as key ingredients in sustainable food production, generating added value to the Andean production chain and more nutritious foods for the population.

## References

1. Sammán, N., Rossi, M., Calliope, S., & Repo-Carrasco-Valencia, R. (2023). Nutritional composition, bioactive and anti-nutritional compounds of Latin American crop grains. In C. M. Haros, M. Reguera, N. Samman, & O. Paredes-López (Eds.), *Latin American crops, agronomic, technology and health aspects* (1st ed., pp. 303–340). CRC Press/Taylor & Francis Group LLC. https://doi.org/10.1201/9781003088424-8
2. Dávalos, J. Z., Tirado, A., Romero, V., Cisneros, G., & Camarra, F. (2023). Structural, thermal and energetic properties of Andean-pseudocereal flours with high nutritional values. *Journal of Thermal Analysis and Calorimetry, 148*, 7207–7215. https://doi.org/10.1007/s10973-023-12224y
3. Lobo, M., Mosso, A., Jiménez, D., & Sammán, N. (2023). Ingredients of high nutritional value obtained from Latin American crops throughout biotechnologies. In C. M. Haros, M. Reguera, N. Samman, & O. Paredes-López (Eds.), *Latin American crops, agronomic, technology and health aspects* (pp. 371–400). CRC Press/Taylor & Francis Group LLC. https://doi.org/10.1201/9781003088424-10
4. Navneet-Martinez, M. M., & Joye, I. J. (2024). Heat-treated bean flour: Exploring techno-functionality via starch-protein structure-function analysis. *Food Hydrocolloids, 157*, 110416. https://doi.org/10.1016/j.foodhyd.2024.110416
5. Jarpa-Parra, M., & Chen, L. (2021). Applications of plant polymer-based solid foams: Current trends in the food industry. Applied Sciences, 11(20), 9605. https://doi.org/10.3390/app11209605
6. Golovianko, M., Terziyan, V., Branytskyi, V., & Malyk, D. (2023). Industry 4.0 vs. Industry 5.0: Co-existence, transition, or a hybrid. Procedia Computer Science, 217, 102–113. https://doi.org/10.1016/j.procs.2022.12.206
7. FAO (Food and Agriculture Organization of the United Nations). (2018). Sustainable food systems: concept and framework. Retrieved January 12, 2025, from https://openknowledge.fao.org/server/api/core/bitstreams/b620989c-407b-4caf-a152-f790f55fec71/content
8. WEF (World Economic Forum). (2024). *How finance can drive the global food industry's sustainability transformation*. Retrieved January 12, 2025, from https://www.weforum.org/stories/2024/09/how-innovative-financial-tools-can-help-the-global-food-industry-complete-a-climate-smart-transformation/
9. FAO. 2022. *Greenhouse gas emissions from agri-food systems. Global, regional and country trends, 2000–2020*. FAOSTAT Analytical Brief Series No. 50. Rome, FAO. Retrieved January 12, 2025, from https://www.fao.org/statistics/highlights-archive/highlights-detail/New-FAOSTAT-data-release-Greenhouse-gas-emissions-from-agrifood-systems/
10. Alvar-Beltrán, J., Marta, A. D., Vivoli, R., Verdi, L., & Orlandini, S. (2022). Greenhouse gas emissions and yield production from an organic and conventional fertilization on quinoa. *Agronomy, 12*(5), 1015. https://doi.org/10.3390/agronomy12051015
11. Balakrishnan, G., & Schneider, R. G. (2022). The role of amaranth, quinoa, and millets for the development of healthy, sustainable food products: A concise review. *Foods, 11*, 2442. https://doi.org/10.3390/foods11162442

12. Kuktaite, R., Repo-Carrasco-Valencia, R., de Mendoza, C. C., Plivelic, T. S., Hall, S., & Johansson, E. (2022). Innovatively processed quinoa (Chenopodium quinoa Willd.) food: Chemistry, structure and end-use characteristics. Journal of the Science of Food and Agriculture, 102(12), 5065–5076. https://doi.org/
13. Huang, K., Shi, J., Li, M., Sun, R., Guan, W., Cao, H., Xiao, G., & Zhang, Y. (2022). Intervention of microwave irradiation on structure and quality characteristics of quinoa protein aggregates. Food Hydrocolloids, 130, 107677. https://doi.org/10.1016/j.foodhyd.2022.107677
14. Xu, F., Chen, J., Wang, L., & Liu, S. (2022). Effect of extruded corn flour addition on the quality characteristics of fine dried noodles. Cereal Chemistry, 99(4), 895–906. https://doi.org/10.1002/cche.10547
15. Patil, S., Kaur, C., Puniya, M. K., Mahapatra, A., Lad, J. D., Jalgaonkar, K., & Mahawar, M. (2021). Functional properties of extruded corn flour. Turkish Journal of Agricultural Engineering Research, 2(1), 167–174. https://doi.org/10.46592/turkager.2021.v02i01.012
16. Menegassi, B., Pilosof, A. M., & Arêas, J. A. (2011). Comparison of properties of native and extruded amaranth (Amaranthus cruentus L.–BRS Alegria) flour. LWT—Food Science and Technology, 44(9), 1915–1921. https://doi.org/10.1016/j.lwt.2011.04.008
17. Muñoz-Pabon, K. S., Roa-Acosta, D. F., Hoyos-Concha, J. L., Bravo-Gómez, J. E., & Ortiz-Gómez, V. (2022). Quinoa snack production at an industrial level: Effect of extrusion and baking on digestibility, bioactive, rheological, and physical properties. Foods, 11(21), 3383. https://doi.org/10.3390/foods11213383
18. Espinosa-Ramírez, J., Rodríguez, A., De la Rosa-Millán, J., Heredia-Olea, E., Pérez-Carrillo, E., & Serna-Saldívar, S. O. (2021). Shear-induced enhancement of technofunctional properties of whole grain flours through extrusion. Food Hydrocolloids, 111, 106400. https://doi.org/10.1016/j.foodhyd.2020.106400
19. Dominguez, N. E., Gimenez, M. A., Segundo, C. N., Gremasqui, I. D. L. A., Lobo, M. O., & Samman, N. C. (2024). Effect of alkaline extrusion temperature on rheological properties of Andean corn dough. Biology and Life Sciences Forum, 37(1), 19. https://doi.org/10.3390/blsf2024037019
20. Enríquez-Castro, C. M., Torres-Chávez, P. I., Ramírez-Wong, B., Quintero-Ramos, A., Ledesma-Osuna, A. I., López-Cervantes, J., & Gerardo-Rodríguez, J. E. (2020). Physicochemical, rheological, and morphological characteristics of products from traditional and extrusion nixtamalization processes and their relation to starch. International Journal of Food Science, 2020(1), 5927670. https://doi.org/10.1155/2020/5927670
21. Beniwal, S. K., Devi, A., & Sindhu, R. (2019). Effect of grain processing on nutritional and physico-chemical, functional and pasting properties of amaranth and quinoa flours. Indian Journal of Traditional Knowledge, 18(3), 500–507. https://www.webofscience.com/wos/woscc/full-record/WOS:000500707300010
22. Cornejo, F., Novillo, G., Villacrés, E., & Rosell, C. M. (2019). Evaluation of the physicochemical and nutritional changes in two amaranth species (Amaranthus quitensis and Amaranthus caudatus) after germination. Food Research International, 121, 933–939. https://doi.org/10.1016/j.foodres.2019.01.022
23. Siwatch, M., Yadav, R. B., & Yadav, B. S. (2017). X-ray diffraction, rheological and FT-IR spectra studies of processed amaranth (Amaranthus hypochondriacus). Journal of Food Measurement and Characterization, 11, 1717–1724. https://doi.org/10.1007/s11694-017-9552-z
24. Liu, S. X., Chen, D., & Xu, J. (2022). Physiochemical properties of jet-cooked amaranth and improved rheological properties by processed oat bran. Future Foods, 5, 100107. https://doi.org/10.1016/j.fufo.2021.100107
25. Singh, M., & Liu, S. X. (2021). Evaluation of amaranth flour processing for noodle making. Journal of Food Processing & Preservation, 45(4), e15270. https://doi.org/10.1111/jfpp.15270
26. Villada, J. A., Sanchez-Sinencio, F., Zelaya-Angel, O., Gutierrez-Cortez, E., & Rodríguez-García, M. E. (2017). Study of the morphological, structural, thermal, and pasting corn trans-

formation during the traditional nixtamalization process: From corn to tortilla. Journal of Food Engineering, 212, 242–251. https://doi.org/10.1016/j.jfoodeng.2017.05.034
27. Chiş, M. S., Păucean, A., Man, S. M., Vodnar, D. C., Teleky, B. E., Pop, C. R., Stan, L., Borsai, O., Kadar, C. B., Urcán, A. C., & Muste, S. (2020). Quinoa sourdough fermented with Lactobacillus plantarum ATCC 8014 designed for gluten-free muffins—A powerful tool to enhance bioactive compounds. Applied Sciences, 10(20), 7140. https://doi.org/10.3390/app10207140
28. Badia-Olmos, C., Sánchez-García, J., Laguna, L., Zúñiga, E., Haros, C. M., Andrés, A. M., & Tarrega, A. (2024). Flours from fermented lentil and quinoa grains as ingredients with new techno-functional properties. Food Research International, 177, 113915. https://doi.org/10.1016/j.foodres.2023.113915
29. Roa-Acosta, D. F., Hoyos-Concha, J. L., & Bravo-Gómez, J. E. (2022). Hydrolysis monitoring of quinoa, soy and rice proteins by using spectroscopy FT-IR technique. Biotecnología En El Sector Agropecuario Y Agroindustrial, 20(2), 76–86. https://doi.org/10.18684/rbsaa.v20.n2.2022.1941
30. Sabbione, A. C., Ibañez, S. M., Martínez, E. N., Añón, M. C., & Scilingo, A. A. (2016). Antithrombotic and antioxidant activity of amaranth hydrolysate obtained by activation of an endogenous protease. Plant Foods for Human Nutrition, 71, 174–182. https://doi.org/10.1007/s11130-016-0540-y
31. Gremasqui, I. D. L. A., Gimenez, M. A., Lobo, M. O., & Sammán, N. C. (2021). Nutritional and functional characterisation of hydrolysates from quinoa flour (Chenopodium quinoa) using two proteases. International Journal of Food Science & Technology, 56(12), 6507–6514. https://doi.org/10.1111/ijfs.15360
32. Zhang, B., Qiao, D., Zhao, S., Lin, Q., Wang, J., & Xie, F. (2021). Starch-based food matrices containing protein: Recent understanding of morphology, structure, and properties. Trends in Food Science and Technology, 114, 212–231. https://doi.org/10.1016/j.tifs.2021.05.033
33. Nkhata, S. G., Ayua, E., Kamau, E. H., & ShingiroJ, B. (2018). Fermentation and germination improve nutritional value of cereals and legumes through activation of endogenous enzymes. Food Science & Nutrition, 6(8), 2446–2458. https://doi.org/10.1002/fsn3.846
34. Thakur, P., Kumar, K., Ahmed, N., Chauhan, D., Rizvi, Q. U. E. H., Jan, S., Pal Singh, T., & Singh Dhaliwal, H. (2021). Effect of soaking and germination treatments on nutritional, anti-nutritional, and bioactive properties of amaranth (Amaranthus hypochondriacus L.), quinoa (Chenopodium quinoa L.), and buckwheat (Fagopyrum esculentum L.). Current Research in Food Science, 4, 917–925. https://doi.org/10.1016/j.crfs.2021.11.019
35. Chauhan, D., Kumar, K., Ahmed, N., Thakur, P., Rizvi, Q. U. E. H., Jan, S., & Yadav, A. N. (2022). Impact of soaking, germination, fermentation, and roasting treatments on nutritional, anti-nutritional, and bioactive composition of black soybean (Glycine max L.). The Journal of Applied Biology & Biotechnology, 10(5), 186–192. https://doi.org/10.7324/JABB.2022.100523
36. Ramirez-Araujo, H., Gaytán-Martínez, M., & Reyes-Vega, M. L. (2019). Alternative technologies to the traditional nixtamalization process. Trends in Food Science and Technology, 85, 34–43. https://doi.org/10.1016/j.tifs.2018.12.007
37. Kamau, E. H., Nkhata, S. G., & Ayua, E. O. (2020). Extrusion and nixtamalization conditions influence the magnitude of change in the nutrients and bioactive components of cereals and legumes. Food Science & Nutrition, 8(4), 1753–1765. https://doi.org/10.1002/fsn3.1473
38. Olšaníková, K., Šárka, E., & Smrčková, P. (2022). Extrusion process of maize grits used for nixtamalization. Czech Journal of Food Sciences, 40(2), 138–146. https://doi.org/10.17221/188/2021-CJFS
39. Cueto, M., Farroni, A., Rodríguez, S. D., Schoenlechner, R., Schleining, G., & del Pilar Buera, M. (2018). Assessing changes in enriched maize flour formulations after extrusion by means of FTIR, XRD, and chemometric analysis. Food and Bioprocess Technology, 11, 1586–1595. https://doi.org/10.1007/s11947-018-2113-6

40. Zhong, L., Fang, Z., Wahlqvist, M. L., Hodgson, J. M., & Johnson, S. K. (2019). Extrusion cooking increases soluble dietary fibre of lupin seed coat. LWT—Food Science and Technology, 99, 547–554. https://doi.org/10.1016/j.lwt.2018.10.018
41. Rueda, J., Lobo, M. O., & Sammán, N. (2020, August). Changes in the antioxidant activity of peptides released during the hydrolysis of quinoa (Chenopodium quinoa willd) protein concentrate. In Proceedings (Vol. 53, No. 1, p. 12). MDPI. https://doi.org/10.3390/proceedings2020053012
42. Pan, M., Liu, K., Yang, J., Liu, S., Wang, S., & Wang, S. (2020). Advances on food-derived peptidic antioxidants—A review. Antioxidants, 9(9), 799. https://doi.org/10.3390/antiox9090799
43. Wang, H., Wang, J., Lv, Z., Liu, Y., & Lu, F. (2013). Preparing oligopeptides from broken rice protein by ultrafiltration-coupled enzymatic hydrolysis. European Food Research and Technology, 236, 419–424. https://doi.org/10.1007/s00217-012-1904-7
44. Xu, E., Wu, Z., Wang, F., Li, H., Xu, X., Jin, Z., & Jiao, A. (2015). Impact of high-shear extrusion combined with enzymatic hydrolysis on rice properties and Chinese rice wine fermentation. Food and Bioprocess Technology, 8, 589–604. https://doi.org/10.1007/s11947-014-1429-0
45. Xu, E., Campanella, O. H., Ye, X., Jin, Z., Liu, D., & BeMiller, J. N. (2020). Advances in conversion of natural biopolymers: A reactive extrusion (REX)–enzyme-combined strategy for starch/protein-based food processing. Trends in Food Science and Technology, 99, 167–180. https://doi.org/10.1016/j.tifs.2020.02.018
46. Dang, T. T., & Vasanthan, T. (2019). Modification of rice bran dietary fiber concentrates using enzyme and extrusion cooking. Food Hydrocolloids, 89, 773–782. https://doi.org/10.1016/j.foodhyd.2018.11.024
47. Odukoya, J. O., De Saeger, S., De Boevre, M., Adegoke, G. O., Audenaert, K., Croubels, S., Antonissen, G., Odukoya, J. O., & Njobeh, P. B. (2022). Influence of nixtamalization cooking ingredients on the minerals composition of nixtamalized maize and sorghum. Journal of Cereal Science, 103, 103373. https://doi.org/10.1016/j.jcs.2021.103373
48. Rincón-Aguirre, A., de Dios Figueroa-Cárdenas, J., Ramírez-Wong, B., Ibba, M. I., & Cruz, E. M. (2022). Nixtamalization of durum wheat and the effect on protein secondary structure, gliadins, dough, and breadmaking quality. Journal of Cereal Science, 107, 103539. https://doi.org/10.1016/j.jcs.2022.103539
49. Luque, G. Y. R., & Luque, M. Y. (2014). Obtención de parámetros tecnológicos para la elaboración de productos nixtamalizados, tipo snacks fritos, a partir de cereales alto andinos: Maíz (Zea mays); Quinua (Chenopodium quinoa); Kiwicha (Amaranthus caudatus). Veritas et Scientia, 3(2), 74–83. https://doi.org/10.47796/ves.v3i2.246
50. Li, W., Gu, Z., Cheng, L., Li, Z., Li, C., Ban, X., & Hong, Y. (2023). Effect of endogenous proteins and heat treatment on the in vitro digestibility and physicochemical properties of corn flour. Food Hydrocolloids, 135, 108220. https://doi.org/10.1016/j.foodhyd.2022.108220
51. Scott, G., & Awika, J. M. (2023). Effect of protein–starch interactions on starch retrogradation and implications for food product quality. Comprehensive Reviews in Food Science and Food Safety, 22(3), 2081–2111. https://doi.org/10.1111/1541-4337.13141
52. Czekus, B., Pećinar, I., Petrović, I., Paunović, N., Savić, S., Jovanović, Z., & Stikić, R. (2019). Raman and Fourier transform infrared spectroscopy application to the Puno and Titicaca cvs. of quinoa seed microstructure and perisperm characterization. Journal of Cereal Science, 87, 25–30. https://doi.org/10.1016/j.jcs.2019.02.011
53. Sadat, A., & Joye, I. J. (2020). Peak fitting applied to fourier transform infrared and raman spectroscopic analysis of proteins. Applied Sciences, 10(17), 5918. https://doi.org/10.3390/app10175918
54. Rodriguez-Saona, L., & Ayvaz, H. (2024). Infrared and Raman spectroscopy. In B. P. Ismail & S. S. Nielsen (Eds.), Nielsen's food analysis. Food science text series. Springer. https://doi.org/10.1007/978-3-031-50643-7_8

55. Cortez-Trejo, M. C., Figueroa-Cárdenas, J. D., Quintanar-Guerrero, D., Baigts-Allende, D. K., Manríquez, J., & Mendoza, S. (2022). Effect of pH and protein-polysaccharide ratio on the intermolecular interactions between amaranth proteins and xanthan gum to produce electrostatic hydrogels. Food Hydrocolloids, 129, 107648. https://doi.org/10.1016/j.foodhyd.2022.107648
56. Rolandelli, G., Gallardo-Navarro, Y. T., Pinilla, S. G., Farroni, A. E., Gutiérrez-López, G. F., & del Pilar Buera, M. (2021). Components interactions and changes at molecular level in maize flour-based blends as affected by the extrusion process. A multi-analytical approach. Journal of Cereal Science, 99, 103186. https://doi.org/10.1016/j.jcs.2021.103186
57. Almeida, M. R., Alves, R. S., Nascimbem, L. B., Stephani, R., Poppi, R. J., & De Oliveira, L. F. C. (2010). Determination of amylose content in starch using Raman spectroscopy and multivariate calibration analysis. Analytical and Bioanalytical Chemistry, 397, 2693–2701. https://doi.org/10.1007/s00216-010-3566-2
58. Jamanca-Gonzales, N. C., Ocrospoma-Dueñas, R. W., Eguilas-Caushi, Y. M., Padilla-Fabian, R. A., & Silva-Paz, R. J. (2024). Technofunctional properties and rheological behavior of quinoa, kiwicha, wheat flours and their mixtures. Molecules, 29(6), 1374. https://doi.org/10.3390/molecules29061374
59. Piga, A., Conte, P., Fois, S., Catzeddu, P., Del Caro, A., Sanguinetti, A. M., & Fadda, C. (2021). Technological, nutritional and sensory properties of an innovative gluten-free double-layered flat bread enriched with amaranth flour. Foods, 10(5), 920. https://doi.org/10.3390/foods10050920
60. Zambelli, R. A., da Rocha, A. P., & de Mendonça, L. G. (2024). Importance of starch characterization. In Standardized procedures and protocols for starch (pp. 1–40). Springer.
61. Roa-Acosta, D. F., Bravo-Gómez, J. E., García-Parra, M. A., Rodríguez-Herrera, R., & Solanilla-Duque, J. F. (2020). Hyper-protein quinoa flour (Chenopodium quinoa wild): Monitoring and study of structural and rheological properties. LWT—Food Science and Technology, 121, 108952. https://doi.org/10.1016/j.lwt.2019.108952
62. Carmona-Garcia, R., Agama-Acevedo, E., Pacheco-Vargas, G., Bello-Perez, L. A., Tovar, J., & Alvarez-Ramirez, J. (2022). Pregelatinised amaranth flour as an ingredient for low-fat gluten-free cakes. International Journal of Food Science and Technology, 57(4), 2346–2355. https://doi.org/10.1111/ijfs.15589
63. Sindhu, R., & Khatkar, B. S. (2018). Thermal, structural and textural properties of amaranth and buckwheat starches. Journal of Food Science and Technology, 55, 5153–5160. https://doi.org/10.1007/s13197-018-3474-6
64. Shotts, M. L., Pujolras, M. P., Rossell, C., & Rodriguez-Saona, L. (2018). Authentication of indigenous flours (quinoa, amaranth and kañiwa) from the Andean region using a portable ATR-infrared device in combination with pattern recognition analysis. Journal of Cereal Science, 82, 65–72. https://doi.org/10.1016/j.jcs.2018.04.005
65. Atudorei, D., Stroe, S. G., & Codină, G. G. (2021). Impact of germination on the microstructural and physicochemical properties of different legume types. Plants, 10(3), 592. https://doi.org/10.3390/plants10030592
66. Grundy, M. M., Momanyi, D. K., Holland, C., Kawaka, F., Tan, S., Salim, M., Boyd, B. J., Bajka, B., Mulet-Cabero, A. I., Bishop, J., & Owino, W. O. (2020). Effects of grain source and processing methods on the nutritional profile and digestibility of grain amaranth. Journal of Functional Foods, 72, 104065. https://doi.org/10.1016/j.jff.2020.104065
67. Cao, H., Huang, Q., Wang, C., Guan, X., Huang, K., & Zhang, Y. (2023). Effect of compositional interaction on in vitro digestion of starch during the milling process of quinoa. Food Chemistry, 403, 134372. https://doi.org/10.1016/j.foodchem.2022.134372
68. Alvarez-Jubete, L., Auty, M., Arendt, E. K., & Gallagher, E. (2010). Baking properties and microstructure of pseudocereal flours in gluten-free bread formulations. European Food Research and Technology, 230, 437–445. https://doi.org/10.1007/s00217-009-1184-z

69. Patole, S., Cheng, L., & Yang, Z. (2022). Impact of incorporations of various polysaccharides on rheological and microstructural characteristics of heat-induced quinoa protein isolates gels. Food Biophysics, 17(3), 314–323. https://doi.org/10.1007/s11483-022-09720-3
70. Li, B., Wu, Y., Ye, A., & Guo, Q. (2025). Emulsifying properties of quinoa protein fibrillar aggregates and their influence on quinoa protein emulsion gel formation. International Journal of Biological Macromolecules, 285, 138234. https://doi.org/10.1016/j.ijbiomac.2024.138234
71. Huang, K., Zhang, S., Guan, X., Li, C., Li, S., Liu, Y., & Shi, J. (2021). Effect of the oat β-glucan on the development of functional quinoa (Chenopodium quinoa wild) milk. Food Chemistry, 349, 129201. https://doi.org/10.1016/j.foodchem.2021.129201
72. Yang, Z., & Cheng, L. (2024). Impact of ultrasound emulsification on the physicochemical properties of emulsions stabilised by quinoa protein isolates at different pHs. Food Biophysics, 19, 160–171. https://doi.org/10.1007/s11483-023-09812-8
73. Zhou, S., Liu, X., Cui, Y., Chen, S., Zhong, F., Lu, J., & Kong, C. (2024). Molecular investigation of soybean protein for improving the stability of quinoa (Chenopodium quinoa willd.) milk substitute. Food Chemistry, 461, 140829.
74. Li, S., Zhang, B., Li, C., Fu, X., & Huang, Q. (2020). Pickering emulsion gel stabilized by octenylsuccinate quinoa starch granule as lutein carrier: Role of the gel network. Food Chemistry, 305, 125476. https://doi.org/10.1016/j.foodchem.2019.125476
75. Dong, J., Huang, L., Chen, W., Zhu, Y., Dun, B., & Shen, R. (2021). Effect of heat-moisture treatments on digestibility and physicochemical property of whole quinoa flour. Foods, 10(12), 3042. https://doi.org/10.3390/foods10123042
76. Zhang, X., Chen, Y., Zhang, R., Zhong, Y., Luo, Y., Xu, S., Liu, J., Xue, J., & Guo, D. (2016). Effects of extrusion treatment on physicochemical properties and in vitro digestion of pre-gelatinized high amylose maize flour. Journal of Cereal Science, 68, 108–115. https://doi.org/10.1016/j.jcs.2016.01.005
77. Yao, T., Wen, Y., Xu, Z., Ma, M., Li, P., Brennan, C., Sui, Z., & Corke, H. (2020). Octenylsuccinylation differentially modifies the physicochemical properties and digestibility of small granule starches. International Journal of Biological Macromolecules, 144, 705–714. https://doi.org/10.1016/j.ijbiomac.2019.12.129
78. Yadav, P., Bharathi, U., Suruthi, K., & Bosco, S. J. D. (2023). Effect of ultrasonic modification on physiochemical, structural, functional properties and in vitro starch digestibility of Amaranthus paniculatus (Rajgeera) starch. Biomass Conversion and Biorefinery, 14, 19017–19024. https://doi.org/10.1007/s13399-023-04134-8
79. Han, L., Qiu, S., Cao, S., Yu, Y., Yu, S., & Liu, Y. (2021). Molecular characteristics and physicochemical properties of very small granule starch isolated from Agriophyllum squarrosum seeds. Carbohydrate Polymers, 273, 118583. https://doi.org/10.1016/j.carbpol.2021.118583
80. Wei, Y., Li, G., & Zhu, F. (2023). Impact of long-term ultrasound treatment on structural and physicochemical properties of starches differing in granule size. Carbohydrate Polymers, 320, 121195. https://doi.org/10.1016/j.carbpol.2023.121195

# Catechin Functionalized Nanomaterials and Their Applications

**Poulami Manna and Sougata Ghosh**

**Abstract** Catechin is a naturally occurring polyphenolic compound found in plants such as, green tea, black tea, black grapes, apricots, broad beans, and strawberries. Catechins include, epicatechin, epigallocatechin, epicatechin gallate, epigallocatechin gallate, gallocatechin, catechin gallate, gallocatechin gallate, and catechin. These compounds exhibit notable biological activities such as anticancer, anti-inflammatory, antioxidative, antibacterial, and antifungal activities. This chapter includes catechin functionalized nanoparticles (NPs) that are synthesized employing various strategies such as precipitation, chemical reduction, and hydrothermal methods. Gold, silver, zinc oxide, and iron oxide nanoparticles showing exotic structural and functional properties after functionalizing with various catechins are elaborated. The resulting nanocomposites are characterized using advanced analytical techniques such as UV-vis spectroscopy, scanning electron microscopy (SEM), high-resolution transmission electron microscopy (HRTEM), dynamic light scattering (DLS), Zeta potential measurements, X-ray diffraction (XRD), and Fourier transform infrared spectroscopy (FTIR). Further, the superior antimicrobial, anticancer, antioxidant, and other therapeutic properties of the catechin functionalized NPs are also covered. Thus, the combination of catechins and nanomaterials can result in promising biomedical applications.

**Keywords** Catechins · Nanoparticles · Functionalization · Characterization · Applications · Nanomedicine

---

P. Manna
Department of Microbiology, School of Science, RK University, Rajkot, Gujarat, India

S. Ghosh (✉)
Department of Microbiology, School of Science, RK University, Rajkot, Gujarat, India

Department of Physics, Faculty of Science, Kasetsart University, Bangkok, Thailand



N. Mazumder et al. (eds.), *Advanced Biophysical Techniques in Biosciences*, Lecture Notes in Nanoscale Science and Technology 7,
https://doi.org/10.1007/978-3-031-94551-9_6

# 1 Introduction

Catechin is a bioactive compound found in plants as well as beverages. It is rich in apples, blueberries, beer, gooseberries, cocoa, grape, green tea, red wine, and strawberry [1]. Eight different catechins include—catechin (C), catechin gallate (CG), gallocatechin (GC), epicatechin (EC), epicatechin gallate (ECG), epigallocatechin (EGC), epigallocatechin gallate (EGCG) and gallocatechin gallate (GCG). Catechin has antibacterial, antifungal, anti-inflammation, antioxidation, anticancer and antiviral activity [1, 2]. The lack of bioavailability, low stability, and fast metabolism are the major disadvantages of catechin and EGCG. Thus, recent developments focus on capping, loading, encapsulating, and conjugating the catechin with nanoparticles to improve its safety and stability [3].

Nanoparticles (NPs) with sizes of 1–100 nm have significant benefits in various sectors like medicine, food, agriculture, textiles, and electronic industries [4]. The NPs can be synthesized by physical, chemical, or biological processes. The physical method involves ball milling, lithography, laser ablation, thermal decomposition, and electron explosion sputtering. Chemical vapor condensation, pyrolysis, spinning, and sol–gel are some of the chemical techniques involved in the synthesis of NPs. The biological process utilizes bacteria, fungi, algae, and plants to synthesize NPs [5].

The utilization of catechin in the synthesis of various NPs offers distinct advantages such as cost-efficiency and eco-friendliness. Catechin-functionalized NPs possess unique physical and chemical properties. Gold, silver, zinc oxide, and iron oxide NPs exhibited enhanced stability and biocompatibility after being conjugated with catechin. Catechin-conjugated NPs have diverse biomedical applications due to their antiviral, antibacterial, and anti-inflammatory activities [6–8]. The metal NPs functionalised with catechin and EGCG can be synthesized using techniques such as chemical reduction and hydrothermal methods [9, 10].

The unique nanoformulation was prepared by utilizing $Mn^{2+}$ ions and EGCG by self-assembly method that significantly inhibited colon tumor cells [11]. The monodispersed selenium nanoparticles (SeNPs) with starch microgel (SM) and EGCG denoted as SM-EGCG-SeNPs exhibited antioxidant activity and anticancer properties [12]. EGCG-loaded lipid NPs and functionalization of EGCG-loaded lipid NPs with folic acid showed cytotoxicity effect against breast cancer cells, MCF-7, MDA-MB-231 and MCF-7TAM. The cell viability decreased by 30% in the MCF-7 cells after being exposed to EGCG-LNPs and EGCG-LNPs-FA whereas cell viability reduced to 41.4 and 57.3% in the case of MCF-7TAM cells after being exposed to EGCG-LNPs and EGCG-LNPs-FA [13].

The aforementioned facts strongly rationalize the immense significance of catechin-conjugated NPs for biomedical applications. Thus, this chapter gives a comprehensive overview of the recent developments in the synthesis of catechin-functionalized nanoparticles, characterization, and their potential biological applications as listed in Table 1.

# 2 Catechin Functionalized Nanoparticles

## 2.1 Gold Nanoparticles

Various attempts are made where catechin is used for the synthesis of nanoscale gold with exotic shapes and sizes. Zhu et al. [14] investigated the synthesis of EGCG functionalized AuNPs (EGCG-AuNPs) for the evaluation of osteoclastogenic activity. EGCG functionalized AuNPs were prepared by mixing 0.1 mL of EGCG solution with 1.0 mL of HAuCl4 solution and continuously stirring at 25 °C. A deep red color appeared within 5 min. The transmission electron microscope (TEM) analysis revealed the spherical shape of EGCG-AuNPs with a size equivalent to 30 nm. The average diameter and zeta potential of EGCG-AuNPs were 35.6 nm and −19.5 mV, respectively, as confirmed by dynamic light scattering (DLS). The Fourier transform infrared (FTIR) spectra of synthesized EGCG-AuNPs showed the stretching vibration of O–H and C–H at 3000 $cm^{-1}$ while the stretching of C=O, benzene ring, C–O and bending of C–H were detected at 1705 $cm^{-1}$, 1450–1620 $cm^{-1}$, 1100–1300 $cm^{-1}$ and 1435 $cm^{-1}$, respectively. Then 15 mg $kg^{-1}$ of EGCG-AuNPs was administered to 5 weeks old female mice for 8 days. The mice were pre-treated with lipopolysaccharide (LPS) at a dose of 10 mg $kg^{-1}$ body weight for 7 days to induce calvarial bone erosion. Anti-osteoclastogenic activity was confirmed as the EGCG-AuNPs showed 4.6% more bone volume (BV)/total volume (TV) ratio than the free EGCG. It also prevented bone loss which was detected by staining the calvarial bone with hematoxylin and eosin (H&E) and tartrate-resistant acid phosphatase (TRAP).

Sanna et al. [15] developed an eco-friendly approach for the fabrication of EGCG-AuNPs, resveratrol conjugated AuNPs (RSV-AuNPs), and fisetin conjugated AuNPs (FS-AuNPs) with antiproliferative and antioxidant activity. The EGCG-AuNPs, FS-AuNPs and RSV-AuNPs were synthesized by reacting 1 mL of 10 mM $HAuCL_4$ with 5 mL of EGCG, FS and RSV, respectively. After incubation at room temperature, the pale yellow colour of the solution changed into purple-red indicating the formation of EGCG-AuNPs, FS-AuNPs and RSV-AuNPs. The solution was stirred for another 30 min and centrifuged at 20,000 rpm for 10 min at 15 °C. The UV-vis spectra of EGCG-AuNPs, FS-AuNPs, and RSV-AuNPs showed an absorption peak at 554 nm, 544 nm, and 540 nm, respectively. X-ray diffraction (XRD) peaks of EGCG-AuNPs, FS-AuNPs, and RSV-AuNPs at $2\theta$ degrees of 38.17°, 44.33°, 64.70°, and 77.70° corresponded to the planes (111), (200), (220), and (311), respectively. TEM images determined various shapes of the EGCG-AuNPs including spheres, triangles and hexagons with a size of 25.55 nm. FS-AuNPs and RSV-AuNPs had spherical shapes with the sizes was 9.76 nm and 14.55 nm, respectively. FTIR spectrum confirmed C=O ester-specific bands at 1700 $cm^{-1}$ for EGCG-AuNPs and free EGCG. It also showed the bending vibration of O–H at 1375 $cm^{-1}$ while C–O, C–C and C=C stretching at 1210 $cm^{-1}$, 1450 $cm^{-1}$ and 1610 $cm^{-1}$, respectively. FTIR spectra of RSV-AuNPs and free RSV showed C–C olefinic, C=C and C-H stretching at 1590 $cm^{-1}$, 1610 $cm^{-1}$ and 1450 $cm^{-1}$, respectively while C–C bending vibration at 1150 $cm^{-1}$. The spectrum of FS-AuNPs and free FS showed the C-OH bending at 1110 $cm^{-1}$ while the stretching of C-C, C=C and C=O at 1260 $cm^{-1}$, 1605 $cm^{-1}$ and 1725 $cm^{-1}$—1690 $cm^{-1}$, respectively. It showed the DPPH scavenging activity of 16% and 60% at 25 μg $mL^{-1}$ and 200 μg $mL^{-1}$, respectively. On the other hand, it scavenged ABTS radical (2,2′-azino-bis(3-ethylbenzothiazoline-6-sulfonic acid) up to 6% and 46% at 25 μg $mL^{-1}$ and 200 μg $mL^{-1}$, respectively. The EGCG-AuNPs showed greater antiproliferative activity against SH-SY5Y-CFP-DEVD-YFP cells. EGCG-AuNPs (50 μg $mL^{-1}$) showed a reduction in the concentration of viable cells to 40% as confirmed from the 3-(4,5-dimethylthiazol-2-yl)-5-(3-carboxymethoxyphenyl)-2-(4-sulfophenyl)-2H-tetrazolium (MTS) activity. The enhanced cleavage of the reporter CFP-DEVD-YFP indicated the activation of apoptosis.

Chavva et al. [16] reported citrate conjugated AuNPs (C-AuNPs) and EGCG-AuNPs with anticancer activity. The EGCG-conjugated AuNPs were synthesized using $HAuCl_4.3H_2O$ (25 mL, 2 mM) and EGCG (25 mL, 1.5 mM) mixed at room temperature. Development of red wine colour indicated the generation of EGCG conjugated AuNPs. Secondly, the citrate-conjugated AuNPs were prepared using $HAuCl_4 \cdot 3H_2O$ (1.25 mL, 10 mM) and trisodium citrate (0.85 mL) mixed at boiling temperature for 30 min. Appearance of red wine colour indicated the generation of C-AuNPs. The absorption band of C-AuNPs and EGCG-AuNPs at 530 nm was determined by the UV-Vis spectroscopy. TEM analysis confirmed the spherical-shaped C-AuNPs and EGCG-AuNPs with a size of 25 nm. The C-AuNPs and EGCG-AuNPs had zeta potential of −81.16 mV, and–72.57 mV, respectively whereas polydispersity index (PDI) was 0.01 and 0.04, respectively. The hydrodynamic size was 41.7 nm and 90.3 nm, respectively as revealed from the DLS analysis. EGCG-AuNPs showed the highest anticancer effect on human cancer cells, MDA-MB-231, A375SM, PC3 and MIA PaCa with the $IC_{50}$ value of 54.7 μg/mL, 67.6 μg/mL, 24.9 μg/mL, and 17.0 μg/mL, respectively. Likewise, C-AuNPs inhibited MDA-MB-231, A375SM, PC3, and MIA PaCa up to 3%, 4.3%, 2.9%, and 4.9%, respectively. The highest apoptosis was observed in MDA-MB-231 and PC3 cells when treated with C-AuNPs and EGCG-AuNPs at the concentration of 50 μg $mL^{-1}$ for 24 h. MDA-MB-231 cells treated with EGCG-AuNPs and C-AuNPs showed a reduction in the nuclear factor-kappaB (NF-κB) transcriptional activity by 73.7% and 10.3%, respectively. PC3 cells showed a reduction in the NF-κB transcriptional activity by 65.5% and 7.7% when treated with EGCG-AuNPs and C-AuNPs, respectively.

Cunha et al. [9] synthesized AuNPs, chitosan-AuNPs, cysteamine-AuNPs, EGCG conjugated chitosan-AuNPs, and EGCG conjugated cysteamine-AuNPs with antioxidant activity and cytotoxicity. The hydrodynamic diameter of the above NPs was 37 nm, 86 nm, 54 nm, 125 nm, and 111 nm, respectively, as detected by DLS. The PDI values were 0.2, 0.2, 0.6, 0.3, and 0.2, respectively. The zeta potential was −33 mV, +22 mV, −37 mV, 36 mV, and −24 mV, respectively. The TEM analysis revealed the spherical shape of the NPs. The ABTS radical scavenging activity of cysteamine-AuNPs, EGCG conjugated chitosan-AuNPs, and EGCG conjugated cysteamine-AuNPs were exhibited at the concentrations of 1.8 mM, 2.5 mM, and 3.5 mM, respectively. The EGCG-conjugated chitosan-AuNPs and EGCG-conjugated cysteamine-AuNPs showed cytotoxicity against pancreatic adenocarcinoma (BxPC3) cells with the $IC_{50}$ value of 3.7 μM and 2.22 μM, respectively.

Safwat et al. [17] reported an eco-friendly method for EGCG-conjugated AuNPs synthesis and further studied their hemocompatibility and anticancer activity. Initially, 2 mL of 0.5 mM $HAuCl_4$ solution was mixed with 8 mL of 0.109 mM to 2.618 mM EGCG solution. The change in the color was observed in 15 min where the pale yellow turned to purple red color. The UV-Vis absorption spectra of EGCG-AuNPs showed a peak at 531 nm. The spherical and discrete EGCG-AuNPs with a diameter of 13.6 nm were identified by TEM. The PDI was between 0.3 and 0.5. The FITR spectra of EGCG-AuNPs (0.4:1 molar ratio of EGCG/$HAuCl_4$) exhibited the absorption peak at 3335.9 $cm^{-1}$ and 1691.3 $cm^{-1}$ and the band of EGCG stretching vibration (1400 $cm^{-1}$—1628.2 $cm^{-1}$). The bands representing stretching of OH and CH groups, carbonyl group, and carbon-carbon bonds were at 3356.9 $cm^{-1}$, 1691 $cm^{-1}$, and 1400–1616 $cm^{-1}$, respectively were detected in the spectrum of EGCG. The EGCG-AuNPs exhibited a hemolytic effect of up to 1.5% at the concentration of 100 μL. The complement component 3 (C3) reduced from 140 mg $dL^{-1}$ to 105 mg $dL^{-1}$ when treated with EGCG-AuNPs. The administration of 1 mg $kg^{-1}$ of EGCG-AuNPs to 4–5 weeks-old male mice for 15 days that were treated with tumor ascites cells (100 μL PBS) to induce Ehrlich's tumor, showed potential anticancer activity.

Reshamwala et al. [18] reported the fabrication of EGCG-AuNPs and RSV-AuNPs with antiviral activity. The eco-friendly synthesis of EGCG functionalized AuNPs was prepared by mixing the solution of EGCG and $HAuCl_4$. The color change was observed in 1 min from bright yellow to red. The mixture was vortexed for 2 h at 500 rpm. The red pellet was recovered after centrifugation at 14,000 rpm (4 °C, 20 min). The RSV-AuNPs were prepared by mixing the solution of EGCG and $HAuCl_4$ followed by vortexing at 500 rpm at room temperature. The yellow color of the solution changed to wine red, indicating the formation of RSV-AuNPs. The mean size of EGCG-AuNPs and RSV-AuNPs were 131.5 nm and 78.6 nm, respectively as identified using TEM analysis. The EGCG-AuNPs and RSV-AuNPs had a zeta potential of −40 mV and −35 mV, respectively as revealed by DLS. The EGCG functionalized AuNPs showed antiviral activity against human *Enterovirus* like CVA9, CVB1, and CVB3 with 50% effective concentration ($EC_{50}$) values of 0.004 μM, 0.237 μM, and 0.051 μM, respectively. The values of $EC_{50}$ of RSV-AuNPs against CVA9, CVB1, and CVB3 were 0.039 μM, 0.068 μM, and 0.004 μM, respectively.

Mukherjee et al. [19] investigated the fabrication of green tea and EGCG-conjugated AuNPs for anticancer activity. The green tea and EGCG conjugated AuNPs were synthesized by chemical reduction technique. The green tea extract was made by brewing green leaves (150 mg) in 10 mL of water. Then 100 μL of $HAuCl_4$ (0.1 M) was added and mixed for 10 min till the color changed into purple-red. EGCG conjugated AuNPs were prepared by mixing the 20 mg/mL EGCG solution with $HAuCl_4$ followed by stirring for 10 min. The formation of EGCG-conjugated

AuNPs and green tea-conjugated AuNPs (GT-AuNPs) was confirmed by the development of a wine-red color after 15 min. The UV-vis spectrum analysis confirmed the absorbance peak at 529 nm and 534 nm of GT-AuNPs and EGCG-AuNPs, respectively. The hydrodynamic sizes were 28.9 nm and 59.8 nm, respectively while the zeta potentials were +43 mV. The PDI were 0.734 and 0.325, respectively. The size of GT-AuNPs and EGCG-AuNPs were 25–30 nm and 45 nm, respectively as revealed by AFM. EGCG-AuNPs exhibited cytotoxic effects against breast cancer cells (MCF-7 cells) that were equivalent to 28.16% and 38.21% at concentrations of 2 μg/mL and 20 μg/mL, respectively. GT-AuNPs showed cytotoxic effects against breast cancer cells (MCF-7 cells) equivalent to 32% and 47.28% at concentrations of 0.15% and 1.5%, respectively.

Tsai et al. [20] synthesized doxorubicin-gelatin and EGCG conjugated AuNPs (DOX-GLT/EGCG AuNPs) and EGCG-AuNPs with anticancer properties. The EGCG-AuNPs were prepared by adding $HAuCl_4$ into 1 mL of EGCG solution followed by continuously stirring at 25 °C for 25 min. The doxorubicin conjugated gelatin-EGCG-AuNPs were synthesized by using the layer-by-layer assembly method. Initially, 2 mL of EGCG-AuNPs was mixed with 2 mg $mL^{-1}$ of DOX conjugated GLT and stirred for 10 min. The DOX-GLT/EGCG AuNPs were obtained after centrifugation of the mixture at 10,000 rpm. The absorbance maxima of EGCG-AuNPs was at 540 nm in the UV-vis spectroscopy. The hydrodynamic size of EGCG-AuNPs was 35–85 nm. TEM image revealed a particle diameter of 30 nm. The presence of elemental carbon, oxygen, and gold at 18.54%, 1.37%, and 80.07%, respectively was confirmed by EDX analysis. The Raman spectrum of EGCG-AuNPs ranged from 1500 $cm^{-1}$ to 1600 $cm^{-1}$. EGCG-AuNPs had a zeta potential of −20 mV. The X-ray photoelectron spectroscope (XPS) analysis showed peaks at 284.7 eV, 289.1 eV, and 537 eV corresponding to the aromatic $sp^2$, C=O, and phenolic hydroxyl, respectively which confirmed the formation of EGCG-AuNPs and DOX-GLT/EGCG AuNPs, respectively. The latter exhibited a significant anticancer property against human prostate cancer cells (PC-3 cells) with an $IC_{50}$ value of 1.4 μM.

## 2.2 Silver Nanoparticles

Seabra et al. [21] synthesized catechin-conjugated AgNPs (catechin-AgNPs), mercaptosuccinic acid functionalized catechin conjugated AgNPs (MSA-catechin-AgNPs), and *S*-nitroso-mercaptosuccinic acid functionalized catechin conjugated AgNPs (*S*-nitroso-MSA-catechin-AgNPs) for bactericidal application. The catechin-AgNPs were prepared by reacting 0.5 mL of catechin solution and 3 mL of $AgNO_3$ at pH 10. The MSA-catechin-AgNPs were fabricated by mixing 10 mL of mercaptosuccinic acid (200 mg) with 10 mL of 20 mg catechin-AgNPs at room temperature for 4 h. Finally, S-nitroso-MSA-catechin-AgNPs was synthesized by mixing MSA-catechin-AgNPs and sodium nitrite ($NaNO_2$). The mixtures were kept under stirred conditions for 30 min. FTIR analysis of catechin-AgNPs revealed the aromatic C=C group at 1563 $cm^{-1}$ and 1456 $cm^{-1}$. XRD peaks were identified at the $2\theta$ degree of 38.10 °, 44.47°, 64.63°, 77.44°, and 81.33° which corresponds to the planes (111), (200), (220), (311), and (222). The spherical shape with a 23.4 nm average size was determined by Field-emission scanning electron microscopy (FE-SEM) whereas a hydrodynamic size of 44 nm was revealed by DLS analysis. The polydispersity index (PDI) and zeta potential were 0.21 and −35.9 mV, respectively. The bactericidal activity of catechin-AgNPs was tested against Gram-positive bacteria (*Staphylococcus aureus*) and Gram-negative (*Escherichia coli, Klebsiella pneumoniae, Pseudomonas aeruginosa* and *Salmonella enterica*) using a microdilution method. The catechin-AgNPs, MSA-catechin-AgNPs, and S-nitroso-MSA-catechin-AgNPs exhibited bactericidal effect against *S. aureus* with a minimum inhibitory concentration (MIC) of 500 μg $mL^{-1}$, 250 μg $mL^{-1}$, and 125 μg $mL^{-1}$, respectively. The MICs of the NPs against *E. coli* and *S. enterica* were 62 μg $mL^{-1}$, 250 μg $mL^{-1}$, and 125 μg $mL^{-1}$, respectively. The MIC values of the aforementioned NPs against *K. pneumoniae* were 1000 μg $mL^{-1}$, 250 μg $mL^{-1}$, and 125 μg $mL^{-1}$, respectively whereas MIC values against *P. aeruginosa* were 62 μg $mL^{-1}$, 125 μg $mL^{-1}$, and 3 μg $mL^{-1}$, respectively.

Meesaragandla et al. [7] checked the anti-biofilm and cytotoxicity activity of EGCG-AgNPs and lysozyme-conjugated EGCG-AgNPs as schematically represented in Fig. 1. The EGCG-AgNPs were synthesized by reacting 10 mM EGCG and 30 mL of $AgNO_3$ under vigorous stirring for 2 h at 27 °C. The colorless solution of $AgNO_3$ changed into a yellowish-brown color which confirmed the synthesis of EGCG-AgNPs. Lysozyme-conjugated EGCG-AgNPs were fabricated by mixing 1 μM EGCG-AgNPs with 100 μg/mL of lysozyme in phosphate-buffered saline (PBS) of pH 6.2. The mixture was incubated for 4 h at 37 °C and centrifuged at room temperature at 17,000 rpm for 30 min. The absorption peak of EGCG-AgNPs and lysozyme-conjugated EGCG-AgNP at 413 nm and 424 nm was noted in their UV-vis spectra. TEM analysis revealed spherical-shaped EGCG-AgNPs of 14 nm

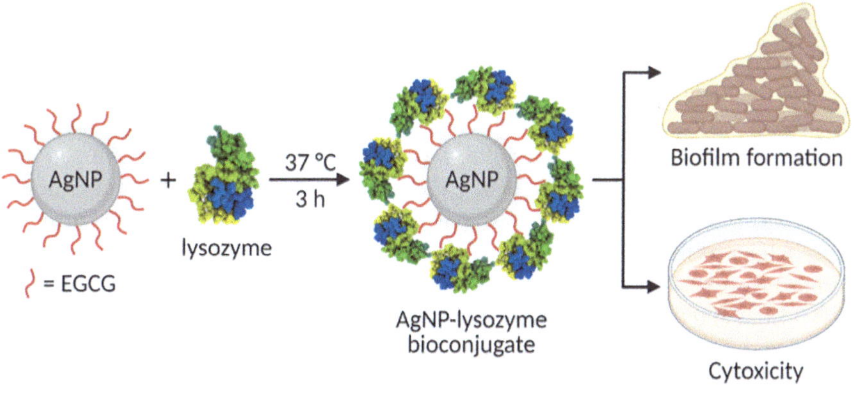


**Fig. 1** Overview of this study illustrating functionalization of AgNPs with EGCG and the formation of a bioconjugate with lysozyme. (Reprinted from Meesaragandla et al. [7]. Copyright © 2022 The Authors. Published by American Chemical Society. (Open Access))

size. The DLS analysis of EGCG-AgNPs revealed a hydrodynamic diameter of 36 nm and 130 nm. The zeta potential of EGCG-AgNPs in water and PBS after 30 min at pH 6.2 were −27 mV and −35 mV, respectively. The EGCG-AgNPs and lysozyme-conjugated EGCG-AgNPs were tested for antibiofilm activity at different concentrations. The culture of wild-type strain of *Bacillus subtilis* (2 μL) and EGCG-AgNPs (10 μL) was added to 88 μL minimal salts glycerol glutamate (MSgg) medium and incubated at 30 °C for 24 h. The biofilm formation was inhibited at a concentration of 40 nM EGCG-conjugated AgNPs. EGCG-AgNPs showed low cytotoxicity against human umbilical vein endothelial cells (HUVEC) at 200 nM whereas the lysozyme-conjugated EGCG-AgNP exhibited cytotoxicity at a concentration of 10 nM.

Krzyzowska et al. [6] reported cytotoxicity and antiviral properties of AgNPs, EGCG-AgNPs, and tannic Acid (TA) functionalized AgNPs (TA-AgNPs). Initially, AgNPs were synthesized by heating a mixture of sodium citrate and deionized water for 15 min at 70 °C. Then $AgNO_3$ and sodium borohydride were mixed and heated for 60 min at 70 °C. EGCG-AgNPs were prepared by mixing AgNPs (1 g) with EGCG (0.015 g) for 60 min. The TA-AgNPs were prepared by mixing $AgNO_3$, sodium citrate, and tannic acid solution followed by stirring for 15 min. The UV-vis spectra of AgNPs, EGCG-AgNPs, and TA-AgNPs detected absorbance maxima at 395 nm, 400 nm, and 407 nm, respectively. High-resolution scanning electron microscope with a transmission detector (HR-STEM) analysis showed the spherical morphology and mean size of 26 nm, 26 nm, and 27 nm for AgNPs, EGCG-AgNPs, and TA-AgNPs, respectively. The hydrodynamic diameters of the aforementioned NPs were 30 nm, 35 nm, and 32 nm, while zeta potentials were −76 mV, −67 mV, and −58 mV, respectively. The AgNPs were tested for cytotoxicity on Vero cell, human vaginal epithelial cells (VK2-E6/E7 cells), and human keratinocytes cells (HaCaT) that showed 50% cellular cytotoxicity ($CC_{50}$) values of 30.8 μg $mL^{-1}$, 27.54 μg $mL^{-1}$ and 33.3 μg $mL^{-1}$, respectively. The EGCG-AgNPs exhibited $CC_{50}$ values equivalent to 39.02 μg $mL^{-1}$, 33.8 μg $mL^{-1}$, and 35.3 μg $mL^{-1}$. The EGCG-AgNPs and TA-AgNPs showed a potent antiviral effect against HSV-1 strain McKrae and HSV-2 strain 333 at the concentration of 72 μg $mL^{-1}$.

Ikram et al. [22] fabricated epicatechin-conjugated AgNPs (EC-AgNPs) and studied its potential for selective gentamicin detection. EC-AgNPs were prepared by mixing 100 μM of $AgNO_3$ with 100 μM of epicatechin and triethylamine as a reducing agent (4–5 drops). The pellet was formed after 12 min of centrifugation at 12,000 rpm. The absorbance band of EC-AgNPs was detected at 412 nm. The PDI value was 0.309 while the mean size was 74.38 nm as determined by DLS. The morphology of the particle was sponge-like with a size ranging between 4 nm and 41 nm as revealed from the SEM and AFM analysis, respectively. FTIR spectrum of epicatechin showed the bands at 3431 $cm^{-1}$ and 2923 $cm^{-1}$ corresponding to OH group and C=C (aromatic) stretching vibrations, respectively. The EC-AgNPs showed a band at 3413 $cm^{-1}$ specific to the hydroxyl group. It showed selective detection for gentamicin in serum, blood plasma, and tap water. The absorbance maximum was 412 nm as determined from the UV-vis spectra which indicated the binding of EC-AgNPs and gentamicin.

## 2.3 *Metal Oxide Nanoparticles*

Zhao et al. [10] reported the preparation of catechin functionalized ZnONPs nanoclusters with antimicrobial activity. The hydrothermal synthesis involved autoclaving a mixture of 0.4 mL $ZnCl_2$ solution and 40 mL catechin solution (pH 8) in a Teflon-lined at 140 °C for 4 h. A brown colloidal solution was observed. The UV-vis spectrum showed the absorption peak at 330–430 nm. The FTIR analysis of catechin-conjugated ZnONPs revealed the C-O and C=C group stretching vibration at 1265 $cm^{-1}$ and 1600 $cm^{-1}$, respectively. The size of nanoclusters ranged from 10 nm to 30 nm while the height was 30 nm as determined by TEM and AFM, respectively. The XRD pattern showed distinct peaks at $2\theta$ values of 31.8°, 34.4°, 36.2°, 47.5°, 56.7°, 62.9°, and 67.8° which were attributed to hexagonal planes (100), (002), (101), (102), (110), (103) and (112), respectively. The two major peaks of XPS were observed at 1026.5 and 1049.5 eV corresponding to Zn2p3/2 and Zn2p1/2, respectively. The catechin-conjugated ZnONPs were evaluated for antimicrobial activity utilizing the disk diffusion method. It showed potent antifungal activity against *Candida albicans* and antibacterial effect against *S. aureus* and *E. coli* with a MIC value of 29.8 μg $mL^{-1}$.

Xiao et al. [23] reported the fabrication of green tea conjugated $Fe_3O_4$NPs (GT-$Fe_3O_4$NPs) and surface functionalization of $Fe_3O_4$NPs with the quinone form of EGCG and EC as schematically represented in Fig. 2. The EGCGq-$Fe_3O_4$NPs were synthesized by mixing the solution of EGCG and $Fe(OH)_3$ for 10 min followed by autoclaving for 3 h. The GT-$Fe_3O_4$NPs were obtained by centrifugation of the mixture at 11,000 rpm for 1 h. The ECq-$Fe_3O_4$NPs were prepared in a similar process using EC instead of EGCG with $Fe(OH)_3$ solution. HRTEM analysis of EGCGq-$Fe_3O_4$NPs and ECq-$Fe_3O_4$NPs revealed a spherical shape with a size of 2.5 nm. The hydrodynamic diameter of GT-$Fe_3O_4$NPs, ECq-$Fe_3O_4$NPs, and EGCGq-$Fe_3O_4$NPs

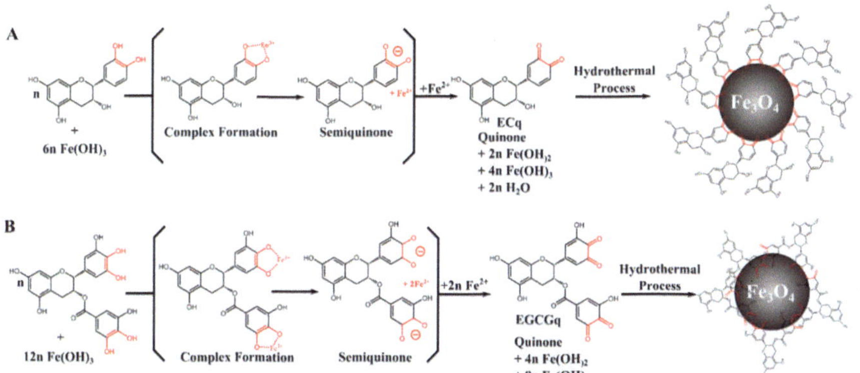


**Fig. 2** Schematic representation for the preparation strategy for (**a**) ECq-$Fe_3O_4$ and (**b**) EGCGq-$Fe_3O_4$ NPs. (Reprinted with permission from Xiao et al. [23]. Copyright © 2015 American Chemical Society)

were 119 nm, 56 nm, and 64 nm, respectively as identified by DLS. The PDI of ECq-$Fe_3O_4$NPs and EGCGq-$Fe_3O_4$NPs were 0.19 and 0.2, respectively. The zeta potential of GT-$Fe_3O_4$NPs, EGCGq-$Fe_3O_4$NPs, and ECq-$Fe_3O_4$NPs were −35.8 mV, −38.3 mV and −29 mV, respectively. The uptake of GT-$Fe_3O_4$NPs, EGCGq-$Fe_3O_4$NPs, and ECq-$Fe_3O_4$NPs by colon cancer cells (CT26 cells) was 0.1 mM, 0.6 mM, and 1 mM, respectively. The cellular uptakes of GT-$Fe_3O_4$NPs, EGCGq-$Fe_3O_4$NPs, and ECq-$Fe_3O_4$NPs in primary human macrophages, J774A cells were 0.3 mM, 0.1 mM, and 1 mM, respectively. The ECq-$Fe_3O_4$NPs and EGCGq-$Fe_3O_4$NPs did not show cytotoxicity against CT26 and J774A cells.

Kader et al. [24] reported a green synthesis for ZnONPs and catechin-conjugated ZnONPs with antibacterial activity. Initially, ZnONPs were synthesized by reacting leaf extract of margosa and zinc chloride hexahydrate ($ZnCl_2 \cdot 6H_2O$) at 75 °C for 2 h. The catechin-conjugated ZnONPs were made by mixing catechin solution with 1 g of ZnONPs and heating at 75 °C in the dark. The formation of catechin-conjugated ZnONPs was confirmed by color change from white to brown. Then it was dried for 2 h at 75 °C. The UV-vis spectrum of catechin, ZnONPs, and catechin-conjugated ZnONPs exhibited the absorption band at 278 nm, 329 nm, and 482 nm, respectively. SEM analysis revealed the particle size of ZnONPs and catechin-conjugated ZnONPs from 20 nm to 80 nm and 60 nm to 120 nm, respectively as evident from Fig. 3. The presence of elements like carbon (8.42%), oxygen (33.17%), and zinc (58.41%) was confirmed by EDX analysis. The XRD analysis of catechin-conjugated ZnONPs revealed the distinct diffraction peaks at the $2\theta$ degree of 31.97°, 34.53°, 37.02°, 47.14°, 55.88°, 62.96°, 65.78, 67.57°, 69.34°, and 72.54° corresponding to the crystalline planes of (100), (002), (101), (102), (110), (103), (200), (112), (201), and (004). The O-H and C=C stretching vibration attributed to

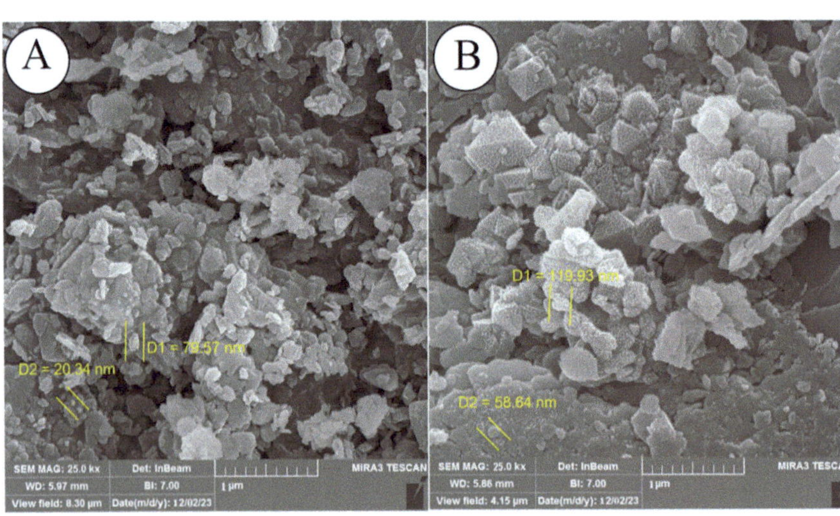


**Fig. 3** Displays scanning electron microscopy (SEM) images for (**a**) the synthesized ZnO nanoparticles and (**b**) the ZnO/catechin nanocomposite. (Reprinted with permission from Kader et al. [24]. Copyright © 2024 Elsevier B.V.)

bands at 3550 $cm^{-1}$—3150 $cm^{-1}$ and 1630 $cm^{-1}$ in the FTIR analysis of catechin conjugated ZnONPs. The bactericidal activity of catechin-conjugated ZnONPs was tested against Gram-positive (*Bacillus cereus* and *S. aureus*) and Gram-negative bacteria (*P. aeruginosa* and *E. coli*). The MIC values against *B. cereus*, *S. aureus*, *E. coli*, and *P. aeruginosa* were 20 μg $mL^{-1}$, 10 μg $mL^{-1}$, 20 μg $mL^{-1}$, and 60 μg $mL^{-1}$, respectively. The zone of inhibition (ZOI) of *Bacillus cereus*, *S. aureus*, *E. coli*, and *P. aeruginosa* were 12 mm, 11 mm, 10 mm, and 10 mm, respectively.

## 2.4 Polymeric and Non-metallic Nanoparticles

Moreno-Vásquez et al. [25] reported the synthesis of chitosan NPs and EGCG-conjugated chitosan NPs employing the precipitation method. The EGCG-conjugated chitosan NPs were synthesized by mixing EGCG and chitosan in the same proportion with acetic acid (0.4%) and stirred for 24 h at 25 °C. The solution was centrifuged for 60 min after adding hydrogen peroxide (1 M, 1 mL) and 0.054 g of ascorbic acid at 3500 rpm at 4 °C. TEM analysis revealed the smooth surface with spherical morphology of NPs. The average diameter of chitosan NPs and EGCG-conjugated chitosan NPs were 139.10 nm and 1375 nm, respectively. The PDI of chitosan NPs and EGCG-conjugated chitosan NPs were 0.844 and 0.523, respectively indicating a monodispersed nature. The chitosan NPs and EGCG conjugated chitosan NPs had zeta potential of 53.03 mV and 40.87 mV, respectively. Chitosan NPs and EGCG-conjugated chitosan NPs showed ABTS radical scavenging activity of 63% and 94.3% at a concentration of 10 mg. $mL^{-1}$ and 2 mg. $mL^{-1}$, respectively. The chitosan NPs and EGCG-chitosan NPs exhibited scavenging activity of DPPH radical with 50% effective concentration ($EC_{50}$) values of 10 mg $mL^{-1}$ and 1 mg $mL^{-1}$, respectively. The EGCG-conjugated chitosan NPs exhibited bactericidal properties against *S. aureus* and *Pseudomonas fluorescens* with MIC values of 0.98 μg $mL^{-1}$ and 15.6 μg $mL^{-1}$, respectively whereas MBC values were 0.98 μg $mL^{-1}$ and 31.2 μg $mL^{-1}$, respectively.

Sistanipour et al. [26] fabricated catechin-functionalized mesoporous hydroxyapatite NPs by conjugating the catechin on the mesoporous hydroxyapatite NPs surface through amide linkage and further studied their antioxidant and osteogenic properties. The mesoporous hydroxyapatite nanoparticles were prepared by mixing 0.15 M $K_2HPO_4 \cdot 3H_2O$ and 0.078 M cetyltrimethylammonium bromide in 100 mL of water at pH 10.5. Then the 0.24 M $CaCl_2$ solution was mixed with it, which resulted in a precipitate that was recovered after 15 min of centrifugation at 6000 rpm. The pellet was dried at 70 °C for 48 h. The 0.1 g of mesoporous hydroxyapatite nanoparticles was mixed with 20 mL of 4.24 M aminopropyltriethoxysilane and stirred for 5 h at 90 °C. The solution was centrifuged at 8000 rpm for 10 min and dried for 24 h at 37 °C to produce 3-aminopropyl conjugated mesoporous hydroxyapatite NPs. Finally, the catechin functionalized mesoporous hydroxyapatite NPs were synthesized by mixing 1.7 mM of catechin with 20 mg of 3-aminopropyl conjugated mesoporous hydroxyapatite NPs for 24 h in the dark. The UV-vis spectra revealed the absorption peaks at 230 nm and 280 nm. The zeta potential of mesoporous hydroxyapatite NPs and catechin functionalized mesoporous hydroxyapatite NPs were −12.95 mV and +16.5 mV, respectively. The presence of elemental carbon, nitrogen, oxygen, phosphorus, calcium, and silicon in the catechin functionalized mesoporous hydroxyapatite NPs was revealed by energy dispersive spectroscopy (EDS). The FTIR spectrum showed the stretching vibrations of C=C at 1492 $cm^{-1}$ and 1387 $cm^{-1}$. The crystalline size of 19 nm was confirmed by XRD analysis. It exhibited the scavenging activity of hydroxyl and DPPH radicals up to 70% and 78.5%, respectively, at a concentration of 500 nM it showed higher superoxide radical scavenging activity at a concentration of 50 nM. The osteogenic differentiation of human bone cancer cells, Saos-2 cells, and mesenchymal stem cells (MSCs) was determined by alkaline phosphatase (ALP) activity grown on the catechin functionalized mesoporous hydroxyapatite NPs film as seen in Fig. 4. The ALP levels in Saos-2 cells and MSCs were 273 ± 9.7 unit $mg^{-1}$ and 1043.28 ± 18 unit $mg^{-1}$, respectively, after an incubation at 37 °C for 14 days. The high level of osteogenic gene expression, namely, OCN and Runx2 was observed in the Saos-2 cells and MSCs.

Liu et al. [27] reported the preparation of folate functionalized chitosan NPs (FC-NPs), and catechin conjugated folate functionalized chitosan NPs (CFC-NPs) with antiproliferative activity. Initially, the FC-NPs were prepared by mixing chitosan

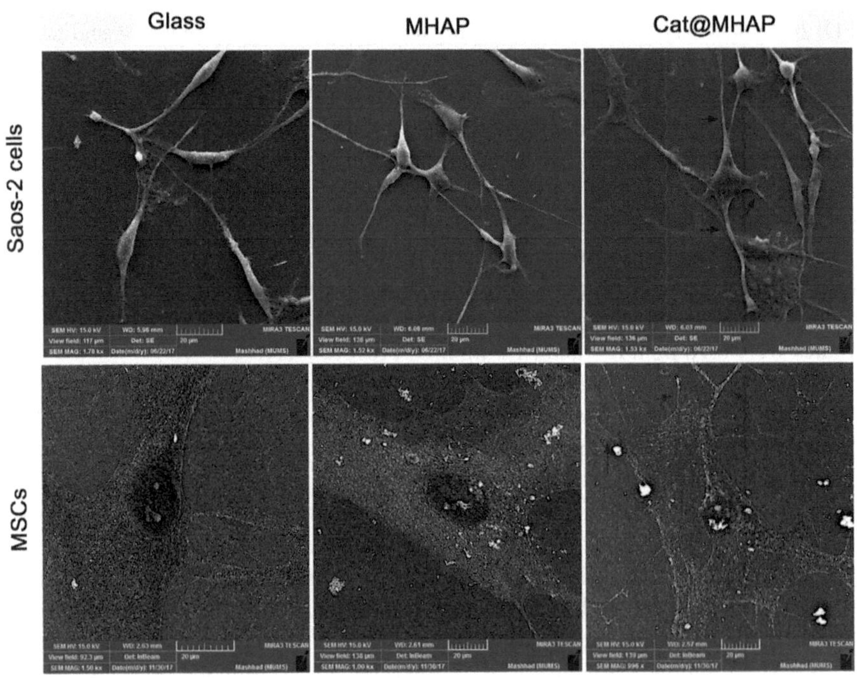


**Fig. 4** Evaluation of cell morphology on the surface of MHAP and Cat@MHAP films. FE-SEM images of cells incubated on different surfaces for 7 days. Black arrows represent the cytoplasmic projections in differentiated cells. The scale bar of all images is 20 μm. (Reprinted with permission from Sistanipour et al. [26]. Copyright © 2018 Elsevier B.V.)

(2 g) and aqacetate (400 mL) followed by the addition of 1-(3-dimethylaminopropyl)-3-ethyl carbodiimide hydrochloride (EDC) and folic acid. The solution was kept under continuous stirring conditions for 16 h at pH 9. Finally, the CFC-NPs were synthesized by mixing catechin with acetic acid (20 mL) and FC-NPs (40 mg). Then, 5 mL sodium tripolyphosphate was added and mixed for 30 min at room temperature. The CFC-NPs were recovered by centrifugation for 30 min at 16,000 rpm. The spherical morphology of CFC-NPs and FC-NPs was observed by TEM analysis. The average diameter of CFC-NPs and FC-NPs was 255 nm and 105 nm, respectively. The zeta potential of CFC-NPs and FC-NPs was 24 mV and 17 mV, respectively. The absorbance maxima of CFC-NPs were at 1630 $cm^{-1}$ as identified from the UV-vis spectra. The CFC-NPs exhibited potential antiproliferative activity against breast cancer cells (MCF-7 cells) and hepatocellular cancer cells (HepG2 cells). It showed cytotoxicity against MCF-7 cells up to 41.36% and 35.22% on incubation with 0.5 mg $mL^{-1}$ CFC-NPs for 24 h and 36 h, respectively, while the cytotoxicity against HepG2 cells was 66.11% and 58.44%, respectively.

Safer et al. [28] reported the fabrication of AuNPs, chitosan NPs, Au-conjugated chitosan NPs, and EGCG conjugated chitosan NPs. The AuNPs were synthesized by mixing 0.25 mM $HAuCl_4$ solution with $NaBH_4$ under continuous stirring at 750 rpm for a few seconds. Then, the $HAuCl_4$ solution was mixed with chitosan solution and was heated for 15 min at 100 °C to produce Au-conjugated chitosan NPs. The EGCG-conjugated chitosan NPs were prepared using EGCG, chitosan, and TPP solution and kept under stirring conditions for 3 h at 1000 rpm. The EGCG-conjugated chitosan NPs were obtained after 10 min of centrifugation at 14,000 rpm. The absorbance maxima of AuNPs were detected by UV-vis spectroscopy at 520 nm. The TEM analysis revealed the size of AuNPs, chitosan NPs, Au-conjugated chitosan NPs, and EGCG-conjugated chitosan NPs as 47 nm, 156 nm, 160 nm, and 414 nm, respectively. The particle sizes were 47.03 nm, 156 nm, 160.03 nm, and 414.8 nm, respectively, while the zeta potentials were 56 mV, −12.9 mV, 63.9 mV, and 67.6 mV, respectively. PDI were 0.328, 0.38, 0.581, and 0.363, respectively, for the aforementioned nanomaterials.

Bae et al. [8] reported cytocompatible EGCG conjugated poly(acrylic acid) (PAA-EGCG) micellar NPs for anti-inflammatory activity. The thiolated PAA was prepared by mixing 4 (4,6-dimethoxy-1,3,5-triazin-2-yl)-4-methylmorpholinium and cysteamine. Then the EGCG was mixed with thiolated PAA for 16 h at pH 7.4. Extensive dialysis against nitrogen-purged water was used to purify the PAA-EGCG conjugates. Finally, the conjugates were self-assembled to form micellar NPs. The UV-vis spectra showed the absorbance maxima of PAA-EGCG micellar NPs at 274 nm. The zeta potential of PAA-EGCG micellar NPs was from −23.2 mV to −47.7 mV. The spherical morphology of the PAA-EGCG micellar NPs was revealed by FE-TEM analysis. PAA-EGCG micellar NPs showed cytocompatibility against human dermal fibroblasts. It showed a decrease in the viable cells up to 62.4% at 120 μM. It exhibited potent anti-inflammatory properties against mouse macrophage (RAW 264.7) activated by lipopolysaccharide at the concentration of 23 μM, which was speculated to be attributed to superior reactive oxygen species (ROS) modulation (Table 1).

**Table 1** Catechin functionalized nanoparticles and their application

| NPs | Functionalized ligand | Size (in nm) | Shape | Application | Reference |
|---|---|---|---|---|---|
| AuNPs | Epigallocatechin-3-gallate | 30 | Spherical | Anti-osteoclastogenic activity | Zhu et al. [14] |
| AuNPs | Epigallocatechin-3-gallate, resveratrol, fisetin | 25 | Spherical | Antioxidant, antiproliferative | Sanna et al. [15] |
| AuNPs | Epigallocatechin-3-gallate, citrate | 25 | Spherical | Anticancer, apoptosis | Chavva et al. [16] |
| AuNPS | Epigallocatechin-3-gallate, chitosan, cysteamine | | Spherical | Antioxidant, cytotoxicity | Cunha et al. [9] |
| AuNPs | Epigallocatechin-3-gallate | 13.6 | Spherical | Anticancer, hemocompatibility | Safwat et al. [17] |
| AuNPs | Epigallocatechin-3-gallate, resveratrol | 131 | – | Antiviral | Reshamwala et al. [18] |
| AuNPs | Epigallocatechin-3-gallate, green tea | 25–30, 45 | – | Anticancer | Mukherjee et al. [19] |
| AuNPs | Epigallocatechin-3-gallate, gelatin, doxorubicin | 30 nm | – | Anticancer | Tsai et al. [20] |
| AgNPs | Catechin, mercaptosuccinic acid | 23.4 | Spherical | Antibacterial | Seabra et al. [21] |
| AgNPs | Epigallocatechin-3-gallate | 14 | Spherical | Antibacterial, cytotoxicity | Meesaragandla et al. [7] |
| AgNPs | Epigallocatechin-3-gallate | 26 | Spherical | Antiviral | Krzyzowska et al. [6] |
| AgNPs | Tannic acid | 27 | Spherical | Antiviral | Krzyzowska et al. [6] |
| AgNPs | Epicatechin | 74.38 | – | Chemosensor for antimicrobial drugs | Ikram et al. [22] |
| ZnONPs | Catechin | 10–30 | Spherical, polydispersed | Antibacterial | Zhao et al. [10] |
| $Fe_3O_4$NPs | Epigallocatechin-3-gallate | 2.5 | Spherical | Cytotoxicity | Xiao et al. [23] |
| $Fe_3O_4$NPs | Epicatechin | 2.7 | Spherical | Cytotoxicity | Xiao et al. [23] |

| | | | | | |
|---|---|---|---|---|---|
| ZnONPs | Catechin | 60–80 | – | Antibacterial | Kader et al. [24] |
| Chitosan NPs | Epigallocatechin-3-gallate | 362.92 | Spherical | Antioxidant, antibacterial | Moreno-Vásquez et al. [25] |
| Mesoporous hydroxyapatite NPs | Catechin | – | – | Antioxidant, osteogenic property | Sistanipour et al. [26] |
| Chitosan NPs | Catechin, folate | – | Spherical | Antiproliferative | Liu et al. [27] |
| Chitosan NPs | Epigallocatechin-3-gallate | 414 | – | – | Safer et al. [28] |
| Poly(acrylic acid) micellar NPs | Epigallocatechin-3-gallate | – | Spherical | Cytocompatibility, anti-inflammatory | Bae et al. [8] |

## 3 Conclusion and Future Perspectives

Catechins are natural polyphenols found in tea and other plants. Catechin-functionalized nanoparticles have demonstrated substantial therapeutic capabilities. These nanocomposites exhibit strong anticancer, antiproliferative, antimicrobial, and antioxidant properties with minimal negative effects. Hence, these nanomaterials can be functionalized with drugs for sustained release. Also, contrast agents and tissue-specific ligands can be functionalized on these nanoparticles for targeted delivery, diagnosis, and bioimaging. The nanohybrids have improved stability and bioavailability, which make them ideal for nanomedicine. Further, they can be used as packaging materials, increasing the shelf-life of vegetables, fruits, and processed food. Therefore, catechin and its derivatives can be rationally functionalized onto various metal, metal oxide, carbon-based, and polymeric nanoparticles for diverse healthcare applications.

**Acknowledgements** Poulami Manna is thankful to the Council of Scientific and Industrial Research (CSIR, Government of India) for a Junior Research Fellowship (JRF) (09/1371(19537)/2024-EMR-I).

## References

1. Vaishnavi, M., Raksha, B., Durga, M., Brindha Banu, B., & Deepikaa, R. (2023). Catechin, as an emerging front-line defense flavonoid against stress-related disorders: An updated review. *International Journal of Pharmacy and Biological Sciences, 13*(3), 122–135.
2. Alavi, M., Yarani, R., Sreedharan, M., & Thomas, S. (2023). Micro and nanoformulations of catechins for therapeutic applications: Recent advances and challenges. *Micro Nano Bio Aspects, 2*(1), 8–19.
3. Sahadevan, R., Singh, S., Binoy, A., & Sadhukhan, S. (2023). Chemico-biological aspects of (−)-epigallocatechin-3-gallate (EGCG) to improve its stability, bioavailability and membrane permeability: Current status and future prospects. *Critical Reviews in Food Science and Nutrition, 63*(30), 10382–10411.
4. Altammar, K. A. (2023). A review on nanoparticles: Characteristics, synthesis, applications, and challenges. *Frontiers in Microbiology, 14*, 1155622.
5. Mekuye, B., & Abera, B. (2023). Nanomaterials: An overview of synthesis, classification, characterization, and applications. *Nano Select, 4*(8), 486–501.
6. Krzyzowska, M., Janicka, M., Chodkowski, M., Patrycy, M., Obuch-Woszczatyńska, O., Tomaszewska, E., Ranoszek-Soliwoda, K., Celichowski, G., & Grobelny, J. (2023). Epigallocatechin gallate-modified silver nanoparticles show antiviral activity against herpes simplex type 1 and 2. *Viruses, 15*(10), 2024.
7. Meesaragandla, B., Hayet, S., Fine, T., Janke, U., Chai, L., & Delcea, M. (2022). Inhibitory effect of epigallocatechin gallate-silver nanoparticles and their lysozyme bioconjugates on biofilm formation and cytotoxicity. *ACS Applied Bio Materials, 5*(9), 4213–4221.
8. Bae, K. H., Chan, K. H., & Kurisawa, M. (2022). Autoxidation-resistant, ROS-scavenging, and anti-inflammatory micellar nanoparticles self-assembled from poly (acrylic acid)–green tea catechin conjugates. *ACS Macro Letters, 11*(7), 835–840.

9. Cunha, L., Coelho, S. C., Pereira, M. D. C., & Coelho, M. A. (2022). Nanocarriers based on gold nanoparticles for epigallocatechin gallate delivery in cancer cells. *Pharmaceutics, 14*(3), 491.
10. Zhao, B., Deng, S., Li, J., Sun, C., Fu, Y., & Liu, Z. (2021). Green synthesis, characterization and antibacterial study on the catechin-functionalized ZnO nanoclusters. *Materials Research Express, 8*(2), 025006.
11. Wang, Q., Liu, P., Wen, Y., Li, K., Bi, B., Li, B. B., Qiu, M., Zhang, S., Li, Y., Li, J., & Chen, H. (2023). Metal-enriched HSP90 nanoinhibitor overcomes heat resistance in hyperthermic intraperitoneal chemotherapy used for peritoneal metastases. *Molecular Cancer, 22*(1), 95.
12. Zhou, J., Liu, Y., Hu, Y., Zhang, D., Xu, W., Chen, L., He, J., Cheng, S., & Cai, J. (2022). Selenium nanoparticles synergistically stabilized by starch microgel and EGCG: Synthesis, characterization, and bioactivity. *Food, 12*(1), 13.
13. Farabegoli, F., Granja, A., Magalhães, J., Purgato, S., Voltattorni, M., & Pinheiro, M. (2022). Epigallocatechin-3-gallate delivered in nanoparticles increases cytotoxicity in three breast carcinoma cell lines. *ACS Omega, 7*(46), 41872–41881.
14. Zhu, S., Zhu, L., Yu, J., Wang, Y., & Peng, B. (2019). Anti-osteoclastogenic effect of epigallocatechin gallate-functionalized gold nanoparticles in vitro and in vivo. *International Journal of Nanomedicine, 14*, 5017–5032.
15. Sanna, V., Pala, N., Dessì, G., Manconi, P., Mariani, A., Dedola, S., Rassu, M., Crosio, C., Iaccarino, C., & Sechi, M. (2014). Single-step green synthesis and characterization of gold-conjugated polyphenol nanoparticles with antioxidant and biological activities. *International Journal of Nanomedicine, 9*, 4935–4951.
16. Chavva, S. R., Deshmukh, S. K., Kanchanapally, R., Tyagi, N., Coym, J. W., Singh, A. P., & Singh, S. (2019). Epigallocatechin gallate-gold nanoparticles exhibit superior antitumor activity compared to conventional gold nanoparticles: Potential synergistic interactions. *Nanomaterials (Basel), 9*(3), 396.
17. Safwat, M. A., Kandil, B. A., Elblbesy, M. A., Soliman, G. M., & Eleraky, N. E. (2020). Epigallocatechin-3-gallate-loaded gold nanoparticles: Preparation and evaluation of anticancer efficacy in ehrlich tumor-bearing mice. *Pharmaceuticals, 13*(9), 254.
18. Reshamwala, D., Shroff, S., Sheik Amamuddy, O., Laquintana, V., Denora, N., Zacheo, A., Lampinen, V., Hytonen, V. P., Tastan Bishop, Ö., Krol, S., & Marjomäki, V. (2021). Polyphenols epigallocatechin gallate and resveratrol, and polyphenol-functionalized nanoparticles prevent *Enterovirus* infection through clustering and stabilization of the viruses. *Pharmaceutics, 13*(8), 1182.
19. Mukherjee, S., Ghosh, S., Das, D. K., Chakraborty, P., Choudhury, S., Gupta, P., Adhikary, A., Dey, S., & Chattopadhyay, S. (2015). Gold-conjugated green tea nanoparticles for enhanced anti-tumor activities and hepatoprotection—Synthesis, characterization and in vitro evaluation. *The Journal of Nutritional Biochemistry, 26*(11), 1283–1297.
20. Tsai, L. C., Hsieh, H. Y., Lu, K. Y., Wang, S. Y., & Mi, F. L. (2016). EGCG/gelatin-doxorubicin gold nanoparticles enhance therapeutic efficacy of doxorubicin for prostate cancer treatment. *Nanomedicine, 11*(1), 9–30.
21. Seabra, A. B., Manosalva, N., de Araujo Lima, B., Pelegrino, M. T., Brocchi, M., Rubilar, O., & Duran, N. (2017). Antibacterial activity of nitric oxide releasing silver nanoparticles. *Journal of Physics: Conference Series, 838*, 012031.
22. Ikram, F., Qayoom, A., & Shah, M. R. (2018). Synthesis of epicatechin coated silver nanoparticles for selective recognition of gentamicin. *Sensors and Actuators B: Chemical, 257*, 897–905.
23. Xiao, L., Mertens, M., Wortmann, L., Kremer, S., Valldor, M., Lammers, T., Kiessling, F., & Mathur, S. (2015). Enhanced in vitro and in vivo cellular imaging with green tea coated water-soluble iron oxide nanocrystals. *ACS Applied Materials & Interfaces, 7*(12), 6530–6540.
24. Kader, D. A., Aziz, D. M., Mohammed, S. J., Maarof, N. N., Karim, W. O., Mhamad, S. A., Rashid, R. M., Ayoob, M. M., Kayani, K. F., & Qurbani, K. (2024). Green synthesis of ZnO/catechin nanocomposite: Comprehensive characterization, optical study, computational

analysis, biological applications and molecular docking. *Materials Chemistry and Physics, 319*, 129408.

25. Moreno-Vásquez, M. J., Plascencia-Jatomea, M., Sánchez-Valdes, S., Tanori-Córdova, J. C., Castillo-Yañez, F. J., Quintero-Reyes, I. E., & Graciano-Verdugo, A. Z. (2021). Characterization of epigallocatechin-gallate-grafted chitosan nanoparticles and evaluation of their antibacterial and antioxidant potential. *Polymers, 13*(9), 1375.
26. Sistanipour, E., Meshkini, A., & Oveisi, H. (2018). Catechin-conjugated mesoporous hydroxyapatite nanoparticle: A novel nano-antioxidant with enhanced osteogenic property. *Colloids and Surfaces B: Biointerfaces, 169*, 329–339.
27. Liu, B., Wang, Y., Yu, Q., Li, D., & Li, F. (2018). Synthesis, characterization of catechin-loaded folate-conjugated chitosan nanoparticles and their anti-proliferative effect. *CyTA-Journal of Food, 16*(1), 868–876.
28. Safer, A. M., Leporatti, S., Jose, J., & Soliman, M. S. (2019). Conjugation of EGCG and chitosan NPs as a novel nano-drug delivery system. *International Journal of Nanomedicine, 14*, 8033–8046.

# Advancements in High-Performance Liquid Chromatography: Transforming Analytical Chemistry Across Biomedical, Pharmaceutical, and Environmental Domains

**Vivek Singh, Mohammad Kashif, Saba Ubaid, Tanvi Singh, Roma Pahwa, Abhishek Kumar Singh, and Anand Singh**

**Abstract** High-performance liquid chromatography (HPLC) is a cornerstone of analytical chemistry, particularly pharmaceutical analysis and disease monitoring. This book chapter underscores the adaptability of HPLC in biomedical, forensic, environmental, and pharmaceutical domains, reassuring the readers about its robustness and versatility. It also highlights the significance of HPLC in handling complex

---

Vivek Singh, Mohmmad Kashif and Saba Ubaid contributed equally with all other contributors.

---

V. Singh · A. Singh (✉)
Thoracic Surgery Branch, National Cancer Institute, National Institutes of Health, Bethesda, MD, USA
e-mail: anand.singh@nih.gov

M. Kashif
Laboratory of Malaria and Vector Research, National Institute of Allergy and Infectious Diseases, National Institutes of Health, Rockville, MD, USA

S. Ubaid
Department of Biochemistry, King George's Medical University (KGMU), Lucknow, Uttar Pradesh, India

T. Singh
Department of Pathology, King George's Medical University (KGMU), Lucknow, Uttar Pradesh, India

R. Pahwa
Urologic Oncology Branch, National Cancer Institute, National Institutes of Health, Bethesda, MD, USA

A. K. Singh (✉)
Manipal Centre for Biotherapeutics Research, Manipal Academy of Higher Education, Manipal, Karnataka, India
e-mail: abhishek.singh@manipal.edu



N. Mazumder et al. (eds.), *Advanced Biophysical Techniques in Biosciences*, Lecture Notes in Nanoscale Science and Technology 7,
https://doi.org/10.1007/978-3-031-94551-9_7

sample matrices, emphasizing its integration with advanced sample preparation techniques like SPE, LLE, QuEChERS, and DLLME. Moreover, the method's utility in biomarker identification, therapeutic monitoring, and environmental safety illustrates its pivotal role in modern scientific research. This article compiles and provides a comprehensive perspective on HPLC's evolving capabilities in addressing analytical challenges by exploring advancements in derivatization, automation, and coupling with mass spectrometry.

**Keywords** HPLC · Therapeutics · Diagnostics · Cancer · Environmental · Food applications

## Abbreviations

| | |
|---|---|
| 17α-E2 | 17α-estradiol |
| 17β-E2 | 17β-estradiol |
| 4-t-OP | 4-octylphenol (OP), 4-tert-octylphenol |
| 6β-OHE | 6β-hydroxycortisone |
| 6β-OHF | 6β-hydroxycortisol |
| 6β-OHP | 6β-hydroxyprednisolone |
| BPA | Bisphenol A |
| DLLME | Dispersive-liquid–liquid extraction |
| E | Cortisone |
| E1 | Estrone |
| E3 | Estriol |
| EE2 | 17α-ethinylestradiol |
| EED | Ethinylestradiol |
| F | Cortisol |
| FFA | Free fatty acids |
| FPSE | Fabric phase sorptive extraction |
| HEX | 4,40-(1,2-diethylethylene)diphenol |
| HF-LPME | Hollow fiber liquid-phase microextraction |
| IF-IHLME | Ionic liquid foam floatation coupled with an ionic liquid-based homogeneous liquid–liquid microextraction |
| LLE | Liquid–liquid extraction |
| MISPE | Molecularly imprinted solid-phase extraction |
| MSPE | Magnetic solid phase extraction |
| NP | 4-nonylphenol |
| PL | Prednisolone |
| PN | Prednisone |
| SPE | Solid phase extraction |
| TBP | 4-tert-butylphenol |
| UA-DLLME | Ultrasonic-assisted dispersive liquid–liquid microextraction |
| VA-DLLME-FOA | Vortex-assisted dispersive liquid–liquid microextraction method based on floating organic acid droplet |

# 1 Introduction

The accurate detection of analytes at trace levels has historically posed significant challenges, particularly in the absence of pretreatment steps. Despite technological advancements in analytical instrumentation, handling samples with complex matrices remains a considerable hurdle. High-performance liquid chromatography (HPLC), as highlighted in this review, often requires sample preparation due to the intricate composition of real-world samples. Over the last two decades, sample preparation has become a pivotal element of the analytical process, evolving significantly to meet modern requirements.

Currently, various sample preparation techniques exist to address three primary objectives:

1. Converting analytes into forms suitable for measurement
2. Simplifying sample matrices
3. Enhancing detection limits through pre-concentration

Advances in fields such as biological therapeutics [1], proteomics [2], lipidomics [3], metabolomics [4], glycan cleanup, and untargeted analysis, and [5] have enabled these techniques to address aqueous and solid sample matrices containing non-volatile analytes. These advancements, when combined with other methodologies, form powerful analytical tools capable of efficient extraction, cleanup, and preconcentration in a single step. This information will keep the audience informed about the latest developments in the field.

Recent trends have introduced offline and automated online coupling of sample preparation techniques to HPLC systems equipped with suitable detectors. This review emphasizes eight prominent methods: solid-phase extraction (SPE), liquid–liquid extraction (LLE), gel permeation chromatography (GPC), Quick, Easy, Cheap, Effective, Rugged, and Safe (QuEChERS), solid-phase microextraction (SPME), ultrasonic-assisted solvent extraction (UASE), and microwave-assisted solvent extraction (MWASE). The principles and applications of these techniques are further explored in the following sections.

Selecting an appropriate sample preparation technique depends on the specific problem or objective being addressed. Regardless of the approach chosen, the ultimate aim remains to develop an analytical method that is accurate, precise, selective, sensitive, and robust. Knowledge of these diverse techniques equips analysts to achieve their targeted separation goals effectively, as shown in Fig. 1 and Table 1.

# 2 Biomedical Applications

Sample preparation continues to be a bottleneck in analytical processes, even with advances in instrumentation. Time-intensive pretreatment steps make efficient methods crucial in biomedical applications. Combining solid-phase extraction (SPE) sorbents such as restricted access materials (RAM) and molecularly imprinted

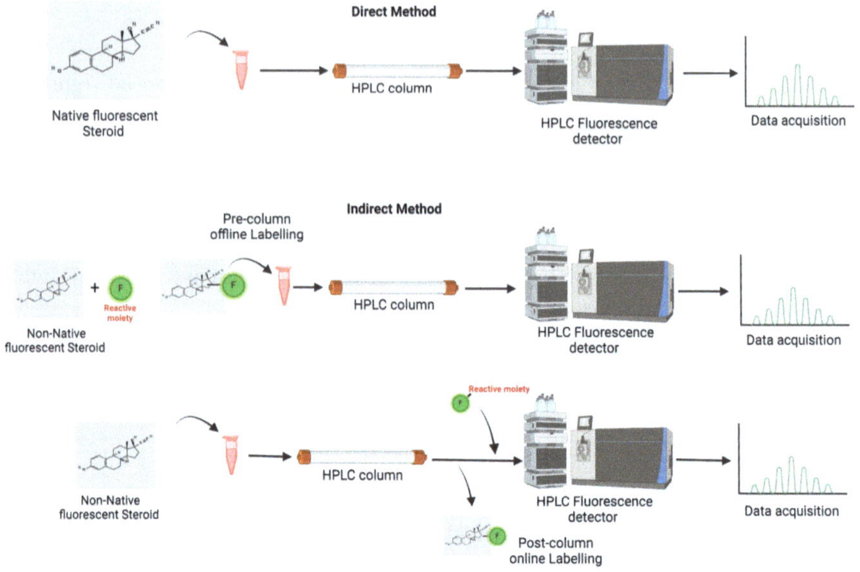


**Fig. 1** Sample preparation methods

polymers (MIP) has proven effective in this field. For example, RAM-MIP SPE sorbents successfully extracted 2-methoxy estradiol from rat plasma, offering reproducibility and efficiency in biological fluid pretreatment [13–16]. This combination enables cleanup and enrichment in one step and supports the separation of large and small molecules. MIP sorbents with HPLC–UV have been applied to detect 3,4-methylenedioxymethamphetamine (MDMA) in plasma and carvedilol in serum [17].

Incorporating magnetic nanoparticles (MMIP) further enhances selectivity and preconcentration. MMIP sorbents enriched cocaine and its metabolites in urine before HPLC–MS/MS analysis, achieving low limits of quantification below urine cut-off levels [18]. Automated online SPE coupled with UHPLC–MS/MS has streamlined drug analysis, as demonstrated in methods identifying and quantifying 42 drugs across various classes [19, 20]. Such advancements, including high-throughput 96-well plate SPE systems, have been employed to analyze aldosterone in plasma [21], antibiotics in animal tissue [22], and Vitamin D metabolites [23].

Other innovations include dispersive micro solid-phase extraction (d-µ-SPE), which enhances efficiency and reduces solvent use. Coupled with HPLC, this technique has been applied to analyze benzodiazepines, bisphosphonates, and acidic or basic drugs [24, 25]. Packed fiber SPE coupled with HPLC–MS/MS has effectively analyzed nucleic acids in human urine [14].

Methods like DLLME, which combine simplicity and eco-friendliness, have applications ranging from pesticide detection in urine [15] to monitoring hypertension drugs in plasma [16] and estrogen levels in human urine [26]. Also, proteomics research has leveraged HPLC combined with laser-induced fluorescence to

**Table 1** Advantages and disadvantages of SPE, LLE, GPC, QUEChERS, UASE, MWASE, SBSE, and SPME

| SP echnique | Advantages | Disadvantages | Coupling to HPLC | References |
|---|---|---|---|---|
| SPE | Simplicity, and convenience, very selective, practical with a variety of matrices, many sorbent choices, less consumption of organic solvents, ability to interface with different instrumentations, high sample throughput, high recoveries, and reproducibility | Greater complexity or difficulty to master, lengthy method development, costly, no optimum sorbent for all applications, matrix constituents can be absorbed, limited selectivity or sensitivity | Offline and online | [6] |
| LLE | High selectivity of separation, high production capacity, low energy consumption, easy continuous operation, inorganic salts easily removed, short method development time, low cost | Difficult to automate, labor-intensive, utilize a large volume of toxic organics, emulsion formation, poor sample throughput, poor reproducibility, and loss of sensitivity | Offline and online | [7] |
| GPC | Good extraction efficiency for large molecules, favorable transfer kinetics, time-saving | Cannot extract small molecules. | Offline and online | [8] |
| QuEChERS | High recoveries for a wide range of analytes, high sample throughput, speed, environmental friendliness, inexpensive, rugged, efficient cleanup | Cannot extract large molecules | | [9] |
| UASE | Simple, inexpensive, reproducible, high extraction yield, improve mass transfer, less extraction time, low energy consumption, low solvent consumption, non-destructive, easily coupled with other extraction techniques | | Offline | |
| MWASE | Short extraction time, simple operation, no organic solvent, high selectivity, versatile & efficient extraction | Difficult to optimize, energy consumption | Offline | [10] |

(continued)

**Table 1** (continued)

| SP echnique | Advantages | Disadvantages | Coupling to HPLC | References |
|---|---|---|---|---|
| SBSE | Simplicity, high sorption capacity, robustness, excellent extraction efficiency | A limited spectrum of analyte polarities for the available stationary phases, strong matrix effects, and the need for heightened control of extraction conditions | Offline and online | [11] |
| SPME | Simple, fast, solvent-free, non-exhaustive extraction technique, reproducible | Fiber breakage, bleeding from conventional coated fibers, sample carry-over problems, pH instability. Note that these drawbacks have been overcome by introducing sol-gel, MIP, and electrochemical-assisted techniques for manufacturing SPME fibers | Offline and online | [12] |

distinguish between normal and cancerous samples [27]. GPC-HPLC systems have been employed in protein profiling and drug stability studies [28].

## 3 Forensic Applications

In forensic chemistry, where drug detection is paramount, recent advancements in sample preparation have improved the analysis of illicit and therapeutic substances. The increasing prevalence of opioids and synthetic cannabinoids has necessitated updates in analytical methods. Techniques such as SPE coupled with HPLC–MS/MS are instrumental in quantifying fentanyl analogs and beta-agonists in milk [29].

Hair has emerged as a critical matrix for drug testing, with SPE-HILIC–MS/MS and MMIP-UHPLC–MS/MS developed to analyze cocaine metabolites [30]. Automated online SPE for post-mortem samples has enhanced the quantification of pharmaceuticals in blood and brain tissue, addressing challenges like sample availability in trauma cases [31]. Advanced techniques like DLLME combined with UHPLC–MS/MS have also been applied to psychoactive substances [32].

Moreover, LLE combined with UHPLC–MS/MS has enabled the detection of gunshot residues, while QuEChERS has been applied to benzodiazepine detection in biological fluids [33]. These innovations demonstrate the importance of coupling versatile preparation techniques with advanced detection systems, enlightening the audience about the synergy between the two.

# 4 Environmental and Industrial Hygiene Applications

Sample preparation methods are extensively applied in environmental analysis, particularly for monitoring pollutants in water. SPE coupled with HPLC–MS/MS has been used to detect endocrine disruptors [34], antibiotics [35], and heavy metals like mercury [36]. Combining SPE sorbents or hybrid methods such as MOF-MEPS–HPLC has further refined the analysis of complex environmental samples [37].

DLLME, often paired with UHPLC–MS/MS, facilitates the detection of trace compounds like fullerene nanoparticles [38] and triazines in wastewater [39]. QuEChERS methods have expanded to analyze pollutants in food and environmental samples, providing efficient cleanup and extraction [40].

Green techniques like UASE and MWASE, often combined with other methods, enable the analysis of hydrocarbons, fungicides, and bioactive plant compounds [41]. These advancements underline the versatility and efficiency of modern sample preparation methods in environmental and industrial hygiene.

# 5 Applications of Chromatographic Methods for Diagnosis

Cancer remains one of the leading causes of mortality worldwide, with over 11 million new cases reported annually—a number that continues to rise. Of the 200 recognized cancer types, lung, prostate, breast, and colon cancers account for over half of all cancer-related deaths [42]. Characterized as a disease of DNA deregulation, cancer arises due to exogenous and endogenous factors. Its progression involves metabolic disturbances that transform normal cells into malignant ones. Early detection and understanding of cancer metabolism are crucial for improving clinical outcomes.

Metabolomics, genomics, and proteomics are pivotal in cancer research, offering insights into normal, precancerous, and cancerous metabolic profiles. Chromatographic techniques such as gas chromatography–mass spectrometry (GC–MS) and liquid chromatography-mass spectrometry (LC–MS) have long been instrumental in biomedical research [43]. These tools are invaluable for identifying potential biomarkers, monitoring therapies, and understanding cellular metabolism, as shown in Fig. 2.

# 6 Chromatography Techniques in Cancer Research

## *6.1 Gas Chromatography–Mass Spectrometry (GC–MS)*

GC–MS, renowned for its sensitivity and rapid analysis capabilities, requires minimal sample volume and excels in human body fluid analysis. It has been widely employed for identifying biomarkers, assessing cancer therapies, and detecting

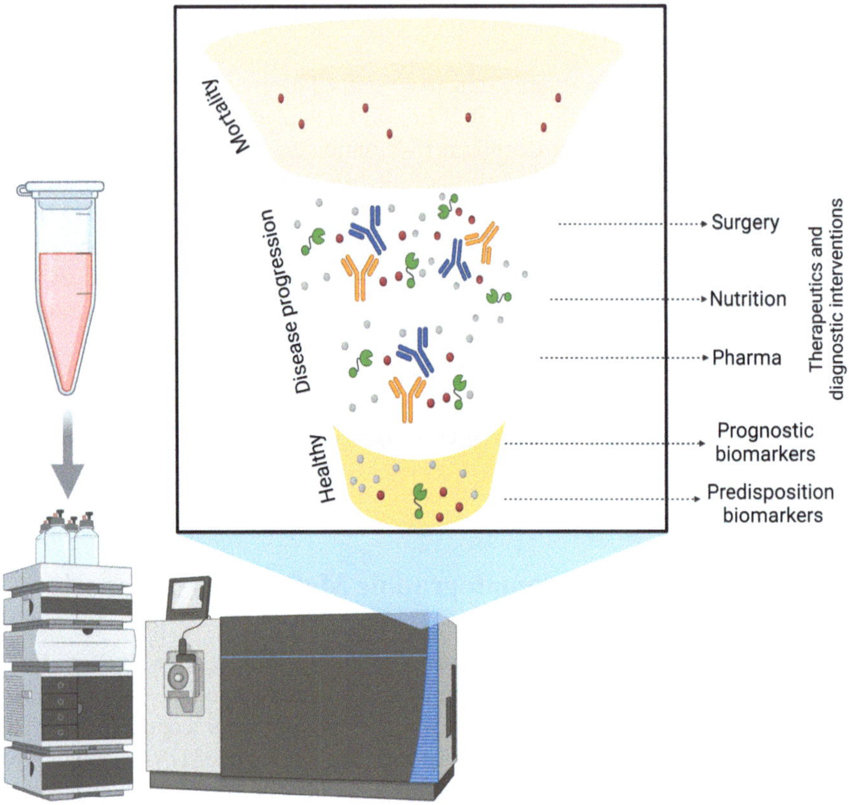


**Fig. 2** Applications of chromatographic methods

DNA damage caused by carcinogens [44]. Its compatibility with other techniques, such as nuclear magnetic resonance (NMR), ensures reliable and comparable results [45]. Additionally, GC-MS has been applied in urinary analysis for over three decades [46].

## 6.2 *High-Performance Liquid Chromatography (HPLC) and Related Techniques*

HPLC enables the detection of polar and thermally labile compounds. When coupled with suitable detectors, HPLC facilitates structural identification and improves resolution and reproducibility [47]. LC–MS and LC–MS/MS, advancements in chromatography, offer high chromatographic peak resolution, rapid analysis, and high-quality mass spectra through ultra-high-pressure liquid chromatography (UHPLC–MS) [48].

These chromatographic methods have been effectively utilized in cancer diagnosis and metabolic disorder research, including acidosis and autism [49]. Validation of these techniques through quality control (QC) and adherence to good laboratory practices ensures their reliability [50].

## 7 Biomarkers in Cancer Diagnosis

The identification of biomarkers is essential for early cancer detection and personalized care. Biomarkers, defined as measurable indicators of biological processes, have diverse applications in disease prediction and therapeutic interventions [51]. For a biomarker to be effective, it must meet specific analytical and clinical validation criteria.

### *7.1 Lung Cancer*

Lung cancer, a leading cause of death globally, is primarily associated with smoking and environmental exposure to carcinogens like radon and asbestos. In exhaled breath, volatile organic compounds (VOCs) are gaining attention as potential diagnostic markers. Techniques such as Solid-Phase Microextraction (SPME) and GC-TOF-MS enable highly sensitive detection of these trace compounds [52]. Recent studies have identified compounds like propane, ethylbenzene, and isopropyl alcohol as significant biomarkers in lung cancer [52].

### *7.2 Prostate Cancer (CaP)*

Prostate cancer ranks among the most common cancers in men. Metabolic profiling has highlighted sarcosine as a key biomarker linked to tumor progression. Techniques like LC–MS/MS and GC–MS facilitate the quantification of sarcosine and other metabolites in biological samples. While urine is a standard matrix for biomarker detection, advancements in LC and GC methods have improved the sensitivity and specificity of prostate cancer diagnosis.

### *7.3 Breast Cancer (BC)*

Breast cancer accounts for a significant percentage of cancer-related deaths in women. Despite diagnostic tools like mammography, early detection remains challenging. Chromatographic methods, including nano-LC and GC–MS, are invaluable for identifying biomarkers like hydroxyestrones in urine and lipophilin B in nipple aspirate fluid [53]. Proteomic techniques such as FD-LC–MS/MS offer quantitative insights into protein expression changes in breast cancer [54].

### *7.4 Colorectal Cancer (CRC)*

Due to late diagnosis, CRC is characterized by high mortality rates. Metabolomics using GC–TOF-MS has identified significant metabolic differences between cancerous and normal tissues, with intermediates of the citric acid cycle and amino acids showing marked variations [57]. Techniques like HILIC and MALDI-TOF-MS have also demonstrated potential in identifying glycosylation changes in CRC tissues [57, 58].

### *7.5 Gastric Cancer*

Gastric cancer's asymptomatic nature often leads to delayed diagnosis. GC–MS and LC–MS techniques have identified metabolites like lactic acid and alanine as potential biomarkers. Metabolic profiling in animal models has revealed distinct metabolic patterns associated with metastasis [50]. These findings underscore the value of metabolomics in gastric cancer diagnosis and treatment.

### *7.6 Hepatocellular Carcinoma (HCC)*

HCC diagnosis relies on tools like ultrasonography, but serology tests often yield false positives. Metabolomics profiling using GC–MS has identified markers like propanoic acid and alanine in urine samples. At the same time, UPLC–MS has detected conjugated bile acids and long-chain fatty acids as stage-specific biomarkers [50, 51].

## 8 Comprehensive Summary of Steroids and HPLC–FLD Techniques

### *8.1 Overview of Steroids*

Steroids are bioactive molecules found in animals, plants, and fungi, as well as synthetic derivatives used in therapeutic applications. Their standard structure, a four-ring core of 17 carbon atoms, supports diverse functional groups and side chains, determining their roles in biological and clinical contexts. Natural steroids, such as sex hormones, mineralocorticoids, glucocorticoids, and cholesterol, influence metabolism, immunity, reproduction, and salt retention as shown in Table 2. Synthetic steroids mimic these roles and are used to treat inflammatory conditions, autoimmune diseases, asthma, and chronic obstructive pulmonary disease (COPD) [127, 128].

### *8.2 Analytical Challenges*

Steroids operate at low concentrations, necessitating sensitive and precise analytical techniques. The complex biological matrices of steroid samples require selective methods to resolve target steroids from similar compounds and interferences. This is especially crucial during sample pretreatment to minimize analyte loss. Steroids' hydrophobicity allows their organic solvents to be extracted, although their metabolism may increase hydrophilicity [129].

## 9 Common Analytical Methods

Several methods exist for steroid analysis:

- **Immunoassays:** Specific but prone to cross-reactivity [102, 130].
- **Gas Chromatography–Mass Spectrometry (GC–MS):** Highly sensitive but demands extensive sample preparation [131].
- **High-Performance Liquid Chromatography–Mass Spectrometry (HPLC–MS):** Offers precision with complex workflows [132].
- **HPLC with Fluorescence Detection (HPLC–FLD) is cost-effective** and sensitive, but it requires derivatization for non-native fluorescent steroids [119].

### *9.1 Importance of HPLC–FLD*

HPLC–FLD is a valuable tool for steroid quantification in clinical, pharmaceutical, and environmental contexts. It leverages fluorescent derivatization to enhance detection, focusing on specific functional groups like alcohols, phenols, and ketones. Key advantages include simplicity, cost-effectiveness, and high sensitivity. Common derivatization agents include coumarins, anthracenes, and benzamidine [119].

**Table 2** HPLC–FLD method conditions used for steroid detection

| Detected steroids | Sample type | Extraction method | LOD | References |
|---|---|---|---|---|
| PLand PN | Human plasma | LLE | 3 ng/mL | [59] |
| F and E | Biological samples | SPE | F: 50 pg<br>E: 70 pg | [60] |
| F | Human serum | – | 0.30 pg/dL | [61] |
| F | Human urine | SPE | 0.26 pg/dL | [61] |
| 7α-hydroxycholesterol | Dog plasma | LLE | 4 pg | [62] |
| Corticosteroids | Urine | Enzyme hydrolysis, extraction with 0.5 M NaOH | NR | [63] |
| F and E | Human plasma | SPE | 3.0 ng/mL | [64] |
| Corticosterone | Rat urine | LLE | 0.5 pmol | [65] |
| F | Human hair | LLE | 1 pg/mg | [66] |
| EE2, E2, and BPA | Human urine and aqueous samples | FPSE | E2: 20 pg/mL<br>EE2: 36 pg/mL<br>BPA: 42 pg/mL | [67] |
| BPA, NP, E2, EE2, and E3 | Human urine | SPE | BPA and E2: 2.7 μg/L<br>NP: 2.9 μg/L<br>E2 and EE2: 4.6 μg/L<br>E3: 8.3 μg/L | [68] |
| E, testosterone, methyltestosterone, bolasterone, testosterone acetate, progesterone | Urine | SPE | Down to 100 pg/mL | [69] |

| | | | | |
|---|---|---|---|---|
| BPA and 8 alkylphenols (4-*sec*-Butylphenol, 2-*tert*-Butylphenol, 3-*tert*-butylphenol, 4-*tert*-butylphenol, 4-*n*-Pentylphenol, 4-*tert*-pentylphenol, 4-*n*-hexylphenol, and 4-*n*-heptylphenol) | Rat plasma and blood | LLE | BPA: 0.1 pg/mL<br>Alkylphenol: 0.7–10 pg/mL | [70] |
| Progesterone and 17-hydroxyprogesterone dehydroepiandrosterone, androstenedione, testosterone and 17-methyltestosterone | Serum from pregnant and non-pregnant women | LLE | 550–3700 fmol per 10 μL | [71] |
| Aldosterone, corticosterone, F, E, dexamethasone, fluocinoloneacetonide, triamcinolone and triamcinolone acetonide | Human plasma | LLE | 0.06–0.3 pg per 100 μL | [72] |
| BPA | Breast milk | SPE then LLE | 0.11 ng/mL | [73] |
| BPA | Human blood serum and ascitic fluid samples | LLE | 0.04 ppb | [74] |
| BPA | Rat brain rat plasma | LLE | 0.3 ppb in 60 μL rat brain<br>4.6 ppb in 50 μL rat plasma | [75] |
| F, E, PL, PN, 6β-OHF, 6β-OHP and 6β-OHE | Human plasma and urine | LLE | F, E, PL and PN: 0.1 ng/mL<br>6β-OHF and 6β-OHP: 0.5 ng/mL | [76] |
| Corticosterone | Rat serum | LLE | 0.1 ng | [77] |

(continued)

**Table 2** (continued)

| Detected steroids | Sample type | Extraction method | LOD | References |
|---|---|---|---|---|
| 18-oxygenated corticosteroids, 18-hydroxycortisol, 18-hydroxycortisone and 18-oxocortisol | Human urine | LLE and SPE | 0.1 pmol | [78] |
| Cholesterol and sitosterol | Saliva and urine biosamples, cow milk, and soybean milk | LLE | 25 nM per 10 μL injected volume | [79] |
| Pregnenolone | Rat brain | SPE | NR | [80] |
| C21 steroids; corticoids | Steroid standards | NR | NR | [81] |
| Triamcinolone | Human plasma | SPE | 1 ng/mL | [82] |
| Butane acid-(5-androsten-17-one-3beta-ol)-diester (A1998) | Rat plasma | LLE | 10 ng/mL | [83] |
| Alfaxalone and pregnanolone | Rat plasma | – | 0.025 μg/mL | [84] |
| EED | Oral contraceptive tablets | – | EED: 0.0538 μg/ml | [85] |
| EED and drospirenone | Oral contraceptive tablets | – | EED: 0.00065 μg/mL<br>DROSP: 0.0774 μg/mL | [86] |
| EED | Coated tablets | – | EED: 0.02 μg/mL | [87] |
| Sodium E1 sulphate, sodium equilin sulphate, E1 and equilin | Raw materials and pharmaceuticals | – | 0.01–1.38 pmol | [88] |
| E1, 17β-estradiol, E3, BPA, NP, OP | Fish, chicken, aquaculture pond water sample | DLLME | 0.02–0.07 μg/L | [89] |
| E1, E2, and E3 | Cow and river buffalo | LLE and SPE | 5–10 ng/kg | [90] |
| α- and β-Trenbolone | Bovine muscle and liver | LLE then SPE | Bovine muscle: 0.2 ng/g<br>liver: 1.0 ng/g | [91] |
| NP, 4-nonylphenolmono-(NP1EO), diethoxylates (NP2EO), BPA, TBP, and OP | Fish and shellfish | LLE | NP NP1EO and NP2EO: 2 ng/g<br>BPA, BP and OP: 1 ng/g | [92] |

| | | | | |
|---|---|---|---|---|
| Nonylphenol and its ethoxylates | Fish tissue | Pressurized fluid extraction | 4–15 ng/mL | [93] |
| E2 and EE2 | Poultry litter | LLE | E2: 4.0 µg/kg<br>EE2: 2.6 µg/kg | [94] |
| E2 and EE2 | Waste water | SPE | 2.5 ng/L | [95] |
| E3, E2, EE2, HEX, mestranol | Water, sediment | SPE | Water: 6–24 ng/L<br>Sediment: 0.1–0.9 ng/g | [96] |
| E2, 17α-EE2, and E1 | Water | SPE | 10–1100 ng/L | [97] |
| OP, NP, BPA, diethylstilbestrol, E1, EE2, E2, and E3 | Wastewatersamples | SPE | 0.3–0.7 ng/L | [98] |
| E2 and EE2 | Tap, surface and waste water | DLLME | E2: 2.0 ng/L<br>EE2: 6.5 ng/L | [99] |
| BPA, 17β-estradiol, and 17α-ethynyl estradiol | Drinking water | PAC | BPA: 201 ng/L<br>E2: 313 ng/L<br>EE2: 284.5 ng/L | [100] |
| 17α- and 17β-Trenbolone | River water | SPE | 4 ng/L | [101] |
| NP, OP, NP polyethoxylates | Municipal wastewater treatment plants | LLE | OP: 2 ng/L<br>NP: 11 ng/L<br>NPE: 52 ng/L | [102] |
| E2, E3, BPA, and 17β-ethinylestradiol | Environmentalwaters | Synthesized in-tube SPME device | 0.006–0.10 ng/mL | [103] |
| E2, E3, EE2, 3-methyl ether EE2, NP, OP, POE(1–2) nonyl phenol and BPA | River water | On-line SPE | 20–50 ng/L | [104] |
| Endocrine disruptors; BPA and EE2 | Environmental water samples | A poly(acrylamide-vinylpyridine) monolithic capillary column | BPA: 0.064 ng/mL<br>17α-ethinylestradiol: 0.12 ng/mL | [105] |

(continued)

**Table 2** (continued)

| Detected steroids | Sample type | Extraction method | LOD | References |
|---|---|---|---|---|
| Brassinosteroids | Plant: (*Vicia faba* L.) | LLE | 50 pg | [106] |
| Brassinolide | *Arabidopsis thaliana*, *Daucus carota* and *Brassica campestris* L. leaves' samples | UA-DLLME | 8.0 ng/L | [107] |
| Brassinolide and castasterone | Pollen of orange (*Citrus sinensis* Osbeck) | LLE | NR | [108] |
| Norgestrel, norethindrone, EE2, gestodene, and norethisterone acetate | Meat samples | MSPE | $1.4 \times 10^3$–$8.7 \times 10^3$ | [109] |
| F and E | Human urine samples | LLE | LLOQ: F: 27.6 nmol/L<br>E: 27.7 nmol/L | [110] |
| FFA | Edible oils and foodstuff | Supercritical $CO_2$ and organic solvent extraction | 0.22–1.06 ng/mL | [111] |
| OP, NP, TBP, BPA, E1, E2, E3 | Milk samples | LLE | 10.5–13.8 ng/L | [112] |
| E3, 2-OHE$_2$, 17β-E2, 17α-E2, EE$_2$, HEX | Dairy products | HF-LPME | 0.23–14.8 μg/kg | [113] |
| Zearalenone | Edible oil | SPE | 10 μg/kg | [114] |
| EE2, E1, E2, E3, and progesterone | Drinking water and wastewater samples | SPE | Drinking water: 1–3.8 ng/L<br>Sewage water: 3.8–7.5 ng/L | [115] |
| E3, E2, E1 | Human urine | VA-DLLME-FOA | E3: 0.01 ng/mL<br>β-E2: 0.01 ng/mL<br>E1: 0.06 ng/mL | [116] |
| 17-α-E2, 17-β-E2 benzoate and quinestrol | Environmental water samples | IF-IHLME | 17-α-estradiol: 0.04 ng/mL<br>E2 and Quinestrol: 0.05 ng/mL | [117] |
| E2 and EE2 | Tap water samples | Nanoparticles of graphene oxide/γ-$Fe_2O_3$ as asorbent for SPE | E2: 2.7 ng/L<br>EE2: 0.8 ng/L | [118] |

| | | | | |
|---|---|---|---|---|
| 17-E2 and E3 | Water samples | Ultrasonication assisted DLLME | DLLME-HPLC/FLD: 7.16–69.22 ng/L | [119] |
| E2, 1,3,5(10)-Estratriene-3,17β-diol | Fish and prawn tissue samples | MISPE | 0.023 mg/L | [120] |
| BPA, EE2, 4-t-OP, 4-OP, and 4-NP | River water | Disposable pipette extraction (DPX) | BPA, EE2, 4-OP and 4-NP: 0.30 μg/L<br>4-t-OP: 0.60 μg/L | [121] |
| E1 and EE2 | Digested sludge | Ultrasonic liquid extraction | E1: 0.305 μg/g<br>EE2: 0.052 μg/g | [122] |
| E2 | Milk sample | SPE | 0.7 ng/mL | [123] |
| Nine BPs | Milk samples | Ultrasonically with acetonitrile and cleaned using the QuEChERS technique. | 1.0–3.1 μg/kg | [124] |
| EE | River water samples | $C_{18}$ SPE disk | 7.4 ng/L | [125] |
| E3, 17β-estradiol glucuronide, 17β-E2, 17α-E2, 17β-E2–3-methyl ether | Wastewater | Molecularly imprinted SPE | 1.4 to 2.5 ng/mL | [126] |

*SPE* solid phase extraction, *LLE* liquid–liquid extraction, *DLLME* dispersive-liquid–liquid extraction, *UA-DLLME* ultrasonic-assisted dispersive liquid–liquid microextraction, *VA-DLLME-FOA* vortex-assisted dispersive liquid–liquid microextraction method based on floating organic acid droplet, *IF-IHLME* ionic liquid foam floatation coupled with an ionic liquid-based homogeneous liquid–liquid microextraction, *MSPE* magnetic solid phase extraction, *FPSE* fabric phase sorptive extraction, *MISPE* molecularly imprinted solid-phase extraction, *HF-LPME* hollow fiber liquid-phase microextraction, *OP* 4-octylphenol, *4-t-OP* 4-tert-octylphenol, *NP* 4-nonylphenol, *TBP* 4-tert-butylphenol, *BPA* bisphenol A, *E1* estrone, *17β-E2* 17β-estradiol, *17α-E2* 17α-estradiol, *E3* estriol, *FFA* free fatty acids, *EE2* 17α-ethinylestradiol, *EED* ethinylestradiol, *F* cortisol, *E* cortisone, *HEX* 4,40-(1,2-diethylethylene)diphenol, *PL* prednisolone, *PN* prednisone, *6β-OHF* 6β-hydroxycortisol, *6β-OH* P6β-hydroxyprednisolone, *6β-OHE* 6β-hydroxycortisone, *RT* room temperature

## 9.2 Applications in Clinical and Environmental Analysis

1. **Disease Diagnostics**
   - Cortisol measurement for stress and Cushing's syndrome [133].
   - Detection of glucocorticoids in hair and plasma using sulfuric acid and fluorescent labeling [66].
   - Derivatization with agents like 9-anthroyl nitrile (9-AN) and dansyl hydrazine enhances the sensitivity and selectivity of corticosteroid quantification [76].
2. **Endocrine Studies**
   - Monitoring estrogen concentrations in urine via methods like solid-phase extraction (SPE) and dispersive liquid-liquid microextraction (DLLME) [67].
   - Detection of testosterone and anabolic steroids in sports and livestock using micellar chromatography [69].
3. **Environmental Monitoring**
   - Analysis of endocrine-disrupting chemicals (EDCs) like bisphenol A (BPA) and alkylphenols in food and biological fluids. Derivatization with reagents such as 4-(4,5-diphenyl-1H-imidazol-2-yl) benzoyl chloride enhances detection sensitivity [75].
4. **Cholesterol and Steroid Hormones:**
   - Quantifying cholesterol and phytosterols in biosamples using HPLC–FLD with naproxen acyl chloride derivations [79].
   - Hormone analysis, including progesterone and 17-hydroxyprogesterone, using reagents like BODIPY FL hydrazide [134].

## 9.3 Advances in Derivatization Techniques

HPLC–FLD has evolved to include derivatization procedures targeting specific steroid functionalities, such as:

- **Primary Alcohols:** Derivatized using agents like 9-AN for corticosteroids in plasma [63].
- **Ketones and Ketolic Groups:** Targeted in Cushing's syndrome diagnosis using benzamidine derivatives [135].
- **Phenolic Groups:** Enhanced detection of estrogens and BPA in environmental samples using p-nitrobenzoyl chloride [68].

### 9.4 Detection Limits and Performance

HPLC–FLD methods achieve remarkable sensitivity:

- Cortisol detection limit: 0.3 pg/dL [61].
- BPA detection in serum: 0.11 ng/mL [74].
- Estrogen detection in urine: 20–42 pg/mL.

## 10 Conclusion

With its robust derivatization strategies, HPLC–FLD is indispensable for steroid analysis across diverse applications. Its ability to detect low concentrations with high precision ensures its utility in clinical diagnostics, environmental safety, and endocrine studies. The growing repertoire of derivatization agents and techniques continues to expand its capabilities, making it a cornerstone of modern analytical biochemistry.

### 10.1 Pharmaceutical Applications

High-Performance Liquid Chromatography (HPLC) has become indispensable in pharmaceutical and biomedical research due to its versatility and precision. It is extensively used for:

- **Drug Discovery and Development Involve Identifying and Quantifying** drugs as active pharmaceutical ingredients (APIs) and within formulations [136].
- **Product Characterization:** Profiling impurities, degradation products, and stability studies [137].
- **Formulation Development:** Assessing dissolution properties, stability, and content uniformity of solid dosage forms [136].

While UV detection is standard in pharmaceutical HPLC applications, Fluorescence Detection (FLD) is increasingly used due to its superior sensitivity and selectivity, particularly in steroid analysis.

### 10.2 Examples of HPLC–FLD in Pharmaceutical Applications

1. **Pharmacokinetic/Pharmacodynamic (PK/PD) Studies**
   - Vesser et al. developed an HPLC–FLD method to quantify plasma neurosteroids (e.g., alfaxalone and pregnanolone). The technique included derivatiza-

tion with dansyl hydrazine and liquid-liquid extraction, achieving a detection limit of 10 ng/mL for a 50 μL plasma sample [83].
- A novel steroid, Butane acid-(5-androsten-17-one-3s-ol)-diester (A1998), was quantified using HPLC–FLD as it is non-ionizable and unsuitable for LC–MS. Derivatization with dansyl hydrazine and sample preparation with ZnSO4 solution improved yield and sensitivity, with a detection limit of 25 ng/mL [62].

2. **Drug Formulation Characterization**
   - A pH-responsive polypeptide-corticosteroid conjugate was characterized using HPLC–FLD to develop a controlled-release treatment for psoriasis. The method used a C18 column with a water/acetonitrile mobile phase and achieved precise quantification [83].
3. **Estrogen Analysis in Tablets**
   - Direct measurement of ethinyl estradiol, a native fluorescent compound, was achieved using reversed-phase HPLC–FLD, avoiding complex sample preparation. Detection limits as low as 0.65 ng/mL were reported [84].

## *10.3 Environmental and Food Applications*

Steroids, including water, soil, and animal products, are widespread in environmental samples. Environmental monitoring often focuses on endocrine-disrupting chemicals (EDCs), which include synthetic steroids and contaminants such as bisphenol A (BPA).

## *10.4 Examples of Environmental Applications*

1. **Water Analysis**
   - HPLC–FLD detected estrogens (e.g., estrone, estradiol, and bisphenol A) in drinking and wastewater using solid-phase microextraction (SPME) and derivatization. Detection limits ranged from 0.006 to 0.10 ng/mL, demonstrating high sensitivity [105].
   - SPE-HPLC–FLD was used to quantify EDCs like nonylphenol and octylphenol in river water. Using optimized SPE cartridges, enhanced detection limits of 20–50 ng/L were achieved [104].
2. **Food and Livestock Monitoring**
   - Steroids in poultry litter were detected after extraction with dichloromethane and methanol. The detection limits for estradiol and ethinyl estradiol were 4.0 μg/kg and 2.6 μg/kg, respectively.

- Natural and synthetic estrogens in dairy products were quantified using hollow-fiber liquid-phase microextraction coupled with HPLC–FLD, achieving detection limits in the low µg/kg range [113].

3. **Toxicology and Bioaccumulation Studies**

   - Trenbolone metabolites in cattle feedlot discharge and river water were quantified using HPLC–FLD with a detection limit of ~4 ng/L. Bovine muscle and liver samples were analyzed with detection limits as low as 0.2 ng/g [91].

4. **Plant Hormone Quantification**

   - Brassinolide, a plant steroid hormone, was analyzed using ultrasonic-assisted dispersive liquid-liquid microextraction coupled with HPLC–FLD, achieving a detection limit of 8.0 ng/L [107].

### *10.5 Advancements in HPLC–FLD Techniques*

1. **Fluorescent Derivatization**

   - Derivatization enhances sensitivity by attaching fluorescent moieties to steroids. Agents like dansyl hydrazine and anthroyl nitrile have been extensively used [83].

2. **Sample Preparation Improvements**

   - Techniques like SPME and SPE optimize extraction efficiency, enabling low detection limits for complex matrices like water and biological samples.

3. **Automation and Integration**

   - Automated SPE systems and online extraction methods streamline sample preparation, reducing processing times while maintaining high precision [99].

## 11 Conclusion

High-performance liquid chromatography has revolutionized the field of analytical chemistry through its precision, sensitivity, and versatility. Its integration with cutting-edge sample preparation techniques has expanded its applications in biomedical research, environmental monitoring, and pharmaceutical development. The ability to detect and quantify trace-level compounds in complex matrices has made HPLC an indispensable tool in disease diagnosis, drug discovery, and forensic investigations. Despite advancements, challenges such as time-intensive sample preparation and the need for method optimization remain. Addressing these barriers will enhance HPLC's utility in routine and specialized analyses.

## 12 Future Perspectives

The future of high-performance liquid chromatography lies in the continued refinement of its techniques and applications. Automation, miniaturization, and the development of eco-friendly sample preparation methods are expected to streamline workflows and reduce environmental impact. Integrating HPLC with advanced analytical tools such as AI-driven data analysis and real-time monitoring systems could revolutionize its efficiency and accuracy. Furthermore, exploring novel derivatization agents and enhanced detection systems will broaden its scope, particularly in emerging fields like personalized medicine, proteomics, and environmental sustainability. Continued interdisciplinary collaborations will drive innovations, ensuring HPLC remains at the forefront of analytical advancements.

## References

1. Allegra, S., Fatiguso, G., De Francia, S., Favata, F., Pirro, E., Carcieri, C., et al. (2018, January). Therapeutic drug monitoring of voriconazole for treatment and prophylaxis of invasive fungal infection in children. *British Journal of Clinical Pharmacology, 84*(1), 197–203.
2. Chen, D., Shen, X., & Sun, L. (2018, July 5). Strong cation exchange-reversed phase liquid chromatography-capillary zone electrophoresis-tandem mass spectrometry platform with high peak capacity for deep bottom-up proteomics. *Analytica Chimica Acta, 1012*, 1–9.
3. Narváez-Rivas, M., Vu, N., Chen, G.-Y., & Zhang, Q. (2017, February 15). Off-line mixed-mode liquid chromatography coupled with reversed phase high performance liquid chromatography-high resolution mass spectrometry to improve coverage in lipidomics analysis. *Analytica Chimica Acta, 954*, 140–150.
4. Jacob, M., Lopata, A. L., Dasouki, M., & Abdel Rahman, A. M. (2019, May). Metabolomics toward personalized medicine. *Mass Spectrometry Reviews, 38*(3), 221–238.
5. Roy, C., Tremblay, P.-Y., Bienvenu, J.-F., & Ayotte, P. (2016, August 1). Quantitative analysis of amino acids and acylcarnitines combined with untargeted metabolomics using ultra-high performance liquid chromatography and quadrupole timeof-flight mass spectrometry. *Journal of Chromatography. B, Analytical Technologies in the Biomedical and Life Sciences, 1027*, 40–49.
6. Płotka-Wasylka, J., Szczepanska, N., de la Guardia, M., & Namiesnik, J. (2016). *Trends in Analytical Chemistry, 27*, 23–43.
7. Silvestre, C. I., Santos, J. L., Lima, J. L., & Zagatto, E. A. (2009). *Analytica Chimica Acta, 652*, 54–65.
8. Barth, H. G., Boyes, B. E., & Jackson, C. (1998). *Analytical Chemistry, 70*, 251R–278R.
9. Gong, X., Wang, H., Zhang, Y., Sun, J., Dong, J., Lin, L., Yu, J., Feng, Z., & Li, K. (2012). *Journal of Chromatography and Separation Techniques, 3*, 125.
10. Sharma, S. C., Verma, N., Saxena, N. P., Singh, & Sinha, A. K. (2006). *Journal of Separation Science, 29*, 613–619.
11. Hasan, C. K., Ghiasvand, A., Lewis, T. W., Nesterenko, P. N., & Paull, B. (2020). *Analytica Chimica Acta, 1139*, 222–240.
12. Ghorbarni, M., Chamsaz, M., & Rounaghi, G. H. (2016). *Journal of Separation Science, 39*, 1082–1089.
13. Kan, X., Zhao, Q., Shao, D., Geng, Z., Wang, Z., & Zhu, J.-J. (2010, March 25). Preparation and recognition properties of bovine hemoglobin magnetic molecularly imprinted polymers. *The Journal of Physical Chemistry. B, 114*(11), 3999–4004.

14. Gan, W., Liu, X.-L., Yu, T., Zou, Y.-G., Li, T.-T., Wang, S., et al. (2018). Urinary 8-oxo-7,8-dihydroguanosine as a potential biomarker of aging. *Frontiers in Aging Neuroscience, 10*, 34.
15. Ramin, M., Khadem, M., Omidi, F., Pourhosein, M., Golbabaei, F., & Shahtaheri, S. J. (2019, October). Development of dispersive liquid-liquid microextraction procedure for trace determination of malathion pesticide in urine samples. *Iranian Journal of Public Health, 48*(10), 1893–1902.
16. Ahmed, S., Atia, N. N., & Bakr Ali, M. F. (2017, March 1). Ultrasound assisted dispersive liquid-liquid microextraction coupled with high performance liquid chromatography designated for bioavailability studies of felodipine combinations in rat plasma. *Journal of Chromatography. B, Analytical Technologies in the Biomedical and Life Sciences, 1046*, 200–210.
17. Azodi-Deilami, S., Abdouss, M., Asadi, E., Hassani Najafabadi, A., Sadeghi, S., Farzaneh, S., et al. (2014, December 5). Magnetic molecularly imprinted polymer nanoparticles coupled with high performance liquid chromatography for solid-phase extraction of carvedilol in serum samples. *Journal of Applied Polymer Science, 131*(23).
18. Sánchez-González, J., Jesús Tabernero, M., Bermejo, A. M., Bermejo-Barrera, P., & Moreda-Piñeiro, A. (2016, January 15). Development of magnetic molecularly imprinted polymers for solid phase extraction of cocaine and metabolites in urine before high performance liquid chromatography—Tandem mass spectrometry. *Talanta, 147*, 641–649.
19. Rogeberg, M., Malerod, H., Roberg-Larsen, H., Aass, C., & Wilson, S. R. (2014, January). On-line solid phase extraction-liquid chromatography, with emphasis on modern bioanalysis and miniaturized systems. *Journal of Pharmaceutical and Biomedical Analysis, 87*, 120–129.
20. Chiuminatto, U., Gosetti, F., Dossetto, P., Mazzucco, E., Zampieri, D., Robotti, E., et al. (2010, July 1). Automated online solid phase extraction ultra high performance liquid chromatography method coupled with tandem mass spectrometry for determination of forty-two therapeutic drugs and drugs of abuse in human urine. *Analytical Chemistry, 82*(13), 5636–5645.
21. Hinchliffe, E., Carter, S., Owen, L. J., & Keevil, B. G. (2013, January 15). Quantitation of aldosterone in human plasma by ultra high performance liquid chromatography tandem mass spectrometry. *Journal of Chromatography. B, Analytical Technologies in the Biomedical and Life Sciences, 913–914*, 19–23.
22. Zhang, Z., Li, X., Ding, S., Jiang, H., Shen, J., & Xia, X. (2016, August 1). Multiresidue analysis of sulfonamides, quinolones, and tetracyclines in animal tissues by ultra-high performance liquid chromatography-tandem mass spectrometry. *Food Chemistry, 204*, 252–262.
23. Duan, X., Weinstock-Guttman, B., Wang, H., Bang, E., Li, J., Ramanathan, M., et al. (2010, March 15). Ultrasensitive quantification of serum vitamin D metabolites using selective solid-phase extraction coupled to microflow liquid chromatography and isotope-dilution mass spectrometry. *Analytical Chemistry, 82*(6), 2488–2497.
24. Kataoka, H., Mizuno, K., Oda, E., & Saito, A. (2016, April 15). Determination of the oxidative stress biomarker urinary 8-hydroxy-2′-deoxyguanosine by automated on-line in-tube solid-phase microextraction coupled with liquid chromatographytandem mass spectrometry. *Journal of Chromatography. B, Analytical Technologies in the Biomedical and Life Sciences, 1019*, 140–146.
25. Rahbar, N., Nazernezhad, L., Asadinezhad, M., Ramezani, Z., & Kouchak, M. (2018, October). A novel micro-extraction strategy for extraction of bisphosphonates from biological fluids using zirconia nanoparticles coupled with spectrofluorimetry and high performance liquid chromatography. *Journal of Food and Drug Analysis, 26*(4), 1303–1311.
26. Merib, J., Spudeit, D. A., Corazza, G., Carasek, E., & Anderson, J. L. (2018, July). Magnetic ionic liquids as versatile extraction phases for the rapid determination of estrogens in human urine by dispersive liquid-liquid microextraction coupled with high-performance liquid chromatography-diode array detection. *Analytical and Bioanalytical Chemistry, 410*(19), 4689–4699.

27. Patil, A., Bhat, S., Pai, K. M., Rai, L., Kartha, V. B., & Chidangil, S. (2015, September 8). Ultra-sensitive high performance liquid chromatography-laser-induced fluorescence based proteomics for clinical applications. *Journal of Proteomics, 127*(Pt A), 202–210.
28. Schäfer, A., von Toerne, C., Becker, S., Sarioglu, H., Neschen, S., Kahle, M., et al. (2012, October 16). Two-dimensional peptide separation improving sensitivity of selected reaction monitoring-based quantitative proteomics in mouse liver tissue: comparing off-gel electrophoresis and strong cation exchange chromatography. *Analytical Chemistry, 84*(20), 8853–8862.
29. Verplaetse, R., & Tytgat, J. (2010, July 15). Development and validation of a sensitive ultra performance liquid chromatography tandem mass spectrometry method for the analysis of fentanyl and its major metabolite norfentanyl in urine and whole blood in forensic context. *Journal of Chromatography. B, Analytical Technologies in the Biomedical and Life Sciences, 878*(22), 1987–1996.
30. Quintela, O., Lendoiro, E., Cruz, A., de Castro, A., Quevedo, A., Jurado, C., et al. (2010, March). Hydrophilic interaction liquid chromatography-tandem mass spectrometry (HILIC-MS/MS) determination of cocaine and its metabolites benzoylecgonine, ecgonine methyl ester, and cocaethylene in hair samples. *Analytical and Bioanalytical Chemistry, 396*(5), 1703–1712.
31. Nielsen, M. K. K., Nedahl, M., Johansen, S. S., & Linnet, K. (2018, January 16). Validation of a fully automated solid-phase extraction and ultra-high-performance liquid chromatography-tandem mass spectrometry method for quantification of 30 pharmaceuticals and metabolites in post-mortem blood and brain samples. *Drug Testing and Analysis*.
32. Sánchez-González, J., Salgueiro-Fernández, R., Cabarcos, P., Bermejo, A. M., Bermejo-Barrera, P., & Moreda-Piñeiro, A. (2017, February). Cannabinoids assessment in plasma and urine by high performance liquid chromatography-tandem mass spectrometry after molecularly imprinted polymer microsolid-phase extraction. *Analytical and Bioanalytical Chemistry, 409*(5), 1207–1220.
33. Licata, M., Rustichelli, C., Palazzoli, F., Ferrari, A., Baraldi, C., Vandelli, D., et al. (2016, July 15). Hair testing in clinical setting: Simultaneous determination of 50 psychoactive drugs and metabolites in headache patients by LC tandem MS. *Journal of Pharmaceutical and Biomedical Analysis, 126*, 14–25.
34. Mei, M., Yu, J., Huang, X., Li, H., Lin, L., & Yuan, D. (2015, March 13). Monitoring of selected estrogen mimics in complicated samples using polymeric ionic liquid-based multiple monolithic fiber solid-phase microextraction combined with highperformance liquid chromatography. *Journal of Chromatography. A, 1385*, 12–19.
35. Dorival-García, N., Zafra-Gómez, A., Cantarero, S., Navalón, A., & Vílchez, J. L. (2013, January). Simultaneous determination of 13 quinolone antibiotic derivatives in wastewater samples using solid-phase extraction and ultra performance liquid chromatography–tandem mass spectrometry. *Microchemical Journal, 106*, 323–333.
36. Yin, Y., Chen, M., Peng, J., Liu, J., & Jiang, G. (2010, June 15). Dithizone-functionalized solid phase extraction-displacement elution high performance liquid chromatography-inductively coupled plasma mass spectrometry for mercury speciation in water samples. *Talanta, 81*(4–5), 1788–1792.
37. Rahimpoor, R., Bahrami, A., Nematollahi, D., Shahna, F. G., & Farhadian, M. (2019, May 1). Facile and sensitive determination of urinary mandelic acid by combination of metal organic frameworks with microextraction by packed sorbents. *Journal of Chromatography. B, Analytical Technologies in the Biomedical and Life Sciences, 1114–1115*, 45–54.
38. Wang, Q., Li, L., Long, C.-L., Luo, L., Yang, Y., Yang, Z.-G., et al. (2020, March). Detection of C60 in environmental water using dispersive liquid-liquid micro-extraction followed by high-performance liquid chromatography. *Environmental Technology, 41*(8), 1015–1022.
39. Aguilar-Arteaga, K., Hernández-Mera, C., Díaz-Batalla, L., Castañeda-Ovando, A., Cruz-Pérez, A. E., Barrado-Esteban, E., et al. (2020, February). Development of a dispersive liquid-liquid microextraction coupled with magnetic dispersive microsolid phase extraction for

determination of triazines in wastewater samples. *Analytical and Bioanalytical Chemistry, 412*(5), 1203–1213.
40. Pan, D., Wang, J., Chen, C., Huang, C., Cai, Q., & Yao, S. (2013, April 15). Ultrasonic assisted extraction combined with titaniumplate based solid phase extraction for the analysis of PAHs in soil samples by HPLC-FLD. *Talanta, 108*, 117–122.
41. Pavlou, M. P., & Diamandis, E. P. (2010, September 10). The cancer cell secretome: A good source for discovering biomarkers? *Journal of Proteomics, 73*(10), 1896–1906.
42. Cho, W. C. S. (2007, April 2). Contribution of oncoproteomics to cancer biomarker discovery. *Molecular Cancer, 6*, 25.
43. Evenson, M. A., & Carmack, G. D. (1979, April). Clinical chemistry. *Analytical Chemistry, 51*(5), 35R–79R.
44. Kałużna-Czaplińska, J. (2011, May). Current medical research with the application of coupled techniques with mass spectrometry. *Medical Science Monitor, 17*(5), RA117–RA123.
45. Hunnewell, M. G., & Forbes, N. S. (2010). Active and inactive metabolic pathways in tumor spheroids: Determination by GC-MS. *Biotechnology Progress, 26*(3), 789–796.
46. Jellum, E., Stokke, O., & Eldjarn, L. (1973, June). Application of gas chromatography, mass spectrometry, and computer methods in clinical biochemistry. *Analytical Chemistry, 46*(7), 1099–1106.
47. Gooding, K. M., & Regnier, F. E. (2002). *Hplc of biological macro- molecules, revised and expanded*. CRC Press.
48. Plumb, R., Castro-Perez, J., Granger, J., Beattie, I., Joncour, K., & Wright, A. (2004). Ultra-performance liquid chromatography coupled to quadrupole-orthogonal time-of-flight mass spectrometry. *Rapid Communications in Mass Spectrometry, 18*(19), 2331–2337.
49. Kauna-Czaplińska, J. (2011, April 29). Current applications of gas chromatography/mass spectrometry in the study of organic acids in urine. *Critical Reviews in Analytical Chemistry, 41*(2), 114–123.
50. Masson, P. (2007, July 27). Quality control techniques for routine analysis with liquid chromatography in laboratories. *Journal of Chromatography. A, 1158*(1–2), 168–173.
51. Strimbu, K., & Tavel, J. A. (2010, November). What are biomarkers? *Current Opinion in HIV and AIDS, 5*(6), 463–466.
52. Amorim, L. C. A., & de L Cardeal, Z. (2007, June 15). Breath air analysis and its use as a biomarker in biological monitoring of occupational and environmental exposure to chemical agents. *Journal of Chromatography. B, Analytical Technologies in the Biomedical and Life Sciences, 853*(1–2), 1–9.
53. Pawlik, T. M., Hawke, D. H., Liu, Y., Krishnamurthy, S., Fritsche, H., Hunt, K. K., et al. (2006, March 16). Proteomic analysis of nipple aspirate fluid from women with early-stage breast cancer using isotope-coded affinity tags and tandem mass spectrometry reveals differential expression of vitamin D binding protein. *BMC Cancer, 6*, 68.
54. Imai, K., Ichibangase, T., Saitoh, R., & Hoshikawa, Y. (2008, November). A proteomics study on human breast cancer cell lines by fluorogenic derivatization-liquid chromatography/tandem mass spectrometry. *Biomedical Chromatography, 22*(11), 1304–1314.
55. Mal, M., Koh, P. K., Cheah, P. Y., & Chan, E. C. Y. (2012, April). Metabotyping of human colorectal cancer using two-dimensional gas chromatography mass spectrometry. *Analytical and Bioanalytical Chemistry, 403*(2), 483–493.
56. Balog, C. I. A., Stavenhagen, K., Fung, W. L. J., Koeleman, C. A., McDonnell, L. A., Verhoeven, A., et al. (2012, September). N-glycosylation of colorectal cancer tissues: A liquid chromatography and mass spectrometry-based investigation. *Molecular & Cellular Proteomics, 11*(9), 571–585.
57. Wu, H., Xue, R., Dong, L., Liu, T., Deng, C., Zeng, H., et al. (2009, August 19). Metabolomic profiling of human urine in hepatocellular carcinoma patients using gas chromatography/mass spectrometry. *Analytica Chimica Acta, 648*(1), 98–104.
58. Dunn, W. B., Bailey, N. J. C., & Johnson, H. E. (2005, May). Measuring the metabolome: current analytical technologies. *The Analyst, 130*(5), 606–625.

59. Yamaguchi, M., Yoshitake, T., Ishida, J., & Nakamura, M. (1989). Determination of 21-hydroxycorticosteroids in human urine by high-performance liquid chromatography with fluorescence detection. *Chemical & Pharmaceutical Bulletin, 37*, 3022–3025.
60. Haegele, A. D., & Wade, S. E. (1991). Ultrasensitive differential measurement of cortisol and cortisone in biological samples using fluorescent ester derivatives in normal phase hplc. *Journal of Liquid Chromatography, 14*, 1133–1148.
61. Nozaki, O., Ohata, T., Ohba, Y., Moriyama, H., & Kato, Y. (1991, September 18). Determination of serum cortisol by reversed-phase liquid chromatography using precolumn sulphuric acid-ethanol fluorescence derivatization and column switching. *Journal of Chromatography, 570*(1), 1–11.
62. Saisho, Y., Shimada, C., & Umeda, T. (1998). Determination of 7α-hydroxycholesterol in dog plasma by highperformance liquid chromatography with fluorescence detection. *Analytical Biochemistry, 265*, 361–367.
63. Neufeld, E., Chayen, R., & Stern, N. (1998, November 6). Fluorescence derivatisation of urinary corticosteroids for high-performance liquid chromatographic analysis. *Journal of Chromatography. B, Biomedical Sciences and Applications, 718*(2), 273–277.
64. Główka, F. K., Kosicka, K., & Karazniewicz-Łada, M. (2010). HPLC method for determination of fluorescence derivatives of cortisol, cortisone and their tetrahydro- and allo-tetrahydro-metabolites in biological fluids. *Journal of Chromatography. B, Analytical Technologies in the Biomedical and Life Sciences, 878*, 283–289.
65. Sudo, A. (1990). Analysis of corticosterone in rat urine by high-performance liquid chromatography and fluorimetry using post-column reaction with sulphuric acid. *Journal of Chromatography. B, Biomedical Sciences and Applications, 528*, 453–458.
66. Gao, W., Xie, Q., Jin, J., Qiao, T., Wang, H., Chen, L., et al. (2010, May). HPLC-FLU detection of cortisol distribution in human hair. *Clinical Biochemistry, 43*(7–8), 677–682.
67. Kumar, R., Gaurav, H., Malik, A. K., Kabir, A., & Furton, K. G. (2014, September 12). Efficient analysis of selected estrogens using fabric phase sorptive extraction and high performance liquid chromatography-fluorescence detection. *Journal of Chromatography. A, 1359*, 16–25.
68. Mao, L., Sun, C., Zhang, H., Li, Y., & Wu, D. (2004, September). Determination of environmental estrogens in human urine by high performance liquid chromatography after fluorescent derivatization with p-nitrobenzoyl chloride. *Analytica Chimica Acta, 522*(2), 241–246.
69. Amin, M., Harrington, K., & von Wandruszka, R. (1993, September 1). Determination of steroids in urine by micellar HPLC with detection by sensitized terbium fluorescence. *Analytical Chemistry, 65*(17), 2346–2351.
70. Katayama, M., Sasaki, T., Matsuda, Y., Kaneko, S., Iwamoto, T., & Tanaka, M. (2001). Sensitive determination of bisphenol A and alkylphenols by high performance liquid chromatography with pre-column derivatization with 2-(4-carboxyphenyl)-5,6-dimethylbenzimidazole. *Biomedical Chromatography, 15*, 403–407.
71. Katayama, M., Nakane, R., Matsuda, Y., Kaneko, S., Hara, I., & Sato, H. (1998). Determination of progesterone and 17-hydroxyprogesterone by high performance liquid chromatography after pre-column derivatization with 4,4-difluoro-5,7-dimethyl-4-bora-3a,4a- diazas-indacene-3-propionohydrazide. *Analyst, 123*, 2339–2342.
72. Katayama, M., Masuda, Y., & Taniguchi, H. (1993). Determination of corticosteroids in plasma by high-performance liquid chromatography after pre-column derivatization with 2-(4-carboxyphenyl)-5,6-dimethylbenzimidazole. *Journal of Chromatography. B, Biomedical Sciences and Applications, 612*, 33–39.
73. Sun, Y., Irie, M., Kishikawa, N., Wada, M., Kuroda, N., & Nakashima, K. (2004). Determination of bisphenol A in human breast milk by HPLC with column-switching and fluorescence detection. *Biomedical Chromatography, 18*, 501–507.
74. Kuroda, N., Kinoshita, Y., Sun, Y., Wada, M., Kishikawa, N., Nakashima, K., et al. (2003, January 15). Measurement of bisphenol A levels in human blood serum and ascitic fluid by HPLC using a fluorescent labeling reagent. *Journal of Pharmaceutical and Biomedical Analysis, 30*(6), 1743–1749.

75. Sun, Y., Nakashima, M. N., Takahashi, M., Kuroda, N., & Nakashima, K. (2002, August). Determination of bisphenol A in rat brain by microdialysis and column switching high-performance liquid chromatography with fluorescence detection. *Biomedical Chromatography, 16*(5), 319–326.
76. Shibata, N., Hayakawa, T., Takada, K., Hoshino, N., Minouchi, T., & Yamaji, A. (1998, March 20). Simultaneous determination of glucocorticoids in plasma or urine by high-performance liquid chromatography with precolumn fluorimetric derivatization by 9-anthroyl nitrile. *Journal of Chromatography. B, Biomedical Sciences and Applications, 706*(2), 191–199.
77. Mason, S. R., Ward, L. C., & Reilly, P. E. B. (1992). Fluorimetric detection of serum corticosterone using highperformance liquid chromatography. *Journal of Chromatography. B, Biomedical Sciences and Applications, 581*, 267–271.
78. Kurosawa, S., Koike, T., Yoshimura, T., Kurosawa, T., Tohma, M., Chiba, H., & Kobayashi, K. (1995). Simultaneous determination of 18-oxygenated corticosteroids by high-performance liquid chromatography with fluorescence detection. *Journal of Liquid Chromatography, 18*, 2383–2396.
79. Lin, Y.-T., Wu, S.-S., & Wu, H.-L. (2007, July 13). Highly sensitive analysis of cholesterol and sitosterol in foods and human biosamples by liquid chromatography with fluorescence detection. *Journal of Chromatography. A, 1156*(1–2), 280–287.
80. Shimada, K., & Nakagi, T. (1996). Studies on neurosteroids. IV. Quantitative determination of pregnenolone in rat brains using highperformanceliquid chromatography. *Journal of Liquid Chromatography and Related Technologies, 19*, 2593–2602.
81. Shimada, K., & Nonaka, M. (1991). Utility of cyclodextrin in mobile phase for high-performance liquid chromatographic separation of C21 steroids. *Journal of Liquid Chromatography, 14*, 2109–2117.
82. Główka, F. K., Karazniewicz, M., & Lipnicka, E. (2006). RP-HPLC method with fluorescence detection for determination of small quantities of triamcinolone in plasma in presence of endogenous steroids after derivatization with 9-anthroyl nitrile; pharmacokinetic studies. *Journal of Chromatography. B, Analytical Technologies in the Biomedical and Life Sciences, 839*, 54–61.
83. Visser, S. A., Smulders, C. J., Gladdines, W. W., Irth, H., van der Graaf, P. H., & Danhof, M. (2000, August 18). High-performance liquid chromatography of the neuroactive steroids alphaxalone and pregnanolone in plasma using dansyl hydrazine as fluorescent label: Application to a pharmacokinetic-pharmacodynamic study in rats. *Journal of Chromatography. B, Biomedical Sciences and Applications, 745*(2), 357–363.
84. Peng, X.-D., Xu, D.-H., Jin, J., Mei, X.-T., Lv, J.-Y., & Xu, S.-B. (2007, June 7). Determination of a new active steroid by high performance liquid chromatography with laser-induced fluorescence detection following the pre-column derivatization. *International Journal of Pharmaceutics, 337*(1–2), 25–30.
85. Arsova-Sarafinovska, Z., Ugrinova, L., Starkoska, K., Djordjev, D., & Dimitrovska, A. (2006). Determination of ethinylestradiol and levonorgestrel in oral contraceptives with HPLC methods with UV detection and UV/fluorescence detection. *Macedonian Pharmaceutical Bulletin, 52*, 9–16.
86. Arsova-sarafinovska, Z., Polozhani, A., & Dimitrovska, A. (2009). Determination of ethinylestradiol and drospirenone in oral contraceptives with HPLC method with UV and fluorescence detection Determination of ethinylestradiol and drospirenone in oral contraceptives with HPLC method with UV and fluorescence detection. *Archives of Public Health, 1*, 67.
87. Silva, V. B., Galdos, A. A. G., Mothe, C. M. A., Pallastrelli, M. B., Prado, M. S. A., Singh, A. K., Kedor-Hackmann, E. R. M., & Santoro, M. I. R. M. (2013). Simultaneous determination of ethinyl estradiol and drospirenone in oral contraceptive by high performance liquid chromatography. *Brazilian Journal of Pharmaceutical Sciences, 49*, 521–528.
88. Gatti, R., Gotti, R., Gioia, M. G., & Cavrini, V. (1998). HPLC analysis of pharmaceutical estrogens in raw materials and dosage forms. *Journal of Pharmaceutical and Biomedical Analysis, 17*, 337–347.

89. Wu, H., Li, G., Liu, S., Hu, N., Geng, D., Chen, G., Sun, Z., Zhao, X., Xia, L., & You, J. (2016). Monitoring the contents of six steroidal and phenolic endocrine disrupting chemicals in chicken, fish and aquaculture pond water samples using pre-column derivatization and dispersive liquid-liquid microextraction with the aid of experimental design metho. *Food Chemistry, 192*, 98–106.
90. Shahbazi, Y., Malekinejad, H., & Tajik, H. (2016). Determination of naturally occurring estrogenic hormones in cow's and river buffalo's meat by HPLC-FLD method. *Journal of Food and Drug Analysis, 24*, 457–463.
91. Yoshioka, N., Akiyama, Y., & Takeda, N. (2000, March 10). Determination of alpha- and beta-trenbolone in bovine muscle and liver by liquid chromatography with fluorescence detection. *Journal of Chromatography. B, Biomedical Sciences and Applications, 739*(2), 363–367.
92. Tsuda, T., Suga, K., Kaneda, E., & Ohsuga, M. (2000). Determination of 4-nonylphenol, nonylphenol monoethoxylate, nonylphenol diethoxylate and other alkylphenols in fish and shellfish by high-performance liquid chromatography with fluorescence detection. *Journal of Chromatography. B, Biomedical Sciences and Applications, 746*, 305–309.
93. Datta, S., Loyo-Rosales, J. E., & Rice, C. P. (2002). A simple method for the determination of trace levels of alkylphenolic compounds in fish tissue using pressurized fluid extraction, solid phase cleanup, and high performance liquid chromatography fluorescence detection. *Journal of Agricultural and Food Chemistry, 50*, 1350–1354.
94. Lu, J., Kong, D., Zhao, L., & Zhou, Q. (2014). Analysis of oestrogenic hormones in chicken litter by HPLC with fluorescence detection. *International Journal of Environmental Analytical Chemistry, 94*, 783–790.
95. De Liz, M. V., Do Amaral, B., Stets, S., Nagata, N., & Peralta-Zamora, P. (2017). Sensitive estrogens determination in wastewater samples by HPLC and fluorescence detection. *Journal of the Brazilian Chemical Society, 28*, 1453–1460.
96. Pérez, R. L., & Escandar, G. M. (2016). Multivariate calibration-assisted high-performance liquid chromatography with dual UV and fluorimetric detection for the analysis of natural and synthetic sex hormones in environmental waters and sediments. *Environmental Pollution, 209*, 114–122.
97. Patrolecco, L., Ademollo, N., Grenni, P., Tolomei, A., Barra Caracciolo, A., & Capri, S. (2013). Simultaneous determination of human pharmaceuticals in water samples by solid phase extraction and HPLC with UV fluorescence detection. *Microchemical Journal, 107*, 165–171.
98. Zhang, S., You, J., Sun, Z., Song, C., Ning, S., Zhao, C., & Suo, Y. (2012). A sensitive method for extraction and determination of endocrine-disrupting compounds from wastewater using 10-ethyl-acridone-2-sulfonyl chloride as pre-column labeling reagent by high-performance liquid chromatography with fluorescence detection. *Microchemical Journal, 103*, 90–96.
99. Lima, D. L. D., Silva, C. P., Otero, M., & Esteves, V. I. (2013, October 15). Low cost methodology for estrogens monitoring in water samples using dispersive liquid-liquid microextraction and HPLC with fluorescence detection. *Talanta, 115*, 980–985.
100. Yoon, Y., Westerhoff, P., Snyder, S. A., & Esparza, M. (2003). HPLC-fluorescence detection and adsorption of bisphenol A, 17β-estradiol, and 17α-ethynyl estradiol on powdered activated carbon. *Water Research, 37*, 3530–3537.
101. Durhan, E. J., Lambright, C. S., Makynen, E. A., Lazorchak, J., Hartig, P. C., Wilson, V. S., Gray, L. E., & Ankley, G. T. (2006). Identification of metabolites of trenbolone acetate in androgenic runoff from a beef feedlot. *Environmental Health Perspectives, 114*, 65–68.
102. Snyder, S. A., Keith, T. L., Verbrugge, D. A., Snyder, E. M., Gross, T. S., Kannan, K., et al. (1999, August 1). Analytical methods for detection of selected estrogenic compounds in aqueous mixtures. *Environmental Science & Technology, 33*(16), 2814–2820.
103. Wen, Y., Zhou, B. S., Xu, Y., Jin, S. W., & Feng, Y. Q. (2006). Analysis of estrogens in environmental waters using polymer monolith in polyether ether ketone tube solid-phase micro-

extraction combined with high-performance liquid chromatography. *Journal of Chromatography. A, 1133*, 21–28.

104. Ying, G.-G., Kookana, R. S., & Chen, Z. (2002, May). On-line solid-phase extraction and fluorescence detection of selected endocrine disrupting chemicals in water by high-performance liquid chromatography. *Journal of Environmental Science and Health. Part. B, 37*(3), 225–234.
105. Fan, Y., Zhang, M., Da, S.-L., & Feng, Y.-Q. (2005, July). Determination of endocrine disruptors in environmental waters using poly(acrylamide-vinylpyridine) monolithic capillary for in-tube solid-phase microextraction coupled to highperformance liquid chromatography with fluorescence detection. *The Analyst, 130*(7), 1065–1069.
106. Gamoh, K., Omote, K., Okamoto, N., & Takatsuto, S. (1989). High-performance liquid chromatography of brassinosteroids in plants with derivatization using 9-phenanthreneboronic acid. *Journal of Chromatography. A, 469*, 424–428.
107. Lv, T., Zhao, X.-E., Zhu, S., Ji, Z., Chen, G., Sun, Z., et al. (2014, December 7). Development of an efficient HPLC fluorescence detection method for brassinolide by ultrasonic-assisted dispersive liquid–liquid microextraction coupled with derivatization. *Chromatographia, 77*(23–24), 1653–1660.
108. Motegi, C., Takatsuto, S., & Gamoh, K. (1994). Identification of brassinolide and castasterone in the pollen of orange (Citrus sinensis Osbeck) by high-performance liquid chromatography. *Journal of Chromatography. A, 658*, 27–30.
109. Li, N., Wu, D., Liu, J., Hu, N., Shi, X., Dai, C., Sun, Z., Suo, Y., Li, G., & Wu, Y. (2018). Magnetic covalent organic frameworks based on magnetic solid phase extraction for determination of six steroidal and phenolic endocrine disrupting chemicals in food samples. *Microchemical Journal, 143*, 350–358.
110. Kosicka, K., Siemiatkowska, A., Szpera-Gozdziewicz, A., Krzyscin, M., Breborowicz, G., & Główka, F. (2019). High-performance liquid chromatography methods for the analysis of endogenous cortisol and cortisone in human urine: Comparison of mass spectrometry and fluorescence detection. *Annals of Clinical Biochemistry, 56*, 82–89.
111. Li, G., You, J., Suo, Y., Song, C., Sun, Z., Xia, L., Zhao, X., & Shi, J. (2011). A developed pre-column derivatization method for the determination of free fatty acids in edible oils by reversed-phase HPLC with fluorescence detection and its application to Lycium barbarum seed oil. *Food Chemistry, 125*, 1365–1372.
112. Liu, J., You, J., Zhang, S., Song, C., Ji, Z., Zhuang, J., & Yu, Y. (2018). New fluorescent labeling reagent Benzimidazo[2,1-b]quinazoline-12(6H)-one-5-ethylimidazole ester and its application in the analysis of endocrine disrupting compounds in milk by high performance liquid chromatography with fluorescence detection. *Microchemical Journal, 138*, 309–315.
113. Socas-Rodríguez, B., Asensio-Ramos, M., Hernández-Borges, J., & Rodríguez-Delgado, M. Á. (2014, April 15). Analysis of oestrogenic compounds in dairy products by hollow-fibre liquid-phase microextraction coupled to liquid chromatography. *Food Chemistry, 149*, 319–325.
114. Drzymala, S. S., Weiz, S., Heinze, J., Marten, S., Prinz, C., Zimathies, A., Garbe, L. A., & Koch, M. (2015). Automated solid-phase extraction coupled online with HPLC-FLD for the quantification of zearalenone in edible oil. *Analytical and Bioanalytical Chemistry, 407*, 3489–3497.
115. Kozłowska-Tylingo, K., Konieczka, P., Gustaw, E., Wasik, A., & Namiesnik, J. (2014). Comparison of high performance liquid chromatography methods with different detectors for determination of steroid hormones in aqueous matrices. *Analytical Letters, 47*, 1449–1464.
116. Wang, P., Qiu, X., & Yang, Y. (2015). Determination of estrogens in human urine by vortex-assisted dispersive liquid liquid microextraction based on floating organic acid droplet combined with high-performance liquid chromatography-fluorescence detection. *Journal of Liquid Chromatography and Related Technologies, 38*, 640–646.
117. Zhang, R., Cheng, X., Guo, J., Zhang, H., & Hao, X. (2017). Comparison of two ionic liquid-based pretreatment methods for three steroids' separation and determination in water samples by HPLC. *Chromatographia, 80*, 237–246.

118. Ferreira, F. N., Benevides, A. P., Cesar, D. V., Luna, A. S., & de Gois, J. S. (2020). Magnetic solid-phase extraction and pre-concentration of 17β-estradiol and 17α-ethinylestradiol in tap water using maghemite-graphene oxide nanoparticles and determination via HPLC with a fluorescence detector. *Microchemical Journal, 157*, 104947.
119. Zhang, G., Yang, Y., Lu, Y., Chen, Y., Li, W., & Wang, S. (2020, February 21). Effect of heavy metal ions on steroid estrogen removal and transport in SAT using DLLME as a detection method of steroid estrogen. *Water, 12*(2), 589.
120. Jiang, T., Zhao, L., Chu, B., Feng, Q., Yan, W., & Lin, J. M. (2009). Molecularly imprinted solid-phase extraction for the selective determination of 17β-estradiol in fishery samples with high performance liquid chromatography. *Talanta, 78*, 442–447.
121. Corazza, G., Merib, J., Magosso, H. A., Bittencourt, O. R., & Carasek, E. (2017). A hybrid material as a sorbent phase for the disposable pipette extraction technique enhances efficiency in the determination of phenolic endocrine disrupting compounds. *Journal of Chromatography. A, 1513*, 42–50.
122. Louros, V. L., Lima, D. L. D., Leitão, J. H., Esteves, V. I., & Nadais, H. G. (2019). Determination of estrone and 17α-ethinylestradiol in digested sludge by ultrasonic liquid extraction and high-performance liquid chromatography with fluorescence detection. *Journal of Separation Science, 42*, 1585–1592.
123. Yuan, Y., Wang, M., Jia, N., Zhai, C., Han, Y., & Yan, H. (2019). Graphene/multi-walled carbon nanotubes as an adsorbent for pipette-tip solid-phase extraction for the determination of 17β-estradiol in milk products. *Journal of Chromatography. A, 1600*, 73–79.
124. Xiong, L., Yan, P., Chu, M., Gao, Y. Q., Li, W. H., & Yang, X. L. (2018). A rapid and simple HPLC–FLD screening method with QuEChERS as the sample treatment for the simultaneous monitoring of nine bisphenols in milk. *Food Chemistry, 244*, 371–377.
125. Ali, M. F. B., Uejo, Y., Kishikawa, N., Ohyama, K., & Kuroda, N. (2020). A selective and highly sensitive high performance liquid chromatography with fluorescence derivatization approach based on Sonogashira coupling reaction for determination of ethinyl estradiol in river water samples. *Journal of Chromatography. A, 1628*, 461440.
126. Guedes-Alonso, R., Santana-Viera, S., Sosa-Ferrera, Z., & Santana-Rodríguez, J. J. (2015). Molecularly imprinted solid-phase extraction coupled with ultra high performance liquid chromatography and fluorescence detection for the determination of estrogens and their metabolites in wastewater. *Journal of Separation Science, 38*, 3961–3968.
127. Rudolph, L. M., Cornil, C. A., Mittelman-Smith, M. A., Rainville, J. R., Remage-Healey, L., Sinchak, K., et al. (2016, November 9). Actions of steroids: New neurotransmitters. *The Journal of Neuroscience, 36*(45), 11449–11458.
128. Cole, T. J., Short, K. L., & Hooper, S. B. (2019, June). The science of steroids. *Seminars in Fetal & Neonatal Medicine, 24*(3), 170–175.
129. Makin, H. L. J., Honour, J. W., Shackleton, C. H. L., & Griffiths, W. J. (2010). General methods for the extraction, purification, and measurement of steroids by chromatography and mass spectrometry. In *Steroid analysis* (pp. 163–282). Springer.
130. Varriale, A., Pennacchio, A., Pinto, G., Oliviero, G., D'Errico, S., Majoli, A., et al. (2015, October 21). A fluorescence polarization assay to detect steroid hormone traces in milk. *Journal of Agricultural and Food Chemistry, 63*(41), 9159–9164.
131. Xiao, X. Y., McCalley, D. V., & McEvoy, J. (2001, July 20). Analysis of estrogens in river water and effluents using solid-phase extraction and gas chromatography-negative chemical ionisation mass spectrometry of the pentafluorobenzoyl derivatives. *Journal of Chromatography. A, 923*(1–2), 195–204.
132. Shackleton, C. (2010, August). Clinical steroid mass spectrometry: A 45-year history culminating in HPLC-MS/MS becoming an essential tool for patient diagnosis. *The Journal of Steroid Biochemistry and Molecular Biology, 121*(3–5), 481–490.
133. Nozaki, O. (2001, November 23). Steroid analysis for medical diagnosis. *Journal of Chromatography. A, 935*(1–2), 267–278.
134. Kawasaki, T., Maeda, M., & Tsuji, A. (1979, June 11). Determination of plasma and urinary cortisol by high-performance liquid chromatography using fluorescence derivatization with dansyl hydrazine. *Journal of Chromatography, 163*(2), 143–150.

135. Seki, T., & Yamaguchi, Y. (1984, January 13). New fluorimetric determination of 17-hydroxycorticosteroids after high-performance liquid chromatography using post-column derivatization with benzamidine. *Journal of Chromatography, 305*(1), 188–193.
136. Levin, S. (2010). High performance liquid chromatography (scpHPLC/scp) in the pharmaceutical analysis. In *Pharmaceutical sciences encyclopedia* (pp. 1–34). Wiley.
137. Bakshi, M., & Singh, S. (2002, June 15). Development of validated stability-indicating assay methods—critical review. *Journal of Pharmaceutical and Biomedical Analysis, 28*(6), 1011–1040.
138. Dolz-Pérez, I., Sallam, M. A., Masiá, E., Morelló-Bolumar, D., Pérez Del Caz, M. D., Graff, P., et al. (2020, February). Polypeptidecorticosteroid conjugates as a topical treatment approach to psoriasis. *Journal of Controlled Release, 318*, 210–222.
139. Strusiak, S. H., Hoogerheide, J. G., & Gardner, M. S. (1982, June). Determination of ethinyl estradiol in solid dosage forms by highperformance liquid chromatography. *Journal of Pharmaceutical Sciences, 71*(6), 636–640.

# RP-HPLC: A Versatile Technique for Identification, Isolation, and Characterization of Therapeutic Antibodies

**Swekcha, T. Meghana, Anshi Agarwal, Krithika Prabhu, and Abhayraj S. Joshi**

**Abstract** Reversed-phase high-performance liquid chromatography (RP-HPLC) has now become a fundamental part of the set of several analytical techniques used in identifying, isolating, and characterizing therapeutic antibodies. It allows effective separation, accurate characterization, and accurate quantification of antibodies based on their hydrophobic characteristics. Thus, this technique has great potential for evaluating a wide range of physicochemical properties, including the chemical nature of the antibodies, charge variants, and post-translational modifications. RP-HPLC is also extremely useful in the quality control process, ensuring that the antibody preparations are highly pure, consistent, and stable. RP-HPLC is of immense importance in the analysis of antibody fragments, the study of drug conjugation with the antibody, and their stability testing. The latest developments in chromatography include the integration of mass spectrometry and ultra-high-performance chromatography systems, which makes its capability of antibody characterization even greater. The methodology has indeed proven to be crucial in developing biosimilar products, enabling comparative analysis with the reference standard products. As therapeutic antibodies continue to advance in biomedical field, RP-HPLC, together with other analytical techniques, will remain a valuable tool for identification, purification, and characterization of those antibodies so that one can bring safer, more effective antibody-based therapeutic or diagnostic agents from bench-side to the patient's bed.

**Keywords** Reversed-phase HPLC · Monoclonal antibodies · Therapeutic antibodies · Diagnostics · Antibody-drug conjugates · Antibody purification

---

Swekcha and T. Meghana contributed equally to this book chapter.

---

Swekcha · T. Meghana · A. Agarwal · K. Prabhu · A. S. Joshi (✉)
Manipal Centre for Biotherapeutics Research, Manipal Academy of Higher Education, Manipal, Udupi, Karnataka, India
e-mail: abhayraj.joshi@manipal.edu



N. Mazumder et al. (eds.), *Advanced Biophysical Techniques in Biosciences*, Lecture Notes in Nanoscale Science and Technology 7,
https://doi.org/10.1007/978-3-031-94551-9_8

# 1 Introduction

## *1.1 Antibodies in Modern Therapeutics: Evolution and Impact*

Antibodies are well established as one of the cornerstones of modern therapeutic medicine and significantly alter the management of many diseases and health problems. With such highly engineered proteins, ranging from cancer to autoimmune diseases, there has been a wide range of possible targets for antibody targeting. The field of therapy has grown in the last two decades, and antibiotic discovery was hastened through the development of biotechnology, molecular biology, and immunological research. Monoclonal antibodies (mAbs), represent a new class of drugs in current medical science, more fundamentally changing the nature of drug discovery and the practice of medicine itself in recent years. Since the first therapeutic monoclonal antibody, OKT4, was approved in 1986, the field has grown exponentially, as highlighted by the approval of 18 additional antibodies in the United States, with many more under developmental phases [1].

The history of antibody therapeutics dates back over a century, beginning with conventional immunoglobulin therapy as the first generation [2]. This technique reflected the diversity inherent in the natural response but was quite limited in practical clinical applications. mAbs have marked the development of the second generation of highly specific drug components that would target antigens associated with particular diseases. During the last 10 years, over 30 mAbs have been commercially launched in the pharmaceutical industry, both with high clinical efficacy and economic feasibility [3]. The third generation of antibody therapeutics, comprised of recombinant polyclonal antibodies, is emerging for challenging and highly mutagenic targets. This generation promises new opportunities for increased therapeutic efficacy. The technological aspects, especially through hybridoma technology and recombinant DNA methodologies, greatly advanced the concept of mAbs [4].

Antibodies were first discovered as therapeutic entities during the 1970s. They were discovered by Georges Köhler and César Milstein [5]. This technological leap has made it possible to generate large quantities of identical antibodies effortlessly. Using hybridoma technology and recombinant DNA technology, several antibodies have been developed that include chimeric antibodies, humanized antibodies, and fully human antibodies [6, 7]. One of the most significant uses of antibodies in modern therapeutics is in the oncology field as novel anticancer agents. The discovery of mAbs targeting specific proteins expressed by cancer cells has significantly altered the course of cancer treatment. A pertinent example is Trastuzumab, commonly known as Herceptin®, which is often used in mainstream therapy of HER2-positive breast cancers; it binds to the HER2 receptor present on the surface of a cancer cell, and thus inhibits cellular proliferation and evokes an immune response toward cancer [8]. Similarly, Rituximab (Rituxan™) works on the CD20 protein of B cells and has revolutionized the treatment of several B-cell lymphomas [9].

Greater success with these early antibody-based therapies has encouraged the development of additional advanced strategies, such as antibody-drug conjugates (ADCs) [10]. ADCs combine the selectivity of antibodies with the strong cytotoxic potency of drugs into one strategy. A specific example is ado-trastuzumab emtansine (Kadcyla™), which is for HER2-positive breast cancers; trastuzumab is conjugated with the chemotherapy agent DM-1 [11]. DM-1 is a highly potent cytotoxin, a derivative of naturally occurring maytansinoid toxin, which helps in inducing death in proliferative cells by inhibiting tubulin polymerization [12]. The conjugation ensures drug delivery to the cancerous cells directly without affecting the healthy tissues around them. Adalimumab (Humira®) and infliximab (Remicade®) are tumor necrosis factor-alpha inhibitors that have revolutionized treatment protocols for diseases such as rheumatoid arthritis, Crohn's disease, and psoriasis [13]. These antibodies neutralize the action of TNF-α, a principal inflammatory mediator, and result in inflammation reduction and relief of symptoms. The success of anti-TNF drugs has also prompted the development of anti-inflammatory antibodies targeted at other pathways, including IL-6 inhibitors like tocilizumab (Actemra®) for rheumatoid arthritis and cytokine storm associated with severe COVID-19 [14]. In infectious diseases, they have been shown to be both preventive as well as therapeutic. Palivizumab, more widely referred to as Synagis®, prevents respiratory syncytial virus (RSV) infections in high-risk infants [15]. More recently, techniques have been developed to engineer antibodies specific to newly emerging infectious diseases. A prime example came with the COVID-19 pandemic when mAbs, such as casirivimab and imdevimab, were rapidly generated and approved for treatment and prophylaxis of severe SARS-CoV-2 infections [16].

The neurological disorders are normally considered a very difficult target for antibody therapies owing to the presence of the blood-brain barrier (BBB). This has become increasingly beneficial due to the development of bispecific antibodies that utilize receptor-mediated transcytosis (RMT) to penetrate the BBB. The recent approval of aducanumab for the treatment of Alzheimer's disease targeting amyloid-beta plaques in the brain, potentially slowing down cognitive decline, reveals how dramatically the frontiers of antibody therapeutics expand in various areas of medicine [17]. Now, with the passing years of antibody engineering and design, bispecific antibodies have been discovered, which can bind two different antigens simultaneously. This came with a great promise for cancer immunotherapy. For example, a bispecific T-cell engager (BiTE), blinatumomab (Blincyto®); binds CD19 on B-cell leukemias and CD3 on T cells, bringing the cells into proximity with activation of the T cells to kill the cancer cells [18]. Amivantamab (Rybrevant®) is another bispecific antibody that targets both EGFR and MET, promoting cell death in non-small cell lung cancer that expresses these receptors [19]. Emicizumab (Hemlibra®), used for treating hemophilia A, is a bispecific antibody that bridges activated factors IX and X, mimicking the function of missing clotting factors in the patients [20]. The leading antibody-based therapeutics in various diseases have been listed in Table 1.

**Table 1** Types of antibodies with their target, disease-associated, and functions

| Antibody Type | Target | Disease | Function | Examples |
|---|---|---|---|---|
| Monoclonal antibodies | Various antigens associated with diseases | Breast cancer, B-cell lymphoma | Target specific antigens to block signaling pathways, induce apoptosis, or recruit immune cells | Trastuzumab (HER2), Rituximab (CD20) |
| Chimeric antibodies | Specific proteins in cancer and autoimmune diseases | Ischemic heart disease, myocardial infarction | Combine murine and human antibody components to reduce immunogenicity | Abciximab (GPIIb/IIIa) |
| Fully human antibodies | Tumor necrosis factor (TNF) | Rheumatoid arthritis, Crohn's disease, and psoriasis | Neutralize TNF-α to reduce inflammation in autoimmune diseases | Adalimumab, Infliximab |
| Bispecific antibodies | Two different antigens | Acute lymphoblastic leukemia (ALL) | Bind two targets simultaneously, enhancing immune response against cancer cells | Blinatumomab (CD19/CD3) |
| Antibody-drug conjugates (ADCs) | Cancer cells | Breast cancer | Deliver cytotoxic drugs directly to tumor cells while sparing healthy tissues | Ado-trastuzumab emtansine |
| Recombinant polyclonal antibodies | Highly mutagenic targets | COVID-19 | Target diverse antigens with a mixture of antibodies for enhanced therapeutic efficacy | Recombinant polyclonal SARS-CoV-2 immunoglobulin (rCIG), SOD1 recombinant polyclonal antibody |
| Neutralizing antibodies | Pathogens or disease-causing molecules | COVID-19 | Block the biological activity of toxins or pathogens | Casirivimab/imdevimab |
| Cytokine inhibitors | Various cytokines | Systemic sclerosis | Target specific inflammatory pathways to manage autoimmune diseases | Tocilizumab (IL-6 inhibitor) |
| Preventive monoclonal antibodies | Infectious agents | Lower respiratory tract disease caused by respiratory syncytial virus (RSV) | Provide prophylaxis against infections, especially in high-risk populations | Palivizumab (RSV prevention) |

With such promises for many therapeutic as well as diagnostic gains in the near future, the field of antibody-based therapeutics/diagnostics will accelerate drug discovery.

At present, the monoclonal and polyclonal antibodies are obtained from different technologies like hybridoma, from recombinant and humanized sources, and to purify them, several methods are widely used. Technologies such as reversed-phase high-performance liquid chromatography (RP-HPLC) have emerged as a powerful analytical tool not only for separation and identification but also for the quantification of antibodies and their fragments [21].

## *1.2 Reversed-Phase High-Performance Liquid Chromatography (RP-HPLC)*

### 1.2.1 Origin and Brief History

Reversed-phase high-performance liquid chromatography (RP-HPLC) is one of the critical analytical techniques widely used in biosciences for the identification, separation, and characterization of biomolecules such as antibodies. In this method, based on hydrophobic interactions, separations are carried out efficiently to make it highly essential for most applications within the fields of biochemistry and biotechnology [22]. Its development has been closely associated with the overall enhancement of liquid chromatography (LC) techniques within the last several decades.

Before the invention of high-performance liquid chromatography (HPLC), traditional techniques in LC were inefficient. The solvent flow relied only on gravity, and it usually took hours or days to separate the samples [23, 24]. In the 1960s, gas chromatography (GC) was a paradigm for the potential automation, rapidity, and detection sensitivity that HPLC could achieve. However, the GC not being ideal for analyzing polar and thermally unstable compounds highlighted the need for LC evolution and progression. Its early applications in amino acid analysis and gel permeation chromatography (GPC) in the early 1960s validated the automation of liquid chromatography techniques [25].

Figure 1 illustrates a timeline representing the brief history of RP-HPLC.

The pioneering work of Archer John Porter Martin and Richard Laurence Millington Synge, who developed partition chromatography in 1941 [26], laid down the basis of HPLC. This research revealed that by reducing the particle size and increasing pressure to enhance solvent flow, liquid chromatography might obtain greater efficiency. Modern HPLC systems were developed as an outcome of various experiments with smaller packing particles and high-pressure systems, administered by researchers such as Calvin Giddings and Ingrid Fogy and others by the 1960s and 1970s [25, 27].

Indeed, it is likely that many individuals considered the potential of utilizing HPLC for separations involving substances other than amino acids or polymers prior to 1965.

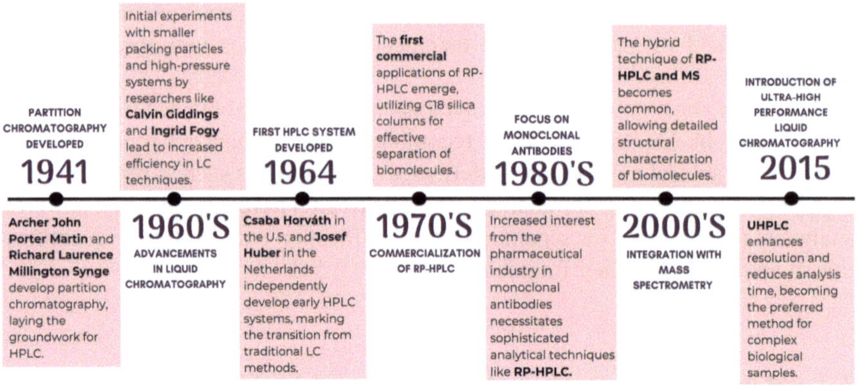


**Fig. 1** Illustration of timeline representation of origin and history of RP-HPLC

Its realization, however, came only through its actual implementation along with commercially available equipment, similar to an amino-acid analyzer and GPC. When considered from this point of view, HPLC was first discovered in the United States in 1964 by Csaba Horváth and at the same time in the Netherlands by Josef Huber [28, 29]. This was the basis for the LCS 1000 Nucleic Acid Analyzer, commercialized by Picker Nuclear and then acquired by Varian. This later became the first general-purpose HPLC known as the Waters ALC-100 [30]. By 1964, Jack Kirkland visited Huber's lab and soon started an HPLC efforts at DuPont [31], which led to Model 820, around the same time as ALC-100.

The first commercial applications of RP-HPLC began in the 1970s and soon advanced to the use of C18 silica columns, which proved to be one of the most highly effective columns in the separation of biomolecules. In the early years, RP-HPLC was largely applied to characterizing small molecules and short peptides. However, by the early 1980s, the pharmaceutical industry directed significant interest toward mAbs. This immediately created a need for highly sophisticated analytical techniques that could provide information on these complex and large biomolecules. Advances in column technology, primarily smaller particle sizes and new improved methods of packing, significantly enhanced the resolution and speed of RP-HPLC separations during the 1990s. In 1995, the gradient elution techniques came into existence to allow even more efficient separation of complex mixtures, further confirming the role of RP-HPLC in analytical laboratories.

It was during the 2000s that the hybrid technique of RP-HPLC and mass spectrometry (MS), became common; it made highly detailed structural characterizations possible for these biomolecules. The credit for this development goes to Tolstikov et al., who collectively studied this hybrid technique (LC–MS) for metabolomics applications [31]. In 2005, sophisticated software pertaining to data analysis and method development was released and greatly improved the analytical processes, leading to an increase in reproducibility and accuracy of the resultant outcomes.

To date, biologics have acquired a significant place in medicine, thereby making RP-HPLC an essential part of quality assurance and characterization of both therapeutic antibodies and other biologics. With the introduction of Ultra-High

Performance Liquid Chromatography (UHPLC) in 2015, researchers have achieved enhanced resolution and reduction in analysis time for the characterization of these complex molecules. Ongoing research is continuously improving the methodologies of RP-HPLC by putting great emphasis on its sensitivity and specificity for different applications.

### 1.2.2 Principle Behind RP-HPLC

RP-HPLC is one of the most significant analytical techniques developed for separating and analyzing molecules based on their hydrophobic properties. The nature of RP-HPLC fundamentals is based upon interactional properties between analyte molecules and the stationary phase. Most often, it consists of silica particles coated with hydrophobic alkyl chains such as C18, C8, or C4. Traditionally, the mobile phase consists of a mixture of water with a polar organic solvent (e.g.: acetonitrile, methanol). RP-HPLC separates the analytes based on differential partitioning between a hydrophobic stationary phase and a relatively polar mobile phase. More hydrophobic molecules interact strongly with the stationary phase and are retained for longer periods when compared with ones that are more hydrophilic [32].

Composition of the mobile phase can be modified during analysis by making a gradient elution in order to improve the separation as well as elution of analytes of varying hydrophobicity levels. Stationary-phase chemistry, mobile-phase composition, column temperature, and flow rate of the system influence the separation efficiency of RP-HPLC [33]. These variables can thus be adjusted to allow for high-resolution separations of complex mixtures. Owing to this virtue, RP-HPLC has become an extremely adaptable technique, useful in biotechnological and pharmaceutical applications, as well as in environmental analysis. The separated analytes are then detected by means of detectors with UV-visible spectrophotometry, fluorescence, or MS capacities, thereby allowing accurate qualitative and quantitative analysis of the constituents of the sample [34].

#### 1.2.2.1 Mechanism of Separation

The separation process begins with the injection of a sample mixture into the chromatographic device. At the start, the analytes are diluted in an aqueous buffer or other relevant solvent that would let them interact with the stationary phase. The mobile phase is introduced to the injection port/sampler and then to the column with a predetermined flow rate. Peristaltic pumps help in pumping the solvent with great precision. Using isocratic elution (where the composition of the solvents is constant) or gradient elution (where the organic composition of the solvent changes with time) the mobile phase moves into the column and then towards the detector and waste outlet [35]. The uniqueness of the properties of the analytes in terms of their hydrophobicity, polarity, solvent compatibility, temperature sensitivity, molecular size, and weight forms the basis of choosing between these methods for the separation of analytes. Gradient elution is especially advantageous in complex mixtures where resolution could be increased by letting less hydrophobic compounds elute earlier while keeping the more hydrophobic compounds for a longer period [34].

#### 1.2.2.2 Factors Affecting Separation

The efficiency of RP-HPLC is dependent on several factors:

**Column Characteristics**: The nature and size of the column have a strong influence on separation performance. Many practical columns are packed with phases of C18, C8, or C4 [34]. C18 columns exhibit the highest level of hydrophobicity among the three types due to their 18-carbon alkyl chain. The C8 column has a shorter 8-carbon alkyl chain, which results in moderate hydrophobicity compared to C18. With only a 4-carbon alkyl chain, C4 columns have the lowest hydrophobicity among the three, making them suitable for larger biomolecules [36].

**Mobile Phase Composition**: The concentration of water to organic solvent is fine-tuned to maximize separation according to the analyte's characteristics. High concentrations of the organic phase typically reduce the retention times [32].

**Temperature**: Increased temperatures tend to decrease the viscosity of the mobile phase and thus increase the mass transfer rates. Therefore, elevated column temperature leads to enhanced resolution and speed of analysis.

**Flow Rate**: The quality of the separation can be affected by modified flow rates; lower flow rates generally facilitate better and prolonged interaction between the analytes and stationary phases [35]. Hence, a slower flow rate often gives better separation quality.

### 1.2.3 Instrumentation

RP-HPLC employs specifically designed apparatuses with all the critical components, each designed to optimize the efficiency as well as the efficacy of the chromatographic process as shown in Fig. 2.

#### 1.2.3.1 Pump

The pump performs the critical function of creating the necessary pressure that enables the movement of the mobile phase in the entire system, thereby allowing the solvent and sample to go into the column. Pumps generate 6000 psi to 9000 psi pressure, i.e. approximately 400–600 bar, required to force the mobile phase through the packed column. It further regulates the flow rate to ensure accurate management

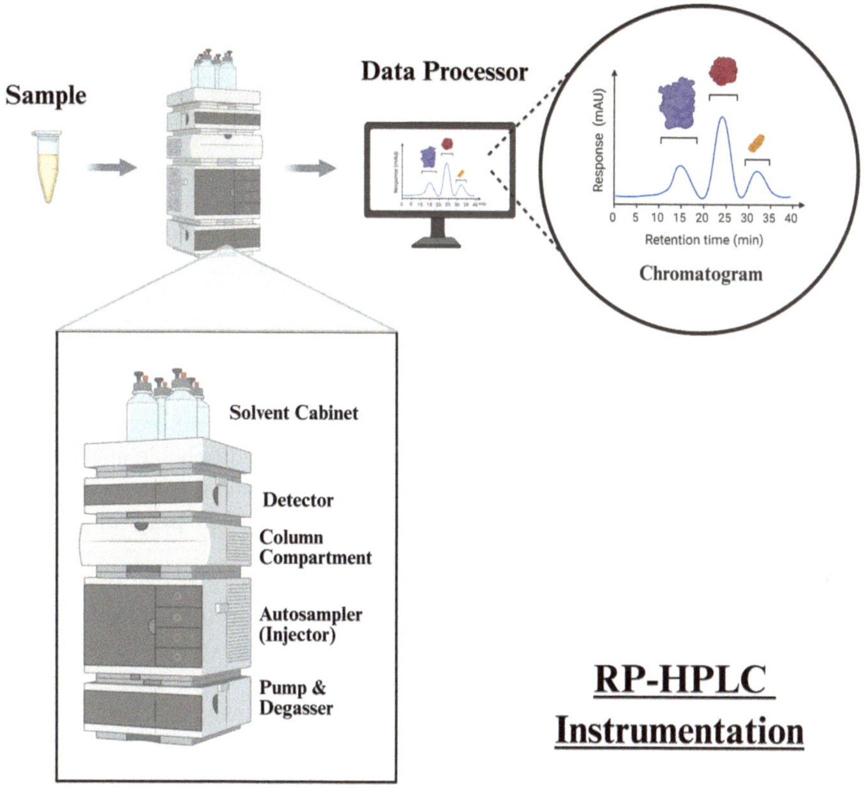


**Fig. 2** Illustration of different components of RP-HPLC instrument

within a range of generally 0.1–10 mL/min. That is very critical for reproducibility and consistency within the chromatographic results. Pumps can maintain a constant mobile phase composition (isocratic) as well as varying composition over time (gradient elution). The gradient elution method is often selected for antibody mixtures because of its ability to enhance the separation and resolution of complex mixtures [37].

#### 1.2.3.2 Injector

The injector needle injects the sample into the mobile phase stream with a defined pressure and volume. Injectors are available in both manual as well as automatic modes. The automated injectors are often preferred over manual ones due to their high reproducibility, minimal chances of errors, and injection accuracy. They are available with various volume capacities that range from 1 to 100 μL. They withstand very high pressures, up to 4000 psi, without leaking or breaking. For quality analysis, a good design of injectors and a sample loop is important to ensure consistent sample volume [38].

#### 1.2.3.3 Column

The actual separation of analytes takes place in the column based on how they interact with the stationary phase. Normally, this phase is made up of non-polar material such as C18 (octadecylsilane), which works with hydrophobic compounds. Silane chemistry involves the use of organosilanes to modify the surface of silica particles, which serves as the stationary phase in RP-HPLC. The end-capping strategy used in C18 reduces undesirable interactions and improves the peak shape of basic analytes [39]. The type of stationary phase influences selectivity and retention times. The efficiency of separations in RP-HPLC is often influenced by different aspects of column packing (like particle size, packing density, and hydrophobic interactions) [39]. They are often between 30 mm to 250 mm in length and 4.6 mm in diameter. The particles of the stationary phase can range between 1.5 and 5 μm, which affects the clarity of the obtained result and the pressure needed [40].

In RP-HPLC different types of columns are utilized to optimize the separation of analytes based on their chemical properties. Specially designed columns for mAbs are available, under the AdvanceBio product line offered by Agilent Technologies, Inc. [41]. The AdvanceBio RP-mAb columns are specially engineered to enhance the performance of therapeutic proteins and reduce mAbs analyses. These columns utilize Poroshell technology for higher resolution, improving the accuracy and reproducibility of results when analyzing complex antibody mixtures. A range of chemistries within the AdvanceBio RP-mAb column series are available, including SB-C8, C4, and diphenyl phases [41].

#### 1.2.3.4 Detectors

In the RP-HPLC system, the two most commonly used detectors are UV-visible and fluorescence detectors. Another common detector is the refractive index detector. UV-visible detectors measure the absorbance at specific wavelengths. They are commonly used since they are sensitive and generally applicable and detect as low as $10^{-6}$ g/ml. One of the types of UV-visible detectors is a photodiode array detector, which is used in modern HPLC units. This detector typically covers the ultraviolet and visible light spectrum (UV-VIS), which ranges from approximately 190 nm to 900 nm [42]. Fluorescence detectors are often used when substances naturally fluoresce or are modified, thus sensitive to fluorescent compounds. Fluorescence detection is more sensitive than UV detection, which allows detection of analytes with as low as 0.1 ng/ml concentrations. Apart from these two, there are other detectors available, including concentration-sensitive universal detectors and molar mass-sensitive detectors [40].

#### 1.2.3.5 Degasser

The main function of the degasser unit in HPLC instrumentation is to remove dissolved gases in the mobile phase and to prevent bubble formation, which may lead to flow disruption, erroneous analysis, and a reduction in instrument efficiency. There are different types of degassers—vacuum degassers use negative pressure to remove all gaseous components from solvents. On the other hand, membrane degassers primarily employ the facility of semi-permeable membranes that let the gases escape while retaining liquid solvents [40].

#### 1.2.3.6 Data Processor

A data processor handles the entire HPLC system and processes signals received from detectors. In modern HPLC, computer software controls the functioning of all units, starting from pumps to the detector. The processor plots the detector response over time to generate the chromatogram, enabling examination of retention times, peak areas, and analyte quantifications. Integration of sophisticated software features like—Peak integration, method development tools, and reporting capabilities; improve the user experience and facilitate better data interpretation [43].

## 2 Advantages of Using RP-HPLC

Dauly et al. have shown through their studies that (non-porous reverse-phase) np-RP-HPLC effectively separates intact proteins depending on their hydrophobicity without breaking its integrity. Proteins as low as 5 ng have been detected by RP-HPLC with a UV detector at 214 nm wavelength, exhibiting outstanding quantitative performance over three orders of magnitude [44]. Its compatibility was also demonstrated by intact protein fractions and their accurate molecular weight using Matrix-Assisted Laser Desorption Ionization-Time of Flight mass spectrometry (MALDI-TOF MS) [45]. In MALDI-TOF MS analyses, sensitivity was greatly increased, identifying as low as picograms (pg) of protein [46]. Separation of complex proteins such as Histones and vimentin is challenging using SDS-PAGE. However, they were identified and separated accurately using np-RP-HPLC to a protein fraction from the membrane of a bovine endothelial cell. The method demonstrated its usefulness in proteomics of specific protein subsets by successfully studying hydrophobic membrane proteins [44].

Patil et al. have discussed Analytical Quality by Design (AQbD) methodology for creating a robust and dependable RP-HPLC method with an emphasis on sensitivity, resolution, and method validation. To guarantee the method's effectiveness, Analytical Target Profile (ATP) was developed, focusing on metrics like resolution and retention time. Method performance was found to be greatly impacted by

Critical Quality Parameter (CQPs), such as the flow rate and the proportion of methanol in the mobile phase. Method variability risks were evaluated in accordance with principles given by International Conference of harmonization of Technical Requirements for Pharmaceuticals for Human Use (ICH Q9 principles), guaranteeing a methodical assessment of quality-influencing elements. The methanol percentage and flow rate were optimized using a Central Composite Design (CCD) inside the Design of Experiments (DoE) framework. To comprehend and illustrate the relationships between variables influencing retention time and theoretical plates (a measure of the efficiency of a chromatography column, representing hypothetical zones where the stationary and mobile phases reach equilibrium during the separation process), polynomial equations and response surface plots were created. The validation as per ICH Q2 (R1) requirements showed excellent linearity with a good correlation coefficient ($R^2 = 0.9998$), Limit of Detection (LOD) of 0.82891 μg/ml, Limit of Quantification (LOQ) of 2.511 μg/ml, and reproducibility is confirmed by low intra-day and inter-day variability with a relative standard deviation (%RSD) of <2%.

The method demonstrated robust and rugged performance, exhibiting minimal impact on analytical outcomes despite slight variations in operating conditions such as mobile phase composition and flow rate; optimal results were achieved with a mobile phase of 43% methanol at a flow rate of 0.9 ml/min using a Phenomenex Luna C18 column, which provided excellent resolution characterized by high theoretical plate values, reduced tailing, and precise, repeatable retention times [47].

Yan et al. have discussed the high-resolution capabilities of RP-HPLC in their paper. It is effective in resolving protein variants with post-translational modifications such as oxidation, glycosylation, and deamidation. This enables the accurate identification and analysis of impurity composition and mAb heterogeneity. As an example, RP-HPLC was used to separate mAb subdomains (which include light and heavy chains, Fab, and Fc) to study specific modifications such as pyroglutamic acid synthesis and its subsequent isomerization. RP-HPLC performs well in the structural and chemical assessment of mAbs. This method is general enough to be employed in a range of analytical applications [48, 49].

The research highlights the advantages of RP-HPLC in the separation of peptides, even if there is a single amino acid variation between them, and some even present as minor variations. It discusses the manner in which RP-HPLC separates two almost identical decapeptides differing owing to the presence of a single amino acid-serine or threonine. Due to this slight structural difference, which featured a methyl group on the threonine side chain, RP-HPLC managed to achieve baseline separation. In the method, 4.6 × 250 mm C18 wide-pore column was utilized along with acetonitrile with 0.08% TFA as mobile phase, and a gradient ranging from 0% to 35% created over a period of 73 min [50]. It was this variability in the hydrophobic bonds that existed between the peptides and the hydrophobic stationary phase of the column that enabled RP-HPLC to achieve this level of resolution [51].

Ambrogelly et al. described the use of RP-HPLC for analyzing mAb variants (IgG1 antibody), detailing its role in differentiation based on hydrophobicity in conjunction with MS and monitoring product-related variants during mAb production. During their study, the variant of mAb1, which eluted late in the chromatogram, was found to be caused by an increase in hydrophobicity as a result of the unprocessed secretion leader peptide at the N-terminus of one heavy chain. This variant was recognized as a post-peak (the behavior and characteristics of the signal detected after the main peak of an analyte has eluted from the chromatographic column), demonstrating the sensitivity and resolution of RP-HPLC in revealing subtle modifications in the primary structure. RP-HPLC with LC-ESI-MS helped to characterize different fractions in detail. The technique revealed that the control mAb and the variant differ in molecular mass, which was construed to indicate modifications on the primary sequence rather than on changes in glycosylation patterns. Based on peptide mapping and MS/MS showing precise sequence identification, truncated forms of the leader peptide bound to the heavy chain were identified using MS. RP-HPLC was used to monitor mAbs variants during the cell culture and purification stages. It gave high-throughput analysis and one more view of product quality. The study showcased the incorporation of RP-HPLC into purification workflows to isolate variants from preparative chromatographic steps like protein A and ion exchange chromatography. It further established the possibility for early identification and characterization of the variant to facilitate process development [52].

## 3 Applications of RP-HPLC

### *3.1 Role of RP-HPLC in Antibodies Therapeutics*

Characterization of antibody therapeutics using RP-HPLC involves studying the physicochemical properties of antibodies [52], separation of diverse types of antibodies [53], and identifying them [54] (Fig. 3). This is essentially very useful for the determination of the heterogeneity of the antibody sample, along with identification of unwanted modifications or degradation products if there are any [55]. Another important application of RP-HPLC is the analysis of antibody fragments obtained after their enzymatic digestion, which include the Fab and Fc regions. These fragments can be feasibly separated and analyzed using RP-HPLC [56].

RP-HPLC is also applied during the quality control of antibody production processes that ensure consistency and purity of the antibody product throughout the produced batch [57].

RP-HPLC techniques have made a significant contribution to the development of antibody-drug conjugates (ADCs). Such complex molecules are difficult to analyze, and the drug-to-antibody ratio becomes an important parameter that affects its efficacy as well as its safety [58]. RP-HPLC is the most frequently applied analytical technique for performing stability studies of antibody therapeutics. For example, the degradation of product formation with time can be tracked using stress conditions such as high temperature and pH alterations. This is necessary for determining

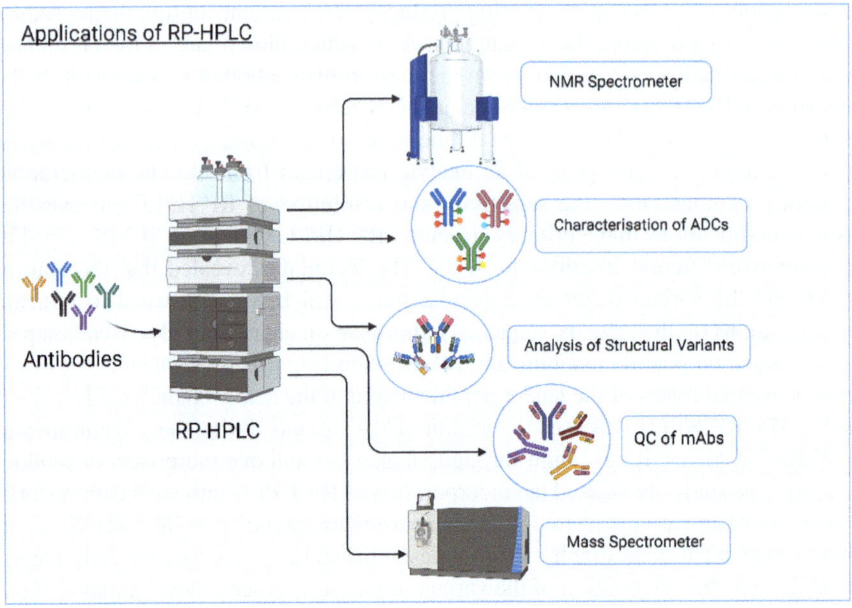


**Fig. 3** Application of RP-HPLC in association with therapeutic antibodies

the shelf life of antibodies and optimization of formulation strategies to obtain the maximum possible stability [59]. Major recent advancements in RP-HPLC technique have led to significant improvements in the field of therapeutic proteins and antibodies. UHPLC systems have brought down analytical times and increased resolution to allow resolution of complex mixtures of antibodies. The integration of RP-HPLC with MS has dramatically enhanced the identification and characterization capacity of molecular modifications of antibodies [60].

RP-HPLC, when used in combination with size-exclusion chromatography (SEC) analysis, provides an excellent explanation related to the nature of antibody aggregates and fragments [52]. This type of collaborative approach allows an enhanced understanding of the quality and stability of the antibodies. RP-HPLC has also contributed significantly to the formulation of biosimilar antibodies [61]. Such studies enable researchers to study similarity in physicochemical properties as well as post-translational modifications by studying the chromatographic profiles of developed antibody product versus reference products. This information is useful for bio-similarity and regulatory approval.

## 3.2 *Characterization of Antibody-Drug Conjugates (ADCs)*

Yiren Xu et al. in their research paper described a robust method for Drug-to-Antibody Ratio (DAR) [62]. Characterization of antibody-drug conjugates (ADCs) using reversed-phase high-performance liquid chromatography (RP-HPLC) was conducted for both reduced and intact forms of the ADCs. Reduction was achieved with guanidine hydrochloride and tris(2-carboxyethyl) phosphine (TCEP). The intact ADCs required minimal sample preparation. The mobile phase A consisted of 0.1% trifluoroacetic acid (TFA) in water, while mobile phase B was acetonitrile, employing a linear gradient from 30% to 45% B over 14 minutes. A Proteomix RP column was optimized specifically for resolving different ADC species. The column temperature was maintained at 80 °C, and detection was performed at an absorption wavelength of 280 nm, which indicated changes in hydrophobicity following the conjugation process. For low ADCs (i.e., ADCs with low DAR), DAR was determined by measuring peak areas of conjugated and unconjugated light chain (LC) and heavy chain (HC) of antibodies. For intact ADCs, the method resolved unconjugated (DAR0), singly conjugated (DAR1), and doubly conjugated (DAR2) species. The relative peak areas were used to calculate the total DAR. Results from RP-HPLC were cross-validated with LC-MS, showing comparable DAR values. The methodology revealed that conjugation gets nearly close to completion under optimal conditions, for example, overnight incubation with a 6:1 drug-to-antibody ratio. Conjugation kinetics studies indicated that 95% was completed within 2 hours. The results also demonstrated a clear indication of increased hydrophobicity in conjugated species, with variation depending on the conjugation site and the drug. RP-HPLC is also effective for the characterization of site-specific ADCs with varied conjugation sites and drugs. It provides complementary views for MS and hydrophobic interaction chromatography (HIC), which are challenged by certain properties of ADCs, such as extreme hydrophobicity [58]. Many research articles [62–65] highlight the importance of RP-HPLC in characterizing ADCs. They show that RP-HPLC is particularly useful for determining the drug-to-antibody ratio, which is essential for evaluating how effective and safe these therapies are. The studies suggest that using site-specific conjugation methods can create more uniform ADCs, potentially leading to better treatment outcomes compared to traditional approaches. Additionally, advancements in analytical techniques, including MS, enhance our understanding of ADC variability and improve quality control processes. Overall, these findings emphasize the need for reliable analytical methods to optimize ADC design and ensure their effectiveness in clinical settings.

### 3.3 Quality Control of mAbs

Rivera et al. have shown the utility of RP-HPLC in quality control checks during mAb production [66]. They showed that RP-HPLC contributes to ensuring the consistency and compliance with regulatory standards for antibody therapeutics. Using 0.1% TFA supplemented water and acetonitrile as mobile phases with a flow rate of 0.5 ml/min the characterization of the post-translational modification of antibodies was performed at around 80 °C. Upon changing the concentration of acidity modifier (TFA) between 0.09–0.11%, the authors observed a significant increase in retention, which is not ideal, especially if the method has a retention time specification.

The robustness of RP-HPLC methods for the analysis of intact protein can be assessed by optimizing the gradient method, focusing on its slope and range (in terms of change in solvent percentage over time) rather than simply its length (in terms of method duration). Adjusting the organic solvent concentration at the start and end of the method provides better control over resolution, as demonstrated by Rivera et al., in case of improved separation of hydrophilic variants with shallower gradients (e.g., 25–45% vs. 30–40% of 0.1% TFA in ACN). Additionally, wide-pore, superficially porous silica-based particles enable higher flow rates (up to 0.8 mL/min on a column with an internal diameter of 2.1-mm) without sacrificing efficiency, supporting shallow gradients at faster linear velocities (e.g., 0.4 cm/s). This approach balanced method sensitivity and minimized degradation while ensuring robust chromatographic performance [66]. Fekete et al. [67] investigated the influence of mobile phase and column temperature on the stability and recovery of commercially available, manufactured IgG molecules of intact IgGs and their fragments: heavy chain, Fc, and Fab using two newly developed wide-pore packing materials: Phenomenex Aeris Widepore (WP) columns feature 3.6 µm core-shell particles with large pores (≥300 Å), available in C18 and C4 phases, with thermal stability up to 90 °C (C18) and 60 °C (C4); whereas Waters Acquity BEH300 columns which use 1.7 µm fully porous particles with similar pore size, offering C18 and C4 bonding and thermal stability up to 80 °C. To minimize the secondary ionic interactions, the authors performed chromatographic separation in the WP column at elevated temperatures of around 90 °C. Trace amounts of a protic solvent such as *n*-butanol were added to the mobile phase at concentrations ranging from 1% to 3% to enhance the peak shape and antibody recovery [67]. Another research group later documented that the coupled-column configuration was useful in enhancing resolution for separations of mAbs [55]. In this report, they demonstrated the Proteomix column (5 µm, 1000 Å, 2.1 × 150 mm) from Sepax Technologies column coupling for improving the resolution of THIOMAB antibodies. The system operated at 0.5 mL/min and 80 °C, using 0.1% TFA in water (A) and 0.1% TFA in acetonitrile (B) as mobile phases, with absorbance monitored at 280 nm. A 20-minute method was used, starting with a 1-minute isocratic hold at 30% B, followed by a gradient to 45% B over 14 min, a 3-min wash at 95% B, and a 2-minute re-equilibration at

30% B. Their main conclusion was that the Xpress CF+ system is highly effective for producing site-specific ADCs with consistently high DARs, potent cell-killing activity, and excellent plasma stability. Additionally, a DAR 4 ADC was developed, showing enhanced therapeutic potential compared to DAR 2 counterparts.

### *3.4 Analysis of Structural Variants*

Glycoforms of antibodies are essentially antibodies attached to sugars such as sialic acids, core fucose, bisecting N-acetylglucosamine, and mannose residues. Glycan profiles influence mAb effector functions (e.g., ADCC, CDC), pharmacokinetics, half-life, and clearance rates. In RP-HPLC, hydrophobic properties of glycoforms of mAbs differentiate them as a function of their retention time. This capability is important for the assessment of how patterns of variations in the glycosylation process may affect therapeutic efficacy and stability. Wide-pore silica columns with short alkyl ligands such as C5 (pentyl), C8 (octyl), and C18 (octadecyl) enhance separation by providing unique selectivity tailored to the specific needs of protein and peptide analyses, while MS-compatible mobile phases i.e., TFA-water/acetonitrile, enable simultaneous detection and identification. It can also be used for stability testing as it identifies glycoforms prone to degradation, aiding stable formulation development. The optimized method parameters included Shim pack GIST C18 column, 2.1 × 150 mm, 3 μm column maintained at 80 °C, gradient elution, and addition of protic solvents like n-butanol for improvement of the peak shape under denaturing conditions. This method ensured the quality standards of mAbs along with their stable desired biological activity, which is vital in biopharmaceutical research and production [57].

A variety of approaches for characterizing therapeutic mAbs by RP-HPLC have been mentioned in the literature. The most common approach is complete digestion of these large glycoproteins using an enzyme like trypsin (also known as the peptide mapping method or bottom-up analysis). The analysis is performed on smaller peptides ranging from 0.5 to 3 kDa. However, the resulting chromatogram remains complex due to the large number of peptides generated from the digestion of the mAbs (mAb). Additionally, the digestion process itself can be time-consuming and inconvenient to carry out. Recently, there has been a new alternative that has been developed based on the limited proteolysis (LP) approach. This approach includes digesting the mAb with enzymes like papain to produce bigger fragments of approximately 50 kDa (Fc and Fab). Other enzymes such as pepsin, Endoproteinase Lys-C, or FabRICATOR® [a modified cysteine protease] may be used to create fragments of different sizes. Finally, the mAb can be analyzed in two ways. First, it can be examined in its intact form to check for impurity levels. Second, it can be reduced to produce two heavy chain fragments of about 50 kDa each and two light chain fragments of approximately 25 kDa each. This helps assess the integrity of the mAb when subjected to denaturing conditions [68].

### 3.5 *Integration with MS and NMR*

Combining reversed-phase high-performance liquid chromatography (RP-HPLC) with MS and nuclear magnetic resonance (NMR) has become a powerful strategy for studying therapeutic antibodies. This integrated approach allows researchers to thoroughly analyze mAbs, gaining valuable insights into their structure, purity, and overall effectiveness.

RP-HPLC is particularly effective at separating different forms of mAbs based on their hydrophobicity. For example, Rivera et al. showed that RP-HPLC can identify various product-related variants during the purification of mAbs, which can then be analyzed by MS to determine their molecular weights and any modifications. This separation is crucial for quality control in the production of biopharmaceuticals, ensuring that only properly processed antibodies are used in treatments [52].

MS complements RP-HPLC by providing detailed information about the molecular weight and structural features of the separated peptides or proteins. It helps identify important post-translational modifications, such as glycosylation patterns, which significantly affect the efficacy and safety of therapeutic antibodies. By combining MS with RP-HPLC, researchers can gain a deeper understanding of the diversity among mAbs, which is essential for their characterization and optimization [69].

NMR adds another layer to this analytical strategy by offering insights into the three-dimensional structure of antibodies. NMR provides information about how mAbs behave at an atomic level, helping researchers understand their mechanisms of action and stability. By integrating NMR with RP-HPLC and MS, scientists can obtain a comprehensive view of therapeutic antibodies, covering their purity, structural integrity, and functional characteristics [55].

In summary, the integration of RP-HPLC with MS and NMR creates a robust analytical framework for developing and ensuring the quality of therapeutic antibodies. This multi-faceted approach helps ensure that mAbs meet the strict regulatory standards necessary for safe and effective use in clinical settings.

## 4 Challenges and Limitations

### 4.1 *Selectivity Issues*

Selectivity is one of the key concerns in the protein purification process, especially for proteins with very similar properties, such as isoforms or variants. Isoforms or variants of closely related proteins have similar physicochemical properties, and therefore, their separation is challenging. Charge, hydrophobicity, and size are usually the characteristics of proteins used for separation. Bispecific antibodies (BsAbs) are biologics that can bind to two different antigens that making them ideal for the therapy of various diseases, primarily cancer. Their unique dual-targeting capability allows them to engage immune cells directly with tumor cells, enhancing the immune response against malignancies. BsAbs have shown significant promise in oncology, with applications in redirecting T cells to attack cancer cells and overcoming therapeutic resistance. Additionally, they are being explored for use in chronic inflammatory diseases, autoimmune disorders, and other conditions such as hemophilia and retinal diseases. Overall, the versatility of BsAbs positions them as a transformative approach in modern therapeutics across multiple medical fields [70, 71]. However, due to their complex structure, the byproducts created include mispaired chains, undesired fragments, and aggregates, which closely resemble the target protein, making purification even more difficult. Impurities such as host cell proteins, acidic proteins, or proteins sharing the same isoelectric points may be co-eluting with the target protein.

Acid elution conditions used in affinity chromatography might destabilize proteins, leading to fragmentation or aggregation. Furthermore, proteins with similar isoelectric points can co-elute during chromatographic processes, necessitating more sophisticated separation techniques to enhance selectivity [72].

### 4.2 Sample Preparation

Sample preparation is seen as the single most important aspect of protein analysis but also introduces various complications of its own. The sample source (such as type of tissue, cell culture), the physical attributes of proteins (size, charge), and their abundance impact the preparation strategy. For instance, high-abundance proteins mask the detection of low-abundance proteins, and thus depletion or enrichment may be necessary to ensure an appropriate sample representation [73]. Besides, the complexity of biological samples typically dictates multi-step protocol designs involving cell lysis, fractionation, and purification to minimize sample loss and enhance reproducibility in subsequent analyses. Each step would need to be optimized carefully to ensure the integrity and functionality of the proteins through the preparation process [74].

### 4.3 Operational Conditions

During the analysis of proteins, the conditions of operation can have a major impact on the results obtained. Temperature is an important factor in keeping the proteins stable, and sensitive proteins might become denatured or aggregate at higher temperatures [72]. In the case of chromatographic techniques, the mobile phase composition affects protein interactions with the stationary phase. Adjusting pH and ionic strength can change protein solubility and retention times on a chromatographic column to be optimal for separation. Therefore, these conditions must be optimized based on the specific properties of the target protein and the analytical method used [75, 76].

# 5 Conclusion

This chapter gives much insight into the contribution of RP-HPLC to the development and characterization of therapeutic antibodies. The importance of evolution in terms of promising and developing therapeutics within the area of mAbs, also including drug conjugates and bispecific antibodies, has been underscored.

RP-HPLC has emerged as one of the most crucial tools in the analytical field of the biopharmaceutical industry, especially for therapeutic antibodies. Its capability to separate and characterize the antibody through hydrophobic interaction provides an understanding of their physicochemical properties, which is critical to product quality and efficacy. The integration of advanced RP-HPLC systems into current leading-edge technologies, such as MS and UHPLC, has further solidified this analytical approach to determine further the accurate characterization in such complex antibody formulations. From here on, therapeutically better safer, and even effective agents can be supported while moving forward towards more enhanced therapeutic gains for a patient under specific treatment across different medical practice areas through RP-HPLC. This method, constantly improved, is sure to open new vistas in drug discovery and therapeutic applications, making this technique of importance in modern biotherapeutics.

## References

1. Bussel, J. B., Giulino, L., Lee, S., Patel, V. L., Sandborg, C., & Stiehm, E. R. (2007). Update on therapeutic monoclonal antibodies. *Current Problems in Pediatric and Adolescent Health Care, 37*. https://doi.org/10.1016/j.cppeds.2007.02.001
2. Kelley, B. The history and potential future of monoclonal antibody therapeutics development and manufacturing in four eras. *MAbs, 16*(1), 2373330. https://doi.org/10.1080/19420862.2024.2373330
3. Haurum, J. S. (2006). Recombinant polyclonal antibodies: The next generation of antibody therapeutics? *Drug Discovery Today, 11*(13), 655–660. https://doi.org/10.1016/j.drudis.2006.05.009
4. Kothari, M., et al. (2024). A comprehensive review of monoclonal antibodies in modern medicine: Tracing the evolution of a revolutionary therapeutic approach. *Cureus, 16*, 6. https://doi.org/10.7759/cureus.61983
5. Köhler, G., & Milstein, C. (1975). Continuous cultures of fused cells secreting antibody of predefined specificity. *Nature, 256*(5517), 495–497. https://doi.org/10.1038/256495a0
6. Morrison, S. L., Johnson, M. J., Herzenberg, L. A., & Oi, V. T. (1984). Chimeric human antibody molecules: Mouse antigen-binding domains with human constant region domains. *Proceedings of the National Academy of Sciences of the United States of America, 81*(21), 6851–6855. https://doi.org/10.1073/pnas.81.21.6851
7. McCafferty, J., Griffiths, A. D., Winter, G., & Chiswell, D. J. (1990). Phage antibodies: Filamentous phage displaying antibody variable domains. *Nature, 348*(6301), 552–554. https://doi.org/10.1038/348552a0
8. Modi, S., et al. (2022). Trastuzumab deruxtecan in previously treated HER2-low advanced breast cancer. *The New England Journal of Medicine, 387*(1), 9–20. https://doi.org/10.1056/NEJMoa2203690

9. Maloney, D. G., et al. (1997). IDEC-C2B8 (rituximab) anti-CD20 monoclonal antibody therapy in patients with relapsed low-grade non-Hodgkin's lymphoma. *Blood, 90*(6), 2188–2195. https://doi.org/10.1182/blood.V90.6.2188
10. Pettinato, M. C. (2021). Introduction to antibody-drug conjugates. *Antibodies, 10*(4), 42. https://doi.org/10.3390/antib10040042
11. Verma, S., et al. (2012). Trastuzumab emtansine for HER2-positive advanced breast cancer. *The New England Journal of Medicine, 367*(19), 1783–1791. https://doi.org/10.1056/NEJMoa1209124
12. Hunter, F. W., et al. (2020). Mechanisms of resistance to trastuzumab emtansine (T-DM1) in HER2-positive breast cancer. *British Journal of Cancer, 122*(5), 603–612. https://doi.org/10.1038/s41416-019-0635-y
13. Hu, S., et al. (2013). Comparison of the inhibition mechanisms of adalimumab and infliximab in treating tumor necrosis factor α-associated diseases from a molecular view. *The Journal of Biological Chemistry, 288*(38), 27059–27067. https://doi.org/10.1074/jbc.M113.491530
14. Emery, P., et al. (2008). IL-6 receptor inhibition with tocilizumab improves treatment outcomes in patients with rheumatoid arthritis refractory to anti-tumour necrosis factor biologicals: Results from a 24-week multicentre randomised placebo-controlled trial. *Annals of the Rheumatic Diseases, 67*(11), 1516–1523. https://doi.org/10.1136/ard.2008.092932
15. . (1998). Palivizumab, a humanized respiratory syncytial virus monoclonal antibody, reduces hospitalization from respiratory syncytial virus infection in high-risk infants. The IMpact-RSV Study Group. *Pediatrics, 102*(3 Pt 1), 531–537. https://publications.aap.org/pediatrics/article/102/3/531/65940/Palivizumab-a-Humanized-Respiratory-Syncytial?autologincheck=redirected
16. Kakinoki, Y., et al. (2022). Impact of antibody cocktail therapy combined with casirivimab and imdevimab on clinical outcome for patients with COVID-19 in a real-life setting: A single institute analysis. *International Journal of Infectious Diseases, 117*, 189–194. https://doi.org/10.1016/j.ijid.2022.01.067
17. Rahman, A., et al. (2023). Aducanumab for the treatment of Alzheimer's disease: A systematic review. *Psychogeriatrics, 23*(3), 512–522. https://doi.org/10.1111/psyg.12944
18. Burt, R., Warcel, D., & Fielding, A. K. (2018). Blinatumomab, a bispecific B-cell and T-cell engaging antibody, in the treatment of B-cell malignancies. *Human Vaccines & Immunotherapeutics, 15*(3), 594–602. https://doi.org/10.1080/21645515.2018.1540828
19. Yahiya Y. Syed. (2021). Amivantamab: First Approval. *Drugs*, 81, 1349–1353. https://link.springer.com/article/10.1007/s40265-021-01561-7
20. *Clinical review report: Emicizumab (Hemlibra): Hoffmann-La Roche Ltd: Indication: Bleeding prevention, hemophilia A.* in CADTH Common Drug Reviews. Ottawa (ON): Canadian Agency for Drugs and Technologies in Health, 2021. Retrieved December 18, 2024, from Available: https://www.ncbi.nlm.nih.gov/books/NBK570372/ [Online].
21. Cook, K., Freeke, J., & Barattini, V. (2013). Reversed-phase analysis of a monoclonal antibody on an Accucore C4-150 HPLC Column. Retrieved November 26, 2024, from https://www.semanticscholar.org/paper/Reversed-Phase-Analysis-of-a-Monoclonal-Antibody-on-Cook-Freeke/5f2b94faa6487d9bfb08d72ccaf3a3bee9c7965a [Online].
22. Mant, C. T., et al. (2007). HPLC analysis and purification of peptides. *Peptide Characterization and Application Protocols, 386*, 3–55. https://doi.org/10.1007/978-1-59745-430-8_1
23. Suh, C., Hoang, L., & Gjerde, D. T. (2010). Method and device for gravity flow chromatography. US20100170852A1, Jul. 08, Retrieved December 25, 2024, from. https://patents.google.com/patent/US20100170852A1/en [Online].
24. Majors, R. (2015). Historical developments in HPLC and UHPLC column technology: The past 25 years. vol. 33, pp. 818–840.
25. Giddings, J. C. (1965). *Dynamics of chromatography: Principles and theory* (1st ed., p. 340). CRC Press. https://doi.org/10.1201/9781315275871
26. Martin, A. J. P., & Synge, R. L. M. (1941). A new form of chromatogram employing two liquid phases. *The Biochemical Journal, 35*(12), 1358–1368.

27. Fogy, I., Schmid, E. R., & Huber, J. F. (1979). Determination of carbamate pesticides in fruits and vegetables by means of high pressure liquid chromatography (author's transl). *Zeitschrift für Lebensmittel-Untersuchung und -Forschung, 169*(6), 438–443. https://doi.org/10.1007/BF01126746
28. Ettre, L. S., & Zlatkis, A., Eds., Csaba Horváth. *Journal of Chromatography Library*, vol. 17, in 75 years of Chromatography a Historical Dialogue, vol. 17, Elsevier, 1979, pp. 151–158. doi: https://doi.org/10.1016/S0301-4770(08)60645-4.
29. Ettre, L. S. and Zlatkis, A., Eds., "J.F.K. Huber," in *Journal of Chromatography Library*, vol. 17, in 75 years of Chromatography a Historical Dialogue, vol. 17, Elsevier, 1979, pp. 159–166. https://doi.org/10.1016/S0301-4770(08)60646-6.
30. Ettre, L. S. (2005). Jim waters: The development of GPC and the first commercial HPLC instruments. *LC-GC North America, 23*(8), 752–758.
31. Tolstikov, V. V., Lommen, A., Nakanishi, K., Tanaka, N., & Fiehn, O. (2003). Monolithic silica-based capillary reversed-phase liquid chromatography/electrospray mass spectrometry for plant metabolomics. *Analytical Chemistry, 75*(23), 6737–6740. https://doi.org/10.1021/ac034716z
32. Shabir, G. A. (2010). Development and validation of a reversed-phase HPLC method for the determination of hydroxybenzene in a cream formulation. *Indian Journal of Pharmaceutical Sciences, 72*(3), 307–311. https://doi.org/10.4103/0250-474X.70475
33. Gupta, S., Kesarla, R., Chotai, N., & Omri, A. (2017). Development and validation of reversed-phase HPLC gradient method for the estimation of efavirenz in plasma. *PLoS One, 12*(5), e0174777. https://doi.org/10.1371/journal.pone.0174777
34. Bhalerao, A., Shelke, S., & Borkar, V. (2023). Advances, applications, and challenges in RP HPLC method development. *International Journal of Advanced Research in Science, Communication and Technology*, 111–123. https://doi.org/10.48175/IJARSCT-9017
35. Aguilar, M.-I. (2004). Reversed-phase high-performance liquid chromatography. In M.-I. Aguilar (Ed.), *HPLC of peptides and proteins: Methods and protocols* (pp. 9–22). Springer New York. https://doi.org/10.1385/1-59259-742-4:9
36. "Reverse phase HPLC basics for LC/MS." Retrieved December 24, 2024. https://www.ionsource.com/tutorial/chromatography/rphplc.htm
37. Moldoveanu, S., & David, V. (2022). Chapter 2—overview of HPLC instrumentation and its use. In S. Moldoveanu & V. David (Eds.), *Essentials in modern HPLC separations* (2nd ed., pp. 21–61). Elsevier. https://doi.org/10.1016/B978-0-323-91177-1.00015-6
38. Papadoyannis, I. N., & Samanidou, V. F. (2004). Validation of HPLC instrumentation. *Journal of Liquid Chromatography and Related Technologies, 27*(5), 753–783. https://doi.org/10.1081/JLC-120029697
39. Ruiz-Angel, M. J., Pous-Torres, S., Carda-Broch, S., & García-Alvarez-Coque, M. C. (2014). Performance of different C18 columns in reversed-phase liquid chromatography with hydro-organic and micellar-organic mobile phases. *Journal of Chromatography. A, 1344*, 76–82. https://doi.org/10.1016/j.chroma.2014.04.011
40. Pasch, H., & Trathnigg, B. (2013). Equipment and materials. In H. Pasch & B. Trathnigg (Eds.), *Multidimensional HPLC of polymers* (pp. 37–53). Springer. https://doi.org/10.1007/978-3-642-36080-0_4
41. "AdvanceBio RP-mAb | Wide Pore Reversed-Phase HPLC Column | Agilent." Retrieved December 24, 2024, from https://www.agilent.com/en/product/biopharma-hplc-analysis/intact-subunit-analysis/advancebio-rp-mab
42. Granite, "Diode Array Detector HPLC-DAD," SCION Instruments. Retrieved December 24, 2024, from https://scioninstruments.com/blog/diode-array-detector-hplc/
43. Zotou, A. (2012). An overview of recent advances in HPLC instrumentation. *Open Chemistry, 10*(3), 554–569. https://doi.org/10.2478/s11532-011-0161-0
44. Dauly, C., Perlman, D. H., Costello, C. E., & McComb, M. E. (2006). Protein separation and characterization by np-RP-HPLC followed by intact MALDI-TOF mass spectrometry and peptide mass mapping analyses. *Journal of Proteome Research, 5*(7), 1688–1700. https://doi.org/10.1021/pr060108z

45. "Characterization of wheat gluten proteins by HPLC and MALDI TOF mass spectrometry," ResearchGate. Retrieved December 24, 2024, from https://pubs.acs.org/doi/10.1016/j.jasms.2008.06.008
46. "Proteomics—an overview | ScienceDirect Topics." Retrieved December 25, 2024, from https://academic.oup.com/clinchem/article/62/1/48/5611769
47. Patil, S., Adhyapak, A., Shetti, P., & Gurao, R. (2023). Development of a robust and reliable reverse-phase high-performance liquid chromatography (RP-HPLC) method using analytical quality by design principles for the accurate determination of esculin in its bulk form. *Future Journal of Pharmaceutical Sciences, 9*(1), 69. https://doi.org/10.1186/s43094-023-00524-x
48. Yan, B., et al. (2007). Analysis of post-translational modifications in recombinant monoclonal antibody IgG1 by reversed-phase liquid chromatography/mass spectrometry. *Journal of Chromatography. A, 1164*(1), 153–161. https://doi.org/10.1016/j.chroma.2007.06.063
49. Alhazmi, H. A., & Albratty, M. (2023). Analytical techniques for the characterization and quantification of monoclonal antibodies. *Pharmaceuticals, 16*(2), 291. https://doi.org/10.3390/ph16020291
50. Carr D. HPLC columns [Internet]. A Guide to the Analysis and Purification of Proteins and Peptides by Reversed-Phase HPLC. p. 2–6. Available from: https://www.hplc.eu/Downloads/ACE_Guide_Peptides.pd.
51. Huang, Y., Pan, L., Zhao, L., Mant, C. T., Hodges, R. S., & Chen, Y. (2014). Structure guided RP-HPLC chromatography of diastereomeric α-helical peptide analogs substituted with single amino acid stereoisomers. *Biomedical Chromatography, 28*(4), 511–517. https://doi.org/10.1002/bmc.3061
52. Ambrogelly, A., et al. (2012). Characterization of antibody variants during process development. *MAbs, 4*(6), 701–709. https://doi.org/10.4161/mabs.21614
53. Haverick, M., Mengisen, S., Shameem, M., & Ambrogelly, A. (2014). Separation of mAbs molecular variants by analytical hydrophobic interaction chromatography HPLC: Overview and applications. *MAbs, 6*(4), 852–858. https://doi.org/10.4161/mabs.28693
54. Identification, characterization and control of a sequence variant in monoclonal antibody drug product: a case study | Scientific Reports. Retrieved December 25, 2024, from https://www.nature.com/articles/s41598-021-92338-1
55. HPLC for characterization and quality control of therapeutic monoclonal antibodies. Retrieved November 24, 2024, from https://www.chromatographyonline.com/view/hplc-characterization-and-quality-control-therapeutic-monoclonal-antibodies
56. Lin S, Avdalovic N, Liu X, Thermo Fisher Scientific. Fast analysis of therapeutic monoclonal antibody fragments using a supermacroporous, Reversed-Phase chromatography column. Application Note 21200. 2021.
57. Štimac, A., Kurtović, T., Pavlović, N., & Halassy, B. (2022). Development of improved high-performance liquid chromatography method for the determination of residual caprylic acid in formulations of human immunoglobulins. *Molecules, 27*(5), 1665. https://doi.org/10.3390/molecules27051665
58. Xu, Y., et al. (2016). RP-HPLC DAR characterization of site-specific antibody drug conjugates produced in a cell-free expression system. *Organic Process Research and Development, 20*(6), 1034–1043. https://doi.org/10.1021/acs.oprd.6b00072
59. Dillon, T. M., Bondarenko, P. V., Rehder, D. S., Pipes, G. D., Kleemann, G. R., & Ricci, M. S. (2006). Optimization of a reversed-phase high-performance liquid chromatography/mass spectrometry method for characterizing recombinant antibody heterogeneity and stability. *Journal of Chromatography. A, 1120*(1), 112–120. https://doi.org/10.1016/j.chroma.2006.01.016
60. Gritti, F., Meyyappan, S., Leveille, W., & Hill, J. (2022). Improved performance of ultrahigh-pressure liquid chromatography–mass spectrometry (UHPLC–MS) hyphenated systems. vol. 35, pp. 285–293.
61. Martínez-Ortega, A., Herrera, A., Salmerón-García, A., Cabeza, J., Cuadros-Rodríguez, L., & Navas, N. (2018). Validated reverse phase HPLC diode array method for the quanti-

fication of intact bevacizumab, infliximab and trastuzumab for long-term stability study. *International Journal of Biological Macromolecules, 116*, 993–1003. https://doi.org/10.1016/j.ijbiomac.2018.05.142
62. RP-HPLC DAR characterization of site-specific antibody drug conjugates produced in a cell-free expression system | Organic Process Research & Development. Retrieved December 24, 2024, from https://pubs.acs.org/doi/full/10.1021/acs.oprd.6b00072
63. Wakankar, A., Chen, Y., Gokarn, Y., & Jacobson, F. S. (2011). Analytical methods for physicochemical characterization of antibody drug conjugates. *MAbs, 3*(2), 161–172. https://doi.org/10.4161/mabs.3.2.14960
64. Tang, Y., et al. (2017). Real-time analysis on drug-antibody ratio of antibody-drug conjugates for synthesis, process optimization, and quality control. *Scientific Reports, 7*(1), 7763. https://doi.org/10.1038/s41598-017-08151-2
65. In-situ reverse phased HPLC analysis of intact antibody-drug conjugates. ResearchGate. Retrieved December 24, 2024, from https://link.springer.com/article/10.2116/analsci.20P424
66. Rivera, B., Lebedev, I., & Eichman, C. (2021). Method robustness and reproducibility for monoclonal antibody purity analysis. 17, 10–16.
67. Fekete, S., Rudaz, S., Veuthey, J.-L., & Guillarme, D. (2012). Impact of mobile phase temperature on recovery and stability of monoclonal antibodies using recent reversed-phase stationary phases. *Journal of Separation Science, 35*(22), 3113–3123. https://doi.org/10.1002/jssc.201200297
68. Fekete, S., & Guillarme, D. (2012). Reversed-phase liquid chromatography for the analysis of therapeutic proteins and recombinant monoclonal antibodies. 25, 540–550.
69. LC-MS-Based Protein Therapeutics Analysis—IN. Retrieved December 24, 2024, from https://www.thermofisher.com/in/en/home/industrial/chromatography/chromatography-learning-center/chromatography-consumables-resources/biopharmaceutical-analysis-u-hplc/lc-ms-based-protein-therapeutics-analysis-columns.html
70. Schofield, D. Applications of bispecific antibodies in therapeutics—evitria. Retrieved December 24, 2024, from https://www.evitria.com/journal/bispecific-antibodies/applications-of-bispecific-antibodies-in-therapeutics/
71. Fan, G., Wang, Z., Hao, M., & Li, J. (2015). Bispecific antibodies and their applications. *Journal of Hematology & Oncology, 8*, 130. https://doi.org/10.1186/s13045-015-0227-0
72. Tang, S., Tao, J., & Li, Y. (2023). Challenges and solutions for the downstream purification of therapeutic proteins. *Antibody Therapeutics, 7*(1), 1–12. https://doi.org/10.1093/abt/tbad028
73. Knecht, S., et al. (2023). An introduction to analytical challenges, approaches, and applications in mass spectrometry–based secretomics. *Molecular & Cellular Proteomics, 22*(9), 100636. https://doi.org/10.1016/j.mcpro.2023.100636
74. Feist, P., & Hummon, A. B. (2015). Proteomic challenges: Sample preparation techniques for microgram-quantity protein analysis from biological samples. *International Journal of Molecular Sciences, 16*(2), 3537–3563. https://doi.org/10.3390/ijms16023537
75. Tips, tricks, and troubleshooting for separations of biomolecules, Part I: Contemporary reversed-phase protein separations. Retrieved November 25, 2024, from, https://www.chromatographyonline.com/view/tips-tricks-and-troubleshooting-separations-biomolecules-part-i-contemporary-reversed-phase-protein
76. Challenges and solutions in purifying recombinant proteins. Retrieved November 25, 2024, from https://blog.mblintl.com/challenges-and-solutions-in-purifying-recombinant-proteins

# Collagen for Health

**Doyal Dey, Samriti Balaji Mudaliar, Nirmal Mazumder, and Alevoor Srinivas Bharath Prasad**

**Abstract** A key structural protein in the extracellular matrix, collagen is essential for preserving the integrity and functionality of tissues. This chapter explores its genesis, classification, hierarchical structure, and biomedical importance. Collagen-based biomaterials (CBBs) have transformed various therapeutic applications, including drug delivery systems, orthopedic implants, and wound healing. The study also briefly describes some autoimmune and hereditary conditions, including Systemic sclerosis, Ehlers-Danlos syndrome, Alport syndrome, etc., connected to collagen disorders. Collagen-based therapeutics continue to face obstacles despite their vast biological inventions. Modern fabrication methods and biomaterial engineering advancements are crucial for maximizing collagen's clinical applications in regenerative medicine and diagnostics.

**Keywords** Collagen · Extracellular matrix · Collagen-based biomaterials · Collagen disorders · Biomedical therapy

## 1 Introduction

Collagen is one of the most abundant structural proteins in the human body. It is extensively present in the extracellular matrix (ECM) of most eukaryotic organisms [1]. The typical structure of a collagen molecule comprises three polypeptide chains, arranged in a triple helix. These molecules further form collagen fibrils and fibers

---

D. Dey · S. B. Mudaliar · A. S. B. Prasad (✉)
Department of Public Health Genomics, Manipal School of Life Sciences (MSLS), Manipal Academy of Higher Education (MAHE), Manipal, Karnataka, India
e-mail: bharath.prasad@manipal.edu

N. Mazumder
Department of Biophysics, Manipal School of Life Sciences (MSLS), Manipal Academy of Higher Education (MAHE), Manipal, Karnataka, India



N. Mazumder et al. (eds.), *Advanced Biophysical Techniques in Biosciences*, Lecture Notes in Nanoscale Science and Technology 7,
https://doi.org/10.1007/978-3-031-94551-9_9

via self-assembly. There are 29 types of collagens that differ from each other in their structure, composition, and morphology. The polypeptide chains for each type of collagen are transcribed from specific *COL* genes in the nucleus and translated inside ribosomes. They then undergo post-translational modifications in the rough endoplasmic reticulum, after which they form the coiled-coil triple helical procollagen. The procollagen molecules then move to the Golgi apparatus to be packaged into secretory vesicles and transported into the extracellular space, where they undergo maturation to form tropocollagen. These tropocollagen molecules then undergo cross-linking, giving rise to different supramolecular morphologies such as networks and fibrils [2, 3]. The structural organization of collagen from polypeptide chains to fibers has been schematically represented in Fig. 1. Collagens have been categorized into seven types based on their composition, distribution, and supramolecular assembly, namely, fibril-forming collagens, fibril-associated collagens with interrupted triple helices (FACITs), network-forming collagens, transmembrane collagens, endostatin-producing collagens, anchoring fibrils, and beaded-filament forming collagens [4].

As a major component of the ECM, collagen forms the basic structural framework of tissues and is thereby responsible for their structural and physico-mechanical properties. Collagen also plays an important role in cell adhesion, cell proliferation, cell differentiation, tissue scaffolding, and tissue morphogenesis. Due to their high stability, fibrillar collagens are mainly responsible for maintaining the structural integrity of body tissues [6]. The occurrence of debilitating disorders, including Fuchs endothelial corneal dystrophy (FECD), dystrophic epidermolysis bullosa, as well as orthopedic conditions like osteoarthritis and osteochondrodysplasia due to abnormalities in the expression and metabolism of collagen, indicates its importance in the healthy functioning of the human body [7, 8]. Recent research trends show the growing interest of the biomedical community in the structural properties of collagen. This has resulted in mounting evidence of the possible correlation between the structural organization of collagen and pathological conditions of the tissue. Furthermore, the heterogeneity in clinical manifestations of certain diseases like Ehlers-Danlos syndrome (EDS) and osteogenesis imperfecta has been attributed to the underlying genetic variation due to mutations in different genes, such as *COL3A1* and *COL11A2,* which encode different types of collagens. Thus, determining the mutated gene will allow more efficient prognosis and management of such collagen-related disorders [9].

## 2 Collagen in Biomedical Therapy

The term 'collagen' originates from the Greek term for glue i.e. 'kolla', since the first known use of collagen was as a source of glue by the Egyptians in 2000 B.C., who obtained the protein by boiling collagen-rich animal tissues [10]. However, its introduction into biomedical therapy occurred centuries later when Joseph Lister, now known as the father of modern surgery, used it to develop a suture material

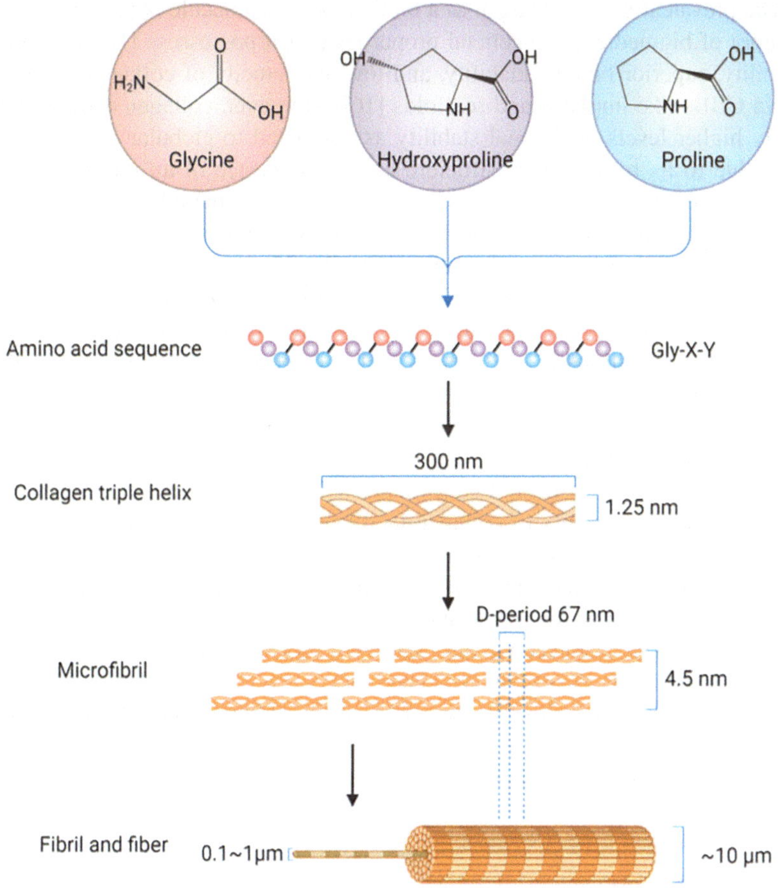


**Fig. 1** Hierarchical structure and organization of collagen [5]

called the 'catgut' in 1881 [11]. Collagen was later used as a matrix for cell culture in 1956, followed by the use of solubilized bovine-derived collagen films for wound healing in 1959 [12]. A few decades later, in 1993, a collagen-based bone implant became the first collagen-based biomaterial (CBB) to be approved for human use by the Food and Drug Administration (FDA) [8]. Currently, several CBBs, including membranes, hydrogels, sponges, microstructures, nanospheres, oral supplements, wound dressings, and injectable solutions, are commercially used in orthopedic medicine, oral therapy, wound healing, drug delivery, tissue repair, and nerve regeneration [13]. Collagen is also a good biomarker for pathological conditions due to its wide distribution and ease of detection in blood and tissue samples. The metabolic products of collagen are employed in the detection and monitoring of a number of diseases, such as osteoarthritis, ankylosing spondylitis, and even certain types of cancer, including lung cancer and breast cancer [14, 15].

The prevalent use of collagen as a biotherapeutic molecule is mainly due to the plethora of biomedically beneficial properties that it possesses. The high biocompatibility, superior biodegradability, and low antigenicity of collagen facilitate its use in CBBs like implants and scaffolds [16]. Moreover, collagen exhibits significantly higher levels of thermal stability as compared to globular proteins, potentially due to its triple helical structure and the presence of an internal ladder of hydrogen bonds. Fibrillar collagens in particular possess incredible mechanical stability, tensile strength, and load-bearing capacity due to the presence of uninterrupted glycine-proline-hydroxyproline (Gly-Pro-Hyp) repeat sequences in their polypeptide chains. This provides rigidity and structural integrity to collagen-based scaffolds, thereby enabling their use in bone regeneration and treating skeletal injuries [17, 18]. Additionally, collagen is highly osteoconductive in nature and promotes bone formation, which aids in the faster recovery of damaged bones [19]. Since collagen promotes cell proliferation and cell migration, it is widely used to accelerate wound healing. It is also used to augment soft tissue integration and angiogenesis in the treatment of gingival and subepithelial wounds [20]. The porous structure and permeability of collagen facilitate its use in controlled drug-delivery systems and as a carrier of bioactive molecules like growth factors and stem cells [21, 22]. While the excellent biotherapeutic characteristics of collagen render it invaluable in biomedicine, most applications require CBBs with highly specific parameters that are achieved via the addition of other biopolymers like chitosan and silk fibroin as well as the use of advanced fabrication strategies such as 3D bioprinting. Precise control over the physical and chemical parameters of CBBs, such as hydrophobicity and porosity, is of utmost importance for their efficient functionality and successful implementation in clinical therapies [23, 24].

## 3 Collagen Disorders

The essential structural elements of all connective tissues, collagens, are also found in the interstitial tissues of almost every parenchymal organ, where they support the rigidity and structural integrity of the tissue. Inborn metabolic errors that impact the structure or metabolism of collagen or autoimmune conditions and some other factors are the causes of collagen disorders. These conditions are still incurable at the biological level, despite advances in our understanding of the structure and manufacturing of collagen. Here, some of the heritable and autoimmune disorders of collagen are described briefly.

## 3.1 Inheritable Disorders of Collagen

1. Alport syndrome is a widespread hereditary condition affecting basement membranes, namely those of glomeruli that contain type IV collagen. The gene on the X chromosome undergoes alterations. An inherited deficiency in typical X-linked Alport syndrome affects the α-5 chain of the collagen type IV collagen gene (COL4A5). Renal impairment, hypertension, abnormalities of the lens and hearing, hematuria, and proteinuria are its most common symptoms. When mutations damage collagen IV, the bound epithelium malfunctions and organ damage ensues [25].
2. Ehlers-Danlos syndrome (EDS) is a connective-tissue illness defined by skin hyperextensibility, tissue fragility, and joint excessive mobility. Mutations in genes such as COL1A1 and COL1A2 impair collagen production, lowering tissue strength. Weak gingiva, pulp stones, mucosal tears, enamel hypoplasia, early-onset periodontitis, and short, malformed roots are all significant oral traits. Patients may exhibit the Gorlin sign (tongue-to-nose contact) and have temporomandibular joint subluxations [26, 27].
3. Inherited epidermolysis bullosa, a class of uncommon genetically transmitted diseases, has many modes of transmission and ranges in severity. Vesicles and bullae that grow on the skin and mucosal membranes are the hallmarks of this multiracial illness. The four forms of epidermolysis bullosa are hemidesmosomal, junctional, simplex, and dystrophic. Dominant simplex type and junctional form have been attributed to specific mutations in the K5 or K14 genes and genes coding for laminin, respectively. Kind VII gene mutations are linked to the dystrophic kind. Mutations in genes linked to several hemidesmosomal attachment proteins, including pleptin, type XVII collagen, and α6α4 integrin, are indicative of the hemidesmosomal type [28, 29].
4. Osteogenesis imperfecta is a set of hereditary illnesses characterized by faulty collagen production or structure, primarily due to alterations in COL1A1 or COL1A2 genes. This affects tissues high in type I collagen, resulting in weak bones, growth deficits, joint hypermobility, hearing loss, blue sclerae, and dentinogenesis imperfecta. Bones have a thin cortical, delicate trabeculation, and osteoporosis, with fractures mending but frequently producing excessive callus [30, 31].
5. Stickler syndrome, an autosomal dominant genetic condition characterized by early osteoarthritis, retinal degeneration, loss of hearing, and orofacial deformities. Variations in the COL2A1, COL11A1, and COL11A2 genes cause the disorder by affecting type II and XI collagen, which are required for connective tissues [32].

## 3.2 Autoimmune Disorders of Collagen

1. Systemic lupus erythematosus (SLE) is a complex autoimmune connective tissue disease that can have both systemic and cutaneous effects on the oral mucosa. Fatigue, malaise, arthritis, weight loss, and fever are typical symptoms. The nose and malar region are covered in a distinctive butterfly-shaped rash. Cardiac involvement is common, accompanied by pericarditis and Libman-Sacks endocarditis (warty vegetations on heart valves). Stomatodynia, candidiasis, xerostomia, erythema, ulceration, discomfort, erythema, hyperkeratosis, and periodontal disease are examples of oral symptoms [28, 33].
2. A connective tissue disease called systemic sclerosis (scleroderma) is characterized by fibrosis of the skin, mucosa, visceral organs, and blood vessels as a result of an overabundance of collagen. Particularly in the distal extremities, the most obvious sign is skin thickening, which starts with pitting edema and progresses to tightening and hardening. The initial indication is frequently Raynaud's phenomenon. Mandibular bone resorption, purse-string furrows around the lips, facial skin hardening, and collagen deposition in the tongue and esophagus are examples of oral manifestations that result in a rigid, immobile tongue and dysphagia [28, 34].
3. Oral submucous fibrosis (OSMF) is a chronic and premalignant condition of the oral mucosa that is currently thought to have an autoimmune basis because of the autoantibodies detected. Due to an imbalance between collagen production and degradation, it is regarded as a collagen-metabolic disorder. Juxta-epithelial inflammation, epithelial atrophy, and fibroelastic alterations in lamina propria are the hallmarks of OSMF, which results in restricted mouth opening (trismus), trouble in speaking, eating, and maintaining oral hygiene. The disorder is caused by abnormal collagen accumulation in the oral tissues, which leads to the formation of dense fibrous bands [35, 36].

# 4 Conclusions

Collagen is a structurally diverse protein with several unique properties, including mechanical stability, biocompatibility, and flexibility that lend themselves to several biomedical applications, including diagnostics and therapeutics. CBBs are widely used in various sectors of medicine and healthcare, ranging from stomatology to neural tissue engineering. This section explores the different aspects of using collagen in healthcare, including an overview of its biotherapeutic properties, its extraction and purification, and the synthesis of different CBBs. It sheds light on the advantages and limitations of collagen as compared to other natural biopolymers like alginate and hyaluronic acid as well as synthetic alternatives such as polyethylene glycol (PEG) and polyglycolic acid (PGA). The forthcoming chapters delve into the clinical applications of collagen-based biomarkers and CBBs in different biomedical sectors, including orthopedics, wound healing, and tissue engineering. Finally, this chapter highlights the need for developing more efficient collagen-based therapies by integrating modern technology in order to overcome the current challenges and facilitate the widespread commercial use of collagen in clinical diagnostics, disease monitoring, and therapeutic biomaterials.

## References


1. Kular, J. K., Basu, S., & Sharma, R. I. (2014). The extracellular matrix: Structure, composition, age-related differences, tools for analysis and applications for tissue engineering. *Journal of Tissue Engineering, 5*, 2041731414557112. https://doi.org/10.1177/2041731414557112
2. Canty, E. G., & Kadler, K. E. (2005). Procollagen trafficking, processing and fibrillogenesis. *Journal of Cell Science, 118*(Pt 7), 1341–1353. https://doi.org/10.1242/jcs.01731
3. Alcaide-Ruggiero, L., Molina-Hernández, V., Granados, M. M., & Domínguez, J. M. (2021). Main and minor types of collagens in the articular cartilage: The role of collagens in repair tissue evaluation in chondral defects. *International Journal of Molecular Sciences, 22*(24), 13329. https://doi.org/10.3390/ijms222413329
4. Onursal, C., Dick, E., Angelidis, I., Schiller, H. B., & Staab-Weijnitz, C. A. (2021). Collagen biosynthesis, processing, and maturation in lung ageing. *Frontiers in Medicine, 8*, 593874. https://doi.org/10.3389/fmed.2021.593874
5. Fan, L., Ren, Y., Emmert, S., Vučković, I., Stojanovic, S., Najman, S., Schnettler, R., Barbeck, M., Schenke-Layland, K., & Xiong, X. (2023). The use of collagen-based materials in bone tissue engineering. *International Journal of Molecular Sciences, 24*(4), 3744. https://doi.org/10.3390/ijms24043744
6. Bella, J., & Hulmes, D. J. (2017). Fibrillar collagens. *Sub-Cellular Biochemistry, 82*, 457–490. https://doi.org/10.1007/978-3-319-49674-0_14
7. Wu, M., Cronin, K., & Crane, J. S. (2023). Biochemistry, collagen synthesis. In *StatPearls*. StatPearls Publishing. https://www.ncbi.nlm.nih.gov/books/NBK507709/
8. Rezvani Ghomi, E., Nourbakhsh, N., Akbari Kenari, M., Zare, M., & Ramakrishna, S. (2021). Collagen-based biomaterials for biomedical applications. *Journal of Biomedical Materials Research. Part B, Applied Biomaterials, 109*(12), 1986–1999. https://doi.org/10.1002/jbm.b.34881
9. Byers, P. H. (2019). Disorders of collagen biosynthesis and structure. In D. L. Valle, S. Antonarakis, A. Ballabio, A. L. Beaudet, & G. A. Mitchell (Eds.), *The online metabolic and molecular bases of inherited disease*. McGraw-Hill Education. https://ommbid.mhmedical.com/content.aspx?bookid=2709&sectionid=225892461
10. Vesentini, S., Redaelli, A., & Gautieri, A. (2013). Nanomechanics of collagen microfibrils. *Muscles, Ligaments and Tendons Journal, 3*(1), 23–34. https://doi.org/10.11138/mltj/2013.3.1.023
11. Lister, J. (1881). An address on the catgut ligature. *British Medical Journal, 1*(1050), 219–221. https://doi.org/10.1136/bmj.1.1050.219
12. Pappas, A. M., & Hyatt, G. W. (1960). The evaluation of collagen film applied to skin defects in mice. *Surgical Forum, 10*, 844–846.
13. Zheng, M., Wang, X., Chen, Y., Yue, O., Bai, Z., Cui, B., Jiang, H., & Liu, X. (2023). A review of recent Progress on collagen-based biomaterials. *Advanced Healthcare Materials, 12*(16), e2202042. https://doi.org/10.1002/adhm.202202042
14. Wahyudi, H., Reynolds, A. A., Li, Y., Owen, S. C., & Yu, S. M. (2016). Targeting collagen for diagnostic imaging and therapeutic delivery. *Journal of Controlled Release: Official Journal of the Controlled Release Society, 240*, 323–331. https://doi.org/10.1016/j.jconrel.2016.01.007

15. Song, K., Yu, Z., Zu, X., Li, G., Hu, Z., & Xue, Y. (2022). Collagen remodeling along cancer progression providing a novel opportunity for cancer diagnosis and treatment. *International Journal of Molecular Sciences, 23*(18), 10509. https://doi.org/10.3390/ijms231810509
16. Meyer, M. (2019). Processing of collagen based biomaterials and the resulting materials properties. *Biomedical Engineering Online, 18*(1), 24. https://doi.org/10.1186/s12938-019-0647-0
17. Shoulders, M. D., & Raines, R. T. (2009). Collagen structure and stability. *Annual Review of Biochemistry, 78*, 929–958. https://doi.org/10.1146/annurev.biochem.77.032207.120833
18. Sorushanova, A., Delgado, L. M., Wu, Z., Shologu, N., Kshirsagar, A., Raghunath, R., Mullen, A. M., Bayon, Y., Pandit, A., Raghunath, M., & Zeugolis, D. I. (2019). The collagen suprafamily: From biosynthesis to advanced biomaterial development. *Advanced Materials (Deerfield Beach, Fla.), 31*(1), e1801651. https://doi.org/10.1002/adma.201801651
19. Kuchler, U., Rybaczek, T., Dobask, T., Heimel, P., Tangl, S., Klehm, J., Menzel, M., & Gruber, R. (2018). Bone-conditioned medium modulates the osteoconductive properties of collagen membranes in a rat calvaria defect model. *Clinical Oral Implants Research, 29*(4), 381–388. https://doi.org/10.1111/clr.13133
20. Caballé-Serrano, J., Zhang, S., Ferrantino, L., Simion, M., Chappuis, V., & Bosshardt, D. D. (2019). Tissue response to a porous collagen matrix used for soft tissue augmentation. *Materials, 12*(22), 3721. https://doi.org/10.3390/ma12223721
21. Ferreira, A. M., Gentile, P., Chiono, V., & Ciardelli, G. (2012). Collagen for bone tissue regeneration. *Acta Biomaterialia, 8*(9), 3191–3200. https://doi.org/10.1016/j.actbio.2012.06.014
22. Lo, S., & Fauzi, M. B. (2021). Current update of collagen nanomaterials—Fabrication, characterisation and its applications: A review. *Pharmaceutics, 13*(3), 316. https://doi.org/10.3390/pharmaceutics13030316
23. Dong, C., & Lv, Y. (2016). Application of collagen scaffold in tissue engineering: Recent advances and new perspectives. *Polymers, 8*(2), 42. https://doi.org/10.3390/polym8020042
24. Patil, V. A., & Masters, K. S. (2020). Engineered collagen matrices. *Bioengineering, 7*(4), 163. https://doi.org/10.3390/bioengineering7040163
25. Sessa, A., & Meroni, M. (2001). *Alport syndrome*. Orphanetencyclop. Available at: http://www.orpha.net/data/patho/GB/uk-alport.pdf..\
26. Yen, J. L., Lin, S. P., Chen, M. R., & Niu, D. M. (2006). Clinical features of Ehlers-Danlos syndrome. *Journal of the Formosan Medical Association = Taiwan yi zhi, 105*(6), 475–480. https://doi.org/10.1016/S0929-6646(09)60187-X
27. Létourneau, Y., Pérusse, R., & Buithieu, H. (2001). Oral manifestations of Ehlers-Danlos syndrome. *Journal (Canadian Dental Association), 67*(6), 330–334.
28. Neville, B. W., Damm, D. D., Allen, C. M., & Bouquot, J. E. (2009). *Oral and maxillofacial pathology (3rd edition)* (pp. 794–802). WB Saunders.
29. Al-Jobeir, A. (2006). Hereditary epidermolysis bullosa: Report of two cases. *Saudi Dental Journal, 18*(3), 155–161.
30. Kierszenbaum, A. L., & Abraham, L. (2002). *Histology and cell biology—An introduction to pathology* (pp. 101–103). Mosby.
31. Gupte, T., Iyer, V., Damle, S. G., Malik, N., & Halbe, A. (2006). Osteogenesis imperfecta. *Journal of the Indian Society of Pedodontics and Preventive Dentistry, 24*(Suppl 1), S44–S46.
32. Rose, P. S., Ahn, N. U., Levy, H. P., Magid, D., Davis, J., Liberfarb, R. M., Sponseller, P. D., & Francomano, C. A. (2001). The hip in stickler syndrome. *Journal of Pediatric Orthopedics, 21*(5), 657–663.
33. Lourenço, S. V., de Carvalho, F. R., Boggio, P., Sotto, M. N., Vilela, M. A., Rivitti, E. A., & Nico, M. M. (2007). Lupus erythematosus: Clinical and histopathological study of oral manifestations and immunohistochemical profile of the inflammatory infiltrate. *Journal of Cutaneous Pathology, 34*(7), 558–564. https://doi.org/10.1111/j.1600-0560.2006.00652.x
34. Cazal, C., Sobral, A. P., Neves, R. F., Freire Filho, F. W., Cardoso, A. B., & da Silveira, M. M. (2008). Oral complaints in progressive systemic sclerosis: Two cases report. *Medicina oral, patologia oral y cirugia bucal, 13*(2), E114–E118.

35. Jha, V. K., Kandula, S., Ningappa Chinnannavar, S., Rout, P., Mishra, S., & Bajoria, A. A. (2019). Oral submucous fibrosis: Correlation of clinical grading to various habit factors. *Journal of International Society of Preventive & Community Dentistry, 9*(4), 363–371. https://doi.org/10.4103/jispcd.JISPCD_92_19
36. Rajalalitha, P., & Vali, S. (2005). Molecular pathogenesis of oral submucous fibrosis—A collagen metabolic disorder. *Journal of Oral Pathology & Medicine, 34*(6), 321–328. https://doi.org/10.1111/j.1600-0714.2005.00325.x

# Polarization-Resolved Second Harmonic Generation Microscopy for Cancer Diagnosis

**Sasha MacArthur, MacAulay Harvey, Richard Cisek, and Danielle Tokarz**

**Abstract** Polarization-resolved second harmonic generation microscopy (PSHG) is becoming an increasingly popular and important technique for differentiating biological tissues based on their disease state, as well as for identifying components related to their ultrastructure that may help in understanding disease progression. In particular, PSHG has shown promise in cancer diagnosis, where PSHG can be used to differentiate the disorganized ultrastructure of collagen from within the extracellular matrix of cancerous tissues from normal tissues. Here, PSHG microscopy techniques and their application in cancer diagnosis over the past 5 years (2020–2024) are described. The PSHG techniques reviewed include polarization-in SHG, a technique where the laser linear polarization is modulated, widefield polarization-in SHG as well as techniques in which the laser polarization is modulated and the polarization state of the outgoing SHG signals is detected, including polarization-in, polarization-out SHG and widefield reduced double Stokes-Mueller SHG. This chapter concludes by describing advances of PSHG to enable the possibility of automated cancer diagnosis and in vivo tissue studies.

**Keywords** Second harmonic generation microscopy · Cancer diagnosis · Collagen ultrastructure · Stokes-Mueller polarization

## 1 Introduction

Second harmonic generation microscopy (SHG) is a nonlinear optical microscopy technique with numerous advantages for biological imaging, including endogenous contrast within samples containing SHG emitters and thus eliminating the need for

S. MacArthur · M. Harvey · R. Cisek · D. Tokarz (✉)
Department of Chemistry, Saint Mary's University, Halifax, NS, Canada
e-mail: Danielle.Tokarz@smu.ca



N. Mazumder et al. (eds.), *Advanced Biophysical Techniques in Biosciences*, Lecture Notes in Nanoscale Science and Technology 7,
https://doi.org/10.1007/978-3-031-94551-9_10

dyes or sample modifications that may introduce artefacts to the sample. SHG also allows intrinsic optical sectioning capabilities without the use of confocal pinholes and the ability to image deeply in highly scattering tissues. Additionally, SHG imaging undergoes a lack of photobleaching or production of phototoxic products because SHG is an energy-conserving process, and any photobleaching that occurs is typically due to an endogenous absorber. Therefore, photobleaching can be alleviated by choosing another laser wavelength for SHG imaging.

Since SHG arises from endogenous SHG emitters in biological structures, changes in the structure of the biological sample can be deduced from measurements of the SHG signal. In particular, applying a polarization-resolved SHG microscopy (PSHG) technique can determine quantitative information related to the molecular ultrastructure. As a result, PSHG has received significant attention due to its ability to provide pixel-by-pixel information about SHG emitters' local molecular structure and hierarchical organization with microscopic spatial resolution.

Recently, PSHG has been used by several research groups to quantify changes in the ultrastructure of the well-known SHG emitter, collagen, and its organization within the extracellular matrix (ECM) of organ tissue during cancer progression and response to treatment. This research has involved a diverse range of sample types, PSHG techniques, and image analysis frameworks. Review papers by Mazumder, et al. [1], James et al. [2], and Cisek et al. [3], and a book chapter by Cisek, et al. [4] have meticulously reviewed PSHG papers for collagen imaging leading up to the year 2020. Therefore, we will review work on PSHG imaging for cancer research from the past 5 years (2020–2024) to provide a brief overview of the current state of the field as well as possible future directions.

## 2 Second Harmonic Generation Microscopy Techniques

### *2.1 Introduction to SHG*

Second harmonic generation (SHG) refers to the frequency doubling that occurs when intense laser radiation interacts with a non-centrosymmetric medium through a spontaneous nonlinear scattering process where two photons are converted into one. In order to produce SHG signal, a sufficiently high photon density is required to interact with the sample. Such a high photon flux is typically achieved using pulsed lasers with picosecond or femtosecond duration. Even with efficient biological SHG emitters, the probability that at least one pair of photons inside that pulse will combine to generate one SHG photon is relatively low. Therefore, the system needs to be optimized to deliver at least ~1–10 nJ of energy per pulse within a small focal volume, typically achieved using a high numerical aperture (NA) microscope objective with NA > 0.6.

Additionally, a material must have non-inversion or non-central symmetry across molecular to microscopic scales in order to generate SHG efficiently. At the molecular scale, the first hyperpolarizability tensor, β, describes the SHG response, which may be due to specific SHG-active bonds or groups of bonds, each with a particular orientation, and having non-inversion symmetry. At the microscopic scale, the second-order nonlinear optical susceptibility tensor, $\chi^{(2)}$, describes the SHG response of a larger structure composed of several SHG active bonds or groups of bonds as well as the non-inversion symmetry of the arrangement of the SHG active subunits and the amount of SHG signal produced in the larger scale structure.

The SHG intensity ($I_{2\omega}$) obtained at each pixel of an image during SHG microscopy is proportional to the second-order nonlinear polarization induced in the material ($I_{2\omega} \propto P^{(2)}$), which itself is an orientational average of all the material hyperpolarizabilities within the focal volume. Therefore, $I_{2\omega}$ is influenced by the ultrastructure of the SHG active bonds or groups of bonds at the nano-scale, described by β, as well as the organization of the SHG active bonds or groups of bonds at the micro-scale, described by $\chi^{(2)}$. Hence, SHG is highly sensitive to the hierarchical size range of a biological sample, such as collagenous tissue, which is composed of many triple helix molecules. This makes SHG a powerful technique for visualizing collagenous tissues and other biological structures.

## 2.2 *SHG Theory*

The theory behind SHG involves the interaction of intense electromagnetic fields, such as those that are produced by ultrafast lasers, and materials at the microscopic scale. When a laser beam is focused onto a sample, the electric field of light induces a polarization in the sample. This induced polarization response is described by a power series, where the induced polarization vector P is related to the applied electric field E by the expression below:

$$\mathrm{P} \propto \chi^{(1)} : \mathrm{E} + \chi^{(2)} : EE + \chi^{(3)} : EEE + \ldots \tag{1}$$

Each term in the above equation corresponds to a different order of interaction between the electric field and the material. $\chi^{(1)}$ is the first-order linear optical susceptibility, which is associated with linear optical effects such as refraction and absorption. $\chi^{(2)}$ is the second-order nonlinear optical susceptibility, which is the key to SHG and other second-order nonlinear optical processes like sum-frequency generation (SFG). $\chi^{(3)}$ is the third-order nonlinear optical susceptibility which is responsible for third-order nonlinear optical interactions such as third harmonic generation (THG) and two-photon absorption.

In Eq. 1, it is assumed that electric dipole interactions dominate, and this seems to work well for PSHG using linear polarization states. However, several recent studies are challenging those assumptions in a method that uses circularly polarized laser excitation (e.g. Eq. 12 below). While higher order interactions, including magnetic dipole or electric quadrupole effects, are usually assumed to be much weaker than electric dipole allowed effects, theoretical modeling has shown that these higher order effects can contribute significantly to SHG from chiral molecules [5]. Recently, in one PSHG study on human cornea by Schmeltz et al., the authors showed that the relative polarity of collagen above and below the imaging plane does not influence the circularly polarized measurement as was previously assumed with the electric dipole approximation [6]. Another study showed that in PSHG of individual collagen fibrils from tendon, a linearly polarized laser measurement induced the expected behavior by correctly scaling SHG with the out-of-plane angle, while a circularly polarized excitation PSHG technique did not [7]. In a different study on PSHG of a section of tendon tissue, the authors showed that the out-of-plane angle did scale with the circular excitation parameter, but only due to a staining procedure [8]. Therefore, further research is needed to uncover the exact source of the extra SHG, and care needs to be taken when interpreting SHG from circular polarization measurements.

SHG exhibits notable advantages over alternative microscopy methodologies, including single-photon excitation fluorescence and multi-photon excitation fluorescence, especially in the context of biological tissue imaging. For example, single- and multi-photon excitation fluorescence entail the absorption of incident energy due to the Stokes shift during relaxation from an excited state. Given that the Stokes energy is necessarily dissipated as heat after each absorption event, single- and multi-photon excitation fluorescence pose a potential risk of photodamage and photobleaching. In contrast, the physical process of SHG avoids energy dissipation because excitation does not occur. Instead, it is termed as an elastic scattering process, where two incident photons combine their energy to generate a single photon at twice the frequency, with no loss in energy. This significantly reduces the risk of photodamage and photobleaching, making SHG an ideal technique for imaging biological tissues, particularly those with highly ordered, non-centrosymmetric structures such as collagenous tissues. It is important to note that the ultrafast laser can still induce sample damage during SHG imaging because other physical processes, such as two-photon absorption, may be simultaneously taking place. Therefore, the SHG imaging laser wavelength is typically chosen outside the absorption window of all absorbers in the sample, or the damage is minimized by choosing lower laser powers and longer repetition rate lasers to allow excited states to relax.

## 2.3 SHG Optical Setup

A simple SHG setup involves an ultrafast laser that is coupled to a confocal laser scanning microscope, which ensures transmission of the laser wavelength and incorporates optical filters to block the laser and fluorescence signals as well as collect SHG signals using a sensitive detector (see Fig. 2). High peak power lasers are required to produce the SHG signal from biological samples, but to avoid damage to biological samples, the average laser power must be low. These two conditions are best fulfilled by employing ultrafast lasers that supply pulses with several nJ of energy or kW of peak power at the biological sample. This is achieved using ultrafast lasers with pulse durations ranging from 10 to 5000 fs and repetition rates ranging from 1 to 100 MHz. To ensure that the laser excitation wavelength is outside the linear absorption region of the sample, and to easily collect SHG signal with detectors that are optimized for light in the visible range, lasers with wavelengths between 700 and 1250 nm are preferred, although using longer wavelengths (1250–1900 nm) is possible albeit at the price of reduced resolution and detector efficiency, and higher dark noise. The SHG signal is only emitted from within the laser focal spot where a sufficiently high photon flux occurs, hence, SHG microscopy features inherent optical sectioning capability, and thus, confocal pinholes are not required in the optical setup.

The typical polarization-resolved laser scanning microscope consists of several elements in addition to the laser, including laser scanners such as galvanometric mirrors which are used to raster scan the laser over the biological sample, a high NA objective which focuses the laser onto the biological sample, and a narrow bandpass filter as well as a photomultiplier tube detector to collect SHG signal. To perform PSHG imaging, the polarization of the laser and sometimes the SHG itself is manipulated via specialized polarization optics in various implementations to create several PSHG techniques. These techniques will be discussed in further detail in Sects. 2.4–2.8.

## 2.4 *Polarization-Resolved SHG Theory*

The SHG properties of a sample can be described using up to 18 unique $\chi^{(2)}$ tensor elements. However, extracting all 18 elements is challenging; therefore simplifying models are typically employed to fit data and extract a reduced number of $\chi^{(2)}$ tensor element values. In many applications, Kleinman symmetry is assumed, where the second-order nonlinear optical susceptibility terms with permuted indices are equal ($\chi^{(2)}_{ZXl} = \chi^{(2)}_{ZlX} = \chi^{(2)}_{lZX}$ where $l$ is $X$, $Y$ or $Z$ in Fig. 1), and occurs when the excitation and SHG wavelengths are far from the absorption bands of the sample, resulting in nine unique tensor elements. Furthermore, when the sample is assumed to have hexagonal $C_{6v}$ symmetry, only two nonzero unique tensor elements remain: $\chi^{(2)}_{ZZZ}$ and $\chi^{(2)}_{ZXX}$ [9, 10]. The ratio of these two unique tensor elements is often denoted as $\rho = \chi^{(2)}_{ZZZ} / \chi^{(2)}_{ZXX}$ [11]. Additionally, if complete permutation symmetry is not assumed, then the Kleinman ratio $K = \chi^{(2)}_{ZXX} / \chi^{(2)}_{XXZ}$ can also be extracted. Allowing $K \neq 1$ means that there is an alternate possibility of defining $\rho^* = \chi^{(2)}_{ZZZ} / \chi^{(2)}_{XXZ}$. Although $K = 1$ typically holds in biological samples, authors may indicate this value as well.

Additionally, if $C_6$ symmetry is assumed instead of $C_{6v}$, then in addition to ρ, $\chi^{(2)}_{XYZ}$ is nonzero, and the nonlinear chirality term can be extracted, defined as $\kappa = \chi^{(2)}_{XYZ} / \chi^{(2)}_{ZXX}$. Also, if $C_{3v}$ symmetry is assumed, then $\chi^{(2)}_{XXX}$ is nonzero, which is accounted for by the parameter $S = \chi^{(2)}_{XXX} / \chi^{(2)}_{ZXX}$. Typically, the angle, δ, representing the projection of

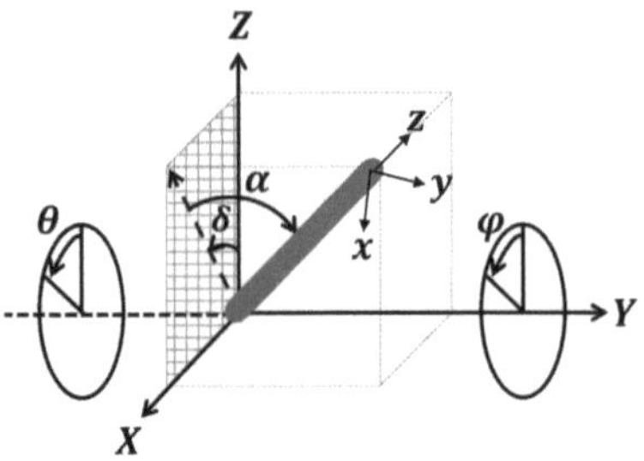


**Fig. 1** The laboratory Cartesian coordinate system is defined as *XYZ* where the principal propagation direction of the laser occurs alongside *Y*, and *X-Z* is the image plane. The gray cylinder represents the average orientation of an SHG emitting structure, which has its own molecular coordinates *xyz*. The molecular coordinates are related to the laboratory Cartesian coordinate system by modified spherical angles including α, the out-of-the-image-plane tilt angle, and *δ*, the in-the-image-plane projection angle from the *Z* axis

the dominant second-order nonlinear optical susceptibility axis into the image (Z-X) plane of the sample, can also be readily extracted from the PSHG data.

## 2.5 *Polarization-In SHG Microscopy*

The classic PSHG technique is referred to here as polarization-in SHG microscopy (PI-SHG), and it remains one of the most commonly used methods. PI-SHG utilizes a polarization state generator (PSG) to rotate the linear polarization state of the laser, θ, see Fig. 1. A half-wave plate (HWP) is commonly used as the PSG, and it is typically placed in a mechanical rotation mount before the excitation objective. The HWP is rotated and an SHG intensity image is recorded at discrete HWP angles. The PSHG dataset consists of SHG images at several HWP angles between 0° to 90°, typically in the range of 8–100.

When $C_{6v}$ symmetry is assumed, the parameters, ρ, δ, and *K* (see Sect. 2.4 for descriptions) can be determined for each pixel or group of pixels by performing nonlinear data fitting using the following equation:

$$I_{2\omega} \propto \left| \left( K\sin^2(\theta-\delta) + \rho\cos^2(\theta-\delta) \right)^2 + \left( \sin 2(\theta-\delta) \right)^2 \right| \tag{2}$$

Note that the terms ρ and *K* are measurements made in the laboratory Cartesian coordinate system. If Kleinman symmetry is assumed ($K = 1$) and the out-of-the-image-plane tilt angle of the SHG emitter is greater than 0° (α > 0°), then the measured ρ can be related to its molecular value (ρ') using the following expression [12]:

$$\rho' = \rho'\cos^2\alpha + 3\sin^2\alpha \tag{3}$$

Interestingly, Eq. (3) reveals that ρ' is the same as $\rho$ when the sample is on the image plane, α=0°. Equation (2) presupposes $C_{6v}$ symmetry; however, when trigonal $C_{3v}$ symmetry is assumed, in addition to ρ, *S* is also obtained [13]:

$$I_{2\omega} \propto \left| \left( K \sin^2(\theta - \delta) + \rho \cos^2(\theta - \delta) \right)^2 + \left( S \sin^2(\theta - \delta) + \sin 2(\theta - \delta) \right)^2 \right| \quad (4)$$

PI-SHG can be easily implemented in both forward and backward detection geometries. Backward PI-SHG is more cost-effective and easier to set up than transmission PI-SHG; however, backward SHG typically has reduced intensity since SHG is emitted mostly in the forward direction due to momentum conservation. Furthermore, a backwards dichroic mirror is typically used to pick off the signal, and the polarization change that the dichroic mirror induces to the excitation beam may need to be accounted for.

## 2.6 Faster Fitting Polarization-In SHG Microscopy

Nonlinear data fitting with PI-SHG to extract ρ and δ can be time-consuming, often taking several hours for images consisting of 500×500 pixels. One method to improve the speed of fitting the PI-SHG data uses a fast Fourier transform (FFT) algorithm. Trigonometric identities can be used to transform Eq. (2) to the following equation [14]:

$$I_{2\omega} \propto \alpha_0 + \alpha_2 \cos 2(\theta - \delta) + \alpha_4 \cos 4(\theta - \delta) \quad (5)$$

where $\alpha_0$, $\alpha_2$, and $\alpha_4$ coefficients can be used to calculate PSHG parameters.

A one-dimensional discrete Fourier transform can be applied to the PI-SHG data along the polarization axis from 0° to 180° to obtain the $\alpha_i$ parameters. Simple algebra can then be used to convert the values into ρ and δ. For instance, the $\alpha_0$, $\alpha_2$, and $\alpha_4$ coefficients can be used to calculate ρ [15]:

$$\rho = \sqrt{\frac{\alpha_0 + \alpha_2 + \alpha_4}{\alpha_0 - \alpha_2 + \alpha_4}} \quad (6)$$

This method of FFT PI-SHG analysis is much faster, resulting in a speed that is five orders of magnitude faster than nonlinear data fitting [14].

To improve data acquisition speed and image size, widefield microscopy can be employed. It should be noted, however, that the technique has reduced axial resolution compared to point scan techniques; therefore thin samples <5 μm are typically utilized.

For the interested reader, Hristu et al. published a comprehensive manuscript that includes all raw and analyzed PSHG measurements using FFT analysis on PI-SHG data. The experimental dataset is potentially very interesting; normal tissue and two diagnosed cancer types of breast and thyroid. The authors analyzed images of PSHG parameters, including δ, ρ, ρ* and the Kleinman coefficient, *K*. Additionally, the authors applied a texture analysis technique based upon the distribution of δ within neighboring pixels via utilizing small windows within δ images, resulting in three additional quantities: Entropy, mean standard deviation, and standard deviation (of δ). In breast tissue, the authors examined three regions: Normal terminal ductal-lobular units (TDLUs), which are the basic functional ductal/glandular structures of the breast, ductal carcinoma in situ (DCIS), which is considered the earliest form of breast cancer, as well as invasive breast carcinoma. In the thyroid, the collagenous capsules surrounding thyroid nodules that were diagnosed as either follicular adenoma or papillary thyroid carcinoma were also included [16].

## 2.7 *Polarization-In, Polarization-Out SHG Microscopy*

Similar to PI-SHG, polarization-in, polarization-out SHG microscopy (PIPO-SHG) often uses a HWP in a mechanical rotation mount as the PSG. However, a polarization state analyzer (PSA) is also used, just after a collection objective and before the filter, designed to measure the polarization state of the emitted SHG signal as a function of the laser orientation. The PSA commonly consists of a linear polarizer in a mechanical rotation mount (at an angle ϕ). As few as 64 SHG images using 8 PSG angles and 8 PSA angles have been used for extracting PSHG parameters. PIPO-SHG acquisition this way takes about 12 minutes to acquire, which is longer than some other PSHG techniques. For instance, it takes about 1.5 minutes to acquire data using PI-SHG with 9 PSG angles [17]. Yet, PIPO-SHG may result in more accurate PSHG parameters due to the extra dimension of the PSHG data, which results in a contour plot rather than a two-dimensional graph.

When hexagonal $C_6$ and Kleinman symmetries ($K = 1$) are assumed, the PSHG parameters, ρ, κ, and δ, are extracted using the following expression by fitting $I_{2\omega}$ as a function of the linear laser electric field polarization orientation determined by the PSG (φ) and the orientation of the PSA analyzer (φ) [18]:

$$I_{2\omega} \propto \left| \begin{array}{c} \sin(\varphi-\delta)\sin 2(\theta-\delta)+\cos(\varphi-\delta)\sin^2(\theta-\delta) \\ +\rho\cos(\varphi-\delta)\cos^2(\theta-\delta)+2\kappa\cos(\varphi-\delta)\sin(\theta-\delta) \end{array} \right|^2 \tag{7}$$

where $\kappa = 0$ when hexagonal $C_{6v}$ and Kleinman symmetries are assumed.

Like in PI-SHG, the measured ρ is related to its molecular value ($\rho'$) using Eq. (3) and similarly, the measured κ is related to its molecular value (κ') by the following expression:

$$\kappa' = \kappa' \sin\alpha \tag{8}$$

Interestingly, Eq. (8) reveals an α behavior of $\kappa$ that is opposite to ρ. Eq. (8) shows that κ' is equal to κ at α=90° and κ = 0 when the sample is on the image plane.

Therefore, κ is typically only measured when the SHG emitting structure is tilted off the plane ($\alpha > 0$), however, utilizing the variation in polarization near the edge of a high NA focus, PIPO-SHG was shown to extract κ from individual collagen fibrils [19].

If, instead of hexagonal $C_{6v}$ symmetry, trigonal $C_{3v}$ symmetry is assumed, then the $S$ parameter can also be extracted [20]:

$$I_{2\omega} \propto \left| \begin{array}{c} \sin(\varphi-\delta)\sin 2(\theta-\delta)+\cos(\varphi-\delta)\sin^2(\theta-\delta) \\ +\rho\cos(\varphi-\delta)\cos^2(\theta-\delta)+S\sin(\varphi-\delta)\sin^2(\theta-\delta) \end{array} \right|^2 \tag{9}$$

In addition to extracting ρ, δ, κ and $S$ parameters, the degree of linear polarization (DOLP) of the SHG signal can be found using PIPO-SHG data:

$$\text{DOLP} = \frac{\sqrt{s_1^2 + s_2^2}}{s_0} \tag{10}$$

where $s_0$, $s_1$, and $s_2$ are Stokes parameters defined as: $s_0 = I_{2\omega,0} + I_{2\omega,90}$, $s_1 = I_{2\omega,0} - I_{2\omega,90}$ and $s_2 = I_{2\omega,45} - I_{2\omega,-45}$ where $I_{2\omega,a}$ refers to the SHG intensity at the $a$ analyzer angle. DOLP values near 1 indicate the polarization state of the emitted SHG is linear, while higher values indicate higher circular or scattered SHG output. In some cases, the DOLP can vary with the input polarization, and therefore, to produce an unbiased DOLP measurement per pixel, DOLP can be performed at a range of evenly spaced linear polarization angles and averaged.

DOLP measurements have been performed by averaging DOLP calculations from 8 input polarizations. Specifically, DOLP calculations were averaged for each dataset, one from analyzer angles at 0°, 45°, 90°, 135° and a second with analyzer angles at 22.5°, 67.5°, 112.5°, 157.5°. In many oriented samples, the DOLP values varied simply because a very low SHG signal was present at one particular laser polarization angle, where noise began to govern the DOLP response [21]. The averaging above works well to reduce the effect of the noise DOLP response in the average, but it is also possible to check Eq. (10) for the particular δ value that gives minimum SHG and then remove all data with similar δ values [22] (Fig. 2).

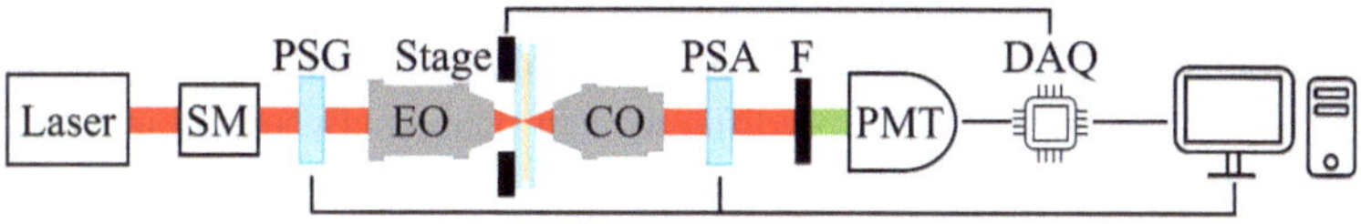


**Fig. 2** Example PSHG microscope schematic for transmission detection. The PSA is present for PIPO-SHG and reduced double Stokes-Mueller SHG microscopy (RDSM-SHG), but not PI-SHG. The abbreviations in the figure refer to: Scanning mirrors (SM), polarization state generator (PSG), excitation objective (EO), collection objective (CO), polarization state analyzer (PSA), filter (F), photomultiplier tube (PMT), and data acquisition (DAQ)

## 2.8 Widefield PSHG Microscopy

Widefield PSHG microscopy allows for faster imaging of larger areas, but it comes at the cost of reduced axial resolution, therefore, thin samples are typically utilized. The technique involves replacing the excitation objective with an achromatic lens which creates a large spot size at the sample, slightly bigger than the desired image size. The illumination area is adjusted by translating the achromatic doublet axially. In the transmission direction, a medium to high NA collection objective is utilized, followed by an imaging lens and then a sensitive camera detector such as an EMCCD or sCMOS. The setup can still use the same PSG and PSA as previously, hence both PI-SHG or PIPO-SHG can be utilized in this modality.

Widefield reduced double Stokes-Mueller SHG microscopy (widefield RDSM-SHG) is emerging as a useful tool for widefield PSHG. The authors of this technique utilized liquid crystal variable retarders (LCVR) and matching polarizers as the PSG and PSA. The polarization states of the PSG and PSA are obtained by employing four retardances on each LCVR; in fractions of a wave, these are: $\lambda/4$, $\lambda/2$, $3\lambda/4$, and $\lambda$. In the PSG, the polarizer is followed by the LCVR, generating four orthogonal polarization states: Left-circularly polarized (LCP), horizontally linearly polarized (HLP), right-circularly polarized (RCP), and vertically linearly polarized (VLP) laser light. The emitted SHG signal is analyzed at each incident laser polarization state. The PSA, which consists of another LCVR and polarizer pair, is set to the same four retardances as the PSG LCVR, allowing for collection of four polarization components of the SHG beam: Two orthogonal linearly polarized states ($I_{2\omega,\,0}$ and $I_{2\omega,\,90}$), and two orthogonal circularly polarized states ($I_{2\omega\,RCP}$ and $I_{2\omega\,LCP}$). The technique obtains 16 SHG intensity images at all combinations of 4 PSG and 4 PSA states, with acquisition taking about 4 minutes to complete a 670 μm × 670 μm imaged area. The exposure time for imaging each polarization state takes 10 s with an additional 5 s delay time for switching polarization states between each exposure time [23].

The 16 combinations of polarization states have been used in widefield RDSM-SHG to determine several useful parameters, including $\delta$, circular dichroism SHG ($SHG_{CD}$), the degree of circular polarization (DCP), linear dichroism SHG ($SHG_{LD}$), linear anisotropy of circular dichroism ($LA_{CD}$), circular anisotropy of circular dichroism of SHG ($CA_{CD}$), linear anisotropy of linear dichroism for SHG ($LA_{LD}$), and circular anisotropy of linear dichroism ($CA_{LD}$) for each pixel of the image. These parameters can then be utilized to obtain hexagonal PSHG parameters, including $\rho$, $\kappa$, as well as $\Delta_{xyz}$, the difference in phase between achiral and chiral second-order nonlinear optical susceptibilities.

To determine these PSHG parameters, a 4-component Stokes vector with $s_3 = I_{2\omega\,RCP} - I_{2\omega\,LCP}$ is used to describe the SHG signal (see Eq. 10 for $s_0 - s_2$). In the following equations, the superscripts on the Stokes components describe the polarization state of the laser. Using circular laser polarizations, the authors calculate the total degree of circular polarization (DCP) as [23]:

$$\mathrm{DCP} = \frac{1}{2}\left( \frac{\left|s_3^{\mathrm{RCP}}\right|}{s_0^{\mathrm{RCP}}} + \frac{\left|s_3^{\mathrm{LCP}}\right|}{s_0^{\mathrm{LCP}}} \right) \tag{11}$$

The circular dichroism SHG is [23]:

$$\mathrm{SHG}_{\mathrm{CD}} = 2\left( \frac{s_0^{\mathrm{RCP}} - s_0^{\mathrm{LCP}}}{s_0^{\mathrm{RCP}} + s_0^{\mathrm{LCP}}} \right) \tag{12}$$

where $\mathrm{SHG}_{\mathrm{CD}}$ ranges from −2 to 2, while the linear dichroism SHG is [23]:

$$\mathrm{SHG}_{\mathrm{LD}} = 2\left( \frac{s_0^{\mathrm{VLP}} - s_0^{\mathrm{HLP}}}{s_0^{\mathrm{VLP}} + s_0^{\mathrm{HLP}}} \right) \tag{13}$$

where $\mathrm{SHG}_{\mathrm{LD}}$ also ranges from −2 to 2. When a circularly polarized laser beam generates SHG signal, and the SHG signal is analyzed with linear polarizations, a measure of the linear anisotropy of circular dichroism ($LA_{CD}$) is obtained [24]:

$$LA_{CD} = 2\frac{s_1^{LCP} - s_1^{RCP}}{s_0^{LCP} + s_0^{RCP}} \tag{14}$$

Similarly, when a circularly polarized laser beam generates SHG signal and that SHG signal is then subjected to a PSA with circular polarizations, a measure of the circular anisotropy of circular dichroism of SHG ($CA_{CD}$) is obtained [24]:

$$CA_{CD} = 2\frac{s_3^{\mathrm{LCP}} - s_3^{\mathrm{RCP}}}{s_0^{\mathrm{LCP}} + s_0^{\mathrm{RCP}}} \tag{15}$$

When linear polarizations are employed for both the incident laser beam and the PSA, the linear anisotropy of linear dichroism for SHG ($LA_{CD}$) can be measured [24]:

$$LA_{LD} = 2\frac{s_1^{HLP} - s_1^{VLP}}{s_0^{HLP} + s_0^{VLP}} \tag{16}$$

When a linearly polarized laser beam generates SHG signal, and that SHG signal is then subjected to a PSA with circular polarizations, the circular anisotropy of linear dichroism ($CA_{LD}$) can be measured [24]:

$$CA_{LD} = 2\frac{s_3^{HLP} - s_3^{VLP}}{s_0^{HLP} + s_0^{VLP}} \tag{17}$$

The above parameters can be used to calculate ρ, $C \sin \Delta_{xyz}$, $\cos 2\delta$ and $I_{2\omega}$. For example, under the assumption that $\kappa \leq 0.1$, ρ can be expressed as [24]:

$$\rho = 1 + \frac{4}{CA_{CD}} \pm 2\sqrt{\frac{4}{CA_{CD}^2} - 1} \tag{18}$$

Similarly, $\kappa \sin \Delta_{xyz}$ can be expressed as a function of $\mathrm{SHG_{CD}}$ [24]:

$$\kappa \sin \Delta_{xyz} = \frac{-\mathrm{SHG_{CD}}\left(\left(\rho - 1\right)^2 + 4\right)}{8\left(\rho + 1\right)} \tag{19}$$

Additionally, $\cos 2\delta$ can be expressed as a function of $\mathrm{SHG_{LD}}$ and $LA_{LD}$ under the assumption that $(\rho + 1) \gg 2\kappa \cos \Delta_{xyz}$ [24]:

$$\cos 2\delta = \frac{-2\left(\frac{\rho - 1}{\rho - 3}\right) - \sqrt{4\left(\frac{\rho - 1}{\rho - 3}\right)^2 - 8\left(\frac{\mathrm{SHG_{LD}}}{LA_{LD}}\right)\left(\left(\frac{\rho + 1}{\rho - 3}\right)\left(\frac{\mathrm{SHG_{LD}}}{LA_{LD}}\right) - 1\right)}}{4\left(\frac{\mathrm{SHG_{LD}}}{LA_{LD}}\right)} \tag{20}$$

Lastly, the in-plane orientation-independent average SHG intensity ($I_{2\omega}$) can be expressed as a function of the incident circularly polarized states [24]:

$$I_{2\omega} = \frac{1}{2}\left(s_0^{\mathrm{LCP}} + s_0^{\mathrm{RCP}}\right) \tag{21}$$

## 3 Applications of PSHG in Cancerous Tissue

PSHG microscopy has been applied to study several biological structures including collagen in cardiac tissue [25], cartilage [26–28], cornea [29–34], skin [35–40], tendon [12, 18, 41–50], reconstituted collagen gels [51–58] as well as myosin filaments [59], microtubules [60], atherosclerotic plaques [61], corpora amylacea [62], astroglial filaments [63], and otoconia [64]. Imaging many of these structures has interesting applications in biomedical research. In particular, PSHG has the advantage of possibly differentiating biological structures based on their disease state.

Recently, the feasibility of diagnosing cancer by measuring PSHG parameters related to the ultrastructure of collagen in tumor tissue has been investigated. Traditionally, cancer diagnosis involves identifying tumor regions in tissue biopsies that have been stained with hematoxylin and eosin dyes (H&E). Hematoxylin stains the nuclear material purple while eosin stains the other tissue regions pink. A pathologist identifies and diagnoses tumor regions by distinguishing variations in tissue coloring as well as nuclear size and shape. Although this technique is considered the gold standard for cancer diagnosis, there are issues with the dyeing process, including dye absorption. The absorption of dye can vary from patient to patient and likely depends on several factors such as age, diet, and genetics. As a result, large cancerous regions (at least a few hundred micrometers in diameter) are needed to identify any patterns, and the result from the pathologist is qualitative, often requiring confirmation from other pathologists. However, PSHG is a promising technique for cancer diagnosis because it provides unbiased quantitative parameters relating to the ultrastructure of tissue at a nanoscale, with high sensitivity to bond angles down to 1°, for example, in wet versus dry collagen [19]. This property has been shown to sensitively differentiate the disorganized ultrastructure of collagen within the extracellular matrix of cancerous tissues from normal tissues. The quantitative characterization that PSHG provides enables the possibility of automated cancer diagnosis and potentially earlier diagnosis since smaller regions of tissue, possibly down to one or several laser focal volumes, can be utilized to diagnose tumors.

In the following sections, results from PSHG microscopy investigations for cancer diagnosis from the past 5 years (2020–2024) are reviewed. To compare and contrast results from different papers, the results are converted to the coordinates described in Fig. 1 for comprehensibility. For instance, the microscope image plane may be defined as *Z-Y* or *X-Y*, rather than *Z-X* as in Fig. 1. In this case, ρ would be $\chi^{(2)}_{ZZZ} / \chi^{(2)}_{ZYY}$ or $\chi^{(2)}_{XXX} / \chi^{(2)}_{XYY}$ in the original paper.

Furthermore, several papers are not included in this chapter that have used PSHG for cancer diagnosis prior to 2020. A thorough review of PSHG for cancer diagnosis prior to 2020 can be found in the book chapter by Cisek et al. [4]. Also, several papers that have been working towards using intensity-based SHG microscopy (as opposed to PSHG) for cancer diagnosis were excluded in this book chapter. A more comprehensive list of SHG microscopy techniques for cancer diagnosis is out of the scope of this chapter, and instead, the focus is placed on recent PSHG results and techniques for cancer diagnosis.

## 3.1 Polarization-In SHG Microscopy Cancer Studies

Tsafas et al. used PI-SHG at 1064 nm to investigate breast cancer tissue samples from eight patients, all older than 60 years. The samples included tissues from two patients who were diagnosed as benign (grade 0, ductal carcinoma) as well as tissues from two patients with grade I tumors, two patients with grade II tumors, and two patients with grade III tumors. The tissue samples were embedded in paraffin and sliced into 5–10 μm thick sections. For each sample, two sections were prepared, one for staining with H&E for histological examination and a second adjacent, unstained section for SHG imaging [15].

Qualitative inspection of the SHG images revealed that the collagen fibers in benign tissues appeared curly and smooth, while the collagen fibers in cancerous tissues showed linearized, thickened, and stiffened collagen surrounding the carcinoma. FFT PI-SHG was performed to extract two PSHG parameters: ρ and a parameter termed the 'ratio parameter'. The ratio parameter was determined by comparing the number of pixels in each image that were successfully fit with $R^2 > 0.9$ divided by the total number of pixels. In general, the mean ρ values of the tissues increased with increasing grade, with statistically significant differences observed between grade 0 versus grade I ($p < 0.01$) and grade I versus grade II ($p < 0.0001$), indicating that ρ values may correlate with tumor progression. The 'ratio parameter' decreased with increasing grade, with statistically significant differences found between grade 0 versus grade II ($p < 0.01$) and grade II versus grade III ($p < 0.0001$). In particular, there was a huge decrease in the 'ratio parameter' from grade II to grade III, which is attributed to the breakdown of collagen's cylindrical symmetric structure. Therefore, PSHG data of grade III tumor tissue may benefit from being fit using different biophysical models, such as one where trigonal symmetry (Eq. (4)) is assumed [15].

Breast cancer tissue was also studied by Mercatelli et al. using PI-SHG at 840 nm. Formalin-fixed and paraffin-embedded tissue sections with 30 μm thickness were investigated. Like Tsafas et al., one tissue slide was stained with H&E dye to select areas for imaging, while a second slide was used for PSHG imaging. However, unlike Tsafas et al., the assumption that collagen fibers have trigonal symmetry was applied in PI-SHG fitting analysis [65].

Initially, two collagen fibril structures were qualitatively observed: Straight-aligned collagen fibrils and wave-like collagen fibrils. It was noted that the curves displayed on a 1D graph of the SHG intensity versus laser polarization angle looked different depending on the qualitatively observed collagen fibril structure. For instance, the straight aligned collagen fibril demonstrated a curve with a highly symmetric profile, while the wave-like collagen fibril displayed a highly asymmetric profile. The corresponding $S$ parameter obtained from the fit of a straight aligned collagen fibril approached zero, while the $S$ parameter obtained from the fit of a wave-like collagen fibril displayed a significantly higher value. This observation was seen as confirmation of different ultrastructural organizations of collagen within the sample, where wave-like collagen structures have a higher prevalence of trigonal symmetry over hexagonal symmetry [65].

It was noted that short, wavy collagen fiber bundles are found in perilesional tissue, while long, straight, and parallel collagen fiber bundles are found in tumor stroma. The $S$ parameter for perilesional tissue was found to be $0.51 \pm 0.14$ while the $S$ parameter for tumor stroma was found to be $0.16 \pm 0.05$. An area between the two different collagen assemblies was found to have an $S$ parameter of $0.30 \pm 0.06$, in line with the observations made above. Therefore, the $S$ parameter may have the ability to discriminate between perilesional tissue and tumor stroma collagen assemblies [65].

Aghigh et al. used PI-SHG at 810 nm to investigate the ECM associated with tumors residing in the mammary gland of mice. Tissue preparation included removing the 4th (inguinal) mammary glands from female BALB/c mice, which are a standard model inbred mouse known to produce plasma cell tumors within soft tissue. These samples served as naïve, non-tumor-bearing mammary gland samples, while murine tumor-bearing samples were obtained using two orthotopic models. One model injected cells from a tumor derived stage 4 breast cancer cell line, 4T1, into nulliparous BALB/c mice. The second model injected another tumor cell line, 66cl4, into BALB/c mice during their postpartum period, resulting in an aggressive form of breast cancer. The tumors were allowed to grow until 14 days post-injection, at which point the mice were euthanized. The primary tumors and surrounding stroma were removed, and the samples were fixed [66].

SHG imaging of the naïve, non-tumor-bearing mammary gland samples revealed well-defined ductal structures. However, in the murine tumor-bearing samples, the tumor edges are more pronounced as the tumors do not generate an SHG signal (Fig. 3a). The authors concluded that a collagen barrier is formed around the tumor as a protection mechanism to constrain the tumor [66].

To obtain ultrastructural information about the collagen fibrils in the murine tumor-bearing samples, FFT analysis of the PI-SHG data was performed. In addition to extracting the δ parameter from the data, the SHG images were also processed using CurveAlign [67], a readily available quantitative tool for fibrillar collagen, and the results were compared. Overall, it was observed that FFT PI-SHG analysis is more detailed, leading to more defined orientations of smaller and finer collagen fibers over CurveAlign. However, PI-SHG analysis led to having some pixel fits be significant even though there is likely no signal there, or, overfilling (Fig. 3b), while CurveAlign was found to under and overfill in areas where the SHG signal was dim (Fig. 3c). The authors argue that overfilling in the PI-SHG analysis results from the chosen goodness of fit parameter, $R^2$. Thus, to ensure that overfilling does not occur, it is important to have a good method of rejecting pixels with low SHG intensity and high noise [66]. Several different pixel rejection methods have been investigated recently by Eftimie et al. [68], including $R^2$, a signal-to-noise ratio (SNR) [32] and an experimental error (Eq. 8 in [14]).

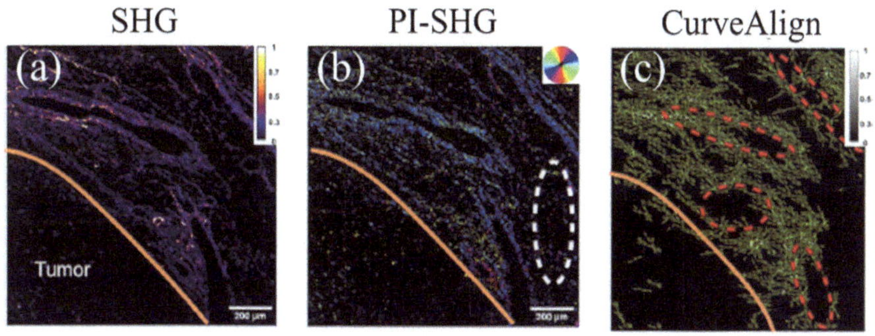


**Fig. 3** SHG intensity (**a**) as well as PI-SHG (**b**) and CurveAlign (**c**) analysis performed on the same region of interest of a sample of the mammary gland ECM. Each colored pixel in (**b**) represents an angle (0–360°) as shown by the color wheel in the top right-hand corner. Overfilling of regions was observed with PI-SHG and is indicated in (**b**) by a white dashed line, while overfilled and underfilling regions were observed with CurveAlign and is indicated in (**c**) by red dashed lines. The scale bar corresponds to 200 μm. Images **a–c** were adapted with permission from [66] © Optica Publishing Group

PI-SHG at 810 nm excitation wavelength was used by Chen et al. to investigate the 4 stages of esophageal squamous cell carcinoma (ESCC) in humans. Although the authors observed that the mean ρ parameter in stage III cancer samples was significantly different that in stages II or IV, the other stages could not be distinguished. The authors concluded that $\chi^{(2)}$ cannot be used to discriminate ESCC cancers because the changes in this particular cancer originate from macromolecular structural changes in collagen fibers rather than molecular scale changes in the collagen triple helix [69].

## 3.2 PIPO-SHG Microscopy Cancer Studies

In 2020, Golaraei et al. applied PIPO-SHG to investigate collagen in human pathological tissue samples of patients who were diagnosed with stage I, -II, and III non-small cell lung carcinoma (NSCLC), and compared to normal lung parenchyma. Collagen fibers present in the ECM of the tissue samples were assumed to have hexagonal $C_6$ symmetry, and several parameters were measured, including ρ, κ , δ and DOLP [70].

The median of the ρ value distribution of the tumor tissue at all the stages of NSCLC was higher than the median of the ρ value distribution of normal tissue (Fig. 4c). Statistically significant differences were observed between normal tissue versus stages II ($p < 0.05$) and -III ($p < 0.01$) NSCLC. A trend between the different tumor stages was observed where the median of the ρ value distribution increased with NSCLC stage; however, a statistically significant difference was not found. The increase in the median of ρ value distributions with stage is indicative of a less organized collagen fiber structure and it was suggested that this may be due to collagen fiber matrix reorganization [70].

Unlike the ρ value analysis, the median absolute deviation (MAD) of the κ value distribution for normal tissues and tumor tissues was compared to one another. The MAD κ value distribution was narrowest in normal tissue and significantly different from normal tissue versus stages I ($p < 0.05$), -II ($p < 0.01$), and -III ($p < 0.05$) NSCLC (Fig. 4d). The larger MAD of the κ value distribution for tumor tissue implies a more disorganized structure since the κ value is related to α (Eq. (8)) and suggests a greater spread of the α angle (the out-of-plane orientation angle). Alternatively, this may indicate changes in the collagen fibril structure, which influences the $\chi^{(2)}_{xyz}$ or $\chi^{(2)}_{zxx}$ components [70].

The median of the δ distribution of stages-I, -II, and -III NSCLC tissues were lower than the median of the δ distribution of normal tissue with stages -II ($p < 0.01$) and -III ($p < 0.05$) having values that are statistically significant from normal tissue (Fig. 4e). This indicates that collagen fibers in tumor tissue have increased fiber alignment compared to normal tissue [70].

The median of the DOLP distribution of stages I ($p < 0.001$), -II ($p < 0.001$), and -III ($p < 0.0001$) NSCLC tissues was statistically significantly lower than the

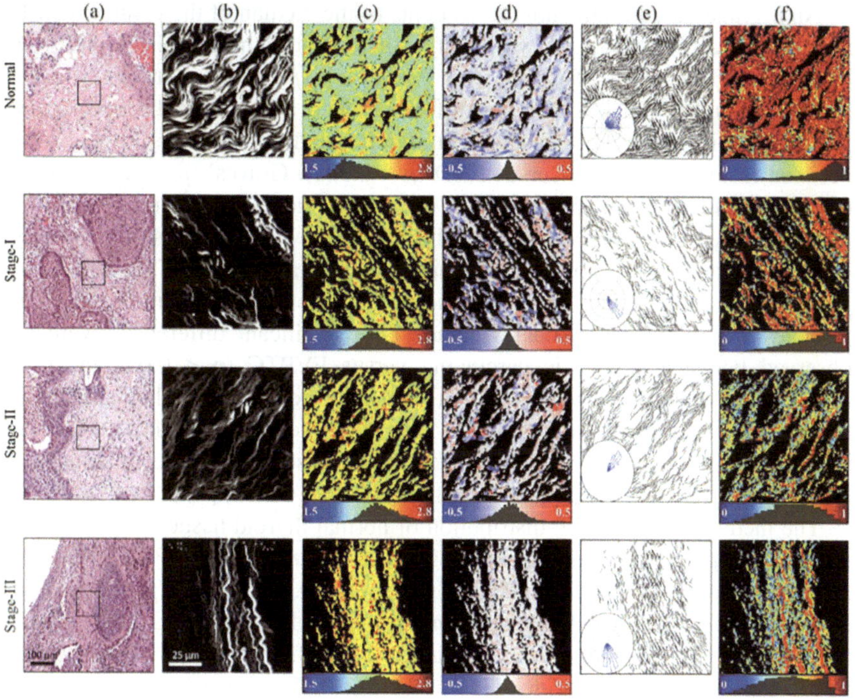


**Fig. 4** H&E stained normal and NSCLC tissue sections displayed using brightfield microscopy (**a**), the average SHG intensity generated from a PIPO-SHG dataset (**b**), $\rho$ color-coded maps (**c**), κ color-coded maps (**d**), the collagen orientation, δ (**e**), and DOLP color-coded maps (**f**). Note that the square in (**a**) indicates the area imaged using PIPO-SHG. The color bars below (**c**), (**d**), and (**f**) correspond to the occurrence pixel histogram of the images above. Images a-f were reprinted with permission from [70] © Optica Publishing Group

median of the DOLP distribution of normal tissue (Fig. 4f). A lower DOLP indicates increased linear depolarization of the outgoing SHG signal, likely due to ultrastructural disorder of collagen fibrils or fragmentation of collagen fibers into smaller segments within the laser focal volume [70].

In 2022, Tokarz et al. similarly applied PIPO-SHG to investigate collagen in human pathological thyroid tissue samples that were diagnosed with follicular nodular disease (FND, benign focal proliferation of the follicular epithelium), Grave's disease (benign diffuse hyperplasia), follicular variant of papillary thyroid carcinoma (FVPTC), classical papillary thyroid carcinoma (cPTC), insular or poorly differentiated carcinoma, and anaplastic or undifferentiated carcinoma versus normal thyroid tissue. Similar to Golaraei et al., the collagen fibers present in the tissue samples were assumed to have hexagonal $C_6$ symmetry and as a result, several parameters were measured, including ρ, the full width at half-maximum (FWHM) of ρ, κ, FWHM of κ and DOLP [71].

An average based on the number patients of the median of the ρ value distribution of normal thyroid tissue was smaller than most pathological thyroid tissues including FVPTC ($p < 0.05$), cPTC ($p < 0.05$), FND ($p < 0.001$), Grave's ($p < 0.01$) and anaplastic ($p < 0.02$). Significant differences were also observed between pathological thyroid tissues. For instance, a significant difference was also observed between insular carcinoma versus FND ($p < 0.001$), Graves' ($p < 0.01$), FVPTC ($p < 0.05$), and anaplastic carcinoma (p < 0.02). An increase in the median ρ values were attributed to increased collagen fiber disorder which may result from differences in the amino acid content, differences in the arrangement of the helices into fibrils, fibers and fascicles, and differences in the distribution of fibrils, fibers and fascicles within the laser focal volume [71]. Significant differences were also observed between anaplastic carcinoma versus FVPTC (p < 0.05) and cPTC ($p < 0.05$) as well as FND versus cPTC ($p < 0.05$). The FWHM of the ρ distribution of normal thyroid tissue was commonly larger than the pathological tissues, including cPTC, insular carcinoma, FND, and Graves' disease; however, a statistically significant difference was only observed for cPTC ($p < 0.05$) [71].

The median of the DOLP distribution of normal thyroid tissue was statistically significantly greater than all the pathological tissues including FVPTC ($p < 0.001$), cPTC ($p < 0.001$), FND ($p < 0.002$), Graves' ($p < 0.01$), insular ($p < 0.001$), and anaplastic carcinoma ($p < 0.001$). Amongst the pathological tissues, the median of the DOLP for insular carcinoma was significantly different than many of the other pathological thyroid tissues including FND ($p < 0.05$), Graves' ($p < 0.02$), cPTC ($p < 0.01$) and anaplastic carcinoma ($p < 0.01$) [71].

On the other hand, the κ values and the FWHM of the κ distribution for the normal thyroid tissue and all the pathological thyroid tissues studied were similar. This is thought to indicate that pathological thyroid tissues may have collagen fibers that display a similar angular distribution to normal thyroid, or the collagen fibers are not extending out of the image plane sufficiently [71].

In 2023, Jeon et al. applied PIPO-SHG to investigate collagen in human gastric tissue pathological samples that were diagnosed with poorly differentiated adenocarcinoma (PDA), moderately differentiated adenocarcinoma (MDA), and well-differentiated adenocarcinoma (WDA) tissues as well as normal gastric tissues. The collagen fibers were assumed to have hexagonal $C_6$ symmetry, which allowed for the extraction of ρ, FWHM of ρ, |κ|, FWHM of κ and DOLP. An additional parameter was obtained, the ratio of disordered pixels (RDP of ρ), which was determined by dividing the number of pixels in a chosen range ($2 \leq \rho \leq 3$) by the total number of fitted pixels per image. Furthermore, trigonal $C_{3v}$ symmetry was assumed, which allowed for the extraction of $|S|$ as well as the FWHM of $S$ [20].

An average based on the median of the ρ value distribution of normal gastric tissue was significantly smaller than PDA ($p < 0.01$) and MDA ($p < 0.01$) when the number of scans was taken into account (Fig. 5b, c) however, a significant difference was not observed when considering the number of patients. Analysis of the RDP of ρ also demonstrated significant differences between normal gastric tissue versus PDA ($p < 0.01$) and MDA ($p < 0.01$). The FWHM of the ρ distribution of

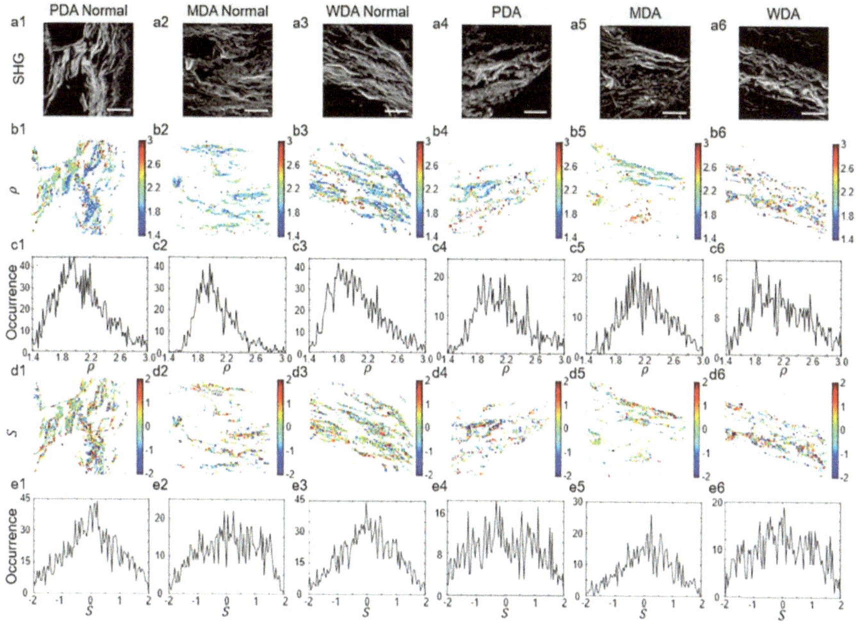


**Fig. 5** PDA, MDA and WDA tissue sections and their corresponding normal tissue sections displayed using the normalized SHG intensity produced by summing the PIPO-SHG data and applying a logarithm (**a**), ρ color-coded maps (**b**), the corresponding occurrence pixel histograms of ρ (**c**), *S* color-coded maps (**d**) and the corresponding occurrence pixel histograms of *S* (**e**). The scale bar in (**a**) corresponds to 25 μm and applies to all images. Images a-e were reprinted with permission from [20] © Optica Publishing Group

normal gastric tissue was significantly different from WDA ($p < 0.05$). As with the lung and thyroid studies above, increases to ρ are likely due to collagen fiber disorder which may result from acid content differences, changes to the assembly of helices into fibrils, fibers and fascicles, and variations in the distributions of fibrils, fibers and fascicles within a focal volume [20].

Significant differences in the |κ| values between the normal gastric tissue and all the adenocarcinoma tissues were not observed, but a statistical difference was found between PDA versus MDA ($p < 0.01$) and MDA versus WDA ($p < 0.02$) when considering the number of scans taken. Statistical differences in the FWHM of the κ distribution for normal gastric tissue versus PDA ($p < 0.001$) and PDA versus WDA ($p < 0.02$) were found when considering the number of scans taken [20].

Different from Golaraei et al. and Tokarz et al., the median of the DOLP distribution and the FWHM of the DOLP distribution for normal gastric tissue versus adenocarcinoma tissues were not significantly different. This was speculated to mean that the collagen fibers present in gastric adenocarcinoma are not misaligned to the same extent as the collagen fibers present in other cancers, including lung and thyroid [20].

Previously, researchers have shown that trigonal symmetry may be a better model to represent collagen structure in cancerous tissue [13, 65]. Since few differences in the PSHG parameters obtained using hexagonal $C_6$ symmetry were observed between normal gastric tissue and gastric adenocarcinoma tissues, fitting with trigonal $C_{3v}$ symmetry was applied to extract the absolute value of the median of the $S$ distribution ($|S|$). In normal gastric tissue $|S|$ was observed to be significantly smaller than PDA ($p < 0.02$), MDA ($p < 0.05$), and WDA ($p < 0.05$) (Fig. 5d, e) when considering the number of scans performed. Numerical simulations were performed to understand the increase in $S$ values in all the adenocarcinoma tissues. It was found that intersecting collagen fibrils of differing diameters or intersecting collagen fibrils of differing $\chi^{(2)}$ values cause a shift in $S$ values from zero, and therefore, it was speculated that more intersecting collagen fibrils are present in the adenocarcinoma tissues. Statistically significant differences were also observed in the FWHM of $S$ for normal gastric tissue versus PDA ($p < 0.05$) and WDA ($p < 0.02$) [20].

## 3.3 Widefield PSHG Microscopy Cancer Studies

Castano et al. developed a widefield RDSM-SHG technique with 1030 nm excitation in which they used to investigate human non-small cell lung carcinoma (NSCLC). The tissue samples were formalin fixed, embedded in paraffin blocks, cut into 5 μm thick sections, mounted on glass slides and stained with H&E [24].

Several PSHG parameters were extracted from the widefield RDSM-SHG data to create maps of the PSHG parameters, including: $SHG_{LD}$, $LA_{LD}$, $CA_{LD}$, $SHG_{CD}$, $LA_{CD}$, $CA_{CD}$, ρ, $C \sin \Delta_{xyz}$, and $\cos 2\delta$, (see Sect. 2.8). The $SHG_{LD}$ color-coded map provides information on the approximate in-image plane orientation of the collagen fibers. It was observed that the collagen fibers in the tumor samples are predominantly oriented along the tumor margin at about 20° from the vertical. Similar to $SHG_{LD}$, $LA_{LD}$ also provides collagen fiber orientation information, hence $LA_{LD}$ also showed that collagen fibers are predominantly oriented along the tumor margin. Therefore, it was speculated that the orientation of collagen fibers may play a part in the metastasis of tumor cells [24].

Interpretation of maps consisting of the $CA_{LD}$ and $LA_{CD}$ was found to be ambiguous. However, it was suggested that these parameters can be used with other PSHG parameters to extract ultrastructural information about collagen fibers. On the other hand, the $CA_{CD}$ parameter is mostly dependent upon the ρ value (see Eq. (15)). Thus, both $CA_{CD}$ and ρ value color-coded maps show a similar distribution, mostly homogeneous with some variability along the tumor margin. The maximum value of the ρ distribution was found to be $2.03 \pm 0.36$, which is consistent with mean ρ values from previous investigations of lung tumor tissue [24].

Similarly, a map consisting of the $SHG_{CD}$ parameter was mostly homogeneous, with only select areas showing negative or positive values. The authors assumed that the $SHG_{CD}$ parameter signifies the polarity of the collagen fibers, and therefore, these regions of negative and positive values were determined to show preferential negative or positive polarity, respectively. Since $SHG_{CD}$ and $\kappa \sin \Delta_{xyz}$ are related through Eq. (19), the color-coded map of the $\kappa \sin \Delta_{xyz}$ parameter is similar to the map of the $SHG_{CD}$ parameter. Like $\kappa \sin \Delta_{xyz}$, $\cos 2\delta$ is a function of $\rho$ as well as $SHG_{LD}$ and $LA_{LD}$ (see Eq. (20)). The color-coded map of the $\cos 2\delta$ parameter is thought to show the in-image-plane orientation of the collagen fibers [24].

In an extension to Castano et al., Mirsanaye et al. used the widefield RDSM-SHG technique to investigate a large area of lung tissue encompassing NSCLC tumor at 1030 nm excitation. The samples were fixed in paraffin blocks, cut into 5 μm thick sections, mounted on glass slides, and stained with H&E [24]. The widefield RDSM-SHG technique used above was also employed to extract several PSHG parameters from widefield RDSM-SHG data of samples where NSCLC was identified by a pathologist. The extracted PSHG parameters were used to create maps of the average SHG intensity from all incident circular polarizations, $\rho$, DCP, $SHG_{CD}$, and $SHG_{LD}$ [72].

The SHG intensity was observed to decrease significantly in the tumor area of the tissue compared to the normal area of the tissue. It was also noted that the normal side of the lung tissue section contained a greater number of brighter pixels than the tumor side. The PSHG parameters, $\rho$, DCP, $SHG_{CD}$, and $SHG_{LD}$, also varied from the tumor side to the normal side of the tissue; however, not as considerably as what was observed in the SHG intensity image [72].

Further, Mirsanaye et al. used the widefield RDSM-SHG technique to investigate a H&E-stained breast tissue microarray. The microarray consisted of thirty-five 0.6 mm diameter breast tissue cores identified as normal or tumor. Fifteen of the cores were deemed normal, while twenty of the cores were tumor containing. The tumor cores were characterized by estrogen receptor (ER), progesterone receptor (PR), and human epidermal growth factor receptor 2 (HER2) expression. The tumor cores studied included: ER+/PR+/HER2+ (triple positive or +/+/+, also known as HER2 positive), ER+/PR+/HER2− (double positive or +/+/−, also known as hormone-receptor positive), and ER−/PR−/HER2− (triple negative or −/−/−). Of the 20 tumor cores, five were triple negative cores, eleven were double negative cores, and four were triple positive cores [23].

Several PSHG parameters were extracted from the widefield RDSM-SHG data to create maps of the PSHG parameters, including: The average SHG intensity from all incident circular polarizations, $\rho$, DCP, $SHG_{CD}$, and $SHG_{LD}$. In general, the tumor cores were found to have larger $\rho$ and DCP values than normal cores. The mean values of $SHG_{CD}$ and $SHG_{LD}$ did not differentiate tumor tissue from normal tissue; however, the mean absolute deviation for both of these parameters was different, where the $SHG_{CD}$ and $SHG_{LD}$ distribution appeared wider in normal tissue than tumor tissue [23].

More recently, Eftimie et al. developed an FFT PI-SHG widefield technique using 1030 nm excitation for investigating the collagen surrounding papillary thyroid carcinoma (PTC) nodules. Formalin-fixed, paraffin-embedded tissue blocks were thin sectioned (4–7 μm thick), mounted on glass slides, and stained with H&E for identification by a pathologist. The pathologist pinpointed PTC nodules, the

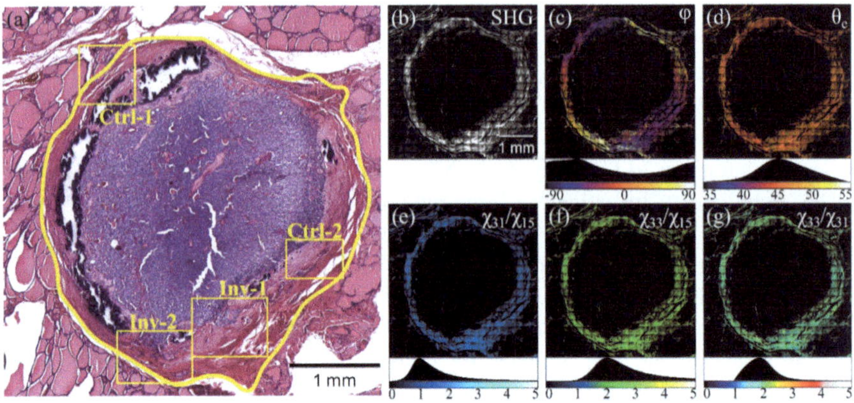


**Fig. 6** An entire thyroid nodule stained with H&E displayed using brightfield microscopy (**a**), the average SHG intensity generated from a PI-SHG dataset (**b**), the collagen orientation, δ (displayed as φ) (**c**), the helical pitch angle (**d**), $K$ (displayed as $\chi_{31}/\chi_{15}$) (**e**), $\rho^*$ (displayed as $\chi_{33}/\chi_{15}$) (**f**) and ρ (displayed as $\chi_{33}/\chi_{31}$) (**g**). Note that the tissue seen outside the yellow curve in (**a**) is considered normal thyroid tissue, while the tissue seen on the inside of the yellow curve contains a collagen capsule encasing the PTC tumor. Yellow rectangles are used to represent invasion sites within the capsule (Inv-1 and Inv-2) and control sites, which are away from the invasion sites (Ctrl-1 and Ctrl-2). The color bars below (**e**), (**f**), and (**g**) correspond to the occurrence pixel histogram of the images above. Images **a**–**g** were reprinted with permission from [68] © Optica Publishing Group

capsules of collagen surrounding them as well as capsular invasion areas and control areas [68].

Individual images of several SHG parameters, including ρ, δ, and the helical pitch angle for collagen, were stitched together to create maps of the entire PTC nodule seen in Fig. 6. From the stitched images of δ, the standard deviation (SD), MAD, and circular standard deviation (circular SD) were calculated to determine the distribution of the collagen orientation. Circular SD is a type of statistical measurement that quantifies the tightness of the grouping of a set of directional data around the mean direction. All three of these parameters show a significant change in regions identified by the pathologist as nodule invasion sites. Therefore, the collagen is suspected of being restructured close to the nodule capsule invasion sites [68].

Subsequently, the authors compared the pixel-by-pixel results of $K$ (Fig. 6e), $\rho^*$ (Fig. 6f) and ρ (Fig. 6g), and noted that statistical differences were mainly observed in ρ and $\rho^*$ when the analyzed dataset was filtered. Data was filtered by selecting collagen-rich areas using a SHG intensity threshold, or by selecting the best data via different fitting assessment methods, including the goodness of fit parameter, $R^2$, the error signal (ERR) or the signal-to-noise ratio (SNR) as defined in [16]. The authors compared fitting assessment methods and indicated the best methods as $R^2 < 0.8 - 0.9$, ERR <0.05–1 and SNR > 2–10, revealing significant differences between regions from control areas and those from nodule capsule invasion areas [68].

## 4 Conclusions and Outlook

Polarization-resolved SHG microscopy shows great promise as a technique for differentiating cancerous tissues from normal tissues and diagnosing cancer. Several improvements to the PSHG instrumentation can be made to enable fast cancer diagnosis, such as using quicker scanning mirrors including spinning mirrors or performing widefield imaging, as well as using faster polarization state generator(s). Faster imaging can also be achieved by adapting liquid crystal variable retarders or electro-optic modulators as the polarization state generators instead of mechanical rotators [73–76]. Employing fewer input laser polarizations could also be considered [77] or alternatively, using a technique by which circularly polarized light of the laser could be used in a single scan to generate several PSHG parameters [17, 78]. Widefield imaging, which expands the field of view can also be used to improve imaging speeds, as has been demonstrated by Mirsanaye et al. and Eftimie et al. [23, 68].

PSHG may also benefit from exploring the use of other laser polarization states, such as radially and azimuthally polarized laser light, which may help to detect a greater variation in the ultrastructure of cancerous tissues. For instance, radially and azimuthally polarized vortex beams used in porcine cornea demonstrated interesting variations in SHG [79].

Furthermore, super-resolution PSHG may provide more ultrastructural differentiation. One type of super-resolution PSHG microscopy technique recently introduced utilizes photonic nanojets where barium titanate glass spheres, 5–63 μm in diameter, are deposited onto the surface of samples. The high refractive index spheres further manipulate the laser excitation to induce higher resolution. The authors demonstrated the PSHG super-resolution technique via imaging of cultured lung spheroids [80].

The super-resolution technique of multifocal structured illumination microscopy was also recently combined with PSHG to image mouse tail tendon. The authors found that the orientation angles of the fibrils within the tendon are regularly arranged in a periodic pattern. This is a new finding that could be used to further differentiate tumor and normal tissues [81]. There are other super resolution SHG microscopy techniques that could be potentially coupled to PSHG to obtain more structural information of tumor tissues [82–90].

Fast data analysis of PSHG, which avoids hours of fitting, has been implemented via Fourier transforms for $C_{6v,}$ allowing to obtaining ρ, δ, and $K$, however, further formulas are needed for $C_{3v}$, κ in $C_6$, and a two-dimensional Fourier implementation is needed for PIPO-SHG. FFT PI-SHG increases the analysis speed of a 100 × 100 pixel image set from ~45 min to ~1 s, and many authors are successfully using this technique for speeding up diagnosis studies [15, 17, 65, 66, 68].

Furthermore, texture analysis techniques using PSHG-derived values can be applied to further quantify the PSHG data to obtain new characterization means for diagnosing cancer. For instance, Golaraei et al. have quantified the spatial relationship between each pixel of an image consisting of a PSHG parameter and its nearest neighbors using the gray-level co-occurrence matrix (GLCM) technique for non-small cell lung carcinoma tissue. GLCM provides a statistical representation of the textural features in the PSHG parameter image. In one instance, significant differences were not observed in the textual features of $\rho$, $\kappa$, $\delta$ and DOLP images; however, trends were observed that could be used to better understand the PIPO-SHG data [70]. Therefore, extracted PSHG parameters and textural features of PSHG parameter maps may reveal meaningful characteristics that can be used to differentiate tumor tissue.

Another technique that may be useful to aid pathologists in diagnosing cancer and enable automated screening is the use of deep learning or machine learning with SHG images. Aghigh et al. extended their PI-SHG study to also use a supervised deep learning model to differentiate SHG images of naïve and tumor-bearing mammary gland tissues. An open-source library from Google that is integrated with TensorFlow was used to create the custom sequential model, and transfer learning was achieved using the MobilenetV2 neural network architecture. The trained deep learning model was able to differentiate naïve mammary gland tissues from tumor-bearing mammary gland tissues with 73% accuracy [66]. Deep learning and machine learning has also been applied to SHG imaging of various cancers including lymphedema [91] and ovarian tissue [92, 93]. However, care must be taken with these methods because it is challenging to achieve a high accuracy when small datasets are used [94].

Machine learning can also be applied with PSHG data. Mirsanaye et al. used PSHG parameters and textural features from widefield PSHG imaging of breast cancer tissue to train a supervised machine learning algorithm (the logistic regression classifier) to differentiate normal breast tissue from tumor tissue with 94.2% accuracy. Further evaluation of the trained logistic regression classifier was performed by predicting the presence of tumor tissue on an independent dataset, resulting in 91.3% accuracy, demonstrating the potential of machine learning to differentiate normal and tumor tissues for cancer diagnosis [23].

Mirsanaye et al. also used an unsupervised machine learning algorithm (K-Means clustering) in a follow-up study using widefield PSHG imaging of NSCLC tumors. Extracted PSHG parameters and textural features were investigated using principal component analysis to classify normal and tumor-like tissue regions [72]. Similarly, Eftimie et al. recently used an unsupervised machine learning algorithm (Cluster analysis algorithm). They generated maps from the PSHG data of $\delta$, $K$, $\rho^*$, and $\rho$ and used cluster analysis to generate a cluster map that highlighted tumor invasion within thyroid nodules and may be used to facilitate diagnosis by identifying questionable regions [68].

In order to potentially translate PSHG into hospitals for use by clinicians, an in vivo PSHG technique is useful, e.g., via endoscope. To do this, backward collection of PSHG signals is needed. Research groups have developed a backward PSHG technique in which they correct the alteration to the linear polarization state of the laser due to the presence of a dichroic mirror by use of a half-wave plate and a quarter-wave plate [95, 96]. However, since backwards SHG occurs primarily due to reflected forward-directed SHG, significant polarization scattering occurs.

Similarly, a resulting decrease in the quality of PSHG signals is expected with depth. However, the polarization contrast in bone has been observed to stay constant with sample depth for up to 70 μm deep [97]. Changes in the PSHG signal response were observed at a similar depth in tendon [98], but when coupled with optical clearing techniques, such as treatment with a 50% glycerol-phosphate buffered saline solution, the PSHG signal response can be preserved in thicker tendon tissue [99], hence, perhaps a non-toxic index matching cream can be developed for patient use with similar clearing properties.

**Acknowledgements** This book chapter was supported by the Natural Science and Engineering Research Council of Canada (NSERC) Discovery Grants Program (RGPIN-2018-05444), Canada's Research Support Fund, and Saint Mary's University. MH was supported by a Nova Scotia Graduate Scholarship and a Scotia Scholars Award (Doctoral).

## References

1. Mazumder, N., Balla, N. K., Zhuo, G.-T., Kistenev, Y. V., Kumar, R., Kao, F.-J., Brasselet, S., Nikolaev, V. V., & Krivova, N. A. (2019). Label-free non-linear multimodal optical microscopy—basics, development, and applications. *Frontiers in Physics*, 7, 170. https://doi.org/10.3389/fphy.2019.00170
2. James, D. S., & Campagnola, P. J. (2021). Recent advancements in optical harmonic generation microscopy: Applications and perspectives. *BME Frontiers*, 2021, 3973857. https://doi.org/10.34133/2021/3973857
3. Cisek, R., Joseph, A., Harvey, M., & Tokarz, D. (2021). Polarization-sensitive second harmonic generation microscopy for investigations of diseased collagenous tissues. *Frontiers in Physics*, 9, 726996. https://doi.org/10.3389/fphy.2021.726996
4. Cisek, R., Harvey, M., Bennett, E., Jeon, H., & Tokarz, D. (2023). Polarization-resolved SHG microscopy for biomedical applications. In N. Mazumder, Y. V. Kistenev, E. Borisova, & S. P. Kabekkodu (Eds.), *Optical polarimetric modalities for biomedical research* (Vol. 1, 1st ed., pp. 215–258). Springer Nature Switzerland AG.
5. Hache, F., Mesnil, H., & Schanne-Klein, M. C. (2001). Application of classical models of chirality to surface second harmonic generation. *Journal of Chemical Physics*, 115, 6707–6715. https://doi.org/10.1063/1.1404983
6. Schmeltz, M., Teulon, C., Pinsard, M., Hansen, U., Alnawaiseh, M., Ghoubay, D., Borderie, V., Mosser, G., Aimé, C., Légaré, F., Latour, G., & Schanne-Klein, M.-C. (2020). Circular dichroism second-harmonic generation microscopy probes the polarity distribution of collagen fibrils. *Optica*, 7(11), 1469–1475. https://doi.org/10.1364/optica.399246
7. Harvey, M., Cisek, R., Tokarz, D., & Kreplak, L. (2023). Effect of out of plane orientation on polarization second harmonic generation of single collagen fibrils. *Biomedical Optics Express*, 14(12), 6271–6282. https://doi.org/10.1364/boe.504304
8. Harvey, M., Lane, B., Cisek, R., Veres, S. P., Kreplak, L., & Tokarz, D. (2024). Histological staining alters circular dichroism SHG measurements of collagen. *Optics Letters*, 49(13), 3705–3708. https://doi.org/10.1364/OL.523689
9. Dailey, C. A., Burke, B. J., & Simpson, G. J. (2004). The general failure of Kleinman symmetry in practical nonlinear optical applications. *Chemical Physics Letters, 390*(1–3), 8–13. https://doi.org/10.1016/j.cplett.2004.03.109
10. Kleinman, D. A. (1962). Nonlinear dielectric polarization in optical media. *Physical Review, 126*(6), 1977–1979. https://doi.org/10.1103/PhysRev.126.1977

11. Roth, S., & Freund, I. (1979). Second harmonic generation in collagen. *The Journal of Chemical Physics, 70*(4), 1637–1643. https://doi.org/10.1063/1.437677
12. Tuer, A. E., Akens, M. K., Krouglov, S., Sandkuijl, D., Wilson, B. C., Whyne, C. M., & Barzda, V. (2012). Hierarchical model of fibrillar collagen organization for interpreting the second-order susceptibility tensors in biological tissue. *Biophysical Journal, 103*(10), 2093–2105. https://doi.org/10.1016/j.bpj.2012.10.019
13. Ambekar, R., Lau, T.-Y., Walsh, M., Bhargava, R., & Toussaint, K. C. (2012). Quantifying collagen structure in breast biopsies using second-harmonic generation imaging. *Biomedical Optics Express, 3*(9), 2021–2035. https://doi.org/10.1364/BOE.3.002021
14. Amat-Roldan, I., Psilodimitrakopoulos, S., Loza-Alvarez, P., & Artigas, D. (2010). Fast image analysis in polarization SHG microscopy. *Optics Express, 18*(16), 17209–17219. https://doi.org/10.1364/OE.18.017209
15. Tsafas, V., Gavgiotaki, E., Tzardi, M., Tsafa, E., Fotakis, C., Athanassakis, I., & Filippidis, G. (2020). Polarization-dependent second-harmonic generation for collagen-based differentiation of breast cancer samples. Journal of Biophotonics, 13(10), e202000180. https://doi.org/10.1002/jbio.202000180
16. Hristu, R., Stanciu, S. G., Dumitru, A., Eftimie, L. G., Paun, B., Tranca, D. E., Gheorghita, P., Costache, M., & Stanciu, G. A. (2022). PSHG-TISS: A collection of polarization-resolved second harmonic generation microscopy images of fixed tissues. Scientific Data, 9(1), 376. https://doi.org/10.1038/s41597-022-01477-1
17. Alizadeh, M., Ghotbi, M., Loza-Alvarez, P., & Merino, D. (2019). Comparison of different polarization sensitive second harmonic generation imaging techniques. Methods and Protocols, 2(49), 1–14. https://doi.org/10.3390/mps2020049
18. Golaraei, A., Mirsanaye, K., Ro, Y., Krouglov, S., Akens, M. K., Wilson, B. C., & Barzda, V. (2019). Collagen chirality and three-dimensional orientation studied with polarimetric second-harmonic generation microscopy. Journal of Biophotonics, 12(1), e201800241. https://doi.org/10.1002/jbio.201800241
19. Harvey, M., Cisek, R., Alizadeh, M., Barzda, V., Kreplak, L., & Tokarz, D. (2023). High numerical aperture imaging allows chirality measurement in individual collagen fibrils using polarization second harmonic generation microscopy. Nanophotonics, 12(11), 2061–2071. https://doi.org/10.1515/nanoph-2023-0177
20. Jeon, H., Harvey, M., Cisek, R., Bennett, E., & Tokarz, D. (2023). Characterization of pathological stomach tissue using polarization-sensitive second harmonic generation microscopy. Biomedical Optics Express, 14(10), 5376–5391. https://doi.org/10.1364/boe.500335
21. Tokarz, D., Cisek, R., Golaraei, A., Asa, S. L., Barzda, V., & Wilson, B. C. (2015). Ultrastructural features of collagen in thyroid carcinoma tissue observed by polarization second harmonic generation microscopy. *Biomedical Optics Express, 6*(9), 3475–3481. https://doi.org/10.1364/BOE.6.003475
22. Golaraei, A., Kontenis, L., Cisek, R., Tokarz, D., Done, S. J., Wilson, B. C., & Barzda, V. (2016). Changes of collagen ultrastructure in breast cancer tissue determined by second-harmonic generation double stokes-mueller polarimetric microscopy. *Biomedical Optics Express, 7*(10), 4054–4068. https://doi.org/10.1364/BOE.7.004054
23. Mirsanaye, K., Castaño, L. U., Kamaliddin, Y., Golaraei, A., Augulis, R., Kontenis, L., Done, S. J., Žurauskas, E., Stambolic, V., Wilson, B. C., & Barzda, V. (2022). Machine learning-enabled cancer diagnostics with widefield polarimetric second-harmonic generation microscopy. Scientific Reports, 12, 10290. https://doi.org/10.1038/s41598-022-13623-1
24. Castaño, L. U., Mirsanaye, K., Kontenis, L., Krouglov, S., Žurauskas, E., Navab, R., Yasufuku, K., Tsao, M.-S., Akens, M. K., Wilson, B. C., & Barzda, V. (2022). Wide field stokes polarimetric microscopy for SHG imaging. Journal of Biophotonics, 16(5), e202200284. https://doi.org/10.1002/jbio.202200284
25. Mirsanaye, K., Golaraei, A., Habach, F., Žurauskas, E., Venius, J., Rotomskis, R., & Barzda, V. (2019). Polar organization of collagen in human cardiac tissue revealed with polarimet-

ric second-harmonic generation microscopy. Biomedical Optics Express, 10(10), 5025–5030. https://doi.org/10.1364/boe.10.005025
26. Brown, C. P., Houle, M.-A., Popov, K., Nicklaus, M., Couture, C.-A., Laliberté, M., Brabec, T., Ruediger, A., Carr, A. J., Price, A. J., Gill, H. S., Ramunno, L., & Légaré, F. (2014). Imaging and modeling collagen architecture from the nano to micro scale. *Biomedical Optics Express, 5*(1), 233–243. https://doi.org/10.1364/boe.5.000233
27. Kumar, R., Grønhaug, K. M., Romijn, E. I., Finnøy, A., Davies, C. L., Drogset, J. O., & Lilledahl, M. B. (2015). Polarization second harmonic generation microscopy provides quantitative enhanced molecular specificity for tissue diagnostics. *Journal of Biophotonics, 8*(9), 730–739. https://doi.org/10.1002/jbio.201400086
28. Mansfield, J. C., Mandalia, V., Toms, A., Peter Winlove, C., & Brasselet, S. (2019). Collagen reorganization in cartilage under strain probed by polarization sensitive second harmonic generation microscopy. *Journal of the Royal Society Interface, 16*(150), 20180611. https://doi.org/10.1098/rsif.2018.0611
29. Alizadeh, M., Merino, D., Lombardo, G., Lombardo, M., Mencucci, R., Ghotbi, M., & Loza-Alvarez, P. (2019). Identifying crossing collagen fibers in human corneal tissues using pSHG images. Biomedical Optics Express, 10(8), 3875–3888. https://doi.org/10.1364/boe.10.003875
30. Ávila, F. J., Del Barco, O., & Bueno, J. M. (2016). Polarization response of second-harmonic images for different collagen spatial distributions. *Journal of Biomedical Optics*, 21(6), 66015. https://doi.org/10.1117/1.JBO.21.6.066015
31. Latour, G., Gusachenko, I., Kowalczuk, L., Lamarre, I., & Schanne-Klein, M.-C. (2012). In vivo structural imaging of the cornea by polarization-resolved second harmonic microscopy. *Biomedical Optics Express, 3*(1), 1. https://doi.org/10.1364/boe.3.000001
32. Lombardo, M., Merino, D., Loza-Alvarez, P., & Lombardo, G. (2015). Translational label-free nonlinear imaging biomarkers to classify the human corneal microstructure. Biomedical Optics Express, 6(8), 2803-2818. https://doi.org/10.1364/boe.6.002803
33. Matteini, P., Cicchi, R., Ratto, F., Kapsokalyvas, D., Rossi, F., De Angelis, M., Pavone, F. S., & Pini, R. (2012). Thermal transitions of fibrillar collagen unveiled by second-harmonic generation microscopy of corneal stroma. *Biophysical Journal, 103*(6), 1179–1187. https://doi.org/10.1016/j.bpj.2012.07.055
34. Raoux, C., Schmeltz, M., Bied, M., Alnawaiseh, M., Hansen, U., Latour, G., & Schanne-Klein, M.-C. (2021). Quantitative structural imaging of keratoconic corneas using polarization-resolved SHG microscopy. *Biomedical Optics Express, 12*(7), 4163–4178. https://doi.org/10.1364/boe.426145
35. Chen, X., Raggio, C., & Campagnola, P. J. (2012). Second-harmonic generation circular dichroism studies of osteogenesis imperfecta. *Optics Letters, 37*(18), 3837–3839. https://doi.org/10.1364/OL.37.003837
36. Erikson, A., Ortegren, J., Hompland, T., Davies, C. D., & Lindgren, M. (2007). Quantification of the second-order nonlinear susceptibility of collagen I using a laser scanning microscope. Journal of Biomedical Optics, 12(4), 044002. https://doi.org/10.1117/1.2772311
37. Hristu, R., Stanciu, S. G., Tranca, D. E., & Stanciu, G. A. (2017). Improved quantification of collagen anisotropy with polarization-resolved second harmonic generation microscopy. *Journal of Biophotonics, 10*(9), 1171–1179. https://doi.org/10.1002/jbio.201600197
38. Miler, I., Rabasovic, M. D., Aleksic, M., Krmpot, A. J., Kalezic, A., Jankovic, A., Korac, B., & Korac, A. (2021). Polarization-resolved SHG imaging as a fast screening method for collagen alterations during aging: Comparison with light and electron microscopy. Journal of Biophotonics, 14(3), e202000362. https://doi.org/10.1002/jbio.202000362
39. Tanaka, Y., Hase, E., Fukushima, S., Ogura, Y., Yamashita, T., Hirao, T., Araki, T., & Yasui, T. (2014). Motion-artifact-robust, polarization-resolved second-harmonic-generation microscopy based on rapid polarization switching with electro-optic Pockells cell and its application to in vivo visualization of collagen fiber orientation in human facial skin. *Biomedical Optics Express, 5*(4), 1099–1113. https://doi.org/10.1364/boe.5.001099

40. Yasui, T., Takahashi, Y., Fukushima, S., Ogura, Y., Yamashita, T., Kuwahara, T., Hirao, T., & Araki, T. (2009). Observation of dermal collagen fiber in wrinkled skin using polarization-resolved second-harmonic-generation microscopy. *Optics Express, 17*(2), 912–923. https://doi.org/10.1364/oe.17.000912
41. Golaraei, A., Kontenis, L., Mirsanaye, K., Krouglov, S., Akens, M. K., Wilson, B. C., & Barzda, V. (2019). Complex susceptibilities and chiroptical effects of collagen measured with polarimetric second-harmonic generation microscopy. Scientific Reports, 9(1), 12488. https://doi.org/10.1038/s41598-019-48636-w
42. Gusachenko, I., Tran, V., Houssen, Y. G., Allain, J. M., & Schanne-Klein, M. C. (2012). Polarization-resolved second-harmonic generation in tendon upon mechanical stretching. *Biophysical Journal, 102*(9), 2220–2229. https://doi.org/10.1016/j.bpj.2012.03.068
43. Mazumder, N., & Kao, F.-J. (2021). Stokes polarimetry-based second harmonic generation microscopy for collagen and skeletal muscle fiber characterization. *Lasers in Medical Science, 36*(6), 1161–1167. https://doi.org/10.1007/s10103-020-03144-6
44. Mazumder, N., Qiu, J., Foreman, M. R., Macias-Romero, C., Hu, C.-W., Tsai, H.-R., Török, P., & Kao, F.-J. (2012). Polarization-resolved second harmonic generation microscopy with a four-channel Stokes-polarimeter. *Optics Express, 20*(13), 14090–14099. https://doi.org/10.1364/OE.20.014090
45. Okoro, C., & Toussaint, K. C. (2017). Second-harmonic patterned polarization-analyzed reflection confocal microscope. Journal of Biomedical Optics, 22(8), 086007. https://doi.org/10.1117/1.JBO.22.8.086007
46. Romijn, E. I., Finnøy, A., & Lilledahl, M. B. (2019). Analyzing the feasibility of discriminating between collagen types I and II using polarization-resolved second harmonic generation. Journal of Biophotonics, 12(1), e201800090. https://doi.org/10.1002/jbio.201800090
47. Roth, S., & Freund, I. (1981). Optical second-harmonic scattering in rat-tail tendon. *Biopolymers, 20*(6), 1271–1290. https://doi.org/10.1002/bip.1981.360200613
48. Silva, D. F. T., Gomes, A. S. L., De Campos Vidal, B., & Ribeiro, M. S. (2013). Birefringence and second harmonic generation on tendon collagen following red linearly polarized laser irradiation. *Annals of Biomedical Engineering, 41*(4), 752–762. https://doi.org/10.1007/s10439-012-0720-3
49. Stoller, P., Reiser, K. M., Celliers, P. M., & Rubenchik, A. M. (2002). Polarization-modulated second harmonic generation in collagen. *Biophysical Journal, 82*(6), 3330–3342. https://doi.org/10.1016/S0006-3495(02)75673-7
50. Tuer, A. E., Krouglov, S., Prent, N., Cisek, R., Sandkuijl, D., Yasufuku, K., Wilson, B. C., & Barzda, V. (2011). Nonlinear optical properties of type I collagen fibers studied by polarization dependent second harmonic generation microscopy. The Journal of Physical Chemistry B, 115(44), 12759–12769. https://doi.org/10.1021/jp206308k
51. Ajeti, V., Nadiarnykh, O., Ponik, S. M., Keely, P. J., Eliceiri, K. W., & Campagnola, P. J. (2011). Structural changes in mixed Col I/Col V collagen gels probed by SHG microscopy: Implications for probing stromal alterations in human breast cancer. *Biomedical Optics Express, 2*(8), 2307–2316. https://doi.org/10.1364/BOE.2.002307
52. Campbell, K. R., & Campagnola, P. J. (2017). Wavelength-dependent second harmonic generation circular dichroism for differentiation of Col i and Col III isoforms in stromal models of ovarian cancer based on intrinsic chirality differences. *Journal of Physical Chemistry B, 121*(8), 1749–1757. https://doi.org/10.1021/acs.jpcb.6b06822
53. Chen, C.-H., Nair, A. V., Chuang, S.-C., Lin, Y.-S., Cheng, M.-H., Lin, C.-Y., Chang, C.-Y., Chen, S.-J., & Lien, C.-H. (2021). Dual-LC PSHG microscopy for imaging collagen type I and type II gels with pixel-resolution analysis. Biomedical Optics Express, 12(5), 3050–3065. https://doi.org/10.1364/boe.416193
54. Fuentes-Corona, C. G., Licea-Rodriguez, J., Younger, R., Rangel-Rojo, R., Potma, E. O., & Rocha-Mendoza, I. (2019). Second harmonic generation signal from type I collagen fibers grown in vitro. Biomedical Optics Express, 10(12), 6449–6461. https://doi.org/10.1364/boe.10.006449

55. Paesen, R., Sanen, K., Smisdom, N., Michiels, L., & Ameloot, M. (2014). Polarization second harmonic generation by image correlation spectroscopy on collagen type i hydrogels. *Acta Biomaterialia, 10*(5), 2036–2042. https://doi.org/10.1016/j.actbio.2014.01.011
56. Rouède, D., Bellanger, J.-J., Bomo, J., Baffet, G., & Tiaho, F. (2015). Linear least square (LLS) method for pixel-resolution analysis of polarization dependent SHG images of collagen fibrils. Optics Express, 23(10), 13309–13319. https://doi.org/10.1364/oe.23.013309
57. Sanen, K., Paesen, R., Luyck, S., Phillips, J., Lambrichts, I., Martens, W., & Ameloot, M. (2016). Label-free mapping of microstructural organisation in self-aligning cellular collagen hydrogels using image correlation spectroscopy. *Acta Biomaterialia, 30*, 258–264. https://doi.org/10.1016/j.actbio.2015.10.047
58. Xydias, D., Ziakas, G., Psilodimitrakopoulos, S., Lemonis, A., Bagli, E., Fotsis, T., Gravanis, A., Tzeranis, D. S., & Stratakis, E. (2021). Three-dimensional characterization of collagen remodeling in cell-seeded collagen scaffolds via polarization second harmonic generation. Biomedical Optics Express, 12(2), 1136–1153. https://doi.org/10.1364/boe.411501
59. Boulesteix, T., Beaurepaire, E., Sauviat, M.-P., & Schanne-Klein, M.-C. (2004). Second-harmonic microscopy of unstained living cardiac myocytes: Measurements of sarcomere length with 20-nm accuracy. Optics Letters, 29(17), 2031–2033. https://doi.org/10.1364/ol.29.002031
60. Campagnola, P. J., Millard, A. C., Terasaki, M., Hoppe, P. E., Malone, C. J., & Mohler, W. A. (2002). Three-dimensional high-resolution second-harmonic generation imaging of endogenous structural proteins in biological tissues. *Biophysical Journal, 82*(1), 493–508. https://doi.org/10.1016/S0006-3495(02)75414-3
61. Suhalim, J. L., Chung, C. Y., Lilledahl, M. B., Lim, R. S., Levi, M., Tromberg, B. J., & Potma, E. O. (2012). Characterization of cholesterol crystals in atherosclerotic plaques using stimulated Raman scattering and second-harmonic generation microscopy. *Biophysical Journal, 102*(8), 1988–1995. https://doi.org/10.1016/j.bpj.2012.03.016
62. Galli, R., Meinhardt, M., Koch, E., Schackert, G., Steiner, G., Kirsch, M., & Uckermann, O. (2018). Optical molecular imaging of corpora amylacea in human brain tissue. *Biomedizinische Technik, 63*(5), 579–585. https://doi.org/10.1515/bmt-2017-0073
63. Fu, Y., Wang, H., Shi, R., & Cheng, J.-X. (2007). Second harmonic and sum frequency generation imaging of fibrous astroglial filaments in ex vivo spinal tissues. Biophysical Journal, 92(9), 3251–3259. https://doi.org/10.1529/biophysj.106.089011
64. Brittain, K., Harvey, M., Cisek, R., Pillai, S., Christie, S. D., & Tokarz, D. (2022). Second harmonic generation microscopy of otoconia. *Biomedical Optics Express, 13*(6), 3593–3600. https://doi.org/10.1364/boe.457967
65. Mercatelli, R., Triulzi, T., Pavone, F. S., Orlandi, R., & Cicchi, R. (2020). Collagen ultrastructural symmetry and its malignant alterations in human breast cancer revealed by polarization-resolved second-harmonic generation microscopy. Journal of Biophotonics, 13(8), e202000159. https://doi.org/10.1002/jbio.202000159
66. Aghigh, A., Preston, S. E. J., Jargot, G., Ibrahim, H., Del Rincón, S. V., & Légaré, F. (2023). Nonlinear microscopy and deep learning classification for mammary gland microenvironment studies. Biomedical Optics Express, 14(5), 2181–2195. https://doi.org/10.1364/boe.487087
67. Bredfeldt, J. S., Liu, Y., Conklin, M. W., Keely, P. J., Mackie, T. R., & Eliceiri, K. W. (2014). Automated quantification of aligned collagen for human breast carcinoma prognosis. *Journal of Pathology Informatics, 5*(1), 28. https://doi.org/10.4103/2153-3539.139707
68. Eftimie, L. G., Padrez, Y., Golubewa, L., Rutkauskas, D., & Hristu, R. (2024). Widefield polarization-resolved second harmonic generation imaging of entire thyroid nodule sections for the detection of capsular invasion. Biomedical Optics Express, 15(8), 4705–4718. https://doi.org/10.1364/boe.523052
69. Chen, W. C., Chen, Y. J., Lin, S. T., Hung, W. H., Chan, M. C., Wu, I. C., Wu, M. T., Kuo, C. T., Das, S., Kao, F. J., & Zhuo, G. Y. (2020). Label-free characterization of collagen fibers in cancerous esophagus tissues using ratiometric nonlinear optical microscopy. *Experimental Biology and Medicine, 245*(14), 1213–1221. https://doi.org/10.1177/1535370220934039

70. Golaraei, A., Mostaço-Guidolin, L. B., Raja, V., Navab, R., Wang, T., Sakashita, S., Yasufuku, K., Tsao, M.-S., Wilson, B. C., & Barzda, V. (2020). Polarimetric second-harmonic generation microscopy of the hierarchical structure of collagen in stage I-III non-small cell lung carcinoma. Biomedical Optics Express, 11(4), 1851–1863. https://doi.org/10.1364/boe.387744
71. Tokarz, D., Cisek, R., Joseph, A., Asa, S. L., Wilson, B. C., & Barzda, V. (2020). Characterization of pathological thyroid tissue using polarization-sensitive second harmonic generation microscopy. *Laboratory Investigation, 100*(2020), 1280–1287. https://doi.org/10.1038/s41374-020-0475-7
72. Mirsanaye, K., Castaño, L. U., Kamaliddin, Y., Golaraei, A., Kontenis, A., Žurauskas, E., Navab, R., Yasufuku, K., Tsao, M.-S., Wilson, B. C., & Barzda, V. (2022). Unsupervised determination of lung tumor margin with widefield polarimetric second-harmonic generation microscopy. Scientific Reports, 12, 20713. https://doi.org/10.1038/s41598-022-24973-1
73. Dewalt, E. L., Sullivan, S. Z., Schmitt, P. D., Muir, R. D., & Simpson, G. J. (2014). Polarization-modulated second harmonic generation ellipsometric microscopy at video rate. *Analytical Chemistry, 86*(16), 8448–8456. https://doi.org/10.1021/ac502124v
74. Dow, X. Y., DeWalt, E. L., Sullivan, S. Z., Schmitt, P. D., Ulcickas, J. R. W., & Simpson, G. J. (2016). Imaging the nonlinear susceptibility tensor of collagen by nonlinear optical stokes Ellipsometry. *Biophysical Journal, 111*(7), 1361–1374. https://doi.org/10.1016/j.bpj.2016.05.055
75. Lien, C.-H., Tilbury, K., Chen, S.-J., & Campagnola, P. J. (2013). Precise, motion-free polarization control in second harmonic generation microscopy using a liquid crystal modulator in the infinity space. *Biomedical Optics Express, 4*(10), 1991–2002. https://doi.org/10.1364/BOE.4.001991
76. Reiser, K., Stoller, P., & Knoesen, A. (2017). Three-dimensional geometry of collagenous tissues by second harmonic polarimetry. Scientific Reports, 7(1), 2642. https://doi.org/10.1038/s41598-017-02326-7
77. Ávila, F. J., Del Barco, O., & Bueno, J. M. (2017). Quantifying external and internal collagen organization from Stokes-vector-based second harmonic generation imaging polarimetry. Journal of Optics, 19(10), 10530.1 https://doi.org/10.1088/2040-8986/aa825d
78. Psilodimitrakopoulos, S., Loza-Alvarez, P., & Artigas, D. (2014). Fast monitoring of in-vivo conformational changes in myosin using single scan polarization-SHG microscopy. *Biomedical Optics Express, 5*(12), 4362–4373. https://doi.org/10.1364/BOE.5.004362
79. Ehmke, T., Nitzsche, T. H., Knebl, A., & Heisterkamp, A. (2014). Molecular orientation sensitive second harmonic microscopy by radially and azimuthally polarized light. Biomedical Optics Express, 5(7), 2231–2246. https://doi.org/10.1364/boe.5.002231
80. Johnson, P. B., Karvounis, A., Singh, H. J., Brereton, C. J., Bourdakos, K. N., Lunn, K., Roberts, J. J. W., Davies, D. E., Muskens, O. L., Jones, M. G., & Mahajan, S. (2021). Superresolved polarization-enhanced second-harmonic generation for direct imaging of nanoscale changes in collagen architecture. Optica, 8(5), 674-685. https://doi.org/10.1364/optica.411325
81. Zhang, Y., Zhang, C., Zhang, R., Xu, R., Yu, B., Lin, D., & Qu, J. (2024). Polarization-resolved super-resolution second-harmonic generation imaging based on multifocal structured illumination microscopy. Optics Letters, 49(6), 1540-1543. https://doi.org/10.1364/ol.514724
82. Barlow, A. M., Mostaço-Guidolin, L. B., Osei, E. T., Booth, S., & Hackett, T. L. (2020). Super resolution measurement of collagen fibers in biological samples: Validation of a commercial solution for multiphoton microscopy. PLoS One, 15(2), 0229278. https://doi.org/10.1371/journal.pone.0229278
83. Gregor, I., Spiecker, M., Petrovsky, R., Groβhans, J., Ros, R., & Enderlein, J. (2017). Rapid nonlinear image scanning microscopy. Nature Methods, 14(11), 1087–1089. https://doi.org/10.1038/nmeth.4467
84. Lee, Y. U., Wisna, G. B. M., Hsu, S. W., Zhao, J., Lei, M., Li, S., Tao, A. R., Tao, A. R., Liu, Z., & Liu, Z. (2020). Imaging of nanoscale light confinement in plasmonic nanoantennas by Brownian optical microscopy. *ACS Nano, 14*(6), 7666–7672. https://doi.org/10.1021/acsnano.0c04019

85. Liu, J., Cho, I. H., Cui, Y., & Irudayaraj, J. (2014). Second harmonic super-resolution microscopy for quantification of mRNA at single copy sensitivity. *ACS Nano, 8*(12), 12418–12427. https://doi.org/10.1021/nn505096t
86. Raanan, D., Song, M. S., Tisdale, W. A., & Oron, D. (2022). Super-resolved second harmonic generation imaging by coherent image scanning microscopy. Applied Physics Letters, 120, 071111. https://doi.org/10.1063/5.0073161
87. Stanciu, S. G., Hristu, R., Stanciu, G. A., Tranca, D. E., Eftimie, L., Dumitru, A., Costache, M., Stenmark, H. A., & Analyzed, E. M. M. M. (2022). Super-resolution re-scan second harmonic generation microscopy. The Proceedings of the National Academy of Sciences, 119(47), e2214662119. https://doi.org/10.1073/pnas
88. Urban, B. E., Dong, B., Zhang, X., Yang, H., & Zhang, H. F. (2018). Patterned-illumination second harmonic generation microscopy of collagen fibrils in rat Scleras. Optics Letters, 43(21), 5190–5193. https://doi.org/10.1364/OL.43.005190
89. Wang, L., Zheng, X., Zhou, J., Wang, M., Chen, J., Zeng, Y., Xu, G., Wang, Y., Qiu, H., Shao, Y., Qu, J., Gao, B. Z., & Gu, Y. (2022). Improvement in resolution of multiphoton scanning structured illumination microscopy via harmonics. Engineering, 16(9), 65–72. https://doi.org/10.1016/j.eng.2021.12.010
90. Wu, W., Chen, J., Wang, M., Wang, L., Zheng, X., Li, J., Qu, J., Gao, B. Z., & Shao, Y. (2023). Comparison of point detection and area detection for point-scanning structured illumination microscopy. Journal of Innovative Optical Health Sciences, 16(4), 2350010. https://doi.org/10.1142/S1793545823500104
91. Kistenev, Y. V., Nikolaev, V. V., Kurochkina, O. S., Borisov, A. V., Vrazhnov, D. A., & Sandykova, E. A. (2019). Application of multiphoton imaging and machine learning to lymphedema tissue analysis. Biomedical Optics Express, 10(7), 3353–3368. https://doi.org/10.1364/boe.10.003353
92. Huttunen, M. J., Hassan, A., McCloskey, C. W., Fasih, S., Upham, J., Vanderhyden, B. C., Boyd, R. W., & Murugkar, S. (2018). Automated classification of multiphoton microscopy images of ovarian tissue using deep learning. Journal of Biomedical Optics, 23(6), 1–7. https://doi.org/10.1117/1.jbo.23.6.066002
93. Wang, G., Sun, Y., Jiang, S., Wu, G., Liao, W., Chen, Y., Lin, Z., Liu, Z., & Zhuo, S. (2021). Machine learning-based rapid diagnosis of human borderline ovarian cancer on second-harmonic generation images. *Biomedical Optics Express*, 12(9), 5658–5669. https://doi.org/10.1364/boe.429918
94. Safonova, A., Ghazaryan, G., Stiller, S., Main-Knorn, M., Nendel, C., & Ryo, M. (2023). Ten deep learning techniques to address small data problems with remote sensing. International *Journal of Applied Earth Observation and Geoinformation*, 125, 103569. https://doi.org/10.1016/j.jag.2023.103569
95. Chou, C.-K., Chen, W.-L., Fwu, P. T., Lin, S.-J., Lee, H.-S., & Dong, C.-Y. (2008). Polarization ellipticity compensation in polarization second-harmonic generation microscopy without specimen rotation. *Journal of Biomedical Optics*, 13(1), 014005. https://doi.org/10.1117/1.2824379
96. Romijn, E. I., Finnøy, A., Kumar, R., & Lilledahl, M. B. (2018). Automated calibration and control for polarization-resolved second harmonic generation on commercial microscopes. *PLoS One*, 13(4), e0195027. https://doi.org/10.1371/journal.pone.0195027
97. Pendleton, E. G., Tehrani, K. F., Barrow, R. P., & Mortensen, L. J. (2020). Second harmonic generation characterization of collagen in whole bone. *Biomedical Optics Express, 11*(8), 4379–4396. https://doi.org/10.1364/BOE.391866
98. Mansfield, J. C., Winlove, C. P., Moger, J., & Matcher, S. J. (2008). Collagen fiber arrangement in normal and diseased cartilage studied by polarization sensitive nonlinear microscopy. *Journal of Biomedical Optics, 13*(4), 044020. https://doi.org/10.1117/1.2950318
99. Nadiarnykh, O., & Campagnola, P. J. (2009). Retention of polarization signatures in SHG microscopy of scattering tissues through optical clearing. *Optics Express, 17*(7), 5794–5806. https://doi.org/10.1364/OE.17.005794

# Nonlinear Optical Microscopy for Cancer Diagnosis

**Gagan Raju, Kausalya Neelavara Makkithaya, Jackson Rodrigues, Ming-Chi Chen, Nirmal Mazumder, Ankur Gogoi, and Guan-Yu Zhuo**

**Abstract** Nonlinear optical microscopy (NLOM) emerged as a powerful imaging modality in cancer diagnosis, offering unparalleled capabilities for label-free, high-resolution visualization of tissue architecture and cellular dynamics with minimal invasion. By leveraging nonlinear optical phenomena, techniques such as two-photon fluorescence (TPF), second harmonic generation (SHG), third harmonic generation (THG), and coherent anti-Stokes Raman scattering (CARS) enable detailed imaging of the tumor microenvironment (TME), extracellular matrix (ECM), and cellular interactions at subcellular resolution without the requirement for external contrast agents. These techniques are particularly effective for assessing the ECM structure, collagen organization, and tumor-associated features critical to cancer progression.

This chapter systematically examines specific applications of NLOM across major cancer types, including breast, skin, brain, and gastrointestinal (GI) cancers. For breast cancer, NLOM excels in detecting tumor-associated collagen signatures (TACS) and monitoring collagen reorganization. In skin cancer, it enables detailed imaging of tissue morphology and assessment of tumor margins, supporting both diagnostic accuracy and surgical precision, while in brain cancers, it delineates tumor boundaries and cellular infiltration patterns. The NLOM also facilitates early detection of GI malignancies and accurate evaluation of surgical margins. A comparative analysis with conventional optical imaging modalities, such as confocal microscopy, optical coherence tomography (OCT), and Raman microspectroscopy, underscores NLOM's superior imaging depth, resolution, and intrinsic label-free imaging capabilities. Challenges related to depth penetration, signal-to-noise ratio

G. Raju · K. N. Makkithaya · N. Mazumder
Department of Biophysics, Manipal School of Life Sciences, Manipal Academy of Higher Education, Manipal, Karnataka, India

J. Rodrigues · M.-C. Chen · G.-Y. Zhuo (✉)
Institute of Biophotonics, National Yang Ming Chiao Tung University, Taipei, Taiwan

A. Gogoi
Department of Physics, Jagannath Barooah University, Jorhat, Assam, India



N. Mazumder et al. (eds.), *Advanced Biophysical Techniques in Biosciences*, Lecture Notes in Nanoscale Science and Technology 7,
https://doi.org/10.1007/978-3-031-94551-9_11

(SNR) optimization, and miniaturization for clinical translation are discussed alongside strategies to overcome them. This chapter highlights advancements in hardware, software, and artificial intelligence (AI), emphasizing their role in real-time, in vivo cancer detection and monitoring. This extensive review is a valuable resource for researchers and clinicians interested in understanding the principles, applications, and future directions of NLOM in cancer diagnosis.

**Keywords** Optical microscopy · Cancer detection · Cancer monitoring · Tumor-associated collagen signature

## 1 Introduction

In recent years, nonlinear optical imaging (NLOI) has emerged as a promising technique in biomedical imaging, with its most impactful applications observed in preclinical and clinical cancer research (Fig. 1) [1]. Several nonlinear optical microscopy (NLOM) modalities, including two-photon fluorescence (TPF), second harmonic generation (SHG), third harmonic generation (THG), and coherent anti-Stokes Raman scattering (CARS), offer label-free imaging that can reveal fine details of the tumor microenvironment (TME), such as the extracellular matrix (ECM) structure

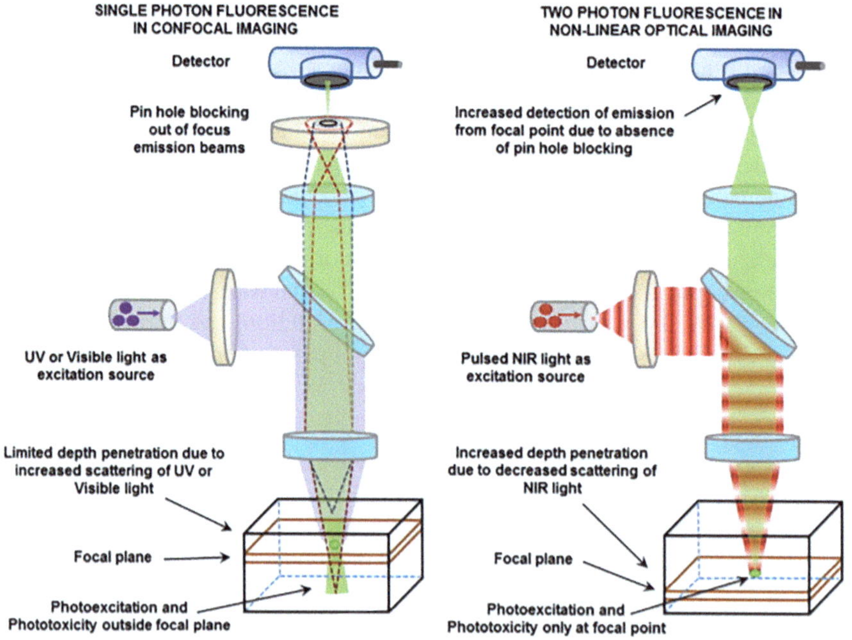


**Fig. 1** Comparison of single-photon excitation fluorescence in confocal microscopy and two-photon fluorescence in NLOM. (Figure and caption used with permission from [1])

and cellular interactions [2]. These modalities are useful for visualizing collagen fibers, lipid droplets, and other key markers involved in cancer progression, without the requirement for external contrast agents, offering a minimally invasive alternative for tumor analysis [3, 4].

Nonlinear optical techniques have been extensively used in biomedical research due to the notable nonlinear optical properties exhibited by many biological tissue structures. These techniques, particularly in the field of bioimaging, can offer quick, label-free, and chemically precise imaging of biological samples. This allows for the analysis of samples and the study of biological processes in ways that are not possible with conventional microscopy techniques [5]. This makes it a useful tool for both early diagnosis and therapy evaluation since it enables the assessment of treatment outcomes and tumor morphology. Additionally, NLOM can detect subtle biochemical and morphological changes in tissues that could be indicative of early-stage cancer [6–8]. Its noninvasive nature may minimize patient risk, offering an alternative to traditional biopsies.

## *1.1 Brief Overview of Cancer Diagnosis Challenges*

Cancer diagnosis is difficult since the disease is highly heterogeneous. Tumors have various genetic and molecular traits [9, 10], making them difficult to detect and classify in their early stages. Early-stage cancers frequently cause nonspecific or no symptoms, complicating detection. Biopsies are frequently required because noninvasive diagnostic tools are unable to distinguish between benign and malignant conditions [11]. In addition, the absence of universal biomarkers for predicting radioresistance in cancer contributes to this challenge [12], as healthcare providers strive to meet the demands of precision medicine, where treatments are increasingly personalized to individual genetic profiles and molecular signatures. This increasing demand for personalized diagnosis compels existing diagnostic modalities, which are often slow and require technological advancements.

Among the world's biggest causes of death is cancer, which varies greatly in incidence and mortality rates among nations and geographical areas. According to GLOBOCAN, out of 183 countries worldwide, 112 have cancer as their top or second leading cause of death. A few factors, such as lifestyle and aging, have contributed to these trends. A rising number of cancer-related deaths in both high- and low-income nations is a major obstacle to increasing lifespans globally. Global data and various cancer registries show a steady increase in the incidence of cancer. In 2020, an estimated 19.3 million new cancer cases (18.1 million excluding nonmelanoma skin cancer) and over 10.0 million cancer deaths (9.9 million excluding nonmelanoma skin cancer) were reported globally [13], putting further pressure on healthcare systems to improve both the early detection and treatment of cancer. Although the incidence of cancer is increasing, some areas have higher survival rates because of early detection and advancements in treatment [14]. According to cancer registries, early detection through screening programs has increased survival rates for some cancers, including breast cancer [15]. However, the worldwide cancer burden continues to climb, and these advancements are not evenly distributed across all cancer types or geographical areas [16]. This emphasizes the necessity of easily available and efficient diagnostic instruments that are applicable to a range of patients and medical environments. Furthermore, early diagnosis is strongly correlated with survival rates [17], as cancers detected at later stages are typically more difficult to treat.

## *1.2 Role of Optical Imaging in Cancer Diagnosis*

Optical imaging has become a critical tool in cancer diagnostics with the use of molecular imaging [18], providing a noninvasive means to quantitatively and qualitatively analyze tumor growth in vivo [19, 20]. Recent developments in this technology have significantly improved its feasibility in clinical settings, particularly through the development of high-resolution and label-free techniques [2, 21]. These advances provide detailed structural, metabolic, and functional information via the interaction of light with biological tissues, which is difficult to obtain with traditional imaging methods such as MRI or CT.

Among the most important developments in optical imaging are NLOM, which encompasses modalities such as TPF, SHG, THG, and CARS. Without the use of exogenous fluorophores, these methods enable deeper tissue penetration and subcellular resolution [22]. A crucial part of the ECM in the TME, fibrillar collagen, is an example of a noncentrosymmetric structure to which SHG microscopy is extremely sensitive [2]. SHG is an effective diagnostic technique for assessing cancer invasion because changes in collagen fiber organization, density, and orientation indicate tumor growth and metastasis. Label-free imaging methods are gaining popularity due to their ability to give intrinsic contrast based on the metabolic and structural properties of tissues. THG microscopy, on the other hand, is an emerging cancer diagnostic tool equipped with high-resolution, label-free imaging of cellular and extracellular structures [23]. THG uses the refractive index difference between cellular interfaces, such as water and lipid or water and protein borders, which provides contrast to the image. This allows label-free visualization of tissue morphology [24]. In cancerous tissues, THG can detect the cellular variations in size, shape, and organization. It also detects alterations in the ECM associated with tumor progression on the basis of collagen information [25]. In breast and pancreatic cancer detection, a combined approach of THG with SHG to differentiate between malignant and benign tissues to obtain structural information associated with tumor growth has been reported [26]. Furthermore, molecule-specific THG microscopy, which is enriched by silver nanoparticles matched to THG wavelengths, provides high-contrast imaging of cancer cell membranes. This technique successfully visualizes HER2/neu oncogene expression for targeted cancer imaging [27]. Therefore, NLOM can be considered a powerful and risk-free technique for real-time diagnosis of tumor cells during surgical procedures (Fig. 2) [28]. This label-free imaging approach offers great promise for developing surgical precision and may provide early-stage cancer detection, potentially improving diagnostic approaches in clinical practice.

CARS microscopy, for example, shows promise as a label-free imaging method for quickly screening many cells in cytopathological samples [29]. This is essential for researching metabolic reprogramming in cancer cells because it uses the vibrational modes of molecules to create high-contrast images of lipid-rich regions in tumors. This method demonstrates the ability to apply quantitative kinetic models and parameters to analyze lipid metabolism in breast cancer cells and to examine

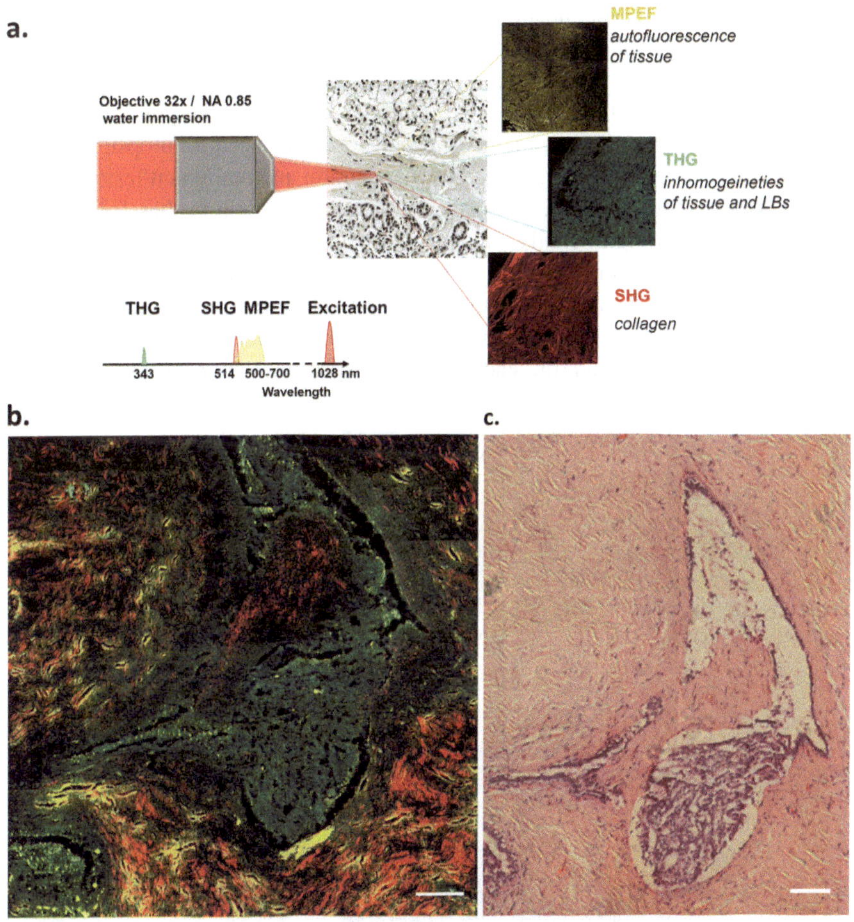


**Fig. 2** Schematic diagram for the collection of multimodal NLO signals from breast tissue (**a**). Characteristic multimodal NLO image (MPEF, THG, SHG) (**b**). THG signal is represented in cyan, SHG signal is represented in red, and MPEF is represented in yellow. The overlap of MPEF (yellow) and THG (cyan) signals generates green color. (**c**) Corresponding H&E image. The scale bar represents 100 μm. (Figure and caption used with permission from [28])

lipid changes in living cells with minimal disruption [30]. The multimodal imaging techniques used to visualize low- and high-grade squamous intraepithelial lesions in Pap smears using CARS and SHG/TPF are shown in Fig. 3.

Furthermore, fluorescence imaging offers a highly sensitive approach to assess endogenous chemicals such as NADH and flavins, providing vital information on the metabolic state of cancer cells [31]. The developments in multiphoton microscopy (MPM), such as two-photon excitation microscopy, have made it possible to image tissues more deeply with less photodamage, which makes it especially helpful for observing intricate tissue structures in vivo [32]. The interactions between cancer cells and the stroma, immune system, and blood arteries around them are critical for tumor formation, and these approaches give unrivaled insights into these relationships [33]. Its diagnostic capabilities have been further improved by the combination of optical imaging and machine learning (ML) algorithms. Artificial intelligence (AI)-powered image analysis can identify minute morphological changes in tissues that the human eye might miss, increasing the precision of early detection and diagnosis. These methods also decrease the need for biopsies by accurately and specifically differentiating between benign and malignant lesions [34].

The development of biomarker-driven molecular imaging probes for precision radiotherapy relies significantly on optical imaging, which allows for real-time observation of biomarker expression in tumors, simplifying the selection of molecular targets for probe design. These methods allow noninvasive evaluation of tumor characteristics and validation of imaging biomarkers, leading to individualized treatment regimens based on each patient's specific tumor biology [35]. By

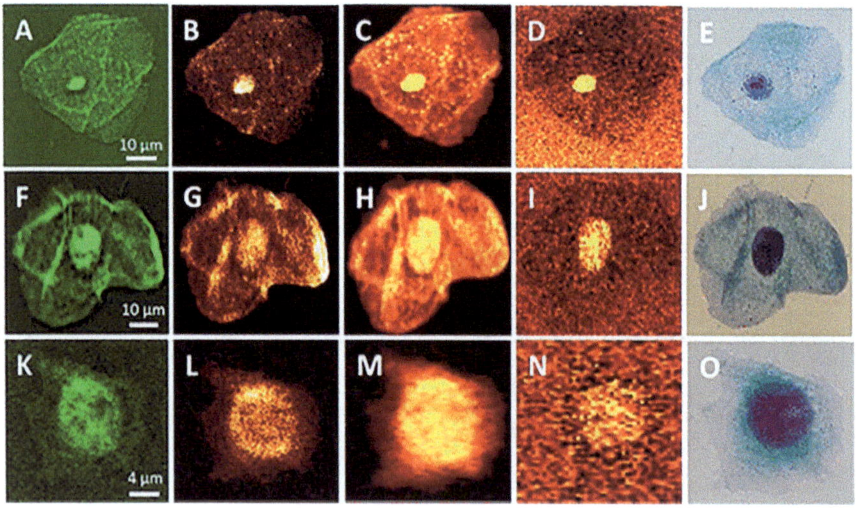


**Fig. 3** Multimodal imaging techniques for normal (**a–e**), low-grade squamous intraepithelial lesion (**f–j**), and high-grade squamous intraepithelial lesion (**k–o**) cells in Pap smears. CARS images (**a**, **f**, **k**). SHG/TPF images (**b**, **g**, **l**). Integrated Raman intensities of cells in the 2850–3050 $cm^{-1}$ region (**c**, **h**, **m**) and in the 785–805 $cm^{-1}$ region (**d**, **i**, **n**). Papanicolaou-stained images (**e**, **j**, **o**). (Figure and caption used with permission from [29])

combining multiple NLOI modalities, researchers have created multimodal NLOM systems that have successfully demonstrated preclinical vertebral imaging. The quantitative information obtained from these nonlinear interactions enables accurate biochemical, morphological, and structural tissue analyses, effectively distinguishing between normal and pathological conditions. Given these capabilities, NLOM shows great potential for integration into standard histopathological analysis in preclinical studies, with promising applications for examining larger tissue sections in clinical settings [36].

## *1.3 Advantages of NLOM Optical Microscopy in Cancer Diagnosis*

NLOM offers several advantages over conventional imaging techniques, establishing itself as a transformative imaging modality in cancer diagnosis. By implementing nonlinear optical techniques such as fluorescence lifetime imaging, SHG, and TPF, NLOM improves the visualization of molecular interactions and cellular structures, thereby increasing diagnostic accuracy and therapeutic monitoring [2].

One of the most significant advantages of NLOM over conventional fluorescence microscopy is its ability to achieve imaging at greater depths within biological tissues [37]. The use of near-infrared (NIR) excitation light minimizes scattering, allowing imaging up to several hundred micrometers deep in tissues [38]. This property is particularly beneficial in oncology, where tumors are often located within dense tissue. The improved penetration depth enables researchers and clinicians to obtain high-resolution images of the TME, facilitating the identification of subtle morphological changes associated with malignancy [39]. The common drawbacks of conventional fluorescence microscopy, phototoxicity, and photobleaching are significantly mitigated by NLOM [40]. The photons are only absorbed at the focal point because NLOM uses two (or three)-photon excitation, which reduces exposure to nearby tissues. In addition to maintaining cellular integrity, this permits extended imaging sessions without endangering the viability of living cells [41]. These properties are essential for cancer research studies, where time-bound tracking of dynamic biological processes is of high priority.

To improve specificity and contrast, NLOM techniques can employ intrinsic fluorophores and contrast agents [42]. SHG is sensitive to the structural features of collagen, which is a critical component of the TME [2]. This property enables visualization of the ECM and its remodeling, which is frequently associated with cancer progression and metastasis. Furthermore, NLOM can be combined with specific contrast agents targeting cancer biomarkers to provide real-time, high-contrast images that help differentiate between malignant and benign tissues. The NLOM provides not only structural information but also functional imaging. Fluorescence lifetime imaging microscopy (FLIM) is a technique for assessing molecular interactions by measuring the fluorescence decay time of the molecules under study. This can be linked to metabolic activity, the cellular environment, and signaling pathways, providing valuable insights into tumor biology for diagnosis and treatment planning. The detailed study of the TME enables a deeper understanding of cancer heterogeneity and treatment resistance [43]. Several benefits of NLOI for the diagnosis of cancer include improved imaging depth and resolution, reduced phototoxicity, improved specificity and contrast, functional imaging capabilities, the ability to observe biological processes in real time, and integration with other imaging modalities. These advantages make NLOM a valuable instrument for enhancing cancer detection and care, opening the door for more personalized treatment strategies. Better patient outcomes are anticipated, as NLOM is used frequently in clinical settings for diagnosis.

## 2 Applications of NLOM in Cancer Diagnosis

### *2.1 Breast Cancer*

Breast cancer is one of the most prevalent diseases affecting women worldwide, with an estimated 2.3 million new cases diagnosed every year [44]. Early detection is crucial for effective treatment, but current diagnostic techniques, such as mammography and ultrasound, have inherent limitations, including poor resolution in dense breast tissues, false positives, and the need for invasive biopsies [45]. Over the past two decades, NLOM, which includes techniques such as TPF, SHG, THG, and CARS, has evolved as a strong imaging technique, providing distinct advantages in breast cancer diagnosis [46]. The application of NLOM tools in breast cancer diagnosis has evolved significantly, offering a noninvasive, label-free method that provides highly detailed imaging of tissue structures at the cellular level. This approach allows scientists to obtain information about the TME by exciting endogenous fluorophores such as NADH and FAD, which are important for assessing metabolic activities within cells. Early work by Wu et al. [47] demonstrated that MPM could differentiate between healthy and malignant breast tissues with high specificity by visualizing collagen fiber structures. Their study provided crucial insights into the potential of MPM for detecting structural changes linked to malignancies, highlighting the unique collagen fiber arrangements in tumors that can be used as diagnostic markers [47].

Further advancements included the use of SHG microscopy to visualize collagen structures more specifically. Due to its ability to detect noncentrosymmetric molecules such as collagen without additional staining, SHG microscopy enables a closer look at collagen reorganization in cancerous tissues [48]. Studies have shown how SHG imaging can reveal detailed structural changes in the ECM, which are indicative of the stromal alterations often observed in breast cancer progression [49]. Additionally, studies by Ambekar et al. [50] quantified collagen orientation and density in breast biopsies, revealing a strong correlation between ECM architecture and cancer invasiveness [50, 51]. A study by Provenzano et al. [52] described three types of tumor-associated collagen signatures (TACS) critical for understanding tumor invasion: TACS-1 shows dense collagen near tumors as a marker for small tumors; TACS-2 involves taut collagen fibers around the tumor, limiting its growth; and TACS-3 features radially aligned fibers that promote invasion, with fiber alignment at approximately 90°, indicating active invasion and near 0°, indicating noninvasiveness (as shown in Fig. 4). This framework helps in

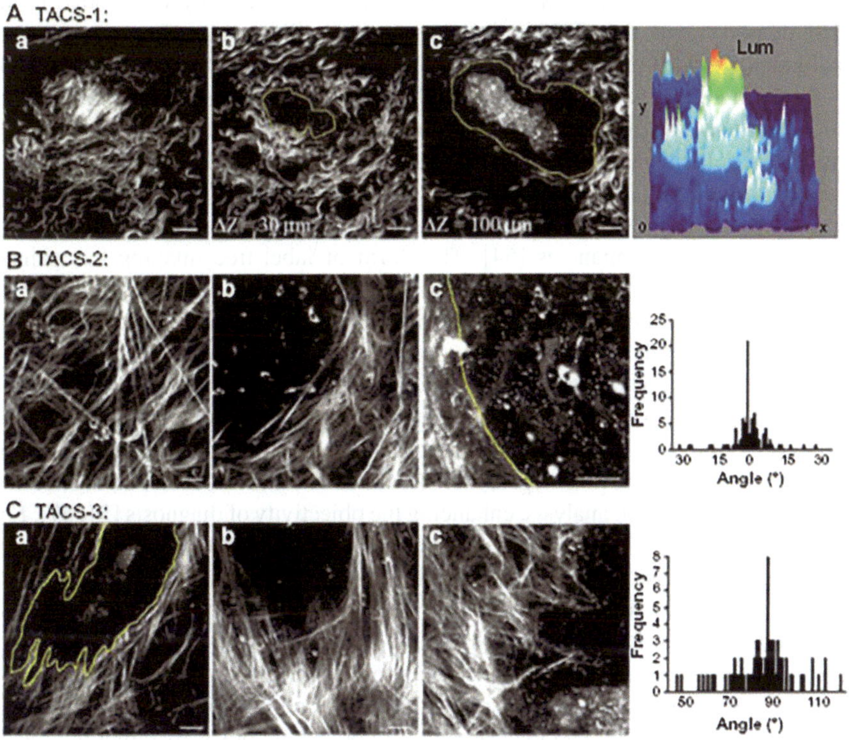


**Fig. 4** The tumor-stromal interaction: representation of the three TACS in Wnt-1 mouse tumors. (**A**–**C**) Micrographs illustrating the identified TACS. (A) TACS-1. MPE/SHG image of TACS-1. Namely, a region of dense collagen (a, and surface map) above a non-palpable tumor (b, c; yellow outline) that is indicative of the presence of a small tumor (also see Additional file 5). The surface map quantifies the intensity of the fluorescent signal relative to x–y location and clearly demonstrates an increased collagen signal. This map is representative of six Wnt-1 tumors and eight PyVT tumors (not shown). (**B**) TACS-2. MPE/SHG image of the second TACS indicated by the presence of straightened (taut) fibers characteristic of a larger Wnt-1 tumor. (**B**) (a–c), MPE/SHG images of collagen fibers in Wnt-1 mice stretched around a relatively smooth tumor boundary (outlined with a yellow line in c) as demonstrated by the fact that majority of the fibers are parallel to the tumor boundary. (**B**) Histogram, the angle of collagen fibers relative to a line tangential to the tumor boundary was measured for 86 regions in six independent tumors and graphed as a frequency distribution resulting in a distribution of fibers around 0°. (**C**) TACS-3. The third TACS: Aligned collagen fibers at regions of cell invasion in Wnt-1 mice. (**C**) (a–c), The irregular tumor boundary associated with local invasion is outlined in yellow (in a) and connected to fibers that are primarily distributed normal to the initial tumor boundary, represented by a frequency distribution around 90° relative to the tumor boundary. (**C**) Histogram, the angle of collagen fibers relative to a line tangential to the tumor boundary was measured for 71 regions in six independent tumors and graphed as a frequency distribution resulting in a distribution of fibers near 90°. Scale bars equal 25 μm. (Figure and caption used with permission from [53])

identifying and assessing the aggressiveness of breast cancer progression [52]. These early findings solidified the role of SHG as a diagnostic tool, allowing pathologists to visualize and analyze changes in the ECM directly, which traditional staining techniques cannot achieve. As NLOM technology progressed, research began to emphasize THG microscopy for higher contrast imaging of cellular boundaries and lipid structures. For example, a study by Gavgiotaki et al. [28] demonstrated that THG microscopy could differentiate between various malignancy grades in breast tissue biopsies, revealing fine structural details such as tissue inhomogeneities, including cellular boundaries, lipid bodies, and structural contrasts within tumor tissues, that are essential for early diagnosis [28]. Another significant advancement is CARS microscopy, which provides chemical-specific imaging of cellular lipids and proteins. Yang et al. [54] utilized CARS to analyze breast tissue without labeling, effectively detecting lipid content changes associated with malignancies [54]. This form of label-free imaging has gained traction, as it enables clinicians to visualize the biochemical environment of breast tissues directly, offering insight into metabolic variations that are often indicative of cancerous growth [55, 56].

Recent works have extended these modalities to quantify collagen fiber alignment and density, which are crucial factors in cancer prognosis. Burke et al. [51] demonstrated that changes in collagen fiber alignment, as visualized by SHG, can indicate tumor progression stages, and Bredfeldt et al. [57] developed algorithms for automated collagen analysis, enhancing the objectivity of diagnosis [51, 57]. For example, Tao et al. [58] examined the effectiveness of NLOM in assessing the pathology of breast cancer, including ductal carcinoma, by integrating various NLOM methods to obtain a multifaceted view of the TME. These studies have provided a foundation for the use of NLOM to distinguish subtle pathological differences that might otherwise be undetected [58]. More recent developments have focused on hyperspectral stimulated Raman scattering (SRS) microscopy. Studies by Shin et al. [59] illustrated that SRS microscopy could precisely target microcalcifications, which are often found in breast cancers. By analyzing these microcalcifications, researchers can evaluate tumor malignancy more accurately, which has substantially improved diagnostic accuracy [59].

Currently, NLOM techniques are at the forefront of providing rapid, high-resolution imaging capabilities essential for breast cancer diagnosis. However, challenges remain in terms of accessibility and integration into clinical practice. While the noninvasive, label-free nature of these techniques offers considerable benefits, their widespread clinical use requires additional validation, especially for rare and complex breast cancer subtypes. Studies such as that of Ren et al. [60] underscore the need for refining these techniques to visualize even finer lymphatic invasion patterns, which are crucial for assessing metastatic potential [60]. The future of NLOM in breast cancer diagnostics lies in further improving image acquisition speed, enhancing algorithm-driven image analysis for real-time diagnostics, and reducing equipment costs to facilitate wider clinical adoption.

## 2.2 Skin Cancer

The history of NLOM in cancer diagnostics began with SHG and TPF imaging techniques in the late 1990s and early 2000s. Researchers initially recognized the potential of these techniques for imaging dense and optically challenging tissues, which allowed for detailed examination of skin layers without invasive biopsies [61]. The appeal of NLOM methods has increased as studies have revealed how they can clearly visualize collagen fibers, elastin, and other structural proteins—key components for identifying pathological changes in skin tissue that are indicative of skin cancers such as melanoma and basal cell carcinoma (BCC) [62].

Over the years, advancements in NLOM, particularly the development of CARS and SRS microscopy, have expanded the scope of skin cancer diagnosis. These Raman-based imaging modalities enable more specific detection of molecular compositions in cells, distinguishing cancerous cells from healthy ones by identifying variations in lipid and protein distributions [63]. Studies have shown the efficacy of CARS imaging in detecting lipid-rich sebaceous glands and inflammatory infiltrates around melanoma cells, offering insights into both cancer progression and the TME [64]. NLOM applications have also been facilitated by improvements in laser technologies and detector sensitivities. Higher laser powers and more efficient detection systems now enable real-time imaging, reducing scanning times and providing clinicians with rapid diagnostic information [65]. Recent studies have leveraged these advancements to assess different stages of melanoma. For example, Delrue et al. [66] demonstrated the ability of SRS imaging to visualize melanin and keratin distributions across various melanoma stages, providing a foundation for staging tumors and guiding treatment. The ability of NLOM to image deeper skin layers without invasive procedures offers an advantage over conventional histopathology and supports more precise, rapid diagnostics in clinical dermatology [66]. Recently, a study by Guo et al. [67] illustrated the ability of spectral- and time-resolved two-photon microscopy (TPM) to characterize BCC. Figure 5 shows detailed imaging results comparing cancerous and noncancerous tissues, emphasizing variations in fluorescence lifetimes and emission spectra. The multidimensional analysis helps identify key differences in biochemical composition and microstructure, which improves the diagnostic accuracy of BCC. These findings demonstrate the potential of TPM for revealing subtle tissue characteristics essential for early and precise skin cancer diagnosis [67].

Current developments in NLOM focus on multiplex imaging and improved resolution to differentiate even subtle cellular changes. Hyperspectral and multimodal approaches combine SHG, TPE, and SRS imaging into a single examination, offering a more comprehensive view of tissue structure and function in one session. This multimodal imaging approach provides dermatologists with simultaneous data on multiple biomarkers, enabling a more precise assessment of skin cancers. Current works highlight this technique's high sensitivity and specificity in identifying both benign and malignant lesions, particularly where early detection is crucial [68].

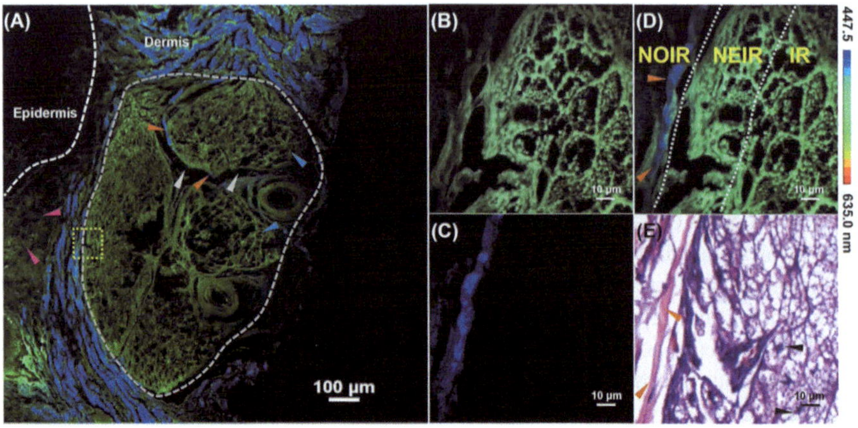


**Fig. 5** Multimodal imaging (TPEF and SHG) of basal cell carcinoma and corresponding histological imaging. (**A**) Largefield two-photon and second harmonic scanning imaging of the lesion area; (**B**) spectrum-coded image of twophoton signals in the yellow box in (**A**); (**C**) spectrum-coded image of multichannel superimposition; (**D**) spectrumcoded image of second harmonic signals; (**E**) histological image of H&E staining. In these images, TPEF signals are shown in green, SHG signals are shown in blue, the epidermis and dermis are distinguished by white dashed lines, the lesion areas are circled by gray dashed lines, the purple arrows indicate areas of fibrolysis, the orange arrows indicate fibers, the blue arrows indicate basal cell lobules, the gray arrows indicate mesenchyme, and the black arrows indicate nuclei [67]).

Despite these clear advantages, some areas still require improvement for broader clinical integration. First, equipment costs remain high, limiting accessibility in some clinical settings. Additionally, the interpretation of NLOM images can be complex, necessitating skilled personnel and potentially ML algorithms for enhanced analysis. Future research aims to simplify these processes, with efforts toward developing user-friendly software to interpret images more efficiently. Increasing portability and accessibility through the miniaturization of NLOM devices could also facilitate point-of-care diagnostics, allowing for accurate screening of suspicious lesions on the spot. In summary, NLOM has significantly advanced skin cancer diagnostics, enabling detailed, noninvasive imaging that reveals both structural and molecular insights. With further advancements in accessibility, automation, and the integration of AI-assisted interpretation, NLOM could become a staple in dermatology clinics for real-time skin cancer screening and monitoring.

## *2.3 Brain Tumors*

The use of NLOM in brain tumor research began with the need for deeper, noninvasive, and high-contrast imaging that conventional microscopy cannot achieve. Over the years, its utility in cancer diagnostics has progressed remarkably, particularly with the adoption of TPF and SHG microscopy, which enable label-free visualization of tumor margins and intricate cellular details [69]. These techniques have evolved into reliable tools for distinguishing cancerous from healthy brain tissue based on cellular and ECM characteristics, which has increased diagnostic accuracy.

An early study by Jiang et al. (2011) demonstrated the feasibility of using TPF and SHG for observing cancer cell morphology and structural irregularities in brain tissue, achieving effective differentiation between healthy and tumorous cells (as shown in Fig. 6) [69]. The depth penetration ability of the TPF and the sensitivity of the SHG to collagen and other fibrillar proteins allow for the precise visualization of tumor borders and the assessment of invasive behavior in brain cancer [70]. These capabilities have transformed the diagnostic environment, shifting from traditional biopsy-based approaches to optical imaging tools that may be employed in vivo. Advances in NLOM have further expanded with the integration of ML algorithms, particularly in texture analysis, which helps differentiate subtle histopathological features in brain tumors. For example, Wu et al. [71] applied ML-enabled texture

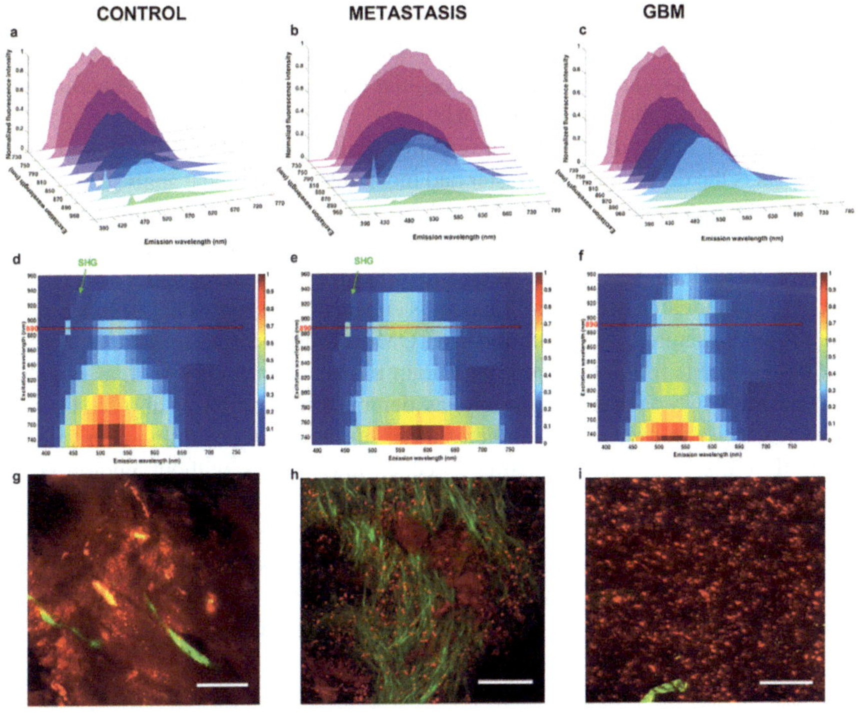


**Fig. 6** Comparative fluorescence analysis of the brain tissue groups. (**a–c**) Topological spectra for the control, metastasis, and GBM groups; (**d–f**) emission–excitation colormaps, highlighting SHG signals (green arrow) and optimal excitation wavelengths (red line); (**g–i**) TPF imaging of selected tissue regions, scale bar = 100 μm. This demonstrates the technique's efficacy in differentiating tissue types through fluorescence characteristics. (Figure and caption used with permission from [70])

analysis in SHG imaging, providing a robust, quantitative method for distinguishing between glioblastoma and less aggressive brain tumor types. This innovation enables clinicians to determine tumor grades and possible growth patterns noninvasively, greatly aiding in treatment planning [71, 72].

Another recent development involves the use of wide-field polarimetric SHG microscopy, as highlighted by Mirsanaye et al. [73], which allows extensive imaging of tumor margins. This technique uses unique polarization properties to further enhance contrast, making it particularly effective for detecting invasive edges of tumors, which are often challenging to identify with other methods. By enhancing the visualization of invasive cancerous boundaries, polarimetric SHG has shown potential for intraoperative use, helping surgeons make more informed decisions during tumor resection [73]. Additionally, innovations in three-photon microscopy are emerging in response to the need for even deeper penetration and greater SNR in brain tissue imaging. The advanced form of NLOM recently explored by Schubert et al. [74] employs tailored three-photon microscopy, providing subcellular resolution even in densely packed tissue environments and facilitating in-depth analysis of brain TMEs. This technology enables researchers to study tumor margins and real-time interactions between cancer cells and their surrounding microenvironments. The method's enhanced SNR and contrast allow for detailed visualization of tumor cell dynamics, vascular structures, and cellular infiltration, paving the way for innovative therapies targeting the TME [74].

In terms of future directions, an ongoing challenge in NLOM for brain cancer diagnosis is the development of faster and more automated image analysis tools that can handle the vast datasets generated by high-resolution imaging. Continued improvements in computational techniques, such as AI and ML, are expected to provide new diagnostic biomarkers and optimize the interpretation of NLOM images. Another area needing refinement is enhancing the sensitivity of imaging for deep-brain structures, as conventional NLOM techniques face limitations in penetration depth due to light scattering in brain tissues. Approaches such as quantum dots-assisted two-photon microscopy, as demonstrated by Ni et al. (2021), have shown promise in overcoming some of these limitations by enabling near-infrared (NIR-II) emission, which penetrates deeper into tissues without phototoxicity, improving imaging depth and resolution [75]. In summary, NLOM has grown from promising imaging methods to invaluable diagnostic tools in brain cancer research. Its ability to visualize cancerous tissues in detail, combined with ongoing advancements in imaging depth and AI integration, suggests that NLOM will remain at the forefront of brain cancer diagnostics, with significant potential for further innovation and refinement.

## 2.4 Gastrointestinal Cancer

The use of NLOM in gastrointestinal (GI) cancer diagnosis began with the exploration of TPM to enhance tissue imaging while preserving cellular integrity. Early studies in GI cancer focused on the use of two-photon fluorescence to differentiate between normal and diseased tissue based on structural and metabolic alterations. Chen et al. [76] revealed that TPM could accurately differentiate the mucosa and submucosa of normal vs. malignant gastric tissues, establishing diagnostic features required for GI cancer diagnosis [76]. Additionally, Perry et al. [77] highlighted the potential of both TPM and SHG imaging in clinical and translational research by emphasizing their application in high-resolution structural imaging and their advantages over conventional imaging techniques [77]. Subsequent advancements in NLOM have allowed for label-free imaging, enabling the visualization of tissue architecture without additional dyes, which is particularly beneficial for patients and clinicians. Recent works, such as that of Li et al. [78], have shown how combining TPM with imaging analysis can facilitate the rapid and label-free detection of GI stromal tumors, thereby improving the diagnostic process. In this study, the TPM demonstrated high sensitivity in detecting stromal tumors, highlighting the ability of the NLOM to differentiate between benign and malignant tissues. This label-free approach has significantly reduced preparation times thus improving procedural efficiency and patient outcomes [78].

A study by Jiang et al. [69] shows TPF and SHG imaging for visualizing collagen within human tissues. These approaches offer excellent resolution, label-free imaging, revealing distinct structural variations in collagen associated with normal and pathological states (as shown in Fig. 7). SHG specifically highlights the fibrillar

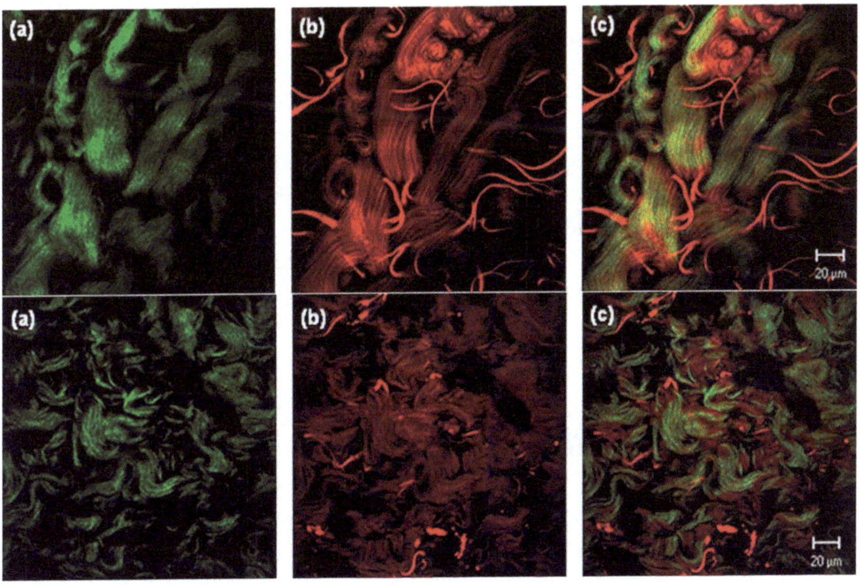


**Fig. 7** Two-photon fluorescence and second-harmonic generation imaging of collagen in human tissue. The top row shows the representative TPEF/SHG images of collagen bundles and elastic fibers of stomach submucosa (**a**) SHG images of collagen fibres, (b) TPEF images of collagen fibres, and (**c**) overlay of SHG/TPEF. The bottom figure shows the TPEF/SHG images of skin dermis representing (**a**) SHG from collagen, (**b**) TPEF from the metabolic markers, and (**c**) overlay of SHG/TPEF. These figures emphasize the power of multiphoton microscopy in biomedical imaging for structural and functional tissue analysis. (Figure and caption used with permission from [69])

organization of collagen, whereas TPF provides complementary information about its molecular environment. These findings underscore the value of multiphoton microscopy for precise histopathological assessment, offering critical insights into structural biomarkers relevant to the early diagnosis and progression of collagen-associated diseases such as skin cancer [69]. The recent integration of enzyme-selective fluorescent probes with TPM represents another innovative improvement. As noted by Noh et al. (2021) [10.3389/fonc.2021.634219], such probes can reveal unique metabolic features of gastric cancer, aiding in more precise diagnosis and expanding the potential applications of TPM. Furthermore, the work by Sarri et al. [79] on SHG and SRS imaging has made NLOM suitable for real-time, intraoperative cancer detection, specifically aiding in the differentiation of tumor boundaries, which is critical for achieving clean surgical margins [79]. The latest advancements in SHG microscopy have also enhanced its effectiveness in GI cancer diagnosis by allowing clinicians to observe collagen and other ECM changes associated with malignancies. Studies have shown that SHG imaging can precisely identify GI stromal tumors through collagen remodeling patterns, making it an invaluable diagnostic aid [80, 81].

The contribution of NLOM to GI cancer diagnostics lies in its ability to visualize microscopic details without relying on traditional stains, which has improved the accuracy and speed of early cancer detection. Studies such as those by Birk et al. [82] highlight the effectiveness of the method, showing how SHG Imaging differentiates high-grade dysplasia and cancer from normal colonic mucosa with high diagnostic accuracy [82]. Furthermore, Yan et al. [83] illustrated how multiphoton microscopy could be used intraoperatively to assess surgical margins in low rectal cancer thus reducing the risk of recurrence due to incomplete excision [83]. The future of NLOM in GI cancer diagnosis looks promising, with ongoing research focusing on enhancing image resolution, reducing imaging times, and integrating AI for automated diagnostic support. However, despite its progress, challenges remain, particularly in standardizing imaging protocols across clinical settings and improving accessibility to NLOM tools. Addressing these areas will be crucial for furthering the impact of NLOM and ensuring that it becomes a staple in routine GI cancer diagnostics.

## *2.5 Other Cancer Types*

NLOM has emerged as a powerful tool for cancer diagnosis, extending beyond its established applications in areas such as skin and GI cancers. Initially, it was used to examine cellular features in cancerous tissues through techniques such as TPE and SHG. The strength of NLOM lies in its ability to deeply penetrate tissues and provide detailed structural and biochemical insights without the need for external dyes, ensuring accurate imaging while maintaining tissue integrity.

In pancreatic cancer, for example, early work by Hu and Fu [84] demonstrated the promise of multiphoton microscopy for examining TMEs, highlighting features of the dense ECM typical of pancreatic malignancies [84]. This ability to capture the ECM structure through SHG has been instrumental in pancreatic cancer diagnosis, as illustrated by Drifka et al. [85], who utilized NLOM to differentiate normal and cancerous pancreatic tissue based on ECM organization. This approach has sbeen essential for identifying pancreatic ductal adenocarcinoma, a particularly aggressive cancer, by allowing pathologists to observe unique collagen patterns and cellular morphology in the ECM [85]. Future developments in polarization-sensitive SHG may further improve the specificity of NLOM in diagnosing such fibrous tumors by detecting the collagen fiber orientation, aiding early pancreatic cancer detection with minimal invasiveness [86].

Similarly, ovarian cancer has benefited from NLOM advancements, particularly through SHG microscopy, as ovarian tumors often exhibit significant ECM remodeling. Ricciardelli and Rodgers [87] reported that ECM alterations are indicative of ovarian tumor invasiveness, suggesting the need for further investigation via SHG microscopy for this cancer type [87]. The application of NLOM in ovarian cancer was expanded by Kirkpatrick et al. [88], who assessed endogenous optical biomarkers in ovarian tissues using multiphoton microscopy, enabling the differentiation of malignant from nonmalignant tissues (as shown in Fig. 8). This technique has facilitated more accurate imaging of tumor boundaries and cellular morphology,

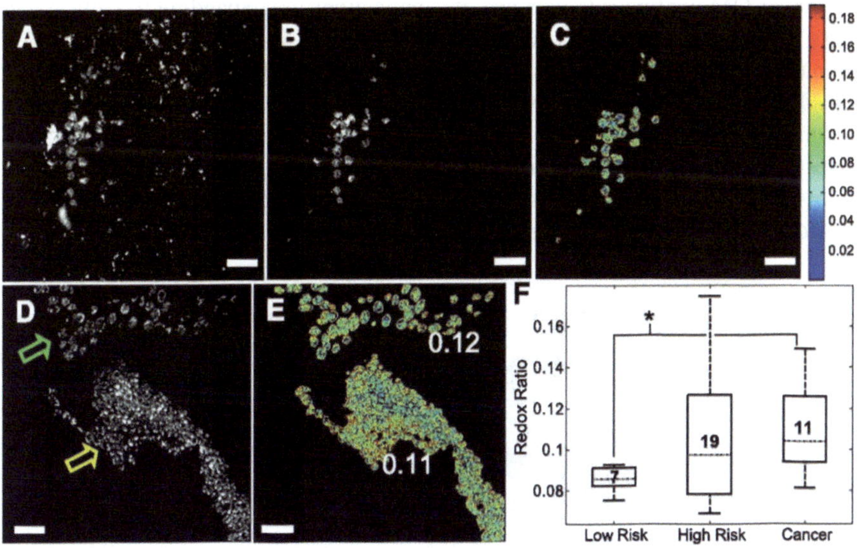


**Fig. 8** Estimation of intracellular redox ratios using TPF imaging in ovarian tissue. (**a**) Combined TPF imaging of cellular and extracellular fluorescence. (**b**) Segmentation of cellular fluorescence based on the FAD-related TPF signals. (**c**) Redox ratio map derived from NAD(P)H and FAD fluorescence, revealing intracellular metabolic variations. (**d, e**) Redox ratios of normal-high-risk tissue showing distinct values between large (green arrow) and small cells (yellow arrow), as calculated from localized fluorescence. (**f**) Box-and-whisker plots comparing redox ratios across the normal-low-risk, high-risk, and cancer groups, highlighting significant metabolic differences. (Figure and caption used with permission from [88])

contributing to better staging and assessment of ovarian cancer [88]. Continued improvements in NLOM resolution and image-processing algorithms will likely enhance our ability to differentiate between benign and malignant ovarian tumors, particularly in early-stage malignancies.

Prostate cancer diagnosis has also progressed through NLOM, where hyperspectral SRS microscopy has shown promise. Chen et al. [89] successfully used this technology to differentiate between low- and high-grade prostate cancer, detecting subtle differences in lipid composition within cancerous cells [89]. These insights are important since prostate cancer often presents challenges in grading and staging. Early applications of CARS microscopy in prostate tissue revealed changes in lipid content due to cancer progression, providing further depth to cancer diagnostics through the analysis of biochemical profiles [52]. Figure 9 illustrates the use of CARS microscopy to analyze lipid distribution in hormone-treated versus control prostate cancer cells. The images compare autofluorescence and CARS signals, revealing alterations in lipid content and organization due to hormonal treatments. This highlights the potential of CARS microscopy to detect biochemical changes in cancer cells, providing insights into lipid metabolism and its role in cancer progression. With improvements in image acquisition speed and spectral sensitivity, NLOM could support personalized treatment planning for prostate cancer patients by accurately identifying molecular markers associated with disease severity.

As NLOM continues to advance, its applications are expanding to include cancers with complex microenvironments and dense ECM structures, where conventional imaging may fail. Future developments are likely to focus on integrating NLOM with ML algorithms to enable automated, real-time analysis, as well as on refining its sensitivity to biochemical markers across various cancer types. Enhanced hardware and software will further push NLOM toward more widespread clinical adoption, offering a tool that not only aids in diagnosis but also tracks treatment response with high specificity and minimal invasiveness.

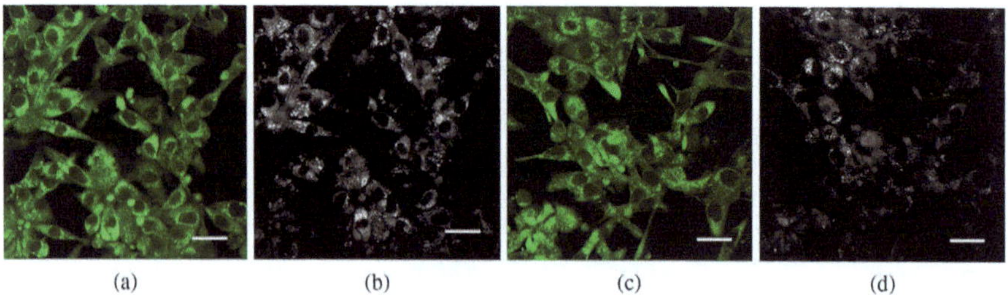


**Fig. 9** Lipid distribution in hormone-treated prostate cancer cells analyzed by CARS microscopy. (**a**) Autofluorescence image of LNCaP-R1881 (hormone-treated) prostate cancer cells. (**b**) CARS microscopy image showing the lipid distribution in the same cells. (**c**) Autofluorescence image of LNCaP-V (vehicle control) cells. (**d**) CARS microscopy image of LNCaP-V cells. The images reveal distinct differences in lipid distribution between treated and control cells, demonstrating the impact of hormone treatments on cellular lipid metabolism. (Figure and caption used with permission from [52])

# 3 Challenges and Limitations of NLOM

## 3.1 Technical Challenges

NLOM can utilize NIR light sources to increase the depth of tissue penetration. However, when light passes through highly scattering or thick biological tissue, its scattering effects and absorption can still affect the resolution of the final image [34, 90]. In a highly scattering environment, light is continuously deflected during propagation, resulting in the attenuation of light signals and an increase in the scattering background, making it difficult to clearly display the structural details of deep tissues. To achieve deeper imaging, the energy of the laser must be increased, but this may cause light damage or thermal effects, further limiting long-term observation of living tissues. Scientists are currently working on developing new optical systems, such as adaptive optics technology [91], to compensate for wavefront distortion caused by scattering, thereby improving imaging depth and resolution.

Another challenge in NLOM is how to improve the signal-to-noise ratio (SNR). Because these techniques rely on ultrashort pulse lasers to excite the sample, the imaging volume and number of photons are very limited, so the signal intensity is usually low. To solve this problem, researchers often use highly sensitive detection technologies, such as time-correlated single-photon counting [3], which can accurately calculate the arrival time of each photon, thereby increasing the accuracy of signal detection. However, this type of technology also slows the imaging speed, which limits the ability to observe dynamic processes or changes in living cells. Moreover, to improve the SNR, it is often necessary to use higher intensity lasers. However, this may cause photodamage to the sample or lead to photobleaching, especially when imaging for long periods of time or observing at high resolution. These problems require researchers to find a balance between the laser intensity and imaging quality, and to develop more precise imaging technology and optical parameter control methods [34].

NLOM systems usually rely on complex and precise optical components and ultrafast pulse lasers. The large size and complex structure of these devices limit their application in clinical settings. In clinical applications, for these technologies to be used in endoscopic or other minimally invasive examinations, miniaturized and flexible devices need to be developed [92]. However, current small nonlinear optical endoscopes still face technical challenges, such as how to maintain sufficient optical resolution and imaging depth while reducing the size of the device and improving its flexibility [93].

## 3.2 Clinical Translation Issues

One of the clinical translation challenges of NLOM is how to effectively apply label-free imaging techniques. The advantage of NLOM is its ability to detect non-linear signals from endogenous molecules such as collagen or lipids, which could theoretically reduce the need for exogenous markers. However, the interpretation of these endogenous signals is often complicated because different tissues and cells may exhibit similar optical properties in different pathological states, making it difficult to accurately distinguish diseased tissues clinically [58, 90, 94, 95]. To achieve clinical diagnostic accuracy, new data processing and analysis methods must be developed to handle and interpret these complex multimodal or multispectral data. This requires professional technicians and advanced computing tools, which not only increases the technical threshold but also slows the clinical application of NLOM [34, 96].

Although the laser systems used by NLOM can achieve deep penetration, these systems rely on high-intensity ultrafast pulse lasers to achieve nonlinear effects. This high-energy light source may still cause thermal effects or light damage to living tissue even at near-infrared wavelengths. Therefore, in clinical applications, imaging parameters must be strictly controlled to ensure patient safety [97]. In addition, the safety of these systems requires compliance with clinical and regulatory standards, which involve lengthy trials and reviews by regulatory agencies, thereby delaying the approval and application of new technologies. To pass these reviews, extensive clinical trials and evidence collection are needed to demonstrate the technology's effectiveness and safety. These processes are often time-consuming and expensive, affecting the clinical translation of NLOM [2].

To use NLOM more flexibly in clinical practice, researchers have attempted to develop miniaturized nonlinear optical endoscopes. However, the design and manufacture of miniaturized systems still face technical challenges. These endoscopes need to reduce the size of the device while maintaining imaging quality and flexibility to perform minimally invasive examinations inside the human body [98]. Such miniaturized devices need to be designed to overcome limitations in optical alignment, imaging resolution, and signal detection while ensuring operational flexibility and safety. Therefore, how to effectively integrate NLOM technology into small clinical devices and respond to different clinical needs remains an ongoing technology research and development challenge [99, 100].

## *3.3 Cost Considerations*

NLOM systems rely on complex optical components and ultrafast pulsed laser sources, such as titanium–sapphire lasers, which are extremely expensive. These high-order light sources must have stable energy outputs and precise pulse widths to stimulate nonlinear effects, so the initial acquisition cost of the entire microscope system is very high. Even for research laboratories, acquiring and maintaining these devices requires significant capital, let alone widespread use in medical institutions with limited funds [101, 102]. In addition, with the development of technology to improve imaging depth, resolution, and system stability, these systems also need to be equipped with additional modules and functions, such as adaptive optics systems and multiphoton detection modules, which further increases the cost.

In addition to the high cost of the equipment itself, the operation and maintenance of nonlinear optical microscopes also require specialized training with a high technical threshold because laser adjustment, system calibration, and control of imaging parameters need to be precisely managed. If medical institutions want to configure this equipment, in addition to the funds to purchase the equipment, they also need to invest a high cost in training technical personnel. In addition, because the optical components of these devices are sophisticated and expensive, maintenance and replacement costs are also very high if they malfunction or become damaged. In daily operations, maintaining these high-precision systems requires engineers with professional backgrounds to conduct regular testing and calibration to ensure the stability of the equipment and imaging quality, which also increases the cost burden of long-term use.

Due to the high cost and high technical threshold of NLOM, even though it has unique advantages in some specific clinical diagnostic fields, its popularity is still limited. Within the healthcare system, cost-effectiveness is an important consideration in determining whether new technologies can be introduced on a large scale. Although NLOM technology can provide high-resolution and depth imaging capabilities, due to its high equipment and operating costs, NLOM technology can currently be used only in some research institutions or large hospitals, and is difficult to use in small clinics or areas with limited resources. For these technologies to become more widely available in the clinic, reducing equipment costs and maintenance requirements will be key. This may involve developing simpler and lower cost laser sources and modular microscopy systems to reduce reliance on specialized personnel and maintenance. However, even with these technological advances, reducing costs enough to enable large-scale clinical application will still require considerable time and significant financial investment.

## 4 Comparison with Other Imaging Techniques

Optical imaging techniques have become indispensable across a wide range of scientific and clinical disciplines, including cell biology, tissue engineering, pharmacology, and materials science. These methods provide powerful insights by enabling high-resolution visualization of biological and material structures, revealing both micro- and nanoscale details that are essential for advancing research and development in these fields [103]. Conventional optical microscopy, confocal microscopy [104], optical coherence tomography (OCT) [105], and Raman spectroscopy [106] represent four widely adopted modalities, each offering unique strengths and catering to distinct imaging needs. Researchers rely on these techniques to investigate the structural, functional, and molecular aspects of samples, with each modality offering advantages that align with specific experimental or diagnostic requirements. While conventional microscopy remains foundational for routine structural observation, confocal microscopy enhances depth resolution, allowing for more detailed three-dimensional analysis. Moreover, OCT extends imaging capabilities into deeper tissue layers, and Raman spectroscopy offers unique chemical specificity that is critical for identifying the molecular composition.

Despite their shared focus on optical imaging, these modalities differ significantly in their technical approaches, resulting in diverse strengths and limitations in terms of resolution, contrast mechanisms, imaging depth, and instrumentation. Conventional optical microscopy, for example, achieves resolution at the diffraction limit but is limited by poor depth sectioning, making it less suitable for thick samples [107]. Confocal microscopy addresses this limitation with optical sectioning, enabling detailed imaging within thicker samples but requiring a more complex setup and higher light intensities, which can pose photodamage risks [108]. OCT advances this further by offering millimeter-level imaging depth with high axial resolution through coherence gating, although it involves sophisticated interferometric setups. Raman spectroscopy provides molecular specificity, mapping the chemical composition of samples with minimal sample preparation but with constraints on spatial resolution and depth penetration [109]. In contrast, NLOM techniques, such as multiphoton and second-harmonic generation microscopy, leverage nonlinear interactions between light and matter to achieve high spatial resolution and deep tissue penetration. Unlike conventional optical microscopy, which is constrained by the diffraction limit and shallow depth imaging, and confocal microscopy, which uses optical sectioning but requires high light intensities, NLOM enables deep tissue imaging while minimizing photodamage and photobleaching by restricting excitation to the focal plane. Compared with OCT, which excels at millimeter-scale imaging depth but lacks molecular specificity, NLOM offers submicron resolution and can simultaneously generate functional and structural insights, making it ideal for studying dynamic biological processes. Additionally, while Raman spectroscopy provides unparalleled chemical specificity, its depth penetration is limited, and it is less suitable for real-time imaging. Nonlinear modalities, in contrast, allow for the visualization of dynamic processes such as tissue metabolism or cancer progression in live samples, complementing the structural insights provided by Raman spectroscopy and confocal imaging.

This section delves into each technique's principles and applications, illustrates how they complement each other in addressing diverse scientific and medical imaging needs, and further compares conventional optical microscopy techniques with NLOM in the field of biological research.

## *4.1 Conventional Optical Microscopy*

Conventional optical microscopy, also known as brightfield microscopy, is one of the most commonly used and accessible imaging modalities. The fundamental concept is the transmission of visible light through a sample, capturing the image based on the contrast between light-absorbing structures within the sample. This technique offers a lateral resolution close to the diffraction limit, approximately 200 nm, and an axial resolution of approximately 500 nm. Owing to these limitations, conventional microscopy excels in imaging thin, transparent samples, such as histological sections and live cells in culture.

For over two centuries, bright-field light microscopy has been the primary method for pathological examination. This technique, which focuses on cellular and tissue morphology, is still considered the "gold standard" for detecting cancer cells and determining cancer type. It works by identifying structural irregularities in tissue architecture and nuclear features in stained samples viewed under a standard bright-field microscope, providing reliable diagnostic indicators that can be applied across numerous tumor types despite the complex molecular diversity present at various stages of cancer development. Surgical resections and small biopsies of tissue samples are the principal sources used to investigate cancer pathology and give diagnostic, prognostic, or therapeutic insights. For clinical uses, tissue samples must be stored in fixatives for short-term usage or processed into frozen or formalin-fixed, paraffin-embedded (FFPE) tissue blocks for long-term preservation. FFPE preservation is regarded as the gold standard in pathology. It also enables the creation of an extensive archive of human tissue linked to clinical outcomes, which enables retrospective analysis across a broad range of diseases, including challenging-to-study precursor lesions that may take years to develop into malignancies [110]. Consequently, imaging techniques adaptable to FFPE tissues offer substantial value, as they can seamlessly integrate into current clinical workflows and support extensive disease research. Sun et al. demonstrated that NLOM effectively mimics traditional H&E staining in analyzing both lung tumors and non-neoplastic pulmonary structures. This imaging technique offers notable benefits over frozen section analysis performed during traditional histopathology and serves as a promising optical platform to support rapid pathological evaluation of resected tissues, particularly in the context of advancing digital pathology [111]. In another work by Tao et al., a blinded analysis of NLOM images by three pathologists yielded a sensitivity of 95.4%, a specificity of 93.3%, and an overall accuracy of 94.1% in differentiating invasive cancer and ductal carcinoma in situ from benign breast tissue compared with histological evaluation. These findings indicate that NLOM holds significant promise as a method for the intraoperative assessment of surgical excised breast samples [58].

While conventional microscopy is straightforward and cost-effective, its inability to optically section thick samples restricts its use in three-dimensional imaging. In thicker samples, out-of-focus light significantly reduces image clarity. The staining or labeling of biological samples is often necessary to achieve adequate contrast, as conventional microscopy does not inherently provide high contrast for transparent or low-contrast samples [112]. Despite its limitations, conventional microscopy is highly practical for routine structural and morphological analyses and remains foundational in many scientific and clinical applications.

## 4.2 Confocal Microscopy

Confocal microscopy advances the capabilities of conventional microscopy by adding optical sectioning, which enhances depth resolution and reduces out-of-focus light, enabling high-contrast imaging of thicker samples. In confocal microscopy (as shown in Fig. 10), a point light source (often a laser) illuminates a small focal point on the sample, and a pinhole is positioned before the detector to let only in-focus light pass through. This setup significantly improves the axial resolution to approximately 500 nm, similar to the lateral resolution, creating sharper, more defined images suitable for three-dimensional reconstructions. Confocal microscopy typically achieves imaging depths of up to 100–200 μm in biological tissues, making it effective for imaging cellular and subcellular structures within thick samples such as tissue sections, organoids, and small model organisms.

Unlike the cross-sectional view typically observed in histological analysis, in vivo confocal microscopy captures images parallel to the tissue surface, referred to as the 'en face' orientation [113]. In its hand-held, miniature format designed for intraoral use, an optical fiber functions as the 'pinhole' within the microscope. Key

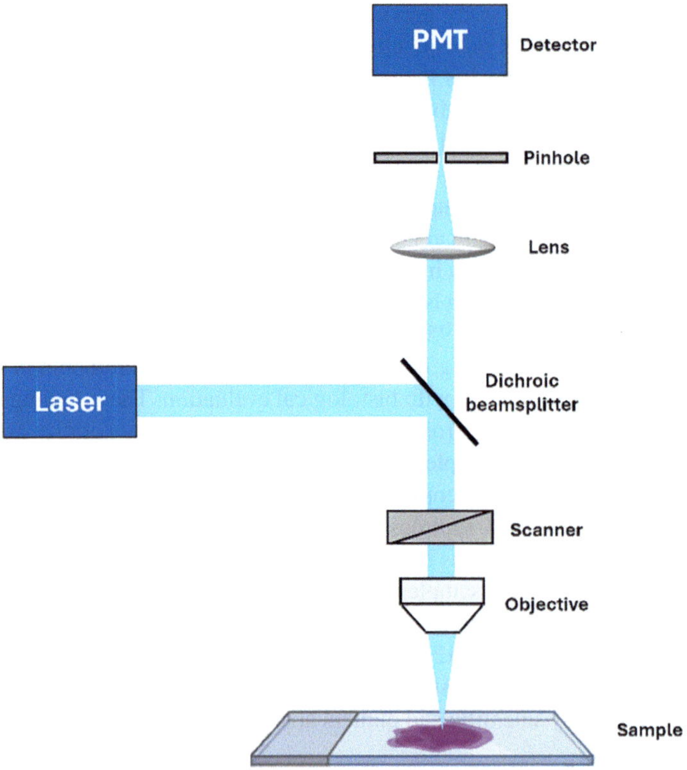


**Fig. 10** A schematic representation of the setup of a confocal laser microscope

challenges facing this portable technology include achieving high resolution, rapid frame rates, and sufficient imaging depth for effective diagnosis [114]. Furthermore, ex vivo confocal microscopy has been proven to be useful for identifying the distinct characteristics of various skin lesions, benign and malignant tumors, and some inflammatory conditions. A primary use of ex vivo confocal microscopy is to evaluate tumor margins in Mohs surgery, which results in a significant time reduction—approximately two-thirds—compared with the time needed for frozen section processing ($p$ <0.001) [115]. Reflectance confocal microscopy is predominantly used to diagnose both neoplastic and non-neoplastic skin lesions. When RCM confirms a lesion as benign, biopsies are often avoided, thereby reducing unnecessary procedures. For skin cancers, the course of treatment—whether surgical or nonsurgical—can be initiated directly on the basis of the cancer stage [116]. RCM offers real-time, cellular-level imaging, confirming its role as an essential diagnostic tool for identifying skin cancers, including both melanocytic and keratinocyte types [117]. Furthermore, fluorescence confocal microscopy could serve as an effective alternative to the use of frozen sections in assessing margin status, particularly in cases where specimen preservation is essential [118].

Despite the diverse applications of confocal microscopy, a common limitation is its potential for photobleaching and phototoxicity, especially when high-resolution images are acquired over time. The high-intensity light required can damage live samples or induce photobleaching, particularly in fluorescently labeled samples. Additionally, confocal microscopy is limited in imaging depth owing to light scattering and attenuation in dense, high-scattering tissues, which restricts its use in applications requiring deep tissue imaging. Nevertheless, confocal microscopy remains invaluable for high-resolution imaging of complex structures and is widely used in neuroscience, cell biology, and pathology.

Confocal microscopy and NLOM are advanced imaging techniques used in biological and medical research, each with distinct capabilities and limitations. Confocal microscopy relies primarily on fluorescent labeling to achieve optical sectioning, offering high-resolution images of cellular structures with a penetration depth of approximately 100–200 μm. However, it requires thin samples and fluorophore staining, limiting its utility for intact tissues or intraoperative applications. NLOM, which includes TPF and SHG, outperforms confocal techniques in several ways. It achieves greater tissue penetration (up to 500–1000 μm) by using near-infrared light, which minimizes photodamage and scattering. Additionally, NLOM can exploit intrinsic tissue signals, such as autofluorescence and SHG, eliminating the need for exogenous labeling. This makes it particularly suitable for real-time imaging of unprocessed tissues in surgical settings [119].

## 4.3 Optical Coherence Tomography (OCT)

OCT combines the principles of confocal microscopy with low-coherence interferometry, achieving high axial resolution and enhanced imaging depth. OCT uses broadband or low-coherence light sources, such as superluminescent diodes or short-pulse lasers, to generate depth-resolved pictures via optical coherence gating. This approach enables high-resolution imaging of subsurface structures without requiring physical sectioning, with an axial resolution of approximately 1–10 μm, making it capable of capturing fine details [120]. Because of these advantages, OCT is now widely used in a variety of biological domains, including ophthalmology, dentistry, dermatology, cancer, and cardiovascular medicine. The setup of an OCT system is depicted in Fig. 11.

OCT has demonstrated considerable promise in imaging tumors within cavity organs through methods such as endoscopy, catheterization, and laparoscopy. Furthermore, using OCT, it is possible to differentiate the four layers of the stomach wall—specifically, the glandular epithelium, mucosal muscle, submucosa, and muscle layer—with blood vessels clearly visible in the submucosa. In a pioneering study, Jansen et al. conducted a prospective analysis on 26 patients with esophageal cancer, calculating plaque contrast within an M-mode scan to discern blood flow regions from surrounding stationary tissues. This research is the first to validate OCT imaging for gastric tissue and in vivo blood flow assessment during esophageal cancer surgery, potentially reducing the incidence of anastomotic leakage post-surgery and enhancing patient outcomes [122]. Furthermore, OCT plays a pivotal role in enabling real-time qualitative analysis and guiding targeted biopsies. In their study, Ding et al. utilized OCT imaging to successfully distinguish dysplastic and cancerous tissue from normal colonic tissue [123]. Efficient and precise intraoperative margin assessment is essential to reduce resection rates in breast cancer surgeries. Singla et al. used OCT imaging and an active reverse-learning, pretrained

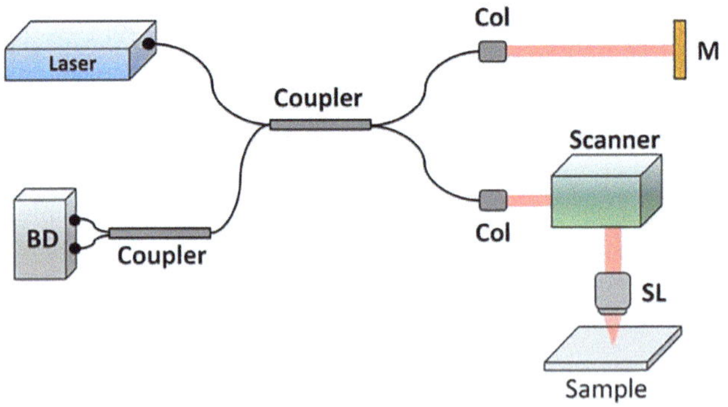


**Fig. 11** A schematic representation of the setup of a swept-source OCT for animal experiments. Col: collimator, M: mirror, SL: scan lens, and BD: balanced detector. (Figure and caption used with permission from [121])

Inception-v3 CNN to distinguish between healthy and malignant breast tissues, with high sensitivity, specificity, and accuracy [124]. Similarly, Mojahed et al. demonstrated that CNN-based algorithms can accurately detect malignant areas within OCT images, supporting their utility in intraoperative cancer detection [125]. OCT also serves as a bronchoscopic adjunct, allowing visualization of the bronchial wall's microstructural layers and providing a near-histopathological assessment without the need for tissue biopsy. This noninvasive approach helps reduce the need for invasive procedures and increases early detection rates, which is crucial for the effective diagnosis and management of lung cancer [126].

Despite the aforementioned applications and advantages, optical coherence microscopy (OCM) complexity, instrumentation requirements, and sensitivity to motion artifacts can present challenges in dynamic or uncontrolled environments. Its relatively high cost also limits accessibility in some settings. Despite these challenges, OCM has become a valuable modality in areas requiring nondestructive imaging at both the cellular and tissue levels, especially in clinical and diagnostic applications. Furthermore, OCT lacks molecular specificity and relies solely on structural contrast. In contrast, NLOM techniques such as TPF and SHG exploit nonlinear optical phenomena to offer molecular-level insights and functional imaging of biological tissues. With depth penetration up to 500–1000 μm and inherent label-free capabilities, NLOM can distinguish cellular and extracellular features through intrinsic signals such as autofluorescence and SHG. While OCT excels in real-time structural imaging, NLOM offers superior biochemical and molecular contrast, making it more suitable for applications requiring detailed functional and compositional analysis. Both modalities are complementary, with studies demonstrating their combined use for comprehensive tissue characterization [127].

### *4.4 Raman Spectroscopy*

Unlike conventional, confocal, and optical coherence microscopy, Raman spectroscopy offers a unique imaging mechanism based on molecular vibrational contrast rather than spatial or optical contrast alone. By detecting inelastically scattered photons following interactions with specific molecular bonds, Raman spectroscopy offers chemical-specific information on a sample's makeup. This method is capable of distinguishing molecular species within samples without the need for labeling, making it highly valuable for applications in biomedical research, pharmacology, and materials science. Raman spectroscopy achieves a lateral spatial resolution similar to that of optical microscopy (~200 nm) but typically offers shallow depth penetration, which is usually limited to less than 100 μm. Figure 12 shows the illustrations of the Raman spectroscopy and Raman probe setups.

Malignant transformation in cells and tissues is marked by substantial biochemical alterations alongside morphological and functional changes. For example, uncontrolled cellular proliferation—a fundamental characteristic of cancer [128]—leads to increased synthesis of DNA, RNA, and proteins while disrupting lipid

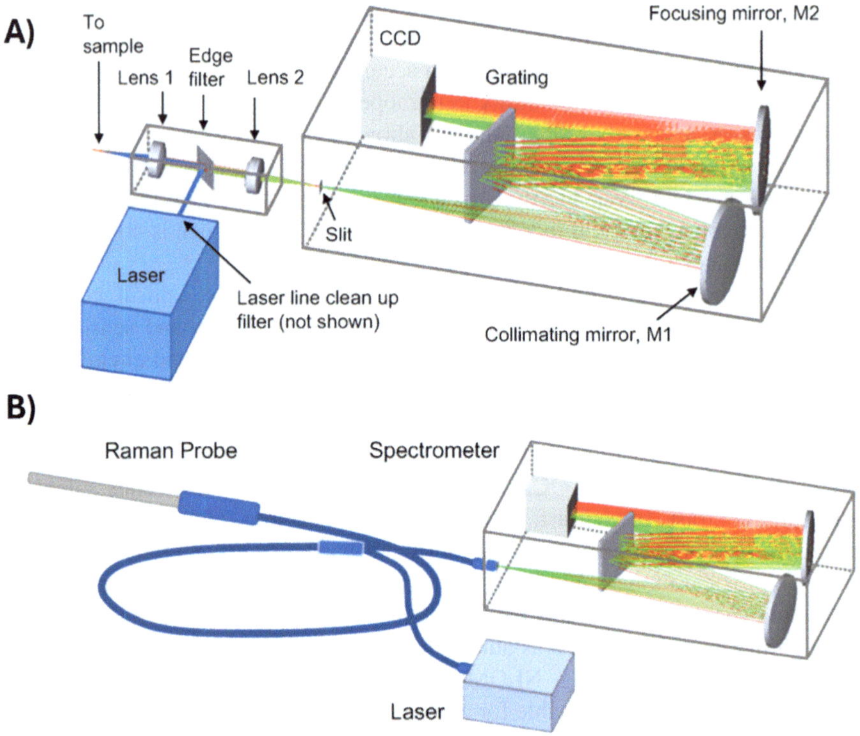


**Fig. 12** (**a**) Typical laboratory Raman spectrometer. (**b**) Raman probe assembly. (Figure and caption used with permission from [63])

metabolism. Notably, these biochemical shifts arise well before clinical symptoms manifest. Consequently, Raman-based techniques are valuable for detecting and quantifying these altered molecular signatures, utilizing spectral profiles as biomarkers for early disease detection and tumor grading. Furthermore, Raman spectroscopy allows for nondestructive analysis, as it requires no extensive sample preparation or labeling. This enables subsequent processing and analyses post-Raman acquisition, which is particularly valuable in cancer diagnostics for constructing a comprehensive profile of individual tumors. Additionally, Raman spectroscopy does not require samples to be in a solid state, leading to growing interest in analyzing circulating biomarkers within various body fluids [129–132] for early cancer detection and diagnosis. The initial Raman investigations on breast tissue focused on qualitative spectral analysis, which identified unique characteristics associated with benign and malignant breast changes. Early findings in human breast tissue revealed that Fourier Transform (FT) Raman spectroscopy, which uses NIR excitation at 1064 nm, could effectively distinguish between normal and abnormal tissue [133]. The main biochemical differences between normal and diseased breast tissues include lipids, carotenoids, and proteins. Spectral analysis of normal, benign, and malignant breast tissues using visual excitation showed a significant drop in peak intensities corresponding to the vibrational modes of β-carotene and lipids in aberrant tissues compared to normal tissues [134].

Although it provides lower spatial resolution and depth capability than OCM does, the specificity of Raman spectroscopy in detecting molecular structures makes it an unparalleled tool for investigating chemical composition, molecular interactions, and tissue diagnostics [135]. A major limitation is the inherently weak Raman scattering signal, which often requires extended acquisition times or signal enhancement techniques, such as surface-enhanced Raman spectroscopy (SERS), for practical imaging. Additionally, fluorescence from certain samples can interfere with the Raman signal, complicating data interpretation [136]. Despite these challenges, Raman spectroscopy is essential for applications requiring precise chemical information, such as cancer diagnosis, metabolic studies, and pharmaceutical quality control.

In comparison with NLOM, Raman spectroscopy typically suffers from low signal intensity, which can lead to slower imaging speeds and limited depth penetration (up to ~100 μm). NLOM techniques, such as TPF and SHG, rely on multiphoton absorption processes to achieve high-resolution imaging of cellular and extracellular structures. With greater imaging depth (up to 500–1000 μm) and label-free imaging capabilities, NLOM excels in visualizing the structural and functional properties of tissues in three dimensions. While Raman spectroscopy is superior for detecting specific molecular fingerprints, NLOM provides faster imaging with greater depth and spatial resolution, making it more suitable for dynamic and in vivo applications. Together, these methods can offer a holistic view of tissue composition and architecture [137].

## *4.5 Comparative Summary*

Each of these four imaging modalities offers distinct advantages that cater to different aspects of biomedical and materials research. Conventional optical microscopy provides a foundational tool for routine structural imaging, particularly in thin samples, although it lacks depth resolution and often requires staining. Confocal microscopy enhances resolution and depth sectioning, enabling detailed imaging of thicker samples, but is limited in depth penetration owing to light scattering and phototoxicity concerns. OCM bridges the gap by extending depth capabilities with high axial resolution and is well suited for subsurface imaging in high-scattering tissues such as skin or corneal layers, although its complex setup and sensitivity to motion artifacts restrict some applications. While not optimized for deep imaging or high spatial resolution, Raman spectroscopy stands out for its chemical specificity, allowing for detailed molecular and compositional analysis that the other modalities cannot provide.

The choice among these modalities depends heavily on the specific application requirements. For example, in cellular imaging, where high resolution and optical sectioning are priorities, confocal microscopy is ideal. In contrast, for noninvasive imaging of deeper tissue layers, OCM is preferred. Moreover, Raman spectroscopy is the tool of choice for applications that demand molecular-level insights, albeit with limitations in spatial resolution and depth. In summary, these optical imaging techniques offer complementary solutions to diverse research and diagnostic challenges, with each modality's unique strengths aligning with particular scientific and medical objectives (Table 1).

## 5 Future Perspectives and Conclusion

The future of NLOM in cancer diagnosis holds immense promise, with several emerging technological frontiers poised to increase its clinical utility. Recent advances in ultrafast laser technology and adaptive optics are enabling deeper tissue penetration and improved resolution, whereas novel combinations of existing modalities such as multiphoton-CARS (MP-CARS) and multicolor multiphoton microscopy are expanding the scope of label-free imaging capabilities [138–141]. These emerging techniques are particularly important for revealing complex cellular dynamics and molecular interactions within the TME in unprecedented detail. The integration of NLOM with complementary diagnostic methods represents another crucial development pathway, where combining NLOM with conventional imaging modalities such as MRI, CT, or ultrasound could provide comprehensive multimodal insights into cancer progression. The ML algorithms are increasingly being employed to analyze the complex datasets generated by these integrated approaches, enabling more accurate and automated diagnoses. The synergy between AI and NLOM is particularly promising for identifying subtle patterns in collagen reorganization and metabolic changes that might escape human detection. Perhaps most significantly, the potential for in vivo real-time diagnosis is being realized through advances in the miniaturization and endoscopic implementation of NLOM. The development of flexible, minimally invasive fiber-based systems and specialized probes is bringing intraoperative guidance and real-time margin assessment closer to clinical reality, potentially revolutionizing surgical oncology procedures.

Building on these future directions, NLOM has demonstrated remarkable progress across various cancer types, offering high-resolution, label-free imaging capabilities that enhance the visualization of cellular and ECM structures. The ability of this technology to provide detailed structural and molecular information without exogenous contrast agents has proven particularly valuable in breast cancer, where tumor-associated collagen signatures can be precisely mapped, and in skin cancer, where real-time assessment of tumor margins is becoming increasingly feasible. The application of techniques such as SHG, CARS, and multiphoton microscopy has proven invaluable for identifying malignancy-associated morphological and biochemical changes, allowing for noninvasive, precise diagnostics. These advances reflect a significant leap toward improved real-time diagnostic methodologies,

**Table 1** Comparison of commonly used imaging techniques

| Feature | Traditional histopathology | Confocal microscopy | Optical coherence tomography | Raman spectroscopy | NLOM |
|---|---|---|---|---|---|
| Process duration | Time-intensive; requires fixation, embedding, and staining | Rapid imaging, limited to thin sections | Fast, provides real-time imaging of tissue structure | Time-intensive for spectral analysis; slower imaging speed | Fast imaging with real-time capabilities |
| Preparation requirements | Tissue sectioning and staining required (e.g., H&E staining) | Minimal; requires staining or fluorescent tagging | No special preparation needed; works on fresh tissue | No staining required; works with unprocessed tissue | No preparation; uses inherent fluorescence and SHG signals |
| Tissue penetration depth | Limited by section thickness | 100–200 μm in biological tissues | 1–3 mm; suitable for epithelial layers | Limited to ~100 μm | 500–1000 μm (varies by setup and tissue type) |
| Image contrast | High contrast with staining | Relies on fluorescent dyes for enhanced contrast | Relatively low contrast; limited to structural imaging | High molecular specificity based on chemical composition | High contrast from intrinsic tissue properties (e.g., SHG) |
| Molecular/ structural information | Provides structural details but limited molecular insights | Structural with some molecular-level resolution | Primarily structural | High chemical specificity at molecular levels | Structural details and some molecular information |
| Real-time imaging capability | Not possible; post-processing required | Limited real-time imaging capabilities | Yes; provides real-time images | No; requires spectral acquisition and processing | Yes; suitable for intraoperative use |
| Specificity to tissue composition | Dependent on staining for cellular and tissue differentiation | Relies on specific fluorescent markers | Limited; provides general tissue morphology | Very high; detects biochemical composition of tissues | High; provides structural and compositional details |
| Digital integration | Requires separate digitization process | Directly generates digital images | Direct digital output; compatible with digital pathology | Data-heavy; spectral data requires further processing | Directly produces high-resolution digital images |

providing unprecedented insights into tumor architecture, ECM remodeling, and lipid distribution, which are pivotal for early detection and treatment planning. Moreover, the integration of artificial intelligence with NLOM data analysis enhances diagnostic accuracy and enables more personalized treatment approaches. However, several difficulties need to be addressed, to fully realize the potential of NLOM in clinical settings. These include reducing system costs, simplifying operational complexity, and developing standardized protocols for image acquisition and interpretation.

The ongoing development of more compact, user-friendly systems and automated analysis tools suggests that these hurdles are gradually being overcome. In the future, the field is moving toward more comprehensive, multimodal imaging platforms that combine the strengths of various nonlinear optical techniques with other imaging modalities, potentially offering a more complete understanding of cancer biology and progression. Future developments aim to refine sensitivity and specificity further, adapting NLOM tools to cover a broader spectrum of cancers and to support personalized diagnostic approaches in oncology. The continued evolution of NLOM technology, coupled with innovations in computational analysis and system miniaturization, positions it as an increasingly important tool in the future of cancer diagnostics, with the potential to significantly impact patient outcomes through earlier detection and more precise guidance for tailored treatment strategies.

## References

1. Thomas, G., van Voskuilen, J., Gerritsen, H. C., & Sterenborg, H. J. C. M. (2014). Advances and challenges in label-free nonlinear optical imaging using two-photon excitation fluorescence and second harmonic generation for cancer research. *Journal of Photochemistry and Photobiology B: Biology, 141*, 128–138.
2. Chakraborty, I., Mazumder, N., Gogoi, A., Chen, M.-C., & Zhuo, G. Y. (2024). Recent advances in label-free imaging techniques based on nonlinear optical microscopy to reveal the heterogeneity of the tumor microenvironment. *Biophysical Reviews, 16*, 581.
3. Mazumder, N., Balla, N. K., Zhuo, G.-Y., Kistenev, Y. V., Kumar, R., Kao, F.-J., et al. (2019). Label-free non-linear multimodal optical microscopy—Basics, development, and applications. *Frontiers in Physics, 7*.
4. Qu, J., Luo, G., Li, S., Qi, K., Hu, R., Li, J., et al. (2024). Label-free in vivo monitoring of lipid droplets for real-time assessment of adipose activator-induced tumor suppression. *Analytical Chemistry, 96*(7), 2754–2758.
5. Zhang, S., Liu, L., Ren, S., Li, Z., Zhao, Y., Yang, Z., et al. (2020). Recent advances in nonlinear optics for bio-imaging applications. *Opto-Electronic Advances, 3*(10), 200003.
6. Kaniyala Melanthota, S., Kistenev, Y. V., Borisova, E., Ivanov, D., Zakharova, O., Boyko, A., et al. (2022). Types of spectroscopy and microscopy techniques for cancer diagnosis: A review. *Lasers in Medical Science, 37*(8), 3067–3084.
7. Pouli, D., Thieu, H.-T., Genega, E. M., Baecher-Lind, L., House, M., Bond, B., et al. (2020). Label-free, high-resolution optical metabolic imaging of human cervical precancers reveals potential for intraepithelial neoplasia diagnosis. *Cell Reports Medicine, 1*(2), 100017.

8. Wang, S., Zhao, Z., Xie, J., Ren, S., Tang, S., Lin, F., et al. (2022). Accurate evaluation of the treatment effects of immunotherapy on subcutaneous ovarian cancer in mice with nonlinear optical imaging and algorithmic analysis. *Biomedical Optics Express, 13*(4), 2266–2277.
9. Henson, D. E., Srivastava, S., & Kramer, B. S. (1999). Molecular and genetic targets in early detection. *Current Opinion in Oncology, 11*(5), 419–425.
10. Kallen, M. E., & Hornick, J. L. (2021). The 2020 WHO classification: What's new in soft tissue tumor pathology? *The American Journal of Surgical Pathology, 45*(1), e1.
11. Neuberger, J., & Cain, O. (2021). The need for alternatives to liver biopsies: Non-invasive analytics and diagnostics. *Hepatic Medicine: Evidence and Research, 13*, 59–69.
12. Larionova, I., Rakina, M., Ivanyuk, E., Trushchuk, Y., Chernyshova, A., & Denisov, E. (2022). Radiotherapy resistance: Identifying universal biomarkers for various human cancers. *Journal of Cancer Research and Clinical Oncology, 148*(5), 1015–1031.
13. Sung, H., Ferlay, J., Siegel, R. L., Laversanne, M., Soerjomataram, I., Jemal, A., & Bray, F. (2021). Global Cancer Statistics 2020: GLOBOCAN estimates of incidence and mortality worldwide for 36 cancers in 185 countries. *CA: A Cancer Journal for Clinicians, 71*(3), 209–249.
14. Whitaker, K. (2020). Earlier diagnosis: The importance of cancer symptoms. *The Lancet Oncology, 21*(1), 6–8.
15. Sasieni, P., Smittenaar, R., Hubbell, E., Broggio, J., Neal, R. D., & Swanton, C. (2023). Modelled mortality benefits of multi-cancer early detection screening in England. *British Journal of Cancer, 129*(1), 72–80.
16. Bray, F., Laversanne, M., Sung, H., Ferlay, J., Siegel, R. L., Soerjomataram, I., & Jemal, A. (2024). Global cancer statistics 2022: GLOBOCAN estimates of incidence and mortality worldwide for 36 cancers in 185 countries. *CA: A Cancer Journal for Clinicians, 74*(3), 229–263.
17. Crosby, D., Lyons, N., Greenwood, E., Harrison, S., Hiom, S., Moffat, J., et al. (2020). A roadmap for the early detection and diagnosis of cancer. *The Lancet Oncology, 21*(11), 1397–1399.
18. Rowe, S. P., & Pomper, M. G. (2022). Molecular imaging in oncology: Current impact and future directions. *CA: A Cancer Journal for Clinicians, 72*(4), 333–352.
19. Ozturk, M. S., Montero, M. G., Wang, L., Chaible, L. M., Jechlinger, M., & Prevedel, R. (2021). Intravital mesoscopic fluorescence molecular tomography allows non-invasive in vivo monitoring and quantification of breast cancer growth dynamics. *Communications Biology, 4*(1), 556.
20. Rodrigues, J., Amin, A., Chandra, S., Mulla, N. J., Nayak, G. S., Rai, S., et al. (2024). Machine learning enabled photoacoustic spectroscopy for noninvasive assessment of breast tumor progression in vivo: A preclinical study. *ACS Sensors, 9*(2), 589–601.
21. Chen, M.-C., Wang, W.-H., Raju, G., Mazumder, N., & Zhuo, G.-Y. (2023). Polarization-resolved second-harmonic generation for tissue imaging. In N. Mazumder, Y. V. Kistenev, E. Borisova, & K. Sharma Prasada (Eds.), *Optical polarimetric modalities for biomedical research* (pp. 259–304). Springer International Publishing.
22. Sun, M., Mu, X., & Li, R. (2024a). The principle, application and imaging of TPEF. In *Linear and nonlinear optical spectroscopy and microscopy. Progress in optical science and photonics* (pp. 87–103). Springer.
23. Gavgiotaki, E., Filippidis, G., Markomanolaki, H., Kenanakis, G., Agelaki, S., Georgoulias, V., & Athanassakis, I. (2017). Distinction between breast cancer cell subtypes using third harmonic generation microscopy. *Journal of Biophotonics, 10*(9), 1152–1162.
24. Harpel, K., Baker, R. D., Amirsolaimani, B., Mehravar, S., Vagner, J., Matsunaga, T. O., et al. (2016). Imaging of targeted lipid microbubbles to detect cancer cells using third harmonic generation microscopy. *Biomedical Optics Express, 7*(7), 2849–2860.
25. Wu, P.-C., Hsieh, T.-Y., Tsai, Z.-U., & Liu, T.-M. (2015). In vivo quantification of the structural changes of collagens in a melanoma microenvironment with second and third harmonic generation microscopy. *Scientific Reports, 5*(1), 8879.

26. Tokarz, D., Cisek, R., Joseph, A., Golaraei, A., Mirsanaye, K., Krouglov, S., et al. (2019). Characterization of pancreatic cancer tissue using multiphoton excitation fluorescence and polarization-sensitive harmonic generation microscopy. *Frontiers in Oncology, 9*, 272.
27. Tai, S. P., Wu, Y., Shieh, D. B., Chen, L. J., Lin, K. J., Yu, C. H., et al. (2007). Molecular imaging of cancer cells using plasmon-resonant-enhanced third-harmonic-generation in silver nanoparticles. *Advanced Materials, 19*(24), 4520–4523.
28. Gavgiotaki, E., Filippidis, G., Tsafas, V., Bovasianos, S., Kenanakis, G., Georgoulias, V., et al. (2020). Third harmonic generation microscopy distinguishes malignant cell grade in human breast tissue biopsies. *Scientific Reports, 10*(1), 11055.
29. Aljakouch, K., Hilal, Z., Daho, I., Schuler, M., Krauß, S. D., Yosef, H. K., et al. (2019). Fast and noninvasive diagnosis of cervical cancer by coherent anti-stokes Raman scattering. *Analytical Chemistry, 91*(21), 13900–13906.
30. Hou, J., Reid, N. E., Tromberg, B. J., & Potma, E. O. (2020). Kinetic analysis of lipid metabolism in breast cancer cells via nonlinear optical microscopy. *Biophysical Journal, 119*(2), 258–264.
31. Cheng, K., Liu, B., Zhang, X.-S., Zhang, R.-Y., Zhang, F., Ashraf, G., et al. (2022). Biomimetic material degradation for synergistic enhanced therapy by regulating endogenous energy metabolism imaging under hypothermia. *Nature Communications, 13*(1), 4567.
32. Xu, C., Nedergaard, M., Fowell, D. J., Friedl, P., & Ji, N. (2024). Multiphoton fluorescence microscopy for in vivo imaging. *Cell, 187*(17), 4458–4487.
33. Perrin, L., Bayarmagnai, B., & Gligorijevic, B. (2020). Frontiers in intravital multiphoton microscopy of cancer. *Cancer Reports, 3*(1), e1192.
34. Parodi, V., Jacchetti, E., Osellame, R., Cerullo, G., Polli, D., & Raimondi, M. T. (2020). Nonlinear optical microscopy: From fundamentals to applications in live bioimaging. *Frontiers in Bioengineering and Biotechnology, 8*, 585363.
35. Li, H., Gong, Q., & Luo, K. (2024). Biomarker-driven molecular imaging probes in radiotherapy. *Theranostics, 14*(10), 4127–4146.
36. Talone, B., Bresci, A., Manetti, F., Vernuccio, F., De la Cadena, A., Ceconello, C., et al. (2022). Label-free multimodal nonlinear optical microscopy reveals features of bone composition in pathophysiological conditions. *Frontiers in Bioengineering and Biotechnology, 10*, 1042680.
37. Ntziachristos, V. (2010). Going deeper than microscopy: The optical imaging frontier in biology. *Nature Methods, 7*(8), 603–614.
38. Hong, G., Antaris, A. L., & Dai, H. (2017). Near-infrared fluorophores for biomedical imaging. *Nature Biomedical Engineering, 1*(1), 0010.
39. Meng, X., Pang, X., Zhang, K., Gong, C., Yang, J., Dong, H., & Zhang, X. (2022). Recent advances in near-infrared-II fluorescence imaging for deep-tissue molecular analysis and cancer diagnosis. *Small, 18*(31), 2202035.
40. Balasubramanian, H., Hobson, C. M., Chew, T.-L., & Aaron, J. S. (2023). Imagining the future of optical microscopy: Everything, everywhere, all at once. *Communications Biology, 6*(1), 1096.
41. Huo, Z., Cao, X., Sun, D., Xu, W., Yang, B., & Xu, S. (2023). Carbonized polymer dot probe for two-photon fluorescence imaging of lipid droplets in living cells and tissues. *ACS Sensors, 8*(5), 1939–1949.
42. Chen, W. C., Chen, Y. J., Lin, S. T., Hung, W. H., Chan, M. C., Wu, I. C., et al. (2020). Label-free characterization of collagen fibers in cancerous esophagus tissues using ratiometric nonlinear optical microscopy. *Experimental Biology and Medicine (Maywood, N.J.), 245*(14), 1213–1221.
43. Baniasadi, A., Das, J. P., Prendergast, C. M., Beizavi, Z., Ma, H. Y., Jaber, M. Y., & Capaccione, K. M. (2024). Imaging at the nexus: How state of the art imaging techniques can enhance our understanding of cancer and fibrosis. *Journal of Translational Medicine, 22*(1), 567.

44. Arnold, M., Morgan, E., Rumgay, H., Mafra, A., Singh, D., Laversanne, M., et al. (2022). Current and future burden of breast cancer: Global statistics for 2020 and 2040. *Breast, 66*, 15–23.
45. Abu Abeelh, E., & AbuAbeileh, Z. (2024). Comparative effectiveness of mammography, ultrasound, and MRI in the detection of breast carcinoma in dense breast tissue: A systematic review. *Cureus, 16*(4), e59054.
46. Tilbury, K., & Campagnola, P. J. (2015). Applications of second-harmonic generation imaging microscopy in ovarian and breast cancer. *Perspectives in Medicinal Chemistry, 7*, 21–32.
47. Wu, X., Chen, G., Lu, J., Zhu, W., Qiu, J., Chen, J., et al. (2013). Label-free detection of breast masses using multiphoton microscopy. *PLoS One, 8*(6), e65933.
48. Okoro, C., Kelkar, V., Sivaguru, M., Emmadi, R., & Toussaint, K. C. (2018). Second-harmonic patterned polarization-analyzed reflection confocal microscopy of stromal collagen in benign and malignant breast tissues. *Scientific Reports, 8*(1), 16243.
49. Ajeti, V., Nadiarnykh, O., Ponik, S. M., Keely, P. J., Eliceiri, K. W., & Campagnola, P. J. (2011). Structural changes in mixed Col I/Col V collagen gels probed by SHG microscopy: Implications for probing stromal alterations in human breast cancer. *Biomedical Optics Express, 2*(8), 2307–2316.
50. Ambekar, R., Lau, T.-Y., Walsh, M., Bhargava, R., & Toussaint, K. C. (2012). Quantifying collagen structure in breast biopsies using second-harmonic generation imaging. *Biomedical Optics Express, 3*(9), 2021–2035.
51. Burke, R. M., Madden, K. S., Perry, S. W., Zettel, M. L., & Brown, E. B., III. (2013). Tumor-associated macrophages and stromal TNF-α regulate collagen structure in a breast tumor model as visualized by second harmonic generation. *Journal of Biomedical Optics, 18*(8), 86003.
52. Provenzano, P.P., Eliceiri, K.W., Campbell, J.M., Inman, D.R., White, J.G. and Keely, P.J., 2006. Collagen reorganization at the tumor-stromal interface facilitates local invasion. *BMC medicine, 4*, pp.1–15.
53. Provenzano, P. P., Eliceiri, K. W., Campbell, J. M., et al. (2006). Collagen reorganization at the tumor-stromal interface facilitates local invasion. BMC Medicine, 4, 38. https://doi.org/10.1186/1741-7015-4-38.
54. Yang, Y., Li, F., Gao, L., Wang, Z., Thrall, M. J., Shen, S. S., et al. (2011). Differential diagnosis of breast cancer using quantitative, label-free and molecular vibrational imaging. *Biomedical Optics Express, 2*(8), 2160–2174.
55. Abramczyk, H., & Brozek-Pluska, B. (2013). Raman imaging in biochemical and biomedical applications. Diagnosis and treatment of breast cancer. *Chemical Reviews, 113*(8), 5766–5781.
56. Xu, J., Yu, T., Zois, C. E., Cheng, J. X., Tang, Y., Harris, A. L., & Huang, W. E. (2021). Unveiling cancer metabolism through spontaneous and coherent Raman spectroscopy and stable isotope probing. *Cancers (Basel), 13*(7).
57. Bredfeldt, J. S., Liu, Y., Conklin, M. W., Keely, P. J., Mackie, T. R., & Eliceiri, K. W. (2014). Automated quantification of aligned collagen for human breast carcinoma prognosis. *Journal of Pathology Informatics, 5*(1), 28.
58. Tao, Y. K., Shen, D., Sheikine, Y., Ahsen, O. O., Wang, H. H., Schmolze, D. B., et al. (2014). Assessment of breast pathologies using nonlinear microscopy. *Proceedings of the National Academy of Sciences of the United States of America, 111*(43), 15304–15309.
59. Shin, K. S., Laohajaratsang, M., Men, S., Figueroa, B., Dintzis, S. M., & Fu, D. (2020). Quantitative chemical imaging of breast calcifications in association with neoplastic processes. *Theranostics, 10*(13), 5865–5878.
60. Ren, L., Li, J., Wang, C., Lou, Z., Gao, S., Zhao, L., et al. (2021). Single cell RNA sequencing for breast cancer: Present and future. *Cell Death Discovery, 7*(1), 104.
61. Narayanamurthy, V., Padmapriya, P., Noorasafrin, A., Pooja, B., Hema, K., Firus Khan, A. Y., et al. (2018). Skin cancer detection using non-invasive techniques. *RSC Advances, 8*(49), 28095–28130.

62. Heuke, S., Vogler, N., Meyer, T., Akimov, D., Kluschke, F., Röwert-Huber, H. J., et al. (2013). Detection and discrimination of non-melanoma skin cancer by multimodal imaging. *Healthcare (Basel), 1*(1), 64–83.
63. Auner, G. W., Koya, S. K., Huang, C., Broadbent, B., Trexler, M., Auner, Z., et al. (2018). Applications of Raman spectroscopy in cancer diagnosis. *Cancer Metastasis Reviews, 37*(4), 691–717.
64. Evans, C. L., & Xie, X. S. (2008). Coherent anti-stokes Raman scattering microscopy: Chemical imaging for biology and medicine. *Annual Review of Analytical Chemistry (Palo Alto Calif), 1*, 883–909.
65. Richards-Kortum, R., Lorenzoni, C., Bagnato, V. S., & Schmeler, K. (2024). Optical imaging for screening and early cancer diagnosis in low-resource settings. *Nature Reviews Bioengineering, 2*(1), 25–43.
66. Delrue, C., Speeckaert, R., Oyaert, M., De Bruyne, S., & Speeckaert, M. M. (2023). From vibrations to visions: Raman spectroscopy's impact on skin cancer diagnostics. *Journal of Clinical Medicine, 12*(23), 7428.
67. Guo, F., Lin, F., Shen, B., Wang, S., Li, Y., Guo, J., et al. (2024). Multidimensional quantitative characterization of basal cell carcinoma by spectral- and time-resolved two-photon microscopy. *Nanophotonics, 13*(2), 217–227.
68. El-Shafai, W., El-Fattah, I. A., & Taha, T. E. (2024). Advancements in non-invasive optical imaging techniques for precise diagnosis of skin disorders. *Optical and Quantum Electronics, 56*(7), 1112.
69. Jiang, X., Zhong, J., Liu, Y., Yu, H., Zhuo, S., & Chen, J. (2011). Two-photon fluorescence and second-harmonic generation imaging of collagen in human tissue based on multiphoton microscopy. *Scanning, 33*(1), 53–56.
70. Poulon, F., Pallud, J., Varlet, P., Zanello, M., Chretien, F., Dezamis, E., et al. (2018). Real-time brain tumor imaging with endogenous fluorophores: A diagnosis proof-of-concept study on fresh human samples. *Scientific Reports, 8*(1), 14888.
71. Wu, J., Wang, T., Uckermann, O., Galli, R., Schackert, G., Cao, L., et al. (2022). Learned end-to-end high-resolution lensless fiber imaging towards real-time cancer diagnosis. *Scientific Reports, 12*(1), 18846.
72. Galli, R., & Uckermann, O. (2024). Vibrational spectroscopy and multiphoton microscopy for label-free visualization of nervous system degeneration and regeneration. *Biophysical Reviews, 16*(2), 219–235.
73. Mirsanaye, K., Uribe Castaño, L., Kamaliddin, Y., Golaraei, A., Kontenis, L., Ẑurauskas, E., et al. (2022). Unsupervised determination of lung tumor margin with widefield polarimetric second-harmonic generation microscopy. *Scientific Reports, 12*(1), 20713.
74. Schubert, M. C., Soyka, S. J., Tamimi, A., Maus, E., Schroers, J., Wißmann, N., et al. (2024). Deep intravital brain tumor imaging enabled by tailored three-photon microscopy and analysis. *Nature Communications, 15*(1), 7383.
75. Ni, H., Wang, Y., Tang, T., Yu, W., Li, D., He, M., et al. (2022). Quantum dots assisted in vivo two-photon microscopy with NIR-II emission. *Photonics Research, 10*(1), 189–196.
76. Chen, J., Zhuo, S., Chen, G., Yan, J., Yang, H., Liu, N., et al. (2011). Establishing diagnostic features for identifying the mucosa and submucosa of normal and cancerous gastric tissues by multiphoton microscopy. *Gastrointestinal Endoscopy, 73*(4), 802–807.
77. Perry, S. W., Burke, R. M., & Brown, E. B. (2012). Two-photon and second harmonic microscopy in clinical and translational cancer research. *Annals of Biomedical Engineering, 40*(2), 277–291.
78. Li, L., Huang, X., Zhang, S., Zhan, Z., Kang, D., Guan, G., et al. (2023). Rapid and label-free detection of gastrointestinal stromal tumor via a combination of two-photon microscopy and imaging analysis. *BMC Cancer, 23*(1), 38.
79. Sarri, B., Canonge, R., Audier, X., Simon, E., Wojak, J., Caillol, F., et al. (2019). Fast stimulated Raman and second harmonic generation imaging for intraoperative gastro-intestinal cancer detection. *Scientific Reports, 9*(1), 10052.

80. Prieto, S. P., Reed, C. L., James, H. M., Quinn, K. P., & Muldoon, T. J. (2019). Differences in colonic crypt morphology of spontaneous and colitis-associated murine models via second harmonic generation imaging to quantify colon cancer development. *BMC Cancer, 19*(1), 428.
81. Zhou, Z. H., Ji, C. D., Xiao, H. L., Zhao, H. B., Cui, Y. H., & Bian, X. W. (2017). Reorganized collagen in the tumor microenvironment of gastric cancer and its association with prognosis. *Journal of Cancer, 8*(8), 1466–1476.
82. Birk, J. W., Tadros, M., Moezardalan, K., Nadyarnykh, O., Forouhar, F., Anderson, J., & Campagnola, P. (2014). Second harmonic generation imaging distinguishes both high-grade dysplasia and cancer from normal colonic mucosa. *Digestive Diseases and Sciences, 59*(7), 1529–1534.
83. Yan, J., Zhuo, S., Chen, G., Milsom, J. W., Zhang, H., Lu, J., et al. (2014). Real-time optical diagnosis for surgical margin in low rectal cancer using multiphoton microscopy. *Surgical Endoscopy, 28*(1), 36–41.
84. Hu, W., & Fu, L. (2013). Simultaneous characterization of pancreatic stellate cells and other pancreatic components within three-dimensional tissue environment during chronic pancreatitis. *Journal of Biomedical Optics, 18*(5), 56002.
85. Drifka, C. R., Tod, J., Loeffler, A. G., Liu, Y., Thomas, G. J., Eliceiri, K. W., & Kao, W. J. (2015). Periductal stromal collagen topology of pancreatic ductal adenocarcinoma differs from that of normal and chronic pancreatitis. *Modern Pathology, 28*(11), 1470–1480.
86. Cisek, R., Joseph, A., Harvey, M., & Tokarz, D. (2021). Polarization-sensitive second harmonic generation microscopy for investigations of diseased collagenous tissues. *Frontiers in Physics, 9*.
87. Ricciardelli, C., & Rodgers, R. J. (2006). Extracellular matrix of ovarian tumors. *Seminars in Reproductive Medicine, 24*(4), 270–282.
88. Kirkpatrick, N. D., Brewer, M. A., & Utzinger, U. (2007). Endogenous optical biomarkers of ovarian cancer evaluated with multiphoton microscopy. *Cancer Epidemiology, Biomarkers & Prevention, 16*(10), 2048–2057.
89. Chen, X., Cui, S., Yan, S., Zhang, S., Fan, Y., Gong, Y., et al. (2021). Hyperspectral stimulated Raman scattering microscopy facilitates differentiation of low-grade and high-grade human prostate cancer. *Journal of Physics D: Applied Physics, 54*(48), 484001.
90. Pallen, S., Shetty, Y., Das, S., Vaz, J. M., & Mazumder, N. (2021). Advances in nonlinear optical microscopy techniques for in vivo and in vitro neuroimaging. *Biophysical Reviews, 13*(6), 1199–1217.
91. Wanjie, W., Zhongya, Q., Junqiang, W., Congping, C., Weitao, C., Kai, L., & Jianan, Y. Q. (2020). In vivo high-resolution multimodal nonlinear optical microscopy of spinal cord in mice. *Proceedings of SPIE, 11252*, 112520P.
92. Zhuo, G.-Y., Tsai, P.-L., Wang, H.-Y., & Chan, M.-C. (2020). Wave-vector-encoded nonlinear endomicroscopy. *Optics Letters, 45*(13), 3713–3716.
93. Liang, W., Hall, G., Messerschmidt, B., Li, M.-J., & Li, X. (2017). Nonlinear optical endomicroscopy for label-free functional histology in vivo. *Light: Science & Applications, 6*(11), e17082.
94. Kretschmer, S., Pieper, M., Hüttmann, G., Bölke, T., Wollenberg, B., Marsh, L. M., et al. (2016). Autofluorescence multiphoton microscopy for visualization of tissue morphology and cellular dynamics in murine and human airways. *Laboratory Investigation, 96*(8), 918–931.
95. Vargas, G., Shilagard, T., Ho, K. H., & McCammon, S. (2009). Multiphoton autofluorescence microscopy and second harmonic generation microscopy of oral epithelial neoplasms. *Annual International Conference of the IEEE Engineering in Medicine and Biology Society, 2009*, 6311–6313.
96. Bower, A. J., Chidester, B., Li, J., Zhao, Y., Marjanovic, M., Chaney, E. J., et al. (2017). A quantitative framework for the analysis of multimodal optical microscopy images. *Quantitative Imaging in Medicine and Surgery, 7*(1), 24–37.

97. Hall, G., Liang, W., & Li, X. (2014). Nonlinear endomicroscopy imaging technology for translational applications. In R. Weigert (Ed.), *Advances in intravital microscopy: From basic to clinical research* (pp. 281–303). Springer Netherlands.
98. Fu, L., & Gu, M. (2007). Fibre-optic nonlinear optical microscopy and endoscopy. *Journal of Microscopy, 226*(3), 195–206.
99. Huang, L., Zhou, X., Liu, Q., MacAulay, C. E., & Tang, S. (2020). Miniaturized multimodal multiphoton microscope for simultaneous two-photon and three-photon imaging with a dual-wavelength Er-doped fiber laser. *Biomedical Optics Express, 11*(2), 624–635.
100. Yu, X., Zhou, L., Qi, T., Zhao, H., & Xie, H. (2023). MEMS enabled miniature two-photon microscopy for biomedical imaging. *Micromachines, 14*, 470.
101. So, P. T. C., Yew, E. Y. S., & Rowlands, C. (2013). High-throughput nonlinear optical microscopy. *Biophysical Journal, 105*(12), 2641–2654.
102. Squier, J., & Müller, M. (2001). High resolution nonlinear microscopy: A review of sources and methods for achieving optimal imaging. *Review of Scientific Instruments, 72*(7), 2855–2867.
103. Demming, A. (2020). Optical microscopy—How small can it go? *Physics World, 33*(10), 41.
104. Elliott, A. D. (2020). Confocal microscopy: Principles and modern practices. *Current Protocols in Cytometry, 92*(1), e68.
105. Irsch, K., Grieve, K., Borderie, M., Vilbert, M., Plamann, K., Ghoubay, D., et al. (2022). Full-field optical coherence microscopy for histology-like analysis of stromal features in corneal grafts. *Journal of Visualized Experiments*, (188).
106. Pezzotti, G. (2021). Raman spectroscopy in cell biology and microbiology. *Journal of Raman Spectroscopy, 52*(12), 2348–2443.
107. Valli, J., Garcia-Burgos, A., Rooney, L. M., Vale de Melo, E. O. B., Duncan, R. R., & Rickman, C. (2021). Seeing beyond the limit: A guide to choosing the right super-resolution microscopy technique. *The Journal of Biological Chemistry, 297*(1), 100791.
108. St Croix, C. M., Shand, S. H., & Watkins, S. C. (2005). Confocal microscopy: Comparisons, applications, and problems. *BioTechniques, 39*(6 Suppl), S2–S5.
109. Ho, C.-S., Jean, N., Hogan, C. A., Blackmon, L., Jeffrey, S. S., Holodniy, M., et al. (2019). Rapid identification of pathogenic bacteria using Raman spectroscopy and deep learning. *Nature Communications, 10*(1), 4927.
110. Baker, M. (2012). Building better biobanks. *Nature, 486*(7401), 141–146.
111. Sun, Y., Weber, T. D., Fujimoto, J. G., Rosen, S., & VanderLaan, P. A. (2024b). Rapid examination of lung tissues by nonlinear microscopy. *American Journal of Clinical Pathology, 162*(4), 369–378.
112. Vennalaganti, P., Kanakadandi, V., Goldblum, J. R., Mathur, S. C., Patil, D. T., Offerhaus, G. J., et al. (2017). Discordance among pathologists in the United States and Europe in diagnosis of low-grade dysplasia for patients with Barrett's esophagus. *Gastroenterology, 152*(3), 564–570.e564.
113. Jabbour, J. M., Saldua, M. A., Bixler, J. N., & Maitland, K. C. (2012). Confocal endomicroscopy: Instrumentation and medical applications. *Annals of Biomedical Engineering, 40*(2), 378–397.
114. Wang, T. D., Mandella, M. J., Contag, C. H., & Kino, G. S. (2003). Dual-axis confocal microscope for high-resolution in vivo imaging. *Optics Letters, 28*(6), 414–416.
115. Bennàssar, A., Vilata, A., Puig, S., & Malvehy, J. (2014). Ex vivo fluorescence confocal microscopy for fast evaluation of tumour margins during Mohs surgery. *The British Journal of Dermatology, 170*(2), 360–365.
116. Yélamos, O., Manubens, E., Jain, M., Chavez-Bourgeois, M., Pulijal, S. V., Dusza, S. W., et al. (2020). Improvement of diagnostic confidence and management of equivocal skin lesions by integration of reflectance confocal microscopy in daily practice: Prospective study in 2 referral skin cancer centers. *Journal of the American Academy of Dermatology, 83*(4), 1057–1063.

117. Shahriari, N., Rabinovitz, H., Oliviero, M., & Grant-Kels, J. M. (2021). Reflectance confocal microscopy: Melanocytic and nonmelanocytic. *Clinics in Dermatology, 39*(4), 643–656.
118. Ragazzi, M., Piana, S., Longo, C., Castagnetti, F., Foroni, M., Ferrari, G., et al. (2014). Fluorescence confocal microscopy for pathologists. *Modern Pathology, 27*(3), 460–471.
119. Lee, J. H., Lee, S., Gho, Y. S., Song, I. S., Tchah, H., Kim, M. J., & Kim, K. H. (2015). Comparison of confocal microscopy and two-photon microscopy in mouse cornea in vivo. *Experimental Eye Research, 132*, 101–108.
120. Hsieh, Y.-S., Ho, Y.-C., Lee, S.-Y., Chuang, C.-C., Tsai, J.-c., Lin, K.-F., & Sun, C.-W. (2013). Dental optical coherence tomography. *Sensors, 13*(7), 8928–8949.
121. Nguyễn-Hoàng, T., Wang, T.-A., Wu, C.-H., & Tsai, M.-T. (2024). Discriminating glioblastoma from normal brain tissue in vivo using optical coherence tomography and angiography: A texture and microvascular analysis approach. *Photonics, 11*(5), 435.
122. Jansen, S. M., Almasian, M., Wilk, L. S., De Bruin, D. M., Van Berge Henegouwen, M. I., Strackee, S. D., et al. (2018). Feasibility of optical coherence tomography (OCT) for intra-operative detection of blood flow during gastric tube reconstruction. *Sensors, 18*, 1331.
123. Ding, J., Li, Q., Lin, J., He, S., Chen, W., He, Q., et al. (2020). Optical coherence tomography for the early detection of colorectal dysplasia and cancer: Validation in a murine model. *Quantitative Imaging in Medicine and Surgery, 11*(1), 371–379.
124. Singla, N., Dubey, K., & Srivastava, V. (2019). Automated assessment of breast cancer margin in optical coherence tomography images via pretrained convolutional neural network. *Journal of Biophotonics, 12*(3), e201800255.
125. Mojahed, D., Ha, R. S., Chang, P., Gan, Y., Yao, X., Angelini, B., et al. (2020). Fully automated postlumpectomy breast margin assessment utilizing convolutional neural network based optical coherence tomography image classification method. *Academic Radiology, 27*(5), e81–e86.
126. Ding, M., Pan, S.-Y., Huang, J., Yuan, C., Zhang, Q., Zhu, X.-I., & Cai, Y. (2022). Optical coherence tomography for identification of malignant pulmonary nodules based on random forest machine learning algorithm. *PLoS One, 16*(12), e0260600.
127. Gubarkova, E. V., Kirillin, M. Y., Dudenkova, V. V., Timashev, P. S., Kotova, S. L., Kiseleva, E. B., et al. (2016). Quantitative evaluation of atherosclerotic plaques using cross-polarization optical coherence tomography, nonlinear, and atomic force microscopy. *Journal of Biomedical Optics, 21*(12), 126010.
128. Hanahan, D., & Weinberg, R. A. (2000). The hallmarks of cancer. *Cell, 100*(1), 57–70.
129. Bilal, M., Bilal, M., Tabassum, S., Saleem, M., Mahmood, H., Sarwar, U., et al. (2017). Optical screening of female breast cancer from whole blood using Raman spectroscopy. *Applied Spectroscopy, 71*(5), 1004–1013.
130. Kim, S., Kim, T. G., Lee, S. H., Kim, W., Bang, A., Moon, S. W., et al. (2020). Label-free surface-enhanced Raman spectroscopy biosensor for on-site breast cancer detection using human tears. *ACS Applied Materials & Interfaces, 12*(7), 7897–7904.
131. Qian, K., Wang, Y., Hua, L., Chen, A., & Zhang, Y. (2018). New method of lung cancer detection by saliva test using surface-enhanced Raman spectroscopy. *Thorac Cancer, 9*(11), 1556–1561.
132. Zhang, K., Liu, X., Man, B., Yang, C., Zhang, C., Liu, M., et al. (2018). Label-free and stable serum analysis based on Ag-NPs/PSi surface-enhanced Raman scattering for noninvasive lung cancer detection. *Biomedical Optics Express, 9*(9), 4345–4358.
133. Liu, C. H., Alfano, R. R., Sha, W. L., Zhu, H. R., Akins, D. L., Cleary, J., et al. (1991). Human breast tissues studied by IR Fourier-transform Raman spectroscopy. In J. G. A. H. T. Bufton & W. Krupke (Eds.), *Conference on lasers and electro-optics* (Vol. 10, p. CWF51). Optica Publishing Group.
134. Albayrak, M., Senol, O., Demirkaya-Miloglu, F., Calik, M., & Kadioglu, Y. (2019). Novel chemometrics-assisted spectroscopic methods for diagnosis and monitoring of invasive ductal carcinoma in breast tissue. *Bratislavské Lekárske Listy, 120*(3), 184–187.

135. Zheng, C., Jia, H. Y., Liu, L. Y., Wang, Q., Jiang, H. C., Teng, L. S., et al. (2020). Molecular fingerprint of precancerous lesions in breast atypical hyperplasia. *The Journal of International Medical Research, 48*(6), 300060520931616.
136. Ma, D., Shang, L., Tang, J., Bao, Y., Fu, J., & Yin, J. (2021). Classifying breast cancer tissue by Raman spectroscopy with one-dimensional convolutional neural network. *Spectrochimica Acta Part A: Molecular and Biomolecular Spectroscopy, 256*, 119732.
137. Cui, S., Zhang, S., & Yue, S. (2018). Raman spectroscopy and imaging for cancer diagnosis. *Journal of Healthcare Engineering, 2018*(1), 8619342.
138. Borile, G., Sandrin, D., Filippi, A., Anderson, K. I., & Romanato, F. (2021). Label-free multiphoton microscopy: Much more than fancy images. *International Journal of Molecular Sciences, 22*(5), 2657.
139. Galli, R., Siciliano, T., Aust, D., Korn, S., Kirsche, K., Baretton, G. B., et al. (2023). Label-free multiphoton microscopy enables histopathological assessment of colorectal liver metastases and supports automated classification of neoplastic tissue. *Scientific Reports, 13*(1), 4274.
140. Rishøj, L., Hernández, I. C., Ramachandran, S., & Jowett, N. (2022). Multiphoton microscopy for label-free multicolor imaging of peripheral nerve. *Journal of Biomedical Optics, 27*(5), 056501.
141. Stringari, C., Abdeladim, L., Malkinson, G., Mahou, P., Solinas, X., Lamarre, I., et al. (2017). Multicolor two-photon imaging of endogenous fluorophores in living tissues by wavelength mixing. *Scientific Reports, 7*(1), 3792.

# An Insight on Photodynamic Therapy for Cancer Treatment

**Manoj Sharma, Rupak Mukhopadhyay, and Aditya Kumar**

**Abstract** Cancer is one of the leading causes of death worldwide with high rate of recurrence even after modern medical intervention. Various treatment approaches, such as chemotherapy, radiation therapy, and targeted gene therapy, have been developed to combat cancer. However, these treatments often come with significant drawbacks, including high costs and severe side effects, which can hinder their effectiveness and reduce patient quality of life. Photodynamic therapy (PDT) is a novel clinical technique that uses photosensitive chemicals known as photosensitizers (PSs) to become cytotoxic when activated by light of a specific wavelength in the presence of oxygen. When activated, the PSs produce reactive oxygen species (ROS), which cause cellular damage and result in the removal of aberrant or malignant cells. Because of its capacity to selectively target cells while avoiding damage to surrounding healthy tissue, PDT is increasingly being used to treat various cancers, particularly surface tumors or benign malignancies, microbial infections, and other medical conditions. This chapter provides an insight into the fundamental chemical and biological concepts that underpin PDT, with a special emphasis on its applications in cancer treatment. It discusses the mechanisms underlying PDT, in which PSs are activated by light to produce ROS capable of selectively destroying cancer cells. This chapter also discusses the current limitations and developments in PDT, particularly in terms of its effectiveness. It examines the issues of improving PS delivery systems and tumor targeting, as well as strategies for overcoming barriers such as tumor hypoxia and poor light penetration. Advances in nanotechnology, such as the invention of nanoparticle-based carriers, are being investigated as methods to increase PS administration, ensure more effective tumor localization, and address oxygen deprivation in tumor microenvironments. Furthermore, efforts to improve light penetration, notably using near-infrared (NIR) light and two-photon

M. Sharma · R. Mukhopadhyay (✉) · A. Kumar (✉)
Department of Molecular Biology and Biotechnology, Tezpur University, Tezpur, Assam, India
e-mail: mrupak@tezu.ernet.in; aditya@tezu.ernet.in



N. Mazumder et al. (eds.), *Advanced Biophysical Techniques in Biosciences*, Lecture Notes in Nanoscale Science and Technology 7,
https://doi.org/10.1007/978-3-031-94551-9_12

excitation, are highlighted, with the goal of improving PDT's efficacy in treating deep-seated cancers.

**Keywords** Photodynamic therapy · Cancer treatment · Photosensitizers · Reactive oxygen species · Cancer treatment

## 1 Introduction

Photodynamic Therapy (PDT) is a minimally invasive treatment that utilizes light-sensitive compounds, called photosensitizers (PSs), and specific wavelengths of light to selectively destroy cancer cells. This technique primarily targets cancerous tissues by inducing a photochemical reaction that generates reactive oxygen species (ROS), such as singlet oxygen, which are toxic to cells and cause cell death through mechanisms like apoptosis, necrosis, or autophagy. PDT requires three key components: a photosensitizer, light, and oxygen. The therapy begins with the administration of a photosensitizer, either systemically or locally, which preferentially accumulates in cancerous tissues due to their abnormal metabolism and structure. After sufficient time for tissue uptake, the target area is exposed to light in the visible or near-infrared (NIR) spectrum, activating the photosensitizer and producing ROS that damage cellular structures, including the cell membrane, mitochondria, and DNA. A major advantage of PDT is its selective cytotoxicity, as the damage is largely confined to the area exposed to light, sparing healthy surrounding tissues, which sets it apart from conventional treatments like chemotherapy and radiotherapy.

## 2 History of Photodynamic Therapy

The ancient history of PDT reveals a rich tapestry of light-based healing practices that date back to early civilizations. The concept of using light for therapeutic purposes, known as heliotherapy, was recognized by ancient Egyptians, Greeks, and Romans, who utilized sunlight to treat various ailments [1, 2]. The invention of the laser by Theodore H. Maiman in 1960, producing photons in a coherent and focused beam, had a transformative impact on phototherapy. Shortly thereafter, lasers were investigated for tumor ablation, leveraging their high energy to produce heat and induce cell damage. However, laser-based tumor ablation faces notable limitations, especially due to its lack of selectivity for targeting only malignant cells. PDT's roots trace back to early twentieth-century discoveries, where it was demonstrated that combining acridine dyes with light exposure could have cytotoxic effects. Around that period, Niels Ryberg Finsen was pioneering the use of red light to treat smallpox and blue light to treat cutaneous tuberculosis (lupus vulgaris) [3]. But it was not until the 1960s and 1970s, with the development of more effective photosensitizers and light sources, that PDT became clinically relevant, particularly through the work of Oscar Raab [4]. The first photosensitizer approved for clinical use was Photofrin, which is still in use today. Over time, PDT has evolved with advances in photosensitizer chemistry and light delivery systems, expanding its applications beyond oncology to other diseases.

## 3 Basic Principles of Photodynamic Therapy

Photodynamic therapy primarily involves three key components: a light source to supply the energy needed for the photodynamic reaction, a photosensitizer (PS) capable of absorbing light and initiating the photodynamic response, and the generation of reactive oxygen species (ROS) through electron or energy transfer from the excited PS, to the components of cells generating ROS, largely singlet oxygen ($^1O_2$) and superoxide. Thus, the effectiveness of PDT relies on the synergistic interaction between light of a specific wavelength, the PS, and ROS production. PDT operates by two basic mechanisms where PS gets activated using light of a specific wavelength and in the process generates ROS, which is capable of damaging cellular components, leading to therapeutic response. ROS generation from photosensitizers occurs by two primary mechanisms: Type I and Type II PDT. Upon activation, PS absorbs the light and gets excited from the groundstate (S0) to the excited singlet short-lived state of higher energy (S1). However, PS eventually loses energy either through fluorescence or heat and enters the groundstate. However, the singlet state can undergo intersystem crossing by spin conversion of electron into higher energy orbitals and enters a more stable, long-lived but reactive triplet electron state (T1). From this state, the energy in PS can either decay to the groundstate through light emission, also known as phosphorescence, or undergo two different types of electron/energy transfer reaction. Electron from the excited state of PS (T1) can either react with nearby substrate, such as unsaturated lipids, proteins, or nucleic acids, in the cells to generate unstable radicals through proton or electron transfer (type I reaction). In the presence of oxygen, this leads to the formation of oxygenated products such as superoxide anion radicals ($O_2^{\bullet -}$), hydroxyl radicals ($OH^{\bullet}$), or hydrogen peroxide ($H_2O_2$), which are highly cytotoxic and induce cellular damage which results in cellular death. Alternatively, the excited PS in its triplet electron state can transfer energy to molecular oxygen ($^3O_2$) present naturally in cells, producing singlet oxygen ($^1O_2$) via a type II reaction (Fig. 1).

During PDT, both type I and type II reactions can occur simultaneously; however, type II reaction predominates in PDT and because of the available oxygen it is effectively able to generate ROS for killing of cells. The balance between type I and type II reactions is influenced by factors such as the type and concentration of the PS, oxygen availability, and the level of irradiation [5, 6]. Since tumor tissue is already highly hypoxic, additionally, oxygen depletion by PS during photodynamic therapy (PDT) can significantly limit its efficacy, as oxygen is critical for the production of reactive oxygen species (ROS), particularly singlet oxygen ($^1O_2$), which drives the cytotoxic effects. Singlet oxygen is a primary cytotoxic agent in PDT,

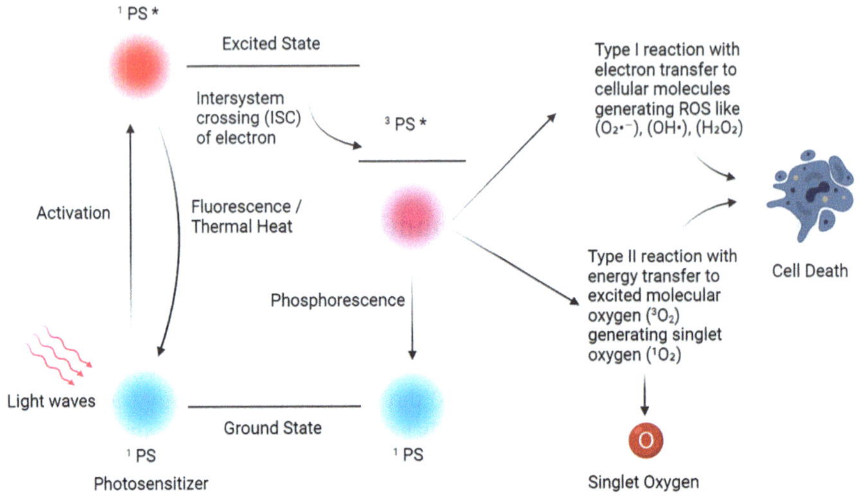


**Fig. 1** A modified Jablonski diagram illustrating the excitation of a photosensitizer (PS) and the subsequent transfer of energy or electrons in type I and type II photodynamic therapy (PDT) mechanisms. These processes are responsible for the generation of reactive oxygen species (ROS) and singlet oxygen ($^3O_2$), ultimately leading to cell death in PDT

particularly under high oxygen conditions. It targets and oxidizes cellular components, leading to apoptosis (programmed cell death), necrosis, or autophagy, depending on the dose and localization. Cells contain a variety of biomolecules that are potential substrates for oxidation by ($^1O_2$) and other reactive species generated by PDT. ROS molecules function by inducing cellular damage in biological molecules like nucleic acid, protein, cell lipid membrane, and cellular structures like mitochondria, thereby causing cell death by various mechanisms. The kind of cell death induced by photodynamic treatment (PDT) is regulated by a number of parameters, including the physicochemical qualities of the photosensitizer (PS), its subcellular localization and concentration, oxygen availability, and the wavelength and intensity of the applied light [7, 8]. When PSs like Z1 and FECTPN that are designed specifically to target mitochondria are used, they normally induce combined pyroptosis and apoptosis in cells at a certain oxidative stress threshold, triggered by photo-generated reactive oxygen species (ROS) and photodamage to electron transport chain components [9]. Certain photosensitizers (PSs), such as Aza-BODIPY and Cationic Amphiphilic Drugs (CADs), can induce lysosomal membrane permeabilization (LMP). This process releases hydrolases into the cytoplasm, disrupting cellular homeostasis and favoring necrosis over apoptosis. PSs that target lysosomes often inhibit or delay the apoptotic pathway, increasing the likelihood of necrotic cell death. Notably, lysosomal disruption and the resultant necrosis can stimulate immune responses, potentially augmenting antitumor immunity [10, 11]. Photosensitizers localized to the plasma membrane can indeed induce apoptosis under modest oxidative stress, while severe damage leading to compromised membrane integrity typically results in necrosis. Photosensitizers like

TBMPEI and Ce6 generate reactive oxygen species (ROS) upon light activation, which can induce apoptosis by damaging cellular components, particularly the plasma membrane [12]. Photosensitizers that target the ER and Golgi apparatus (GA) impair vesicle transport and cause stress in cellular metabolic pathways, and ROS generation that cause Golgi fragmentation disturb membrane integrity and increase necrotic cell death by overloading the cellular repair processes [13]. Overall, the manner of cell death in PDT is determined by the site, type, and amount of ROS created, all of which influence the extent and kind of oxidative damage [14].

## 4 Mechanism of PDT-Induced Tumor Death

Photodynamic therapy (PDT) was initially understood to induce tumor cell death solely through the direct destruction of tumor cells. However, it is now recognized that PDT exerts its therapeutic effects through three interconnected mechanisms. First, it directly induces tumor cell death via apoptosis, necrosis, or autophagy. Second, it causes indirect cytotoxicity by disrupting and damaging the blood vessels and capillaries that supply the tumor, thereby depriving it of essential nutrients and oxygen. Finally, PDT activates acute inflammatory and immune responses, which further contribute to tumor destruction and inhibit tumor growth. These combined mechanisms enhance the overall efficacy of PDT in cancer treatment [15]. New studies have also shown that PDT-mediated direct cytotoxicity can employ multiple pathways, including ferroptosis, necroptosis, pyroptosis, mitotic catastrophe, paraptosis, and parthanatosis [16, 17].

### *4.1 Apoptosis*

Apoptosis is a process of regulated cell death characterized by specific cell modifications such as chromatin condensation and membrane blebbing which occurs with the help of various endonucleases and caspases. In PDT depending on the treatment circumstances, photosensitizers can cause several types of cell death. High levels of photodamage, such as those caused by a high photosensitizer concentration and/or a high light dosage, generally result in necrosis. Moderate damage promotes regulated cell death processes, such as apoptosis. Low degrees of damage, on the other hand, typically induce autophagy, most likely adaptive autophagy, which, depending on the severity of the damage, can then activate other controlled cell death pathways. Oxidative stress is expected to be more severe around blood vessels and at the tumor surface, where photon flux is higher, resulting in cells undergoing necrosis. In contrast, cells located deeper within the tumor, farther from blood vessels, are more likely to undergo regulated forms of cell death, such as apoptosis, whereas Autophagy usually is known to occur in regions where less reactive oxygen species (ROS) are generated [18].

Apoptosis occurs mainly via two primary pathways: the intrinsic pathway, which is triggered by internal stresses such as reactive oxygen species (ROS) or DNA damage and involves caspase-9 activation via CAR-CARD interactions, cytochrome *c* release, and mitochondrial damage; and the extrinsic pathway, which is initiated by external stimuli and activates caspase-8 followed by caspase-3 [19]. The most frequent mechanism of apoptotic cell death mentioned in relation to PDT is intrinsic apoptosis, which is crucially regulated by the B cell lymphoma 2 (BCL-2) family. Photosensitizers, such as ALA, Pc4, SnEt2, and CPO, are some of the commonly used PSs that are known to induce cell death by intrinsic pathways of apoptosis. PDT treatments with various photosensitizers cause changes in BCL-2 family protein levels, often increasing the pro- vs. antiapoptotic protein ratio (e.g., BAX/BCL-2). The location of the photosensitiser (PS) within the cell determines whether the cells follow apoptosis or necrosis pathway in PDT. The increased oxidation damages proteins, leading to their accumulation and activating an endoplasmic reticulum (ER) stress response. Photosensitizers accumulating in mitochondria destroy antiapoptotic proteins like BCL-2, as seen with ALA, Pc4, SnEt2, and others. This destruction upregulates BAX at the mitochondrial membrane, triggering cytochrome *c* release and apoptosis. Photosensitizers targeting the ER-Golgi, such as hypericin and redaporfin, activate the intrinsic apoptosis pathway through ER stress, increased cytosolic $Ca^{2+}$, and CHOP-induced expression of proapoptotic proteins (e.g., BIM, NOXA). PS such as polyamines-targeting derivatives, which are known to accumulate in the cell cytoplasm, also mediate cell death by activating apoptosis via the intrinsic pathway [20]. However, in some cases, inhibition of pro-apoptotic proteins BAX and BAK can shift cell death from apoptosis to autophagy.

## *4.2 Autophagy*

Autophagy is a sequential process that destroys and recycles cytoplasmic components, which is crucial for cellular homeostasis and development. Cells undergoing apoptosis often follow autophagy to remove and recycle the cellular debris. In PDT, autophagy can act as either a protective mechanism, by repairing photodamaged components, or a cell death pathway, particularly in apoptosis-resistant cells, depending on the photosensitizers, cell types, and light dose used [21, 22]. When PDT-induced damage is significant, both autophagy and apoptosis are known to occur simultaneously. In response to PDT, apoptosis often dominates due to its definitive nature, but Bcl-2 proteins can modulate the balance between apoptosis and autophagy. Antiapoptotic proteins from the Bcl-2 family, such as Bid, Bim, Bad, Noxa, and Puma, located in the ER and mitochondria, are key targets in PDT. Their loss can facilitate the initiation of either apoptosis or autophagy. Acting as stress sensors, these proteins lack BH domains 1 and 2, inhibit antiapoptotic proteins, and activate proapoptotic proteins to induce mitochondrial outer membrane permeabilization (MOMP) [23]. They can also reduce autophagy by attaching to Beclin-1. However, under PDT-induced stress, Bcl-2 proteins can be phosphorylated, releasing Beclin-1 and increasing autophagy flux [24]. This increases autophagy as a protective mechanism for managing photodamage by recycling damaged components, particularly when apoptosis is inhibited or delayed, suggesting a potential avenue for enhancing cancer treatment outcomes. However, the complexity of Bcl-2 family interactions necessitates further research to fully understand their roles in therapy resistance.

### *4.3 Necrosis*

Necrosis is a type of cell death brought on by serious injury that damages the cell membrane, allowing cellular contents to flow out and inciting inflammation. Necrosis is caused by acute stressors, including radiation, heat, toxins, or hypoxia, and is energy-independent, in contrast to apoptosis. Necrosis is usually caused by high-dose photodynamic therapy (PDT) that uses high concentrations of photosensitizers or light doses, particularly when photosensitizers build up in the plasma membrane. Short incubation times encourage the localization of photosensitizers to the plasma membrane, which leads to necrosis when photoactivation occurs. Cell death can be changed to apoptosis by longer incubation, which permits photosensitizer buildup in organelles such as the Golgi apparatus (GA). For instance, phthalocyaninato zinc (II) and photofrin cause necrosis after brief incubations, but apoptosis when they target the GA after longer incubations [25, 26]. Similarly, following a brief contact, glycosylated chlorin (GLc4) confines to the plasma membrane and induces necrosis [18]. Some of the commonly used photosensitizers such as Talaporfin, Hypericin, Photolon, Phthalocyanines, HypB-Ac, and DTHe demonstrate the ability to localizing near plasma membrane destroying the integrity of membrane leading to necrosis [27–30].

### *4.4 PDT Effect on Tumor Vascular System*

Photodynamic therapy (PDT) targets tumor vasculature, disrupting the blood supply essential for delivering oxygen and nutrients to tumor cells, thereby inducing their death. This approach has been heavily exploited to develop photosensitizers (PS) that, when activated by light, generate reactive oxygen species (ROS), causing damage to endothelial cells and vascular structures. This further stops the supply of nutrients, blood, and oxygen to tumor tissue, making them an easy target to eliminate tumor. Studies by Henderson et al. [31] and others have shown that PDT induces vascular shutdown, hypoxia, and even anoxia, significantly impacting tumor physiology and leading to delayed tumor growth [31, 32].

Key photosensitizers like Photofrin, benzoporphyrin derivatives, and MV6401 have been shown to induce immediate vasoconstriction and subsequent thrombus formation, cutting off the tumor's supply lines [15]. Short drug–light intervals enhance the vascular effects by concentrating the PS in the vasculature, minimizing systemic toxicity and photosensitivity. These vascular effects make PDT a highly selective and effective method for cancer treatment, often achieved in a single session. However, PDT-induced ROS and hypoxia can paradoxically upregulate angiogenic factors like VEGF and COX-2, which might promote vascular regrowth [33]. Despite this, the destruction of tumor microvasculature remains a key mechanism for long-term PDT efficacy, requiring further research to optimize therapeutic protocols and manage potential compensatory angiogenesis. Vascular PDT offers advantages over traditional PDT methods by rapidly clearing PS, targeting tumor blood vessels directly, and reducing the risk of side effects.

## *4.5 PDT Effect on Immune Cells*

PDT's effect on immune cells is the third most critical element, after direct cell killing and damage to the tumor's vascular system. Apart from apoptosis PDT also stimulate an acute inflammatory response around locally treated tumors thus encouraging antitumor immunity by releasing secondary inflammatory mediators like cytokines and interleukins. PDT-treated tumor tissue produces chemoattractant signals, which attract circulating immune cells such as monocytes, macrophages, neutrophils, and lymphocytes toward the site of tumor. In the late 1980s and 1990s, studies revealed an increase in the infiltration of these cells into PDT-treated tissues, indicating that immunological responses were activated. This immune response is partially mediated by inflammatory cytokines such as interleukin IL-6, IL-1, IL-8, and IL-10. PDT-induced inflammation is caused by oxidative stress, which causes the release of vasoactive molecules, proteinases, reactive oxygen species (ROS), and leukocyte chemoattractant. These components lead to acute inflammation, which is an essential component of the antitumor immune response.

Neutrophils plays a crucial role in the success of PDT. It was discovered that neutrophil increase in PDT-treated tissues inhibited tumor growth, and that reducing neutrophils reduced PDT efficacy [34]. Photodamage to tumor vasculature promotes contraction of endothelial cells, allowing neutrophils to adhere to subendothelial matrix via β2 integrin receptors [35]. PDT causes enhanced expression of adhesion molecules. E-selectin and ICAM1 are crucial for neutrophil adherence to tumor microvessels and penetration into tumor tissue [36]. Additionally, PDT activates macrophages, enhancing their capacity to phagocytize tumor cells and engage in Fc receptor-mediated ingestion. These activated macrophages secrete cytokines and prostaglandins, such as PGE2, which amplify the inflammatory response and recruit T cells to strengthen their antitumor effects.

PDT's effects on cells of adaptive immunity, like dendritic cells and T cells are very noteworthy. Photodynamic treatment (PDT) promotes Dendritic cells (DC) maturation by inducing an inflammatory response and producing damage-associated molecular patterns (DAMPs). DAMPs are released or exposed when tumor cells undergo immunogenic cell death (ICD) caused by PDT. DCs' pattern recognition receptors (PRRs) recognize these DAMPs, prompting signal transduction that results in DC activation and maturation [37]. PDT induces dendritic cells to digest tumor antigens and activate CD4 helper T cells. These helper T cells, in turn, prepare CD8 cytotoxic T lymphocytes to detect and kill tumor cells [38]. T cells play a critical role in preventing tumor recurrence following PDT by mediating long-term immune surveillance and tumor destruction. This was evidenced in a study where immunodeficient mice showed a significantly higher risk of tumor recurrence after PDT compared to immunocompetent mice. Remarkably, this effect was reversed when immunodeficient mice received bone marrow transplants from immunocompetent donors, highlighting the essential role of functional T cells in sustained antitumor immunity [39]. Experimental evidences have also supported the role of T cells in tumor suppression, e.g., immunocompetent (BALB/c) mice were cured by Photofrin-based PDT, while immunosuppressed ones (SCID/Nude mice) were not, emphasizing the role of host lymphoid populations in PDT success [40]. T cells also play a role in the immunostimulatory effects of PDT-produced cancer vaccines. One key mechanism involves using tumor lysates derived from PDT-treated tissues as therapeutic vaccines. These lysates contain tumor antigens released during PDT-induced cell death, which can be recognized by the immune system. The antigens in the lysate are processed by antigen-presenting cells (APCs), such as dendritic cells. This process leads to the activation of cytotoxic T lymphocytes (CTLs), which target and kill tumor cells expressing those specific antigens. Furthermore, PDT lysates promote the production of interleukin-12 (IL-12), a cytokine that enhances the activity of CTLs and natural killer (NK) cells while supporting the development of a long-lasting immune response [41]. PDT-induced necrosis of tumor cells and vasculature elicits a variety of immune responses. Damaged cells release pro-inflammatory chemicals, attracting neutrophils, macrophages, and dendritic cells to the affected area. Macrophages devour PDT-damaged cells and transmit their antigens to CD4 helper T cells, which in turn activate CD8 cytotoxic T cells. These cytotoxic T lymphocytes circulate throughout the body, identifying and eliminating tumor cells on a systemic basis. This creates a long-lasting immunological memory, which can help prevent tumor recurrence.

PDT's ability to trigger both innate and adaptive immune responses makes it more successful as a cancer treatment. It not only directly destroys tumor cells, but it also improves systemic antitumor immunity by recruiting immune cells, activating tumor-specific responses, and establishing immunological memory. Photodynamic treatment (PDT) regulates cytokine levels, balancing pro-inflammatory signals like VEGF and CXCL9 that activate immune cells and anti-inflammatory factors like PD-L1 and HIF-1$\alpha$ that repress them. These alterations represent the local inflammatory response triggered by PDT and have a major impact on immunological function. Monitoring cytokine variations before, during, and after PDT provides essential information for optimizing treatment procedures. For example, combining PDT with immune checkpoint inhibitors may offset immunosuppressive effects, increasing overall treatment efficacy. Understanding these dynamics is critical for enhancing PDT outcomes [42].

# 5 Molecular Targets of Photodynamic Therapy

## 5.1 Proteins

Since proteins are abundant and highly reactive with singlet oxygen ($O_2$) and other reactive oxygen species (ROS), they are the main targets of photodynamic treatment (PDT). Particularly susceptible amino acids include cysteine, methionine, histidine, tyrosine, and tryptophan [43]. These amino acids undergo specific oxidative changes, including the creation of sulfoxides, endoperoxides, and disulfide bridges [44–46]. Different ROS types cause different damage pathways. For instance, hydroxyl radicals react at diffusion-controlled rates, producing secondary effects such as protein cross-linking and tyrosine hydroperoxide production. The amino acid composition, location within the protein structure, and proximity to the photosensitizer all affect protein susceptibility [44]. Such oxidative damage impairs signalling, catalysis, and other vital processes, which ultimately results in cell death.

## 5.2 Lipids

In photodynamic treatment (PDT), lipids—especially unsaturated fatty acids in bio membranes—are the main targets of reactive oxygen species (ROS) and singlet oxygen ($O_2$). Photosensitizers in lipid environments are more likely to produce $^1O_2$ because oxygen is highly soluble in lipids [47]. This $^1O_2$ then interacts with unsaturated fatty acids to form lipid peroxides. When transition metals are present, these peroxides break down, releasing radicals that set off chain reactions that harm the lipid bilayer and produce reactive by-products [48]. In addition to impairing membrane integrity, this lipid peroxidation also has an impact on metabolism, cell signalling, and secondary protein and DNA changes, which ultimately result in cell death [14, 49].

## 5.3 Nucleic Acid

Oxidative damage in the DNA induced by one-electron oxidation, hydroxyl radicals ($HO^•$), and singlet oxygen ($^1O_2$) are the main reason by which DNA damage in photodynamic treatment (PDT) leads to cell death. Among all the five nucleotide bases, guanine because of its low ionization potential is a prime target that produces oxidative products including 8-oxo-7,8-dihydro-2′-deoxyguanosine and guanine radicals [50, 51]. While $^1O_2$ preferentially oxidizes guanine without cleaving the DNA backbone, hydroxyl radicals react indiscriminately with DNA, resulting in base alterations and strand breakage [52]. Compared to other cancer treatments, PDT causes less extensive DNA damage and carries a lower risk of subsequent tumor development. Furthermore, PDT-induced genetic damage is amplified by subsequent oxidation and DNA-protein cross-links.

## 6 Treatment Procedure of PDT

The treatment begins with the administration of a photosensitizer, either systemically (e.g., intravenous injection) or locally (e.g., topical application). These compounds are preferentially taken up or retained by cancerous tissues due to their abnormal metabolic and structural characteristics. After a sufficient period to allow for the photosensitizer to accumulate in the target tissue, the area is exposed to a specific wavelength of light, typically in the visible or near-infrared spectrum, with the help of a probe designed to reach the superficial skin surface and deeper tumor tissues. This light activates the photosensitizer. Upon activation, the photosensitizer undergoes a photochemical reaction in the presence of oxygen, leading to the production of ROS. Since molecular oxygen are critical component of PDT, various methods are utilized to keep in check the percentage of oxygen in blood and tissue like pO2 Histography, Phosphorescent probe compounds (surface imaging), Electron paramagnetic resonance (EPR) particulate, Nuclear magnetic resonance-blood oxygenation level dependent (NMR-BOLD), MRI, Reflectance spectroscopy, Frequency domain photon migration spectroscopy, Fourier transformed-spectral imaging, etc. [53]. These highly reactive molecules damage cellular components such as the cell membrane, mitochondria, and DNA, ultimately causing cell death through apoptosis, necrosis, or autophagy. The cytotoxic effects of PDT are largely confined to the area where the light is applied thus minimizing damage to surrounding healthy tissues. This selectivity is a significant advantage over traditional cancer treatments like chemotherapy and radiotherapy. After photodynamic therapy (PDT), a patient can go home within few hours; however, it usually takes about 2–6 weeks to fully heal after PDT, depending on the area of your body that is being treated. Patients must also take steps to safeguard their skin and the treated area after PDT. Post-treatment patients taking photosensitizing medications should stay out of direct sunshine and bright indoor lights for at least 48 h to avoid burns or irritation since these medications can make the skin and eyes extremely sensitive to light. Proper post-treatment care includes keeping the treated area clean and dry to reduce the risk of infection, covering it to shield it from light, and avoiding scratching to prevent irritation or disruption of healing tissues. These steps are essential for optimizing recovery and achieving the best therapeutic outcomes. Photodynamic therapy (PDT) plays a crucial role in cancer treatment due to its ability to precisely target tumors, making it especially effective for localized cancers such as those of the skin, lungs, oesophagus, bladder, head and neck, gastric (stomach), cervical, colorectal, ovarian, vaginal, vulvar, and liver cancers. It is particularly beneficial for patients who are not candidates for surgery or those seeking to avoid the side effects of traditional therapies. Beyond cancer treatment, PDT is also used to address various conditions, including certain skin disorders like acne, psoriasis, nonmelanoma skin cancer, and precancerous skin conditions. Due to its targeted and efficient approach, PDT is widely used to treat various bacterial, fungal, and viral infections as well.

# 7 Components of PDT

## *7.1 Light Source*

PDT utilizes various light sources, including lasers, incandescent light, and laser-emitting diodes (LED's) for the activation of photosensitisers [54]. The light source used in photodynamic therapy plays a crucial role in determining the penetration depth of photons during in vivo optical imaging. Lasers are commonly used in PDT due to their ability to produce highly focused and monochromatic light. However, they are expensive and require an optical system to expand the beam for larger tissue coverage. Despite their precision, their high cost and limited scalability can be restrictive. Conventional lamps can also serve as light sources for PDT, particularly when paired with optical fibers to deliver specific wavelengths of light. However, these lamps may produce unwanted thermal effects that need to be carefully managed to avoid interference with the therapeutic process. LEDs have gained popularity as an alternative light source for PDT. They are affordable, safe, thermally nondestructive, and available in flexible arrays, making them suitable for a range of applications. Their cost-effectiveness and ease of use make them a practical choice in PDT setups (Fig. 2).

Photon penetration depth is primarily influenced by tissue absorption, scattering, and wavelength of light used (600–800 nm). Traditional light sources, which typically emit light in the visible spectrum, face significant limitations. Their penetration into tissues is shallow and often hindered by interference from endogenous substances. Consequently, these light sources are not suitable for effectively targeting deep-seated tumors in photodynamic therapy. Light penetration in biological tissues depends on wavelength, with longer wavelengths penetrating more effectively. Wavelengths below 600 nm are readily absorbed by biological molecules like haemoglobin and melanin, limiting their tissue penetration and increasing skin photosensitivity. Conversely, wavelengths above 850 nm penetrate deeper but lack the energy needed to excite oxygen into its reactive singlet state, which is crucial for photodynamic therapy (PDT). The optimal range, known as the "tissue optical window" or "phototherapeutic window," lies between 600 and 850 nm, making it an ideal range for PDT. In the NIR-I range (650–950 nm), photon penetration in tissues is limited to 1–2 cm due to autofluorescence noise, impacting imaging quality. The NIR-II windows (1000–1350 nm) offer improved penetration, with the lowest absorption and highest penetration depth around 1125 nm. This range outperforms NIR-I for skin imaging, making it the best option for deeper and clearer bioimaging applications [55, 56].

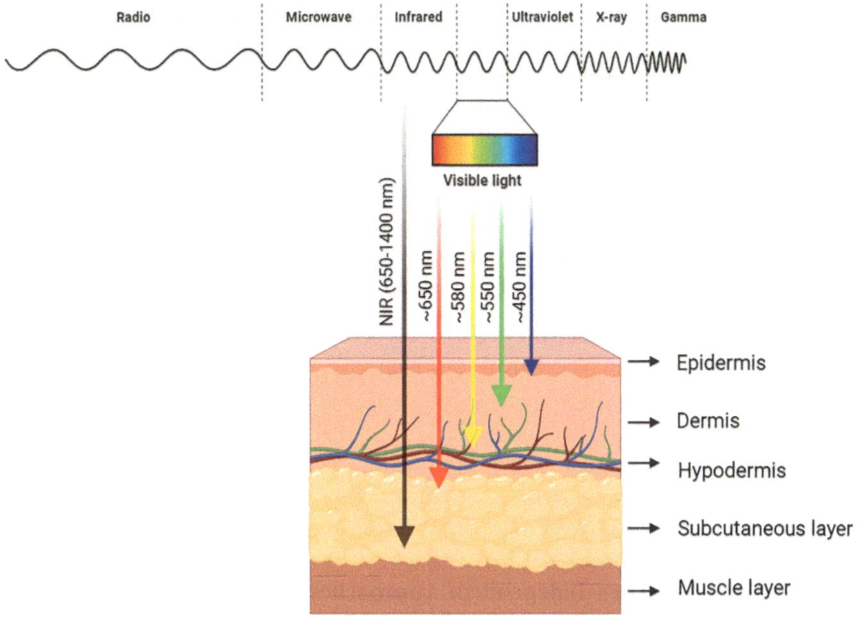


**Fig. 2** The Light penetration in skin in PDT: in PDT, different wavelengths of light interact with various molecules in the skin, resulting in different penetration depths in the skin

## 7.2 *Source of Oxygen*

Oxygen plays a key role in the generation of ROS for killing of cells in tumor. The level of oxygen plays an important role in the effectiveness of PDT. The level of oxygen significantly varies depending on the location of tissue based on the density of blood vessels. In case of a solid tumor, the tumor microenvironment is seriously devoid of oxygen because of a lesser number of blood vessels, and hence oxygen becomes a limiting factor. Hence, oxygen is supplied to tumor patients by two approaches:

1. **Indirect $O_2$ Introduction**: To address tumor hypoxia in photodynamic therapy (PDT), catalase is commonly employed as an $O_2$ carrier. Catalase catalyzes the breakdown of intracellular hydrogen peroxide ($H_2O_2$), which is typically elevated in tumors, into oxygen ($O_2$) and water ($H_2O$) [57]. This not only generates $O_2$ to support PDT but can also reduce $H_2O_2$ levels, potentially reversing the malignant phenotype of cancer cells. More stable catalase-like materials have also been developed to sustain long-term catalytic activity, enhancing their effectiveness in relieving tumor hypoxia.
2. **Direct Introduction of Oxygen**: Direct oxygen delivery to tumors involves carriers like **perfluorocarbons (PFCs)** and haemoglobin. PFCs, known for their high oxygen solubility and biocompatibility, can significantly increase tissue

oxygen levels and alleviate hypoxia [58]. Liquid PFCs have even been used as blood substitutes. In cancer treatment, PFC-based **nanodroplets** have shown promise; these droplets deliver oxygen directly to cancer cells, resulting in reduced cell viability and improved outcomes in photodynamic therapy (PDT).

In case of PDT, treatment with light intensity leads to transient depletion of local oxygen levels, causing a halt in ROS production and reduced treatment effectiveness since it suppresses the ability of PS to generate ROS and singlet oxygen responsible for cellular damage and cell death. Hence, to overcome this situation, various approaches are followed during PDT to avoid oxygen depletion like:

(a) **Enhancing Oxygen Delivery**: Use of Hyperbaric oxygen chambers to enhance oxygen saturation in blood, using artificial oxygen-carrying agents in blood like perfluorocarbon emulsions, and using supplementary oxygen delivery in the targeted tissues.
(b) **Modulating PDT Parameters**: optimizing PS dosage to minimize the oxygen depletion, using low intensity of irradiated light to minimize rapid oxygen consumption, using a pulse of light instead of continuous light to allow time for oxygen levels in tissue.
(c) **Developing Oxygen-Independent Approaches**: using type I photosensitizers which primarily use free radicals instead of singlet oxygen to produce ROS, enhancing oxygen carrying capacity of RBC using RBC's nanoparticles which helps in oxygen supply as well as accumulation of PS in tumor tissue [59]. Incorporating materials that release oxygen upon activation, e.g., perfluoro carbon nanoparticles, covalent organic nanoparticles, fluorinated nanoparticles, Zirconium (IV)-based metal organic framework (UiO-66) nanoparticles, etc., and using in situ oxygen generation using $H_2O_2$-activated, $O_2$-evolving PDT nanoparticles (HAOP NPs), exfoliated manganese oxide ($MnO_2$) nanomaterials, $CaO_2$ decomposition, photo-splitting water, etc., to generate oxygen at the tumor site.

These strategies are designed to address oxygen depletion, sustain effective ROS production, and enhance the therapeutic efficacy of PDT by overcoming the challenge of oxygen consumption during treatment, thereby improving its overall efficiency.

## 7.3 *Photosensitizers*

Photosensitizers are a crucial component of photodynamic therapy (PDT). These chemically inert compounds remain unreactive with biological molecules under normal conditions. However, when exposed to light of a specific wavelength, they interact with molecular oxygen to initiate photodynamic reactions. This process generates reactive oxygen species (ROS) and singlet oxygen, which react with cellular components, ultimately leading to the destruction of tumor cells. During photodynamic therapy (PDT), a photosensitizer is administered topically, intraperitoneally, or intravenously.

The distribution of photosensitizers (PS) in tumor cells is greatly influenced by their charge and solubility. Malignant cells are able to selectively endocytose the low-density lipoprotein (LDL)-photosensitizer complex because of physiological differences between tumors and healthy tissues, particularly in the distribution of LDL receptors. Hydrophobic photosensitizers attach to LDL and enable their quick passage into intracellular membranes like the mitochondria and endoplasmic reticulum (ER), whereas hydrophilic photosensitizers bind to albumin and globulins. Anionic PSs target lysosomes and perinuclear areas, while cationic PSs build up in mitochondria. Lipophilic PSs easily cross lipid barriers, while polar PSs depend on slower endocytosis and hence require longer drug-light intervals (DLIs) [60]. The Enhanced Permeability and Retention (EPR) effect, which arises from leaky blood arteries and inadequate lymphatic drainage in tumors, aids in tumor localization [61]. Through LDL binding, tumor cells with overexpressed LDL receptors also improve PS uptake. This preferential uptake localizes the photosensitizer in tumor cells rather than in normal tissues can also be allotted to lower intracellular pH and the presence of large amounts of collagen. After administration, the photosensitizer typically takes 3–96 h to accumulate in cancer cells and achieve an optimal tumor-to-normal cell concentration ratio. Once localized, the tumor can be identified and closely monitored using the fluorescence of the photosensitizer through techniques such as spectrophotometry, MRI, PET, and fluorescence imaging techniques [62–64]. Once internalized, the patient is treated with photosensitizers specific wavelength of the radiation creates singlet oxygen, which reacts with the tumor and kills it either through necrosis or apoptosis. Some rare instances of PDT side effects are red, swollen, blistered, and itchy skin, coughing, difficulty swallowing, stomach pain, chest pain, etc.

### 7.3.1 Types of Photosensitizers

The majority of photosensitizers used in PDT now belong to the cyclic tetrapyrrolic structural family, which also includes phthalocyanines, chlorins, bacteriochlorins, porphyrins, and their derivatives. When triggered by light, these molecules' distinct chemical and photophysical properties allow them to effectively form reactive oxygen species (ROS), which is an important step in the therapeutic action process. Over the years, the molecular design and functional properties of PSs have evolved, leading to their classification into three distinct generations, each reflecting advancements aimed at overcoming limitations of their predecessors.

**The first generation of photosensitizers**, including hematoporphyrin derivative (HpD) and its commercial equivalent, Photofrin®, laid the groundwork for PDT's clinical applications. These PSs have been used successfully for over three decades to treat malignancies of the lungs, bladder, oesophagus, and brain. However, they have substantial disadvantages, such as low chemical purity, prolonged skin sensitivity to light, and limited activation at wavelengths less than 640 nm, which limits the depth of tissue penetration.

To overcome these restrictions, **second-generation photosensitizers** were created. These PSs, which include derivatives of chlorins, bacteriochlorins, and phthalocyanines, absorb more light in the deep red portion of the spectrum. This characteristic enables deeper tissue penetration and stronger therapeutic effects in tumor areas, significantly improving PDT efficacy while lowering off-target effects.

Following on these advancements, **third-generation photosensitizers** were produced, with a focus on precision and selectivity when targeting tumor cells. These PSs are designed to attain high specificity using functionalization techniques. They can be conjugated with biomolecules such as antibodies, proteins, or carbohydrates that have an affinity for cancer cells. They can also be enclosed in carriers such as gold nanoparticles, silica nanoparticles, quantum dots, and carbon nanotubes. These carriers improve PS stability, transport, and accumulation at tumor locations thus increasing therapeutic effects. The distinct advantages and disadvantages of each class of PS have been thoroughly investigated, some of which are listed in Table 1.

## 8 Limitations of Photodynamic Therapy

Photodynamic therapy (PDT) is often considered a less invasive alternative to traditional cancer treatments like surgery or radiation therapy, which are often very costly and comes with a lot of side effects and are painful. One of the key advantages of PDT is that it does not involve incisions or radiation, making it a more targeted and minimally invasive approach. Additionally, the adverse reactions associated with PDT are generally mild and transient, particularly compared to the more severe and long-lasting side effects of surgery and radiation therapy. However, PDT is not without potential side effects, which can vary depending on the photosensitizer (PS) used and the specific treatment protocol, and include minimal side effects like pain, erythema, edema, fever, itching, pleural effusion, constipation, anaemia, problem with breathing, etc. While PDT has shown promise in treating various cancers, its effectiveness is limited by several factors, particularly in the context of deep-seated or large tumors.

### *8.1 Penetration Depth of Light*

A significant drawback of PDT is the limited ability of light to penetrate biological tissues, typically only reaching a depth of a few millimeters. This restricts PDT's effectiveness in treating large or deep-seated tumors, as these areas often do not receive adequate light exposure. Consequently, the reactive oxygen species (ROS) generated during PDT are confined to the localized region where the photosensitizer

**Table 1** List of first, second, and third generation of clinically approved and experimental photosensitizers in use for the treatment of various cancers and skin diseases

| Generation | Subcategory | Name | Advantages | Drawbacks |
|---|---|---|---|---|
| First generation | Hematoporphyrin derivative (HpD) | Photofrin® (Porfimer sodium) | High singlet oxygen quantum yield; FDA-approved for several cancers, including non-small cell lung cancer and oesophageal cancer | Low $\varepsilon_{max}$ (630 nm); limited tissue penetration (2–3 mm); long-term skin phototoxicity. Requires a long post-treatment period to avoid light exposure due to photosensitivity |
| | | Photogem®, Photosan-3® | This analogue to Photofrin is commonly used in PDT to treat cancer (human glioma, squamous carcinoma, gynecological cancers, head and neck cancers, and pancreatic cancers) | Low tissue penetration |
| Second generation | Porphyrins | m-THPP (*meta*-tetra(hydroxyphenyl) porphyrin) | Higher potency than HpD; effective for tumor photonecrosis at 648 nm | Causes skin phototoxicity |
| | | TPPS4 (5,10,15,20-tetrakis(4-sulfanatophenyl)-21H,23H-porphyrin) | Potential candidate for basal cell carcinoma treatment; irradiated at 645 nm | Lower photochemical efficiency compared to other porphyrins |
| | | PpIX (protoporphyrin IX) | Metabolizes within 48 h; FDA-approved for actinic keratosis | Limited absorption maximum (630–635 nm) |

(continued)

**Table 1** (continued)

| Generation | Subcategory | Name | Advantages | Drawbacks |
|---|---|---|---|---|
| | Chlorins | BPD-MA (benzoporphyrin derivative monoacid ring A) (Verteporfin) | Deeper tissue penetration; reduced skin photosensitivity; approved for age-related macular degeneration | High cost; limited approved cancer treatments |
| | | m-THPC (*meta*-tetra(hydroxyphenyl) chlorin) (Foscan®) | Shorter skin sensitivity and better tissue penetration due to absorption in longer wavelengths<br>Requires small doses; effective for advanced squamous cell carcinoma of the head and neck, breast, and pancreatic cancers | High skin photosensitivity in some patients |
| | | Purlytin™ SnET2 (tin ethyl etiopurpurin) | Deeper penetration; evaluated for metastatic breast cancer and basal cell carcinoma | Dark toxicity; skin photosensitivity; not yet FDA-approved |
| | | Talaporfin (NPe6) (*N*-aspartyl chlorin e6) | Shorter skin sensitivity and better tissue penetration due to absorption in longer wavelengths<br>Minimal skin photosensitivity; approved in Japan for lung cancer | Limited clinical applications outside Japan |
| | Bacteriopheophorbides | Tookad® (WST09) Palladium bacteriopheophorbide | Deeper penetration; quick excretion; and minimal skin photosensitivity | Limited FDA approvals; under evaluation for prostate cancer |
| | | Stakel® (WST11) | Water-soluble; undergoing trials for prostate cancer | Still in early-stage clinical trials |
| | Texaphyrins | Lutrin® (Lu-Tex) Motexafin lutetium | Deeper penetration; water-soluble; evaluated for prostate, breast, and cervical cancers | Requires further studies for safety and efficacy |

| | | | | |
|---|---|---|---|---|
| | Phthalocyanines | Photosens (AlPcS4) Aluminum phthalocyanine tetrasulfonate | High εmax (~200,000 $M^{-1}$ $cm^{-1}$); used in Russia for multiple cancers | Prolonged skin phototoxicity |
| | | (Pc4) Silicon phthalocyanine | Undergoing trials for cutaneous cancers and mycosis fungoides | Limited approved applications; pharmacokinetics not fully established |
| | Non-porphyrin | Hypericin | ROS generation targets cancer cells; evaluated for basal and squamous cell carcinoma; clinically used for specific cancers and microbial infections | Clinical results are unsatisfactory; requires further optimization |
| | | Methylene Blue | Effective for melanoma and Kaposi's sarcoma; positive PDT action in vitro for other cancers. Clinically used for specific cancers and microbial infections | Requires additional clinical validation for widespread application |
| | | Rose Bengal | Water-soluble; experimental agent for breast carcinoma and melanoma. Clinically used for specific cancers and microbial infections | Still in experimental stages for most applications |
| | | Curcumin | Natural dye; potential for bacterial disinfection via PDT. Clinically used for specific cancers and microbial infections | Low absorption maximum (420 nm); limited clinical studies |
| Third generation | Targeted PSs | Antibody-PS conjugates | Higher specificity for cancer cells; reduced off-target effects | Complex synthesis; high cost; early-stage clinical studies |
| | Nanoparticle-based PSs | PS-loaded nanoparticles | Enhanced tumor delivery; better solubility and stability; and deeper tissue penetration | Potential toxicity of nanoparticles; requires further clinical trials |
| | Multifunctional PSs | Theranostic PSs | Combines diagnostic imaging and therapy; tailored photodynamic action | High cost; complex development process |

is activated by light. As a result, PDT is most commonly used for treating superficial tumors or precancerous lesions rather than deeply located or extensive tumors.

## 8.2 Singlet Oxygen Short Half-Life

PDT depends on the production of singlet oxygen, a highly reactive chemical produced when the photosensitizer absorbs light. However, singlet oxygen has a very short half-life (usually in the nanosecond range), which means it decays quickly after production. Because of its brief duration, PDT's effects are limited to the area immediately surrounding the photosensitizer. While this is useful for targeting specific locations, it becomes a serious drawback when treating large or deep-seated cancers. Furthermore, singlet oxygen's transient presence makes it difficult to give long-term therapeutic benefits, which are required for treating large or advanced-stage malignancies.

## 8.3 Tumor Hypoxia

Tumor microenvironment is highly hypoxic because of poorly developed blood vessels. Since PDT requires oxygen to generate ROS to carry out its function, under hypoxic conditions, the effectiveness of PDT can be severely reduced. Hypoxic tumors are particularly challenging for PDT because oxygen is required to convert the photosensitizer to its active state and generate the ROS necessary for cell killing. Not only does the hypoxic condition reduce PDT's effectiveness, but PDT itself can exacerbate tumor hypoxia by consuming oxygen in the treatment area.

## 8.4 Skin Toxicity

Many photosensitizers used in PDT tend to accumulate in the skin, which can cause skin toxicities such as light sensitivity (photosensitivity) and burns after treatment. Photosensitizers can stay in the skin for long periods of time, increasing the risk of sunburn and other skin damage. This side effect is especially significant in clinical settings where the patient's skin may be exposed to ambient light, even if PDT is not directly administered. This constraint reduces the overall utility and safety of PDT, particularly for whole-body treatment.

### 8.5 Drug-Light Interval Variability

The "drug-light interval (DLI)" is the time between administering the photosensitizer and applying the light. This interval has a significant impact on the amount of photosensitizer that accumulates in tumor tissue versus surrounding normal tissue, making PDT effectiveness largely dependent on it. The ideal drug-light interval can change greatly across patients and even between tumor lesions in the same patient. For example, a short drug-light interval may result in preferential targeting of tumor vasculature, whereas a longer interval may directly increase tumor cell killing. This diversity makes it challenging to develop uniform treatment methods.

### 8.6 Vascular Targeting in PDT

PDT can also target tumor blood vessels, altering their structure and cutting off the tumor's blood supply. The efficiency of PDT can thus be determined not only by photosensitizer localization within tumor cells, but also by its distribution in the vascular endothelium. In other circumstances, the primary therapeutic effect may be the breakdown of tumor blood vessels (by short drug-light intervals), resulting in tumor necrosis via ischemia (a lack of blood flow), whereas longer drug-light intervals may enable the photosensitizer to act directly on tumor cells, resulting in cell death.

## 9 Advances and Innovations in Photodynamic Therapy

One of the key strategies for advancing PDT is the development of improved and more efficient compounds that address the limitations of first- and second-generation photosensitizers (PSs). This includes the synthesis of PSs with strong absorption in the near-infrared (NIR) region of the spectrum, which helps overcome the insufficient treatment depth associated with currently approved PSs. To improve PDT treatment depth, several excitation sources have been developed, including NIR light-based techniques, external sources such as X-rays and acoustic waves, and internal self-excitation mechanisms like chemiluminescence and bioluminescence. Internal self-excitation shows particular promise, as it bypasses the need for external light and enables deeper tissue treatment. For example, NanoLuc luciferase and the phototoxic flavoprotein miniSOG, which generates ROS upon luciferase-substrate injection, exemplify a bioluminescence resonance energy transfer (BRET)-activated PDT system. This method combines an internal light source with a photosensitizer (PS) within a single genetic construct, allowing targeted delivery to tumors at virtually any depth and activation through substrate injection [65]. PS accessibility and distribution to tumor tissue is a major challenge associated with the success of PDT therapy. Using nanoparticles in PDT has several advantages over free photosensitizers, including increased solubility, stability, and targeted cancer cell uptake. Nanocarrier photosensitizers can be created via non-covalent (e.g., self-aggregation, core-shell embedding) or covalent (e.g., chemical immobilization) techniques, which improve PS stability and target specific malignancies. Metal nanoparticles, in particular, have great stability, tunable size, and functionalization potential, making them more biocompatible for biological purposes, e.g., Gold Nanoparticles (AuNPs), Copper Nanoparticles (CuNPs), Iron Oxide Nanoparticles ($Fe_3O_4$), Titanium dioxide ($TiO_2$), etc. [66].

Solid tumors provide challenges for PDT because of their hypoxic tumor microenvironment (TME), which is caused by aberrant vascularization and high oxygen consumption by rapidly reproducing cells. This lowers the efficiency of PDT, as standard photosensitizers (PSs) rely on singlet oxygen to destroy tumors. To address this, advanced nanoplatforms and strategies have been developed, including oxygen delivery systems (e.g., haemoglobin, RBCs, PFCs, MOFs), in situ oxygen generation via biochemical reactions such as $H_2O_2$ decomposition catalyzed by catalase (CAT) or CAT-like enzymes, and Type I PSs that reduce oxygen dependence [67]. Hydrogen peroxide ($H_2O_2$) is essential for treating tumor hypoxia in advanced photodynamic treatment (PDT). Tumors naturally have high $H_2O_2$ levels, which can be converted into oxygen ($O_2$) by catalysts like catalase or iron nanoparticles. In situ $O_2$ generation improves PDT efficacy by increasing ROS production. Fe-TBP nanoframeworks and $CaO_2$ nanoparticles use $H_2O_2$ to create $O_2$, targeting hypoxic tumor sites and enhancing therapy with minimal adverse effects [68]. PDT, when paired with immunotherapy, can improve antitumor immune responses by generating immunogenic cell death, which enhances immune responses to tumor antigens. PDT can work in tandem with dendritic cell treatment, cancer vaccines, and chemotherapy to increase immunological memory and promote tumor regression. PDT has been shown in studies to produce effective tumor vaccines by creating tumor cell lysates or dendritic cell-based vaccines, which can be further boosted by therapies such as cyclophosphamide or vinorelbine. Furthermore, combining PDT with immune checkpoint inhibitors like anti-PD-L1 or anti-PD-1 has shown promise in overcoming immune suppression inside the tumor microenvironment, increasing tumor growth suppression, and perhaps treating distant metastases [69]. This combination technique could lead to more effective cancer therapy by strengthening the immune response and increasing the body's ability to fight tumors.

Photodynamic therapy (PDT) has advanced tremendously in terms of efficacy and usefulness in cancer treatment. Type I photosensitizers that work efficiently in hypoxic circumstances, as well as nanotechnology-driven approaches to targeted oxygen delivery and in situ oxygen generation, are examples of innovation. Furthermore, combining PDT with immunotherapy, photothermal treatment (PTT), and chemotherapy has shown synergistic advantages in overcoming constraints, including tumor resistance and metastasis. These developments pave the way for cancer treatments that are more accurate, efficient, and tailored to each individual patient.

## 10 Conclusion

Photodynamic therapy (PDT) has demonstrated immense potential as a minimally invasive, highly targeted therapeutic technique with applications for a variety of cancers, notably superficial and skin cancers, due to its minimally invasive nature and ability to specifically target aberrant tissues. However, issues such as low photosensitizer (PS) transport efficiency, restricted light penetration, and hypoxia-induced therapeutic restrictions limit its broad use, particularly in deep-seated or large tumors. Nanotechnology advancements have addressed some of these constraints by increasing PS delivery, allowing for controlled medication release, improving reactive oxygen species (ROS) production, and facilitating oxygen delivery via novel nanocarrier systems. Future advancement in PDT is dependent on the discovery of improved PS compounds with stronger photochemical reactivity, deeper tissue penetration via near-infrared (NIR) stimulation, and low cytotoxicity. Furthermore, combining PDT with other modalities like as chemotherapy and immunotherapy has yielded excellent results, increasing the therapeutic potential.

## References


1. Daniell, M. D., & Hill, J. S. (1991). A history of photodynamic therapy. *The Australian and New Zealand Journal of Surgery, 61*(5), 340–348.
2. Brodsky, M., et al. (2017). Revisiting the history and importance of phototherapy in dermatology. *JAMA Dermatology, 153*(5), 435.
3. Li, X., et al. (2020). Clinical development and potential of photothermal and photodynamic therapies for cancer. *Nature Reviews Clinical Oncology, 17*(11), 657–674.
4. Kolarikova, M., et al. (2023). Photodynamic therapy: Innovative approaches for antibacterial and anticancer treatments. *Medicinal Research Reviews, 43*(4), 717–774.
5. Sai, D. L., et al. (2021). Tailoring photosensitive ROS for advanced photodynamic therapy. *Experimental & Molecular Medicine, 53*(4), 495–504.
6. Yao, Q., et al. (2022). The concept and examples of type-III photosensitizers for cancer photodynamic therapy. *Chem, 8*(1), 197–209.
7. Akbar, A., et al. (2023). Unleashing the power of porphyrin photosensitizers: Illuminating breakthroughs in photodynamic therapy. *Journal of Photochemistry and Photobiology B, 248*, 112796.
8. Robertson, C. A., Evans, D. H., & Abrahamse, H. (2009). Photodynamic therapy (PDT): A short review on cellular mechanisms and cancer research applications for PDT. *Journal of Photochemistry and Photobiology B, 96*(1), 1–8.
9. Wang, B., et al. (2024). A mitochondria-targeted photosensitizer for combined pyroptosis and apoptosis with NIR-II imaging/photoacoustic imaging-guided phototherapy. *Angewandte Chemie International Edition, 63*(39), e202408874.
10. He, T., et al. (2024). Peptide-driven proton sponge nano-assembly for imaging and triggering lysosome-regulated immunogenic cancer cell death. *Advanced Materials, 36*(19), e2307679.
11. Hu, M., Dong, X., & Zhao, W. (2024). Lysosome-targeted Aza-BODIPY photosensitizers for anti-cancer photodynamic therapy. *Bioorganic & Medicinal Chemistry, 99*, 117583.
12. Wei, R., et al. (2021). Bioorthogonal pretargeting strategy for anchoring activatable photosensitizers on plasma membranes for effective photodynamic therapy. *ACS Applied Materials & Interfaces, 13*(12), 14004–14014.

13. Moserova, I., & Kralova, J. (2012). Role of ER stress response in photodynamic therapy: ROS generated in different subcellular compartments trigger diverse cell death pathways. *PLoS One, 7*(3), e32972.
14. Benov, L. (2015). Photodynamic therapy: Current status and future directions. *Medical Principles and Practice, 24 Suppl 1*(Suppl 1), 14–28.
15. Gunaydin, G., Gedik, M. E., & Ayan, S. (2021). Photodynamic therapy-Current limitations and novel approaches. *Frontiers in Chemistry, 9*, 691697.
16. Mishchenko, T., et al. (2022). Which cell death modality wins the contest for photodynamic therapy of cancer? *Cell Death & Disease, 13*(5), 455.
17. Chang, Q., et al. (2024). A review on ferroptosis and photodynamic therapy synergism: Enhancing anticancer treatment. *Heliyon, 10*(7), e28942.
18. Donohoe, C., et al. (2019). Cell death in photodynamic therapy: From oxidative stress to anti-tumor immunity. *Biochimica Et Biophysica Acta: Reviews on Cancer, 1872*(2), 188308.
19. Wu, C. C., & Bratton, S. B. (2013). Regulation of the intrinsic apoptosis pathway by reactive oxygen species. *Antioxidants & Redox Signaling, 19*(6), 546–558.
20. Taba, F., et al. (2018). Mitochondria-targeting polyamine-protoporphyrin conjugates for photodynamic therapy. *ChemMedChem, 13*(1), 15–19.
21. Zhu, B., et al. (2019). Inhibition of autophagy with chloroquine enhanced sinoporphyrin sodium mediated photodynamic therapy-induced apoptosis in human colorectal cancer cells. *International Journal of Biological Sciences, 15*(1), 12–23.
22. Song, C., et al. (2020). Photodynamic therapy induces autophagy-mediated cell death in human colorectal cancer cells via activation of the ROS/JNK signaling pathway. *Cell Death & Disease, 11*(10), 938.
23. Kessel, D., & Oleinick, N. L. (2009). Initiation of autophagy by photodynamic therapy. *Methods in Enzymology, 453*, 1–16.
24. Aniogo, E. C., George, B. P. A., & Abrahamse, H. (2020). Role of Bcl-2 family proteins in photodynamic therapy mediated cell survival and regulation. *Molecules, 25*(22), 5308.
25. Dellinger, M. (1996). Apoptosis or necrosis following photofrin photosensitization: Influence of the incubation protocol. *Photochemistry and Photobiology, 64*(1), 182–187.
26. Fabris, C., et al. (2001). Photosensitization with zinc (II) phthalocyanine as a switch in the decision between apoptosis and necrosis. *Cancer Research, 61*(20), 7495–7500.
27. Miki, Y., et al. (2015). Photodynamic therapy using talaporfin sodium induces concentration-dependent programmed necroptosis in human glioblastoma T98G cells. *Lasers in Medical Science, 30*, 1739–1745.
28. Lavie, G., et al. (1999). A photodynamic pathway to apoptosis and necrosis induced by dimethyl tetrahydroxyhelianthrone and hypericin in leukaemic cells: Possible relevance to photodynamic therapy. *British Journal of Cancer, 79*(3), 423–432.
29. Santin, G., et al. (2013). Regulated forms of cell death are induced by the photodynamic action of the fluorogenic substrate, Hypocrellin B-acetate. *Journal of Photochemistry and Photobiology B: Biology, 125*, 90–97.
30. de Almeida, D. R., et al. (2023). Necroptosis activation is associated with greater methylene blue-photodynamic therapy-induced cytotoxicity in human pancreatic ductal adenocarcinoma cells. *Photochemical & Photobiological Sciences, 22*(4), 729–744.
31. Henderson, B. W., et al. (1985). Tumor destruction and kinetics of tumor cell death in two experimental mouse tumors following photodynamic therapy. *Cancer Research, 45*(2), 572–576.
32. Henderson, B. W., & Fingar, V. H. (1989). Oxygen limitation of direct tumor cell kill during photodynamic treatment of a murine tumor model. *Photochemistry and Photobiology, 49*(3), 299–304.
33. Dolmans, D. E., Fukumura, D., & Jain, R. K. (2003). Photodynamic therapy for cancer. *Nature Reviews. Cancer, 3*(5), 380–387.
34. de Vree, W. J., et al. (1996). Evidence for an important role of neutrophils in the efficacy of photodynamic therapy in vivo. *Cancer Research, 56*(13), 2908–2911.

35. Castano, A. P., Mroz, P., & Hamblin, M. R. (2006). Photodynamic therapy and anti-tumour immunity. *Nature Reviews Cancer, 6*(7), 535–545.
36. Dudzik, T., Domanski, I., & Makuch, S. (2024). The impact of photodynamic therapy on immune system in cancer—An update. *Frontiers in Immunology, 15*, 1335920.
37. Falk-Mahapatra, R., & Gollnick, S. O. (2020). Photodynamic therapy and immunity: An update. *Photochemistry and Photobiology, 96*(3), 550–559.
38. Etminan, N., et al. (2011). Heat-shock protein 70-dependent dendritic cell activation by 5-aminolevulinic acid-mediated photodynamic treatment of human glioblastoma spheroids in vitro. *British Journal of Cancer, 105*(7), 961–969.
39. Preise, D., et al. (2009). Systemic antitumor protection by vascular-targeted photodynamic therapy involves cellular and humoral immunity. *Cancer Immunology, Immunotherapy, 58*(1), 71–84.
40. Korbelik, M., et al. (1996). The role of host lymphoid populations in the response of mouse EMT6 tumor to photodynamic therapy. *Cancer Research, 56*(24), 5647–5652.
41. Jalili, A., et al. (2004). Effective photoimmunotherapy of murine colon carcinoma induced by the combination of photodynamic therapy and dendritic cells. *Clinical Cancer Research, 10*(13), 4498–4508.
42. Thiruppathi, J., et al. (2024). Enhancing cancer immunotherapy with photodynamic therapy and nanoparticle: Making tumor microenvironment hotter to make immunotherapeutic work better. *Frontiers in Immunology, 15*, 1375767.
43. Davies, M. J. (2004). Reactive species formed on proteins exposed to singlet oxygen. *Photochemical & Photobiological Sciences, 3*(1), 17–25.
44. Benov, L., Craik, J., & Batinic-Haberle, I. (2012). Protein damage by photo-activated Zn(II) N-alkylpyridylporphyrins. *Amino Acids, 42*(1), 117–128.
45. Davies, M. J., & Truscott, R. J. (2001). Photo-oxidation of proteins and its role in cataractogenesis. *Journal of Photochemistry and Photobiology B, 63*(1–3), 114–125.
46. Kim, J., et al. (2008). Oxidative modification of cytochrome c by singlet oxygen. *Free Radical Biology & Medicine, 44*(9), 1700–1711.
47. Girotti, A. W., & Kriska, T. (2004). Role of lipid hydroperoxides in photo-oxidative stress signaling. *Antioxidants & Redox Signaling, 6*(2), 301–310.
48. Yoshida, Y., & Niki, E. (1992). Oxidation of methyl linoleate in aqueous dispersions induced by copper and iron. *Archives of Biochemistry and Biophysics, 295*(1), 107–114.
49. Niki, E. (2009). Lipid peroxidation: Physiological levels and dual biological effects. *Free Radical Biology & Medicine, 47*(5), 469–484.
50. Cadet, J., Douki, T., & Ravanat, J. L. (2008). Oxidatively generated damage to the guanine moiety of DNA: Mechanistic aspects and formation in cells. *Accounts of Chemical Research, 41*(8), 1075–1083.
51. Cadet, J., et al. (2012). Photoinduced damage to cellular DNA: Direct and photosensitized reactions. *Photochemistry and Photobiology, 88*(5), 1048–1065.
52. Cadet, J., & Wagner, J. R. (2013). DNA base damage by reactive oxygen species, oxidizing agents, and UV radiation. *Cold Spring Harbor Perspectives in Biology, 5*(2), a012559.
53. Woodhams, J. H., Macrobert, A. J., & Bown, S. G. (2007). The role of oxygen monitoring during photodynamic therapy and its potential for treatment dosimetry. *Photochemical & Photobiological Sciences, 6*(12), 1246–1256.
54. Chen, D., et al. (2011). Light-emitting diode based illumination system for in vitro photodynamic therapy. *Photodiagnosis and Photodynamic Therapy, 2*(8), 200.
55. Allison, R. R., et al. (2024). Tools of clinical photodynamic therapy (PDT): A mini compendium. *Photodiagnosis and Photodynamic Therapy, 46*, 104058.
56. Li, B., & Lin, L. (2022). Internal light source for deep photodynamic therapy. *Light: Science & Applications, 11*(1), 85.
57. Zhao, J., et al. (2020). Insight into the efficiency of oxygen introduced photodynamic therapy (PDT) and deep PDT against cancers with various assembled nanocarriers. *Wiley Interdisciplinary Reviews: Nanomedicine and Nanobiotechnology, 12*(1), e1583.

58. Palumbo, F., et al. (2014). Perfluorocarbon functionalized hyaluronic acid derivatives as oxygenating systems for cell culture. *RSC Advances, 4*(44), 22894–22901.
59. Wei, F., et al. (2021). Oxygen self-sufficient photodynamic therapy. *Coordination Chemistry Reviews, 432*, 213714.
60. Hamblin, M. R., & Newman, E. L. (1994). New trends in photobiology: On the mechanism of the tumour-localising effect in photodynamic therapy. *Journal of Photochemistry and Photobiology B: Biology, 23*(1), 3–8.
61. Abrahamse, H., & Michael, R. (2016). Hamblin, new photosensitizers for photodynamic therapy. *Biochemical Journal, 473*(4), 347–364.
62. Sarbadhikary, P., George, B. P., & Abrahamse, H. (2021). Recent advances in photosensitizers as multifunctional theranostic agents for imaging-guided photodynamic therapy of cancer. *Theranostics, 11*(18), 9054.
63. Foster, T. H., et al. (2010). Intratumor administration of the photosensitizer pc 4 affords photodynamic therapy efficacy and selectivity at short drug-light intervals. *Translational Oncology, 3*(2), 135–141.
64. Efendiev, K., et al. (2024). Tumor fluorescence and oxygenation monitoring during photodynamic therapy with chlorin e6 photosensitizer. *Photodiagnosis and Photodynamic Therapy, 45*, 103969.
65. Shramova, E. I., et al. (2022). Genetically encoded BRET-activated photodynamic therapy for the treatment of deep-seated tumors. *Light: Science & Applications, 11*(1), 38.
66. Udrea, A. M., et al. (2023). Photosensitizers-loaded nanocarriers for enhancement of photodynamic therapy in melanoma treatment. *Pharmaceutics, 15*(8), 2124.
67. Zuo, T., et al. (2024). Engineering tumor-oxygenated nanomaterials: Advancing photodynamic therapy for cancer treatment. *Frontiers in Bioengineering and Biotechnology, 12*, 1383930.
68. Lan, G., et al. (2018). Nanoscale metal-organic framework overcomes hypoxia for photodynamic therapy primed cancer immunotherapy. *Journal of the American Chemical Society, 140*(17), 5670–5673.
69. Gunaydin, G., Gedik, M. E., & Ayan, S. (2021). Photodynamic therapy for the treatment and diagnosis of cancer—A review of the current clinical status. *Frontiers in Chemistry, 9*, 686303.

# Photobiomodulation

**Trishna Deka, Prerana Kalita, Khyati Goswami, and Venkata Satish Kumar Mattaparthi**

**Abstract** Photobiomodulation (PBM) is a promising new therapy that uses gentle, non-damaging light—like low-level lasers or LEDs—to help the body heal and function better. This chapter takes a closer look at how PBM works, starting with how cells absorb light and how this sparks a chain reaction of biological processes that support healing and repair. The authors delve into how PBM influences key cellular functions, mainly how it boosts energy production in mitochondria, reduces harmful oxidative stress, and even affects how genes behave. The chapter also explores the application of PBM for the regeneration of damaged areas in the brain, muscle and nervous system. The chapter also emphasises the importance of various parameters, such as wavelength, intensity, and duration, in the effectiveness of PBM therapy. In the final sections, the authors also elucidate real-world applications of PBM to manage diabetes, neurological conditions, and inflammatory disorders, and to ease side effects after radiation treatment. Drawing from a wide body of current research, this chapter highlights how PBM is not just a scientific curiosity but a versatile tool with real potential to transform patient care.

**Keywords** Photobiomodulation · Light therapy · Phototransduction pathway · Biophotonic signalling

Trishna Deka, Prerana Kalita and Khyati Goswami contributed equally with all other contributors.

T. Deka · P. Kalita · K. Goswami · V. S. K. Mattaparthi (✉)
Molecular Modelling and Simulation Laboratory, Department of Molecular Biology and Biotechnology, Tezpur University, Tezpur, Assam, India
e-mail: venkata@tezu.ernet.in



N. Mazumder et al. (eds.), *Advanced Biophysical Techniques in Biosciences*, Lecture Notes in Nanoscale Science and Technology 7,
https://doi.org/10.1007/978-3-031-94551-9_13

# 1 Introduction

Photobiomodulation (PBM), also known as low-level light treatment (LLLT, laser phototherapy), is a nondestructive, nonthermal method of treating a variety of soft tissue and neurologic diseases. It employs monochromatic and quasimonochromatic light in the optical region of around 600–1000 nm. The Medical Subject Headings (MeSH) Identifier "D028022," Established Data 2002–01–01, defines LLLT as *"treatment using irradiation with light of low power intensity so that the effects are a response to the light and not due to heat. A variety of light sources, especially low-power lasers are used."*

We can plainly see how light affects biological systems and functions in our circadian rhythms, sleep–wake cycles, and vitamin absorption. Researchers now have the ability to obtain and employ high-powered light (at specified wavelengths) in biology, opening up new possibilities for its application in tissue engineering and healing [1].

Photobiomodulation originated in Hungary in 1960 when Dr. Endre Mester at the Semmelweis Medical University tried to replicate an experiment by Paul McGuff of Boston, USA. At that time, it was known as low-level laser therapy. This was because, in comparison to other forms of laser treatment used for tissue ablation, cutting, and thermal coagulation, the optimal levels of energy density given are low. Generally speaking, LLLT uses power densities that are less than 100 $mW/cm^2$, depending on the wavelength and type of tissue, which is less than what is required to heat the tissue [2].

McGuff used ruby laser to cure malignant tumor in rats. But the laser Mester used was not as strong as the laser used by McGuff. Mester did not observe any significant improvement in tumor; rather, he observed some other effects like heightened hair growth, better wound healing at the site of tumor implantation. The experiment was later called "laser biostimulation." The biostimulation effect was first used in therapeutic settings in 1974.

The term low-level laser therapy was used as the process mainly involved the use of ruby laser and HeNe laser. But soon after this term was obsolete as there was no significant parameter that could define low level.

Their nomenclature was evolving as research on the use of light for medicinal purposes progressed. The term "low-level laser therapy," "low-level laser irradiation," "non-thermal laser," "low-level laser," "cold laser," "soft laser," "low energy laser irradiation," "laser therapy," and, more recently, "photobiomodulation"(PBM) have all been used to characterize the effect reported by Mester [3].

# 2 Mechanisms of Photobiomodulation

Biological mediums and tissues exhibit varying degrees of absorption, scattering, and reflection of various light wavelengths. A variety of chromophores, including water, myoglobin, oxyhemoglobin (HbO2), deoxyhemoglobin (Hb), melanin, cytochromes, and flavins, are typically found in biological tissues. Melanin, hemoglobin, and flavins absorb light energy at wavelengths shorter than 600 nm, but water molecules absorb light energy at wavelengths larger than 970 nm. Due to these factors, there is an optical window for photobiomodulation (PBM) therapy in the red to near-infrared (NIR) spectrum [4], and multiple studies have established the therapeutic efficacy of this wavelength range.

While PBM has garnered the attention of numerous research teams lately, elucidating the fundamental mechanisms of its functioning could facilitate a more comprehensive idea of PBM's effects on cells and their multiplicity of advantages for enhanced healing. Studies on PBM have produced experimental and clinical data that have served as the scientific foundation for treatment protocols in a variety of fields, including medicine, dentistry, nursing, physiotherapy, and esthetics [5].

## *2.1 Molecular Mechanism*

### 2.1.1 Chromophores

According to the fundamental law of photobiology, light photons cannot affect biological processes unless they are absorbed by a molecule found within the tissue known as a chromophore [6].

#### 2.1.1.1 Cytochrome c Oxidase

Many molecules in biological system are naturally photosensitive. One such molecule in eukaryotic cells is the terminal enzyme of the mitochondrial respiratory chain, cytochrome c oxidase (Cox). Cox transfers electrons from cytochrome c to molecular oxygen, which is the terminal electron acceptor of the respiratory chain. Cox has two copper centers and two heme centers. The copper centers are $Cu_A$ and $Cu_B$, and the heme centers are a and a3. All four centers are responsible for absorbing light and pumping protons. Cox accepts photosignals in the red and near-infrared regions of the light spectrum. It has been observed that low-level light irradiation increases oxygen consumption in Cox, which promotes photodissociation of nitric oxide (NO) from Cox. This leads to increased enzyme activity due to non-covalent binding of NO between the heme centers a and a3 and inhibits Cox activity [7]. It is thought that PBM increases the mitochondrial membrane potential (MMP) along

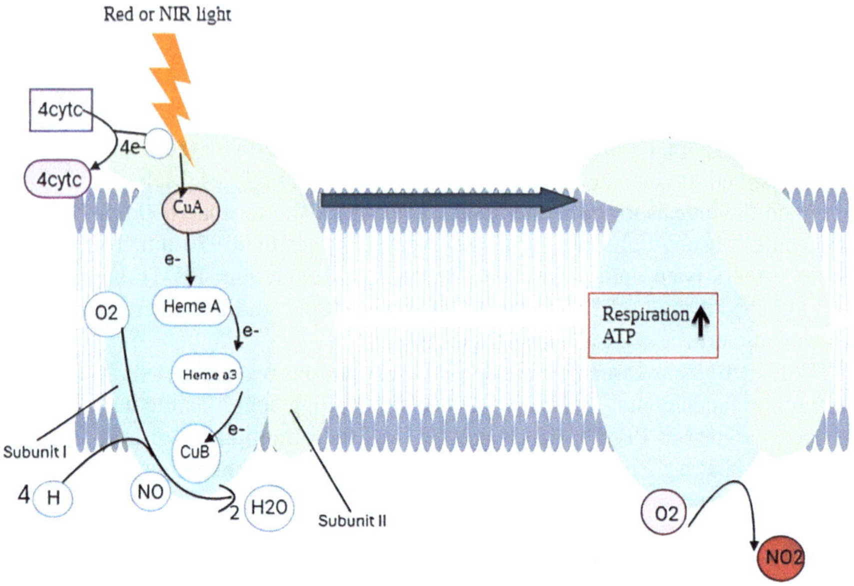


**Fig. 1** NO photodissociation from cytochrome c oxidase (Cox) is proposed. A multisubunit enzyme called Cox, which has two copper centers and two heme cofactors, oxidizes four reduced cytochrome c molecules while also converting oxygen to water and generating four protons that ATP synthase uses to create ATP. It has been suggested that red or NIR light can dissociate the non-covalently bound nitric oxide (NO), and hence enhancing the rate of respiration and ATP synthesis. Nitric oxide can block this process by attaching to CuB [6]

with increasing levels of cyclic adenosine monophosphate (cAMP), adenosine triphosphate (ATP), and reactive oxygen species (ROS) (Fig. 1).

#### 2.1.1.2 Retrograde Mitochondrial Signaling

For light signal to pass to mitochondria and eventually to cytochrome c oxidase there must be a cellular signalling pathway between the mitochondrial respiratory chain and the nucleus [8]. This is known as retrograde mitochondrial signalling. It starts with light activation in the visible and infrared regions. The aforementioned interaction results in an elevation of the mitochondrial membrane potential ($\Delta\psi m$), which in turn leads to a rise in ATP production and modifications in the quantities of ROS, $Ca^{2+}$, and NO. Additionally, alterations in the fission-fusion equilibrium in a changing mitochondrial network, or modifications in the ultrastructure of the mitochondria, are responsible for the communication that occurs between the nucleus and the mitochondria. Variations in the ultrastructure of the mitochondria lead to modifications in the amounts of cyclic adenosine monophosphate (cAMP), pH, intracellular redox potential, and ATP production. Activator protein-1 (AP1) and nuclear factor kappa-light-chain-enhancer of activated B cells (NF-κB) activity are impacted by changes in membrane permeability and ion flux at the cell membrane. Additionally, Karu suggested a number of complementary routes, such as the direct upregulation of specific genes (Fig. 2).

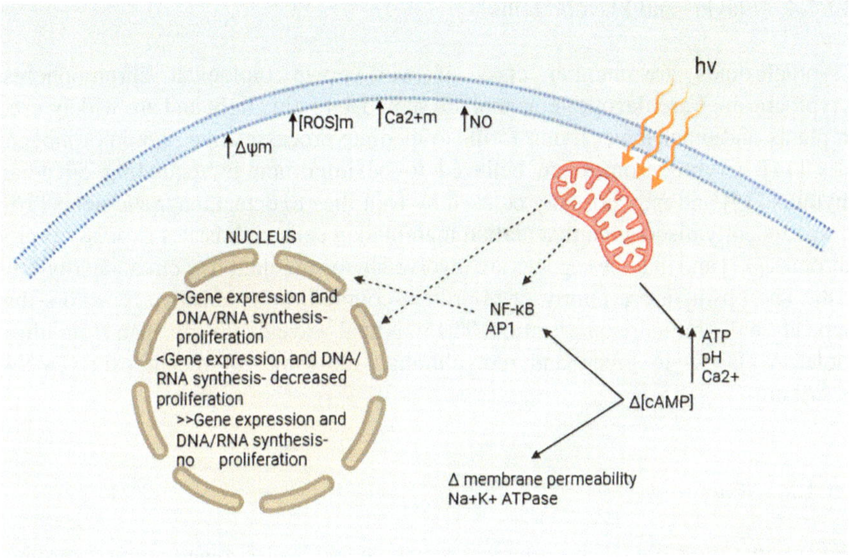


**Fig. 2** Diagram illustrating Karu's suggested mitochondrial retrograde signaling channels. Segmented arrows depict the complementary pathways, whereas continuous arrows represent the main pathway [9]

#### 2.1.1.3 Light-Sensitive Ion Channels

Apart from cytochrome c oxidase, another class of photoreceptors is found to be involved in PBM that is particularly sensitive to blue and green light [6]. These mostly include the ion channels that are directly gated by light. It has been suggested that these photoreceptors belong to the family of light-sensitive G-protein coupled receptors known as opsins (OPN). Opsins mediate vision in the rod and cone photoreceptor cells of the mammalian retina by photoisomerizing a cis-retinal cofactor, which results in a conformational change in the protein. Other members of the opsin family (OPN2–5) are expressed in many other tissues of the body, including the brain [10].

On the other hand, it has recently been discovered that a large class of ion channels is light-sensitive. These channels, which are responsible for insect vision, are known as "transient receptor potential" (TRP) channels initially found in a Drosophila mutant [11]. The TRPC ('Canonical') subfamily, the TRPV ('Vanilloid'), the TRPM ('Melastatin'), the TRPP ('Polycystin'), the TRPML ('Mucolipin'), the TRPA ('Ankyrin'), and the TRPN ('NOMPC') subfamilies are the seven subfamilies among which there are currently at least 50 distinct known TRP isoforms [12]. Many stimuli, including heat, cold, light, sound, toxic substances, mechanical forces, hormones, neurotransmitters, spices, and voltage, can alter the activity of distinct TRP channels. Calcium channels called TRP are influenced by phosphoinositides [13].

#### 2.1.1.4 Flavins and Flavoproteins

Cryptochromes are another class of well-known biological chromophores. Cryptochromes are flavoproteins that are sensitive to blue light and are widely used in plants and other lower living forms to mediate processes like photomorphogenesis [14]. Cryptochromes are believed to be important in regulating circadian rhythms [15] and may possibly be used by fruit flies to detect magnetic fields [16]. It was recently discovered that certain mammalian cells and tissues express cryptochromes [17] and that these genes are involved in the regulation of circadian rhythms [18]. The opsins are a family of G-protein-coupled, light-sensitive receptors that depend on cis-retinal isomerization. The maximal wavelength can span from ultraviolet A (UVA) to green and red, although the kmax of melanopsin (OPN4) is 479 nm.

#### 2.1.1.5 Water

Water is the most prevalent biological molecule and is also a well-studied chromophore. The absorption spectrum of water peaks at 980 nm, and also at most wavelengths longer than 1200 nm [6]. Though the possibility of water acting as a chromophore is still speculative, it is challenging to come up with a different explanation for the advantages of PBM at wavelengths ranging from 1000 nm to 10,000 nm.

### 2.1.2 Signaling Molecules

#### 2.1.2.1 Adenosine Triphosphates

One of the most common and important results of PBM, both in vitro and in vivo, is a rise in intracellular ATP [19]. Cox's enhanced activity upon activation by light is the reason for the stimulated generation of ATP. It was stated that when PBM is performed prior to a variety of activities, for example, the mechanism underlying improved muscle function is higher Cox activity [20].

#### 2.1.2.2 Reactive Oxygen Species

It is established that reactive oxygen species (ROS) are physiological byproducts of the mitochondria's regular aerobic metabolism. Mitochondria consume almost 90% of the molecular O2 that is respired by mammalian tissues and cells [21]. The function of the antioxidant systems opposes the generation of reactive oxygen species (ROS) by the mitochondrial respiratory chain under healthy conditions. Nevertheless, after PBM, there may be an imbalance in the way that ROS generation and antioxidant system activity are balanced, or an improper PBM may cause the mitochondria to uncouple, increasing ROS within the cell. The molecules that make up ROS are classified as Giano Bifronte, which are extremely toxic at large quantities and can cause cellular damage, but positive and stimulating at low concentrations when it comes to redox signaling from the mitochondria to the rest of the cell. Actually, ROS can interact with the nucleus through activating the protein complex known as nuclear factor kappa-light-chain-enhancer of activated B cells (NF-κβ), which regulates DNA transcription, cytokine synthesis, promotes cell survival, and initiates a series of intracellular events in PBM [22].

#### 2.1.2.3 cAMP

Another effect of PBM is the increase in levels of cAMP. This could be due to the increase in levels of ATP. It was shown that PBM raised cAMP in alveolar macrophages [6]. Three distinct types of sensors are activated by cAMP to produce its effects on cells: cAMP response element-binding protein (CREB) is phosphorylated and activated by cAMP-dependent protein kinase A (PKA). CREB subsequently binds to the CRE domain on DNA, activating the gene [23]. Taniguchi et al. [24] proposed that after exposure to a low-power red laser, PBM-induced synovial fibroblast proliferation is mediated via cAMP.

### 2.1.3 Transcription Factors and Signaling Pathways

Transcription factors and signaling pathways can be activated by several of the secondary mediators of PBM, including ROS, NO, and cAMP. One of the transcription factors induced by PBM is hypoxia-inducible factor (HIF-1α). HIF-1a stimulates the expression of genes related to the cellular response to hypoxia, including phosphoglycerate kinase (PGK), glucose carrier (GLUT-1), vascular endothelial growth factor (VEGF), and VEGF-receptor. The phosphatidylinositol 3-kinase (PI3K)/Akt and mitogen-activated protein kinase (MAPK) signaling pathways may be stimulated to cause HIF-1a activation [25].

PBM may activate the AKT/GSK3β/β-catenin pathway, which in turn may have a prosurvival effect on cells. Fundamentally, PBM has the ability to activate protein kinase B, or AKT, which subsequently interacts with glycogen synthase kinase 3β (GSK3β) to decrease its activity. The serine-threonine kinase GSK3β is implicated in both oncogenesis and cell death [6].

Akt/mTOR/cyclin D1 is an additional pathway that can be triggered by PBM-mediated activation of Akt. Proteins, including Akt, HSP90, pS6ser240/244, and cyclin D1, that are linked to the progression and invasion of cancer were expressed at increased levels.

In a study published by Lima and coauthors, rats were exposed to 660 nm light (5.4 J) through skin radiation across the bronchus (chest). They noticed a significant increase in peroxisome proliferator-activated receptors (PPAR) mRNA expression following PBM and elevated PPAR-y activity in the bronchoalveolar membrane

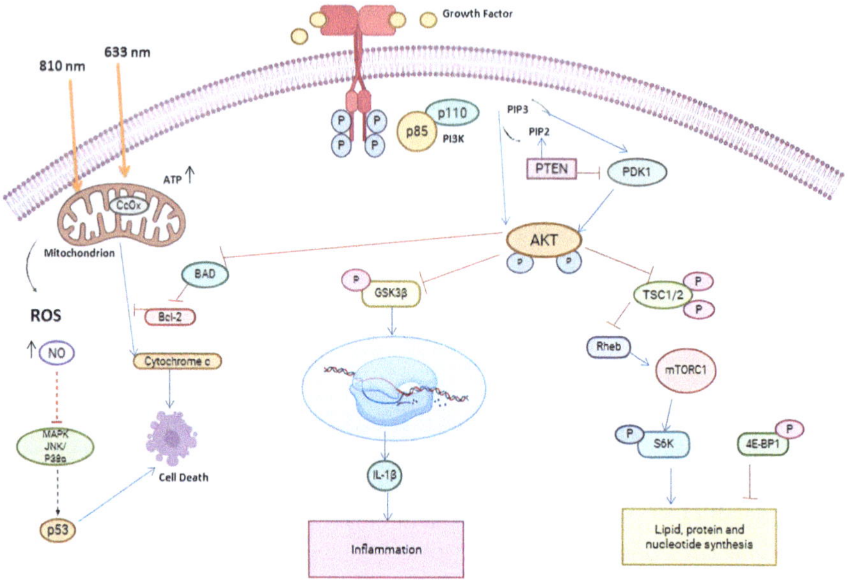


**Fig. 3** Molecular mechanism of PBM: PBM has the potential to activate PI3K kinase, which in turn recruits PDK1 kinase by phosphorylating phosphatidyl-4,5-bisphosphate (PIP2) to produce phosphatidylinositol-3,4,5-triphosphate (PI3P). Once phosphorylated, PDK1 triggers AKT, which in turn triggers a transducing cascade signal that inhibits the pro-apoptotic protein BAD, the TSC1/2 complex, and glycogen synthase kinase 3 beta (GSK3β). After becoming active, AKT increases mTORC1 activity, which in turn increases the synthesis of lipids, proteins, and nucleotides to promote tissue regeneration. This alleviates the inhibitory effect of the TSC1/2 complex on Rheb. In addition to mTOR, photoactivated AKT may decrease BAD and GSK3β to cause anti-apoptotic and anti-inflammatory signals, respectively. Additionally, PBM causes an increase in NO, which modulates pro-survival routine by reducing MAPK/JNK/p38α. The principal photoreceptor in the red and near-infrared spectrum is the enzyme cytochrome c oxidase (COx), which is responsible for mildly producing ROS and increasing ATP production [27]

animal lavage (BALF) cells from laser-treated animals. Lima concluded by suggesting that PBM may function as a homeostatic facilitator, elevating a transcription factor's expression, triggering the production of HSP70 and other antiproteins that cause inflammation [26] (Fig. 3).

## 2.2 *Cellular Mechanism*

The molecular impacts of transcription factors, effector and signaling molecules, and trigger agents ultimately result in the cellular consequences associated with PBM. It has been reported that exposure to low-power laser and LED radiation causes cellular effects such as migration, apoptosis, differentiation, viability, and proliferation [28, 29].

Anti-inflammatory effect of PBM was reported in the study conducted by Lim and his coworkers. They found that 635 nm light irradiation at low power inhibits prostaglandin E2 (PGE2) production and COX-1 and COX-2 mRNA expression. The cellular signaling mechanism's regulation by NF-κB is necessary for PGE2 production. Lim demonstrated that PGE2 production was decreased by exposure to 635 nm light, possibly through the inhibition of the NF-κB pathway. In human gingival fibroblasts (hGFs) silenced by HSP27, it did not change the phosphorylation of IκB, IKK, or NF-κB, suggesting that NF-κB modulation by 635 nm light via HSP27 is necessary for the downregulation of pro-inflammatory gene expression in these fibroblasts [30].

PBM has shown its beneficial effects in cellular protection by treatment with different toxins. For example, methanol produces formic acid, a hazardous metabolite that prevents cytochrome c oxidase from functioning. PBM increases cell survival following formic acid poisoning by stimulating cytochrome c oxidase, which in turn increases mitochondrial activity [31]. Another harmful substance whose effects PBM could reduce is cyanide. The death of neurons produced by potassium cyanide was reduced when 670 nm light was applied beforehand [32]. Low-power laser radiation induces Yes-associated protein (YAP) cytoplasmic translocation and amyloid-βpeptide (Aβ) inhibition, which makes PBM potentially helpful in treating Alzheimer's disease.

PBM can boost the rates of proliferation of a number of different cell types. For instance, following 660 nm light irradiation, keratinocytes demonstrated increased proliferation, which was accompanied by an increase in Cyclin D1 expression and a faster maturation of keratinocytes during migration to the wound sites, via the expression of proteins involved in the process of epithelial proliferation, specifically p63, CK10, and CK14. This is beneficial for enhancing epithelial repair [33]. When exposed to 635 nm radiation, vascular endothelial cells grow more quickly than unirradiated cells and exhibit a lower VEGF content. This implies that a drop in VEGF levels is connected to laser-induced cell growth. The endothelial cells' release of TGF-β was reduced during exposure to 830 nm light [34].

Migration of tenocytes to the wounded location, followed by their proliferation and the creation of extracellular matrix, is necessary for tendon repair. When Tsai and colleagues examined the impact of 660 nm light on tenocytes generated from rat Achilles tendon, they discovered that in vitro migration was promoted and dynamin-2 expression was increased. The stimulatory impact of PBM was diminished when dynamin-2 was inhibited with dynasore, suggesting that dynamin-2 upregulation mediated the tenocyte migration induced by low-level light [35]. Light exposure also affects other types of cells. For example, after being exposed to blue and red light, melanocytes demonstrated increased viability and proliferation. Lower concentrations of UV, blue, and red light increased the migration of melanocytes; greater doses of light had a non-stimulatory impact. When compared to UV and red lasers, blue light appeared to be more efficient [36].

The proteins p63, CK10, and CK14, which are involved in the growth and maturation of epithelial cells, were shown to be more highly expressed when exposed to PBM [33]. Low-light conditions can actually make a number of other proteins more expressed. As shown by Frozanfar and colleagues in 2013, fibroblasts' increased production of collagen I 2 days after exposure to 810 nm light is a good example [37]. Additionally, osteoblasts exposed to 830 nm light showed an increase in the expression of mimecan and osteoglycin, two examples of proteoglycans. Previously known as osteoinductive factor, osteoglycin is a leucine-rich proteoglycan that is readily present in connective tissues, cartilage cells, and bone matrix. In conjunction with certain growth factors and matrix proteins, PBM administered to osteoblasts during their early proliferation stage is crucial for the stimulation of bone formation because they play a regulatory role in the cell proliferation, differentiation, and adhesion of osteoblastic cells [38].

Stem cells seem to be especially light-sensitive. PBM activates protein expression and increases cell motility, differentiation, proliferation, and viability in stem cells [39]. Mesenchymal stem cells multiply more when exposed to light (generally at wavelengths between 600 and 700 nm). These stem cells are often derived from bone marrow, tooth pulp, periodontal ligament, and adipose tissue. PBM provides a workable substitute to encourage the translation of stem cell research into the clinical setting because undifferentiated stem cells exhibit a decreased rate of proliferation, which may be a limiting factor for the clinical efficacy of stem cell therapies [40].

The ability of nonionizing radiation emitted from low-power sources, such as low-power therapeutic lasers and LEDs, to alter cellular responses—which are the outcome of those radiation-induced molecular effects—such as cell proliferation, viability, differentiation, apoptosis, and migration—can therefore be interpreted as PBM at the cellular level based on the data collected from the studies.

## 2.3 *Tissue Mechanism*

### 2.3.1 Muscle

The primary chromophore in cells that can detect red and near-infrared light is cytochrome c oxidase (Cox). While Cox activity plays a crucial role in the absorption of photon, assessing the activity of Cox by itself might not be enough to show whether light treatment can cause "extra" ATP generation. Therefore, assessing mitochondrial membrane potential (MMP) alongside ATP synthesis can offer insights into the speed of changes within the electron transport chain (ETC) and the dynamics of proton ($H^+$) movement between the intermembrane space and mitochondrial matrix ([20, 41, 42 62]. The findings of Ferraresi and colleagues align with those of Xu et al., who found that light therapy had no immediate effects on mitochondrial membrane potential (MMP). Although Xu et al. did not evaluate ATP content, the results indicated no significant increase in ATP levels immediately following light therapy, in contrast to the control group. Additionally, the combined effects of both red and near-infrared light treatment on MMP and ATP synthesis were observed in C2C12 myotubes (mouse muscle cells). The data suggest that a time frame of 3–6 h post treatment may be optimal for enhancing muscle metabolism [42, 43].

Additionally, studies have shown that in intact gastrocnemius muscle, low-power laser irradiation affects the presence of receptors for acetylcholine (AChRs) and creatine kinase (CK) activity. Studies show that photobiomodulation (PBM) notably elevates activity of CK and levels of AChR over 1–2 months as compared to control subjects. These biochemical changes in muscle cells may reflect a trophic signal promoting CK activity, keeping a supply of high-energy phosphates on hand to facilitate quick ATP synthesis [9].

### 2.3.2 Brain

In the neurological field, photobiomodulation has been associated with improved memory and cognitive benefits following brain damage brought on by controlled cortical impact (CCI) [9]. Khuman and colleagues found that a 500 mW $cm^{-2}$ laser irradiation (60 J $cm^{-2}$) given for 2 min enhanced memory and spatial learning in mice with CCI, alongside a reduction in microgliosis within 48 h [9, 44]. Near-infrared (NIR) light appears to have a neuroprotective impact, though the underlying mode of action remains partially understood. One proposed mechanism involves NIR light facilitating brain repair, lowering inflammation, and improving the function of mitochondria [9]. Xuan et al. reported that in mice with CCI, transcranial NIR light promotes neurogenesis in the subventricular zone (SVZ) and the hippocampus region [9, 45, 46]. Newly formed neuroprogenitor cells may migrate to damaged cortical regions, aiding in repair [9]. Furthermore, BDNF levels are elevated in the SVZ and hippocampus 1 week after TBI, with synaptogenesis occurring in the cortex by 4 weeks, indicating that light can stimulate new connections between existing neurons [9, 46].

Another proposed mechanism suggests that NIR light may trigger a systemic response through circulating molecules or cell types. Evidence indicates that NIR irradiation of specific sites, such as skin wounds, can elicit remote effects on tissues. Studies have shown neuroprotection in mice after NIR light was applied to the dorsum, without direct head irradiation. This might involve macrophages and mast cells that may aid in safeguarding brain tissue while modulating inflammatory mediators, such as downregulating pro-inflammatory cytokines and upregulating anti-inflammatory cytokines. Additionally, in the bone marrow, NIR light may enhance the growth of c-kit-positive stem cells, which are subsequently drawn to injured tissues, especially in myocardial infarctions. These progenitor cells are able to secrete trophic and pro-survival factors, including nerve growth factor (NGF) and vascular endothelial growth factor (VEGF). Mitochondria themselves might release an unidentified extracellular signal termed a "mitokine," facilitating communication with remote cells [9, 44].

### 2.3.3 Nerves

The analgesic effects of low-level light therapy arise from various mechanisms, including the absorption of light by mitochondrial chromophores, primarily Cox, leading to an increase in the levels of nitric oxide (NO), cell division, vasodilation, and biomodulation [9]. Recent investigations by Serra and Ashmawi explored whether serotonin influences analgesia induced by PBM; however, their findings suggested that peripheral opioid receptors, not serotoninergic pathways, are the mechanisms underlying this action [9, 47]. PBM is effective in pain reduction and addressing pathological conditions related to the nervous system, potentially by modulating synaptic transmission to dorsal horn neurons and suppressing afferent fiber signaling, which includes inhibition of substance P and promoting the alleviation of pain [9, 48]. Additionally, PBM exhibits strong anti-inflammatory properties that help in the peripheral nervous system to heal and regenerate after damage, and the improvement of neurological deficits following strokes and traumatic brain injuries [9, 44].

### 2.3.4 Healing

In the context of bone healing, low-power laser irradiation creates favorable conditions that accelerate recovery. PBM promotes the development of osteoblasts and proliferation in vitro and in vivo, which increases osteocalcin expression, alkaline phosphatase (ALP) activity, and enhance bone formation. This implies that bone growth can be directly enhanced by laser irradiation, potentially mediated by the expression of insulin-like growth factor (IGF) [9, 49]. Additionally, PBM has been shown to stimulate mineralization through increased expression of bone morphogenetic proteins (BMPs) and Runx-2, along with osteoblast differentiation into MC3T3-E1 cells. BMPs, critical mediators of bone induction, may partially explain the formation of bone nodules following PBM treatment [9, 49]. When there is a flaw in the process of hemostasis, PBM may help to promote healing [9, 50].

### 2.3.5 Hair

Photobiomodulation has also demonstrated efficacy in stimulating hair growth. Although the mechanisms remain under investigation, it is believed that even minimal heat generated by PBM may activate follicular stem cells to divide and multiply as a result of an increase in the heat shock protein levels [9, 51]. Furthermore, PBM can cause the secretion of factors that affect the cell cycle and promote angiogenesis [9, 51]. Hair growth occurs in three phases: anagen (growth), catagen (transition), and telogen (resting) [9]. PBM has been shown to encourage follicles of telogen hair to transition into the anagen phase and prolong its duration, potentially by enhancing protein synthesis via activated transcription factors, which lead to cell migration and proliferation [9, 51]. Increased levels of NO in treated tissues, a known vasodilator, may also support hair growth by improving blood flow to hair follicles. Moreover, PBM can inhibit the expression of pro-inflammatory mediators, which is crucial since inflammation can adversely affect hair follicles, suggesting its therapeutic potential in treating conditions like alopecia areata [9, 51].

## 3 Parameters of PBM

The fundamentals of PBM consist of light and its administered dosages. The parameters regarding the source of light and the doses of light are shown in Tables 1 and 2, respectively. In low-level light therapy, red or near-infrared light is used, having wavelengths between 600 and 700 nm and 780 and 1100 nm. The laser or LEDs used in this therapy normally have an irradiance or power density of 5 mW $cm^{-2}$ to 5 W $cm^{-2}$. Red and near-infrared light is used because tissue chromophores scatter and absorb less of it in this wavelength range, allowing for maximum penetration through tissue [9]. PBM effectiveness in promoting in vitro cell proliferation is significantly influenced by both light wavelength and photon number (fluence) [52]. Compared to NIR (808 nm = 1.53 eV), red light, having a frequency of 600–700 nm, has a greater photon quantum energy (2.07–1.77 eV) and can more readily induce electrochemical alterations in tissues. However, NIR light mostly increases the molecular vibrational state, which may increase metabolic activity and a transient thermal effect of a minimum of 2 °C in tissue ranging in thickness from 3.0 to 5.0 mm.

**Table 1** Description of the irradiation parameters [9]

| Irradiation Parameter | Measurement unit | Description |
|---|---|---|
| Wavelength | Nm | Light behaves like a wave and is an electromagnetic kind of energy. It is visible in the 400–700 nm range, and its wavelength is expressed in nanometers (nm) |
| Irradiance | W $cm^{-2}$ | It is equivalent to the power (in W) divided by the area (in $cm^{-2}$) and is also known as power density or intensity |
| Pulse structure | Peak power (W), pulse frequency (Hz), pulse width (s), duty cycle (%) | The power should be referred to as average power if the beam is pulsed, and it is computed as follows: Peak power (W) times pulse width (s) times pulse frequency (Hz) equals average power (W) |
| Coherence | Coherence length depends on spectral bandwidth | Laser speckle, which is produced by coherent light, is thought to be crucial for photobiomodulation's interaction with cells and organelles. |
| Polarization | Linear polarized or circular polarized | This characteristic is rarely taken into account when analyzing the effects of PBM because polarized light is known to lose its polarity in highly scattering media, such as biological tissues |

**Table 2** Description of the light dose parameters [9]

| Irradiation parameter | Measurement unit | Description |
|---|---|---|
| Energy | Joules (J) | The inverse relationship between power and time, known as reciprocity, prevents it from being mistaken for a dose. The formula is as follows: Energy (J) = power (W) x time (s) |
| Energy density | J $cm^{-2}$ | Although this is a significant dosage description, it may not be accurate because it presumes a reciprocal relationship between time and irradiance |
| Irradiation time | s | Determining the four factors in Table 1 and then using the irradiation time as the actual "dose" may be the most effective method for prescribing and documenting PBM |
| Treatment interval | Hours, days, or weeks | Although different time periods may produce distinct results, more information is required to determine how much they differ from one another |

The wavelength of light has a significant impact on tissue penetration. When treating superficial tissue, shorter wavelengths (600–700 nm) are thought to be the most effective; for treating deeper tissues, longer wavelengths (780–950 nm) are selected. Unlike red wavelengths, which travel 0.5–1 mm, near-infrared radiation travels 2 mm prior to losing 37% of its intensity [53, 54]. Red light has a lower penetration rate than infrared laser light, which is why infrared wavelengths have a greater effect on bone regeneration than red light [55].

PBM therapy uses both noncoherent LEDs and coherent lasers. Collimated laser beams have a higher probability of being forward-scattered within the tissue compared to non-collimated LED light. This indicates that, if all other parameters remain the same, the penetration depth with lasers is probably going to be deeper. Furthermore, LED light is noncoherent, whereas lasers produce coherent light. When the bandwidth is reduced, the coherence length increases. For PBM applications, a recent review paper found no discernible differences between lasers and LEDs as long as all other light characteristics stayed the same [56].

There must be sufficient photon intensity or irradiance ($W - cm^2$ or spectrum irradiance). At lower intensities, photon absorption will not be enough to achieve the intended result; at higher intensities, photon energy will be transformed into excessive heat in the target tissue [55]. Also, the dosage needs to be sufficient ($J - cm^2$). Achieving the optimal fluence or dose by using low irradiance and extending the irradiation time will not yield a satisfactory outcome. Known as the second law of photobiology, the Bunsen-Roscoe law of reciprocity is not valid at low incident power densities [57].

The Arndt-Schulz law, which states that "weak stimuli slightly accelerate vital activity, stronger stimuli raise it further until a peak is reached, and even stronger stimuli suppress it until a negative response is achieved," is based on the biphasic response [58]. The biphasic dose response curve, also referred to as hormesis, shows that excessive light delivery can occasionally have unintended inhibitory effects. Alternatively, very low or very high doses (fluence ($J/cm^2$), irradiance ($mW/cm^2$), delivery time, or number of repetitions) can have no apparent impact [59]. Thus, it is well established that the treatment will probably be ineffective if the wrong conditions are used.

## 4 Effects of Photobiomodulation

Photobiomodulation has been shown to offer various clinical benefits, exhibiting biological effects that are dose-dependent. At lower to moderate doses, PBM stimulates positive changes, while higher doses can have inhibitory effects [59–61]. The effectiveness of PBM is likely influenced by total fluence and exposure duration [61, 62]. Research indicates that applying PBM at a fluence of 0.5–5 J/cm$^2$ can enhance the physiological processes of dermal fibroblasts in vitro, contributing significantly to wound healing [61, 63–67]. In these dosage ranges, cellular proliferation, viability, and migration improve compared to controls that are not irradiated [61, 68]. The cutaneous wound healing process is complex and relies on interplay of several regulated factors that work together to restore skin barrier function [69]. Ongoing research continues to unveil the intricate mechanisms underlying physiological wound repair [70, 71]).

In the case of wakefulness and sleep, numerous studies over the past 70 years have demonstrated that PBM, which can improve neuronal survival and function in both healthy and pathological conditions by applying red to near-infrared light to cells. These benefits are primarily attributed to PBM's effects on mitochondrial and non-mitochondrial mechanisms across various cell types, including neurons. PBM has been shown to improve neuronal function throughout the brain and protect against damage, yielding neuroprotective outcomes in multiple animal disease models [72]. While the effects of PBM during wakefulness have been extensively studied, its impact on brain function during sleep remains underexplored. It appears that the cellular effects of PBM can vary based on the level of arousal, potentially following a circadian rhythm. During wakefulness, PBM may primarily stimulate neuronal activity, activate mitochondrial functions, enhance gene expression, and influence brain wave patterns (such as alpha, beta, and gamma waves). It also provides neuroprotection against illness and stress by fostering neuronal survival. Conversely, PBM may be less successful at promoting neuronal activity during sleep, but it may improve the brain's ability to eliminate waste and fluids by triggering the glymphatic system [72–76].

PBM is a promising method for triggering cellular metabolic pathways that influence viability, differentiation, proliferation, and migration, ultimately enhancing regenerative capacities in mesenchymal cells [77]. Light is known to induce significant biological effects, including vision, circadian hormone regulation, vitamin D synthesis, and melanin production [78]. Recent studies have indicated that specific wavelengths targeting the cytochrome C receptor in mitochondria can substantially maintain organ function and lessen the extent of myocardial infarcts and strokes [78].

Numerous studies have demonstrated that PBM can speed up and improve the quality of gingival wound healing through a variety of methods [79]. For instance, Keskiner et al. indicated increases in transforming growth factor-beta 1 (TGF-β1), palatal wound fluid (PWF), interleukin-8 (IL-8), and platelet-derived growth factor-BB (PDGF-BB) levels, suggesting that PBM stimulates the secretion of key mediators involved in healing [79, 80]. Enhanced collagen production, increased growth factor levels, and improved extracellular matrix remodeling have also been observed in a dose-dependent manner [79, 81–84]. PBM is a useful adjunct in boosting second-intention wound healing in periodontal soft tissue surgeries. These effects may contribute to tissue thickness, re-epithelialization, and improvement in pain relief [79].

## 5 Applications of Photobiomodulation

Over the last ten years, there has been an increase in the quantity of studies highlighting the application of laser treatment in dentistry and medicine [85]. Cell growth is modulated by the photobiomodulatory effect, which is mostly caused by diode lasers operating in the red and near-infrared region (630–940 nm). Using lasers with wavelengths in the "optical window" (600–1100 nm) allows for deeper penetration, which in turn causes a broader response from cells to light [86]. In photobiomodulation (PBM), low-energy density monochromatic light is applied to cells, resulting in nonthermal photochemical effects. It has been suggested that this novel approach improves primary stability while also facilitating the process of bone repair [87]. When used in conjunction with certain medical therapies, PBM has been shown to enhance their effects.

### *5.1 Diabetes Treatment*

Diabetes is a prevalent illness and a major global public health issue. Research on PBM use in patients with diabetes revealed that PBM could lower insulin levels by about 75% or permit patients with Type 1 and Type 2 diabetes to stop taking their medications for 6 months, respectively [88]. Knowledge gathered from PBM treatment over many years shows that it is beneficial in treating diabetes. Microcirculation and myocardial contractility appear to be enhanced by PBM's antioxidant and immunomodulating effects.

## 5.2 Neural Disease Treatment

Transcranial PBM enhances regional cerebral blood flow in Parkinson's disease patients, severe depressive patients, and traumatic brain injury patients [89]. Accelerated restoration of the injured nervous system may be possible with increased ATP synthesis and enhanced regional cerebral blood flow. It has been proposed that NIR therapy prevents the death of a large number of dopaminergic cells. This PBM effect was applied to Parkinson's disease treatment.

Additionally, PBM has been applied to spinal cord recovery. Starting from the tip of the brain stem and extending to the end of the spine, the spinal cord is a thin, long, tail-shaped tissue. Nerves in the spinal cord carry signals from the brain to the body's other organs [1]. Huang et al. reported a beneficial effect of PBM for spinal cord repair (SCI). The results of their study indicate that intravascular laser irradiation of blood might effectively reduce oxidative stress and mitochondrial dysfunction in patients with chronic SCI [90]. Yamany et al. then applied PBM to 30 patients suffering from excruciating diabetic neuropathy and saw changes in the skin's microcirculation on the feet, a decrease in pain, and certain electrophysiological markers of peripheral nerve activity [91].

## 5.3 Dermatology

PBM's effectiveness has been verified and demonstrated in a number of dermatological treatments, including procedures for fat reduction, hair loss treatment, and skin rejuvenation. PBM's effectiveness was used to treat skin issues such as wrinkles and skin laxity [92]. Wikramanayake et al. used a 655-nm laser wavelength for 20 s three times a week for a total of 6 weeks to validate the hair growth findings of PBM on the C3H/HeJ mouse model [93].

## 5.4 Inflammatory Disorders

PBM study was done for chronic autoimmune disease thyroiditis. A pilot study was conducted from 2010 to 2017, and finally it indicated that LLLT can help enhance thyroid function and has long-lasting advantages [94]. Also, PBM for muscles seeks to improve training and athletic performance, lessen delayed onset muscle soreness (DOMS), and improve muscle damage indicators (creatine kinase) following strenuous or extended exercise. Furthermore, PBM can be used to treat open muscle injury brought on by trauma or stress [6]. Additionally, PBM effects were also seen in treating arthritis. Rheumatoid arthritis (RA) and osteoarthritis (OA) are the two most prevalent types of arthritis. Rheumatoid arthritis (RA) affects approximately 0.24% of the population, but osteoarthritis (OA) affects over 3.8% of people. PBM has been used to treat both kinds with success. Cochrane systematic reviews discovered some evidence in the case of OA [95] and strong evidence for its efficacy in RA [96]. Since it is challenging to measure inflammation in human subjects, the majority of clinical investigations have tested the effectiveness using range of motion scores and pain scales.

## *5.5 Management of Secondary Complications Following Radiation*

Fungal mucositis is a common side effect of oral radiation therapy that severely reduces a patient's quality of life. It follows that effective management of this issue is essential. Nevertheless, randomized clinical trials of dental care are insufficient in addressing this issue. In 1985, Ciais et al. reported that PBM with 10–150 mW He/Ne or diode laser was effective in reducing the severity of oral mucositis lesions in vivo. A randomized clinical trial by Cowenet et al. in patients receiving bone marrow transplantation also demonstrated the efficacy of PBM in preventing chemotherapy-induced oral mucositis. Officially, in 2014, the International Society of Oral Oncology (ISOO) and the Multinational Association of Supportive Care in Cancer (MASCC) advised the use of PBM for patients receiving high-dose chemotherapy during hematopoietic stem cell transplantation [97].

## *5.6 Dentistry*

In nearly all aspects of clinical dentistry, lasers and light-emitting diodes (LEDs) are being employed to improve patient care and quality of life. The effectiveness of the therapy has grown as a result of PBM's involvement in reducing pain and encouraging healing. It can be useful for treating many oral diseases, promoting tissue repair, and reducing inflammation through a variety of physiological mechanisms, making it applicable to many areas of dentistry. Encouraging cell migration, proliferation, and differentiation, it also contributes to improved therapeutic outcomes [98]. PBM's dental applications rely on the use of biophotonics therapy at a low dose. PBM is utilized in orthodontic treatment to lessen post-orthodontia appliance implantation pain [99, 100] Additionally, it expedites bone remodeling and enhances collagen deposition and osseointegration [101, 102]. It was demonstrated to support implant stability in surgically assisted therapy, such as implant and mini-implant implantation [103], healing time was shortened, and swelling was decreased, all of which improved postoperative comfort.

## 6 Conclusion and Future Prospects

Alterations at the tissue, cellular, and molecular levels can result from interactions between cells and low concentrations of red and near-infrared (NIR) light [9]. Photobiomodulation (PBM) influences a variety of biological and pathological processes. [104]. It can regulate essential biological functions, including cell survival, proliferation, and apoptosis, which depend on key molecules such as Src tyrosine kinases, protein kinase C (PKC), protein kinase B (Akt/PKB), and interleukin-8/1a (IL-8/1a) [9]. Notably, at low doses, light can have a stimulatory effect on cell proliferation—beneficial for wound healing—while higher doses may inhibit proliferation, potentially aiding in specific types of scar formation [9]. The outcomes are influenced by the parameters employed during treatment [9, 105].

Although PBM operates through a common mechanism involving the absorption of light by mitochondrial chromophores, primarily cytochrome c oxidase (Cox), the resulting effects vary based on the parameters that are employed, the specific tissue being treated, and the activated signalling pathways [9, 106]. For successful clinical application of PBM, clinicians must understand the appropriate parameters and their implications for each targeted tissue [9].

PBM is a versatile technology that has found applications across various medical disciplines, including dermatology, psychiatry, oncology, and, more recently, ophthalmology. This noninvasive approach has demonstrated promising results in treating a variety of ocular conditions, from the surface of the eye to deeper structures, garnering increasing interest from the ophthalmology research community [107]. In the future, the establishment of standardized protocols regarding the number and frequency of sessions, treatment parameters, and types of light, alongside strict patient selection criteria, will enhance our understanding of PBM's role in ocular disease management [107].

Despite significant emphasis on the photoelectrical and photochemical mechanisms of PBM, its complex functions remain only partially understood [104]. Additionally, the photophysical aspects of PBM have received limited attention. One notable effect of PBM involves the nonvisual phototransduction pathway, which encompasses micro-oscillatory cellular interactions, mechanotransduction, biophotonic signaling, and modulation of cytoskeletal structures [104]. Gaining insight into these photophysical mechanisms will be instrumental in guiding future experimental and clinical studies on photobiomodulation therapy [104].

## References

1. Dompe, C., Moncrieff, L., Matys, J., Grzech-Leśniak, K., Kocherova, I., Bryja, A., & Dyszkiewicz-Konwińska, M. (2020). Photobiomodulation—Underlying mechanism and clinical applications. *Journal of Clinical Medicine, 9*(6), 1724.
2. Hamblin, M. R. (2018). Mechanisms and mitochondrial redox signaling in Photobiomodulation. *Photochemistry and Photobiology, 94*(2), 199–212. https://doi.org/10.1111/php.12864

3. Mester, A. (2013). Laser biostimulation. *Photomedicine and Laser Surgery, 31*(6), 237–239. https://doi.org/10.1089/pho.2013.9876
4. Hamblin, M. R. (2019). Mechanisms of photobiomodulation in the brain. In *Photobiomodulation in the brain* (pp. 97–110). Academic Press.
5. Hamblin, M. R. (2016). Photobiomodulation or low-level laser therapy. *Journal of Biophotonics, 9*, 1122–1124.
6. Hamblin, M. R. (2017). Mechanisms and applications of the anti-inflammatory effects of photobiomodulation. *AIMS Biophysics, 4*(3), 337.
7. da Silva, T. G., Ribeiro, R. S., Mencalha, A. L., & de Souza Fonseca, A. (2023). Photobiomodulation at molecular, cellular, and systemic levels. *Lasers in Medical Science, 38*(1), 136.
8. Karu, T. I. (2013). Cellular and molecular mechanisms of photobiomodulation (low-power laser therapy). *IEEE Journal of Selected Topics in Quantum Electronics, 20*(2), 143–148.
9. De Freitas, L. F., & Hamblin, M. R. (2016). Proposed mechanisms of photobiomodulation or low-level light therapy. *IEEE Journal of Selected Topics in Quantum Electronics, 22*(3), 348–364.
10. Fernandes, A. M., Fero, K., Driever, W., & Burgess, H. A. (2013). Enlightening the brain: Linking deep brain photoreception with behavior and physiology. *BioEssays: News and Reviews in Molecular, Cellular and Developmental Biology, 35*(9), 775–779. https://doi.org/10.1002/bies.201300034
11. Hardie RC. Photosensitive TRPs. 2014:795–826.
12. Nilius, B., & Voets, T. (2005). TRP channels: a TR(I)P through a world of multifunctional cation channels. *Pflugers Archiv : European Journal of Physiology, 451*(1), 1–10. https://doi.org/10.1007/s00424-005-1462-y
13. Rohacs, T. (2014). Phosphoinositide regulation of TRP channels. *Handbook of Experimental Pharmacology, 223*, 1143–1176. https://doi.org/10.1007/978-3-319-05161-1_18
14. Liu, B., Yang, Z., Gomez, A., Liu, B., Lin, C., & Oka, Y. (2016). Signaling Mechanisms of Plant Cryptochromes. *Annual Review of Plant Biology*, 67, 445–476.
15. Emery, P., So, W.V., Kaneko, M., Hall, J.C., & Rosbash, M. (1998). CRY, a Drosophila clock gene, controls circadian rhythms. Cell, 95(5), 669–679.
16. Yoshii, T., Ahmad, M., & Helfrich-Förster, C. (2009). Cryptochrome mediates light-dependent magnetosensitivity of Drosophila's circadian clock. *PLoS Biology*, 7(4), e1000086.
17. Liu, B., & Zhang, E.E. (2016). Cryptochrome in mammals: A magnetosensor? Science, 354(6316), 1119–1120.
18. Chiou, Y.W. (2016). Mammalian cryptochromes: implications for circadian regulation and beyond. *Journal of Biomedical Science*, 23, 4.
19. Farivar, S., Malekshahabi, T., & Shiari, R. (2014). Biological effects of low level laser therapy. *Journal of Lasers in Medical Sciences, 5*(2), 58–62.
20. Ferraresi, C., Hamblin, M. R., & Parizotto, N. A. (2012). Low-level laser (light) therapy (LLLT) on muscle tissue: Performance, fatigue and repair benefited by the power of light. *Photonics & Lasers in Medicine, 1*(4), 267–286. https://doi.org/10.1515/plm-2012-0032
21. Srinivasan, S., & Avadhani, N. G. (2012). Cytochrome c oxidase dysfunction in oxidative stress. *Free Radical Biology & Medicine, 53*(6), 1252–1263. https://doi.org/10.1016/j.freeradbiomed.2012.07.021
22. Hamblin, M. R., & Demidova, T. N. (2006). Mechanisms of low level light therapy. *Mechanisms for low-light therapy, 6140*, 614001.
23. Taskén, K., & Aandahl, E. M. (2004). Localized effects of cAMP mediated by distinct routes of protein kinase A. *Physiological Reviews, 84*(1), 137–167. https://doi.org/10.1152/physrev.00021.2003
24. Taniguchi, D., Dai, P., Hojo, T., Yamaoka, Y., Kubo, T., & Takamatsu, T. (2009). Low-energy laser irradiation promotes synovial fibroblast proliferation by modulating p15 subcellular localization. *Lasers in Surgery and Medicine, 41*(3), 232–239. https://doi.org/10.1002/lsm.20750

25. Cury, V., Moretti, A. I., Assis, L., Bossini, P., deS Crusca, J., Neto, C. B., Fangel, R., de Souza, H. P., Hamblin, M. R., & Parizotto, N. A. (2013). Low level laser therapy increases angiogenesis in a model of ischemic skin flap in rats mediated by VEGF, HIF-1α and MMP-2. *Journal of Photochemistry and Photobiology B, Biology, 125*, 164–170. https://doi.org/10.1016/j.jphotobiol.2013.06.004
26. de Lima, F. M., Albertini, R., Dantas, Y., Maia-Filho, A. L., deL Santana, C., Castro-Faria-Neto, H. C., França, C., Villaverde, A. B., & Aimbire, F. (2013). Low-level laser therapy restores the oxidative stress balance in acute lung injury induced by gut ischemia and reperfusion. *Photochemistry and Photobiology, 89*(1), 179–188. https://doi.org/10.1111/j.1751-1097.2012.01214.x
27. Felician, M. C. P., Belotto, R., Tardivo, J. P., Baptista, M. S., & Martins, W. K. (2023). Photobiomodulation: Cellular, molecular, and clinical aspects. *Journal of Photochemistry and Photobiology, 17*, 100197.
28. Tam, S. Y., Tam, V. C., Ramkumar, S., Khaw, M. L., Law, H. K., & Lee, S. W. (2020). Review on the cellular mechanisms of low-level laser therapy use in oncology. *Frontiers in Oncology, 10*, 1255.
29. Zhu, Q., Xiao, S., Hua, Z., Yang, D., Hu, M., Zhu, Y. T., & Zhong, H. (2021). Near infrared (NIR) light therapy of eye diseases: a review. *International Journal of Medical Sciences, 18*(1), 109.
30. Lim, W., Kim, J., Kim, S., Karna, S., Won, J., Jeon, S. M., & Kim, O. (2013). Modulation of lipopolysaccharide-induced NF-κB signaling pathway by 635 nm irradiation via heat shock protein 27 in human gingival fibroblast cells. *Photochemistry and Photobiology, 89*(1), 199–207.
31. Eells, J. T., Henry, M. M., Summerfelt, P., Wong-Riley, M. T. T., Buchmann, E. V., Kane, M., & Whelan, H. T. (2003). Therapeutic photobiomodulation for methanol-induced retinal toxicity. *Proceedings of the National Academy of Sciences, 100*(6), 3439–3444.
32. Liang, H. L., Whelan, H. T., Eells, J. T., Meng, H., Buchmann, E., Lerch-Gaggl, A., & Wong-Riley, M. (2006). Photobiomodulation partially rescues visual cortical neurons from cyanide-induced apoptosis. *Neuroscience, 139*(2), 639–649.
33. Sperandio, F. F., Simões, A., Corrêa, L., Aranha, A. C. C., Giudice, F. S., Hamblin, M. R., & Sousa, S. C. (2015). Low-level laser irradiation promotes the proliferation and maturation of keratinocytes during epithelial wound repair. *Journal of Biophotonics, 8*(10), 795–803.
34. Szymanska, J., Goralczyk, K., Klawe, J. J., Lukowicz, M., Michalska, M., Goralczyk, B., & Rosc, D. (2013). Phototherapy with low-level laser influences the proliferation of endothelial cells and vascular endothelial growth factor and transforming growth factor-beta secretion. *Journal of Physiology and Pharmacology, 64*(3), 387–391.
35. Tsai, W. C., Hsu, C. C., Pang, J. H. S., Lin, M. S., Chen, Y. H., & Liang, F. C. (2012). Low-level laser irradiation stimulates tenocyte migration with up-regulation of dynamin II expression. *PLoS One, 7*(5), e38235.
36. AlGhamdi, K. M., Kumar, A., Ashour, A. E., & AlGhamdi, A. A. (2015). A comparative study of the effects of different low-level lasers on the proliferation, viability, and migration of human melanocytes in vitro. *Lasers in Medical Science, 30*, 1541–1551.
37. Frozanfar, A., Ramezani, M., Rahpeyma, A., Khajehahmadi, S., & Arbab, H. R. (2013). The effects of low level laser therapy on the expression of collagen type I gene and proliferation of human gingival fibroblasts (Hgf3-Pi 53): In vitro study. *Iranian Journal of Basic Medical Sciences, 16*(10), 1071.
38. Hamajima, S., Hiratsuka, K., Kiyama-Kishikawa, M., Tagawa, T., Kawahara, M., Ohta, M., & Abiko, Y. (2003). Effect of low-level laser irradiation on osteoglycin gene expression in osteoblasts. *Lasers in Medical Science, 18*, 78–82.
39. Abrahamse, H. (2012). Regenerative medicine, stem cells, and low-level laser therapy: Future directives. *Photomedicine and Laser Surgery, 30*(12), 681–682.

40. Ginani, F., Soares, D. M., Barreto, M. P. E. V., & Barboza, C. A. G. (2015). Effect of low-level laser therapy on mesenchymal stem cell proliferation: a systematic review. *Lasers in Medical Science, 30*, 2189–2194.
41. Borsa, P. A., Larkin, K. A., & True, J. M. (2013). Does phototherapy enhance skeletal muscle contractile function and postexercise recovery? A systematic review. *Journal of Athletic Training, 48*(1), 57–67.
42. Ferraresi, C., Kaippert, B., Avci, P., Huang, Y. Y., De Sousa, M. V., Bagnato, V. S., & Hamblin, M. R. (2015). Low-level laser (light) therapy increases mitochondrial membrane potential and ATP synthesis in C2C12 myotubes with a peak response at 3–6 h. *Photochemistry and Photobiology, 91*(2), 411–416.
43. Xu, X., Zhao, X., Liu, T. C. Y., & Pan, H. (2008). Low-intensity laser irradiation improves the mitochondrial dysfunction of C2C12 induced by electrical stimulation. *Photomedicine and Laser Surgery, 26*(3), 197–202.
44. Khuman, J., Zhang, J., Park, J., Carroll, J. D., Donahue, C., & Whalen, M. J. (2012). Low-level laser light therapy improves cognitive deficits and inhibits microglial activation after controlled cortical impact in mice. *Journal of Neurotrauma, 29*(2), 408–417.
45. Bouvet-Gerbettaz, S., Merigo, E., Rocca, J. P., Carle, G. F., & Rochet, N. (2009). Effects of low-level laser therapy on proliferation and differentiation of murine bone marrow cells into osteoblasts and osteoclasts. *Lasers in Surgery and Medicine: The Official Journal of the American Society for Laser Medicine and Surgery, 41*(4), 291–297.
46. Xuan, W., Agrawal, T., Huang, L., Gupta, G. K., & Hamblin, M. R. (2015). Low-level laser therapy for traumatic brain injury in mice increases brain derived neurotrophic factor (BDNF) and synaptogenesis. *Journal of Biophotonics, 8*(6), 502–511.
47. Serra, P. E. A., & Ashmawi, A. H. (2010). Influence of naloxone and methysergide on the analgesic effects of low-level laser in an experimental pain model. *Revista Brasileira de Anestesiologia, 60*(3), 302–310. [PubMed: 20682161].
48. Yan, W., Chow, R., & Armati, P. J. (2011). Inhibitory effects of visible 650-nm and infrared 808-nm laser irradiation on somatosensory and compound muscle action potentials in rat sciatic nerve: Implications for laser-induced analgesia. *Journal of the Peripheral Nervous System, 16*(2), 130–135.
49. Fujimoto, K., Kiyosaki, T., Mitsui, N., Mayahara, K., Omasa, S., Suzuki, N., & Shimizu, N. (2010). Low-intensity laser irradiation stimulates mineralization via increased BMPs in MC3T3-E1 cells. *Lasers in Surgery and Medicine, 42*(6), 519–526.
50. Hoffman, M., & Monroe, D. M. (2012). Low intensity laser therapy speeds wound healing in hemophilia by enhancing platelet procoagulant activity. *Wound Repair and Regeneration, 20*(5), 770–777.
51. Avci, P., Gupta, G. K., Clark, J., Wikonkal, N., & Hamblin, M. R. (2014). Low-level laser (light) therapy (LLLT) for treatment of hair loss. *Lasers in Surgery and Medicine, 46*(2), 144–151.
52. Wang, Y., Jiang, J., Zhang, H., Dong, X., Wang, L., Ranjan, R., & Zomaya, A. Y. (2017). A scalable parallel algorithm for atmospheric general circulation models on a multi-core cluster. *Future Generation Computer Systems, 72*, 1–10.
53. Henderson, T. A., & Morries, L. D. (2015). Near-infrared photonic energy penetration: Can infrared phototherapy effectively reach the human brain? *Neuropsychiatric Disease and Treatment, 11*, 2191–2208.
54. Padalkar, M. V., & Pleshko, N. (2015). Wavelength-dependent penetration depth of near infrared radiation into cartilage. *Analyst, 140*(7), 2093–2100.
55. Zein, R., Selting, W., & Hamblin, M. R. (2018). Review of light parameters and photobiomodulation efficacy: Dive into complexity. *Journal of Biomedical Optics, 23*(12), 120901–120901.
56. Heiskanen, V., & Hamblin, M. R. (2018). Photobiomodulation: Lasers vs. light emitting diodes? *Photochemical & Photobiological Sciences, 17*(8), 1003–1017.

57. Bolton, J. R., Mayor-Smith, I., & Linden, K. G. (2015). Rethinking the concepts of fluence (UV dose) and fluence rate: The importance of photon-based units–a systemic review. *Photochemistry and Photobiology, 91*(6), 1252–1262.
58. Lanzafame, R. J., Stadler, I., Kurtz, A. F., Connelly, R., Brondon, P., & Olson, D. (2007). Reciprocity of exposure time and irradiance on energy density during photoradiation on wound healing in a murine pressure ulcer model. *Lasers in Surgery and Medicine: The Official Journal of the American Society for Laser Medicine and Surgery, 39*(6), 534–542.
59. Huang, Y. Y., Sharma, S. K., Carroll, J., & Hamblin, M. R. (2011). Biphasic dose response in low level light therapy–an update. *Dose-Response, 9*(4), dose-response.
60. Huang, Y. Y., Chen, A. C. H., Carroll, J. D., & Hamblin, M. R. (2009). Biphasic dose response in low level light therapy. *Dose-Response, 7*(4), 1.
61. Tripodi, N., Corcoran, D., Antonello, P., Balic, N., Caddy, D., Knight, A., & Feehan, J. (2021). The effects of photobiomodulation on human dermal fibroblasts in vitro: A systematic review. *Journal of Photochemistry and Photobiology B: Biology, 214*, 112100.
62. Leal-Junior, E. C. P., Vanin, A. A., Miranda, E. F., de Carvalho, P. D. T. C., Dal Corso, S., & Bjordal, J. M. (2015). Effect of phototherapy (low-level laser therapy and light-emitting diode therapy) on exercise performance and markers of exercise recovery: a systematic review with meta-analysis. *Lasers in Medical Science, 30*, 925–939.
63. Ayuk, S. M., Houreld, N. N., & Abrahamse, H. (2018). Effect of 660 nm visible red light on cell proliferation and viability in diabetic models in vitro under stressed conditions. *Lasers in Medical Science, 33*, 1085–1093.
64. Jere, S. W., Houreld, N. N., & Abrahamse, H. (2018). Photobiomodulation at 660 nm stimulates proliferation and migration of diabetic wounded cells via the expression of epidermal growth factor and the JAK/STAT pathway. *Journal of Photochemistry and Photobiology B: Biology, 179*, 74–83.
65. Li, W., Hu, X., Lu, X., Liu, J., Chen, Z., Zhou, X., & Liu, S. (2020). RNA-Seq analysis revealed the molecular mechanisms of photobiomodulation effect on human fibroblasts. *Photodermatology, Photoimmunology & Photomedicine, 36*(4), 299–307.
66. Sekhejane, P. R., Houreld, N. N., & Abrahamse, H. (2011). Irradiation at 636 nm positively affects diabetic wounded and hypoxic cells in vitro. *Photomedicine and Laser Surgery, 29*(8), 521–530.
67. Zungu, I. L., Hawkins Evans, D., & Abrahamse, H. (2009). Mitochondrial responses of normal and injured human skin fibroblasts following low level laser irradiation—An in vitro study. *Photochemistry and Photobiology, 85*(4), 987–996.
68. Ammann, K. R., DeCook, K. J., Li, M., & Slepian, M. J. (2019). Migration versus proliferation as contributor to in vitro wound healing of vascular endothelial and smooth muscle cells. *Experimental Cell Research, 376*(1), 58–66.
69. Paquette, D., & Falanga, V. (2002). Leg ulcers. *Clinics in Geriatric Medicine, 18*(1), 77–88.
70. Han, G., & Ceilley, R. (2017). Chronic Wound Healing: A Review of Current Management and Treatments. *Advances in therapy*, 34(3), 599–610.
71. Sen, C. K., Gordillo, G. M., Roy, S., Kirsner, R., Lambert, L., Hunt, T. K., & Longaker, M. T. (2009). Human skin wounds: a major and snowballing threat to public health and the economy. *Wound Repair and Regeneration, 17*(6), 763–771.
72. Moro, C., Valverde, A., Dole, M., Hoh Kam, J., Hamilton, C., Liebert, A., & Mitrofanis, J. (2022). The effect of photobiomodulation on the brain during wakefulness and sleep. *Frontiers in Neuroscience, 16*, 942536.
73. Semyachkina-Glushkovskaya, O., Abdurashitov, A., Dubrovsky, A., Klimova, M., Agranovich, I., Terskov, A., & Kurths, J. (2020a). Photobiomodulation of lymphatic drainage and clearance: Perspective strategy for augmentation of meningeal lymphatic functions. *Biomedical Optics Express, 11*(2), 725–734.
74. Semyachkina-Glushkovskaya, O., Klimova, M., Iskra, T., Bragin, D., Abdurashitov, A., Dubrovsky, A., & Kurths, J. (2021a). Transcranial photobiomodulation of clearance of beta-amyloid from the mouse brain: Effects on the meningeal lymphatic drainage and blood

oxygen saturation of the brain. In *Oxygen transport to tissue XLII* (pp. 57–61). Springer International Publishing.
75. Semyachkina-Glushkovskaya, O., Penzel, T., Blokhina, I., Khorovodov, A., Fedosov, I., Yu, T., & Kurths, J. (2021b). Night photostimulation of clearance of beta-amyloid from mouse brain: New strategies in preventing Alzheimer's disease. *Cells, 10*(12), 3289.
76. Semyachkina-Glushkovskaya, O., Postnov, D., Lavrova, A., Fedosov, I., Borisova, E., Nikolenko, V., & Tuchin, V. (2020b). Biophotonic strategies of measurement and stimulation of the cranial and the extracranial lymphatic drainage function. *IEEE Journal of Selected Topics in Quantum Electronics, 27*(4), 1–13.
77. Pinto, H., Goñi Oliver, P., & Sánchez-Vizcaíno Mengual, E. (2021). The effect of photobiomodulation on human mesenchymal cells: a literature review. *Aesthetic Plastic Surgery, 45*, 1826–1842.
78. Streeter, J., De Taboada, L., & Oron, U. (2004). Mechanisms of action of light therapy for stroke and acute myocardial infarction. *Mitochondrion*, 4(5-6), 569–576.
79. Ebrahimi, P., Hadilou, M., Naserneysari, F., Dolatabadi, A., Tarzemany, R., Vahed, N., & Gholami, L. (2021). Effect of photobiomodulation in secondary intention gingival wound healing—a systematic review and meta-analysis. *BMC Oral Health, 21*(1), 258.
80. Keskiner, I., Lutfioğlu, M., Aydogdu, A., Saygun, N. I., & Serdar, M. A. (2016). Effect of photobiomodulation on transforming growth factor-β1, platelet-derived growth factor-BB, and interleukin-8 release in palatal wounds after free gingival graft harvesting: a randomized clinical study. *Photomedicine and Laser Surgery, 34*(6), 263–271.
81. Heidari, M., Paknejad, M., Jamali, R., Nokhbatolfoghahaei, H., Fekrazad, R., & Moslemi, N. (2017). Effect of laser photobiomodulation on wound healing and postoperative pain following free gingival graft: A split-mouth triple-blind randomized controlled clinical trial. *Journal of Photochemistry and Photobiology B: Biology, 172*, 109–114.
82. Kohale, B. R., Agrawal, A. A., & Raut, C. P. (2018). Effect of low-level laser therapy on wound healing and patients' response after scalpel gingivectomy: A randomized clinical split-mouth study. *Journal of Indian Society of Periodontology, 22*(5), 419–426.
83. Ustaoglu, G., Ercan, E., & Tunali, M. (2017). Low-level laser therapy in enhancing wound healing and preserving tissue thickness at free gingival graft donor sites: a randomized, controlled clinical study. *Photomedicine and Laser Surgery, 35*(4), 223–230.
84. Yildiz, M. S., & Gunpinar, S. (2019). Free gingival graft adjunct with low-level laser therapy: a randomized placebo-controlled parallel group study. *Clinical Oral Investigations, 23*, 1845–1854.
85. Matys, J., Świder, K., & Flieger, R. (2017). Laser instant implant impression method: A case presentation. *Dental and Medical Problems, 54*(1), 101–106.
86. AlGhamdi, K. M., Kumar, A., & Moussa, N. A. (2012). Low-level laser therapy: a useful technique for enhancing the proliferation of various cultured cells. *Lasers in Medical Science, 27*, 237–249.
87. Pires Oliveira, D. A., de Oliveira, R. F., Zangaro, R. A., & Soares, C. P. (2008). Evaluation of low-level laser therapy of osteoblastic cells. *Photomedicine and Laser Surgery, 26*(4), 401–404.
88. Bodnar, P. M., Peshko, A. O., Prystupiuk, O. M., Voronko, A. A., Kyriienko, D. V., Mykhal'chyshyn, H. P., & Naumova, M. I. (1999). Laser therapy in diabetes mellitus. *Likars' ka Sprava, 6*, 125–128.
89. Zhang, J., Xing, D., & Gao, X. (2008). Low-power laser irradiation activates Src tyrosine kinase through reactive oxygen species-mediated signaling pathway. *Journal of Cellular Physiology, 217*(2), 518–528.
90. Huang, S. F., Tsai, Y. A., Wu, S. B., Wei, Y. H., Tsai, P. Y., & Chuang, T. Y. (2012). Effects of intravascular laser irradiation of blood in mitochondria dysfunction and oxidative stress in adults with chronic spinal cord injury. *Photomedicine and Laser Surgery, 30*(10), 579–586.
91. Yamany, A. A., & Sayed, H. M. (2012). Effect of low level laser therapy on neurovascular function of diabetic peripheral neuropathy. *Journal of Advanced Research, 3*(1), 21–28.

92. Barolet, D., Roberge, C. J., Auger, F. A., Boucher, A., & Germain, L. (2009). Regulation of skin collagen metabolism in vitro using a pulsed 660 nm LED light source: Clinical correlation with a single-blinded study. *Journal of Investigative Dermatology, 129*(12), 2751–2759.
93. Wikramanayake, T. C., Rodriguez, R., Choudhary, S., Mauro, L. M., Nouri, K., Schachner, L. A., & Jimenez, J. J. (2012). Effects of the Lexington LaserComb on hair regrowth in the C3H/HeJ mouse model of alopecia areata. *Lasers in Medical Science, 27*, 431–436.
94. Hofling, D., Chavantes, M. C., Buchpiguel, C. A., Cerri, G. G., Carneiro, P. C., Marui, S., & Chammas, M. C. (2017). Long-term follow-up of patients with hypothyroidism induced by autoimmune thyroiditis submitted to low-level laser therapy. *Lasers in Surgery and Medicine, 49*, 36.
95. Brosseau, L., Robinson, V., Wells, G. A., deBie, R., Gam, A., Harman, K., & de Bie, R. (2004). Low level laser therapy (classes I, II and III) for treating osteoarthritis. *Cochrane Database of Systematic Reviews, 2004*(3), CD002046.
96. Brosseau, L., Welch, V., Wells, G. A., de Bie, R., Gam, A., Harman, K., & Tugwell, P. (2005). Low level laser therapy (classes I, II and III) for treating rheumatoid arthritis. *Cochrane Database of Systematic Reviews, 2005*(4), CD002049.
97. Lalla, R. V., Bowen, J., Barasch, A., Elting, L., Epstein, J., Keefe, D. M., & Mucositis Guidelines Leadership Group of the Multinational Association of Supportive Care in Cancer and International Society of Oral Oncology (MASCC/ISOO). (2014). MASCC/ISOO clinical practice guidelines for the management of mucositis secondary to cancer therapy. *Cancer, 120*(10), 1453–1461.
98. Soni, S., & Thakar, S. (2022). A review of Photobiomodulation and its application in dentistry. *Indian Journal of Dental Sciences, 14*(4), 209–212.
99. Artés-Ribas, M., Arnabat-Dominguez, J., & Puigdollers, A. (2013). Analgesic effect of a low-level laser therapy (830 nm) in early orthodontic treatment. *Lasers in Medical Science, 28*, 335–341.
100. Genc, G., Kocadereli, I., Tasar, F., Kilinc, K., El, S., & Sarkarati, B. (2013). Effect of low-level laser therapy (LLLT) on orthodontic tooth movement. *Lasers in Medical Science, 28*, 41–47.
101. Faria Amorim, J. C., Sousa, G. R. D., Silveira, L. D. B., Prates, R. A., Pinotti, M., & Ribeiro, M. S. (2006). Clinical study of the gingiva healing after gingivectomy and low-level laser therapy. *Photomedicine and Laser Therapy, 24*(5), 588–594.
102. Kim, S. J., Kang, Y. G., Park, J. H., Kim, E. C., & Park, Y. G. (2013). Effects of low-intensity laser therapy on periodontal tissue remodeling during relapse and retention of orthodontically moved teeth. *Lasers in Medical Science, 28*, 325–333.
103. Matys, J., Świder, K., Grzech-Leśniak, K., Dominiak, M., & Romeo, U. (2019). Photobiomodulation by a 635nm diode laser on Peri-implant bone: Primary and secondary stability and bone density analysis—A randomized clinical trial. *BioMed Research International, 2019*(1), 2785302.
104. Liebert, A., Capon, W., Pang, V., Vila, D., Bicknell, B., McLachlan, C., & Kiat, H. (2023). Photophysical mechanisms of photobiomodulation therapy as precision medicine. *Biomedicine, 11*(2), 237.
105. Chu, J., Wu, S., & Xing, D. (2010). Survivin mediates self-protection through ROS/cdc25c/CDK1 signaling pathway during tumor cell apoptosis induced by high fluence low-power laser irradiation. *Cancer Letters, 297*(2), 207–219.
106. Choi, M., Kim, J. E., Cho, K. H., & Lee, J. H. (2013). In vivo and in vitro analysis of low level light therapy: a useful therapeutic approach for sensitive skin. *Lasers in Medical Science, 28*, 1573–1579.
107. Cannas, C., Pintus, B., Corgiolu, L., Borrelli, E., Boscia, G., Toro, M. D., & Giannaccare, G. (2024). Current applications and future perspectives of Photobiomodulation in ocular diseases: A narrative review. *Applied Sciences, 14*(6), 2623.

# Raman Spectroscopy

**Babli Sharma, Chainee Das, and Venkata Satish Kumar Mattaparthi**

**Abstract** Raman spectroscopy is a versatile analytical technique that provides detailed molecular information through the inelastic scattering of monochromatic light, usually from a laser. When an incident light interacts with a molecule's vibrational, rotational, or other low-frequency modes, most of the scattered light remains at the same wavelength (Rayleigh scattering). However, a small fraction is scattered at different wavelengths due to energy exchanges, known as Raman scattering. The resulting spectrum acts as a molecular fingerprint, enabling the identification of compounds and insights into molecular structures, phase transitions, and chemical environments. A key advantage of Raman spectroscopy is its minimal sample preparation and resistance to water interference, making it ideal for studying biological samples, aqueous solutions, and various materials. Its nondestructive nature ensures sample preservation, valuable in the field of forensics. Raman spectroscopy is applicable to gases, liquids, and solids, providing high specificity for molecular identification. Raman Spectroscopy basically examines changes in the polarizability of a chemical bond. Interaction of light with a molecule can cause its electron cloud to distort, and this deformation is known as polarizability change. Raman active modes are generated when polarizability varies due to unique energy transitions that occur in molecular bonds. Widely used across disciplines such as chemistry, biology, materials science, and medical diagnostics, it offers critical insights into molecular structures and compositions.

**Keywords** Raman scattering · Vibrational modes · Stokes shift · Anti-Stokes shift · Polarizability change

Babli Sharma and Chainee Das contributed equally with all other contributors.

B. Sharma · C. Das · V. S. K. Mattaparthi (✉)
Molecular Modelling and Simulation Laboratory, Department of Molecular Biology and Biotechnology, Tezpur University, Tezpur, Assam, India
e-mail: venkata@tezu.ernet.in



N. Mazumder et al. (eds.), *Advanced Biophysical Techniques in Biosciences*, Lecture Notes in Nanoscale Science and Technology 7,
https://doi.org/10.1007/978-3-031-94551-9_14

# 1 Introduction

The discipline of spectroscopy studies the interactions between electromagnetic radiation and matter, encompassing various phenomena like emission, absorption, fluorescence, and scattering [1]. Spectroscopic techniques are frequently used for analyzing and characterizing a broad range of samples. These methods serve two primary purposes: qualitative analysis to identify a sample's composition and quantitative analysis to determine the concentration of analyte in the sample [2, 3]. While some spectroscopic methods, such as UV-Vis Spectrophotometry, provide preliminary identification and are used as screening tools, others like Infrared (IR) Spectroscopy and Mass Spectrometry offer definitive identification and are employed as confirmatory tests due to their high specificity and reliability [4, 5].

When charged molecules or electrons interact with an incident light of frequency ($\nu_i$) or a beam of monochromatic light (a laser), which typically has a frequency in the visible, near-infrared, or ultraviolet range, a phenomenon known as scattering occurs (scattered light having frequency, $\nu_s$), resulting in the re-emission of photons in various directions. The light interacts with the electron cloud and vibrational states of the molecule. This scattering process can take two forms: elastic and inelastic. Rayleigh scattering, or elastic scattering, occurs when light interacts with a molecule without changing its wavelength. In contrast, Raman scattering, or inelastic scattering, involves an energy exchange between light and the molecule, resulting in a shift in the light's wavelength [6]. In Rayleigh scattering, the scattered light retains its original energy and wavelength, while in Raman scattering, the energy transfer between light and molecule alters the wavelength, providing valuable information about the molecular structure and composition [1]. Unlike Rayleigh scattering, where the molecule returns to its original state, in Raman scattering, the molecule either absorbs or releases energy, resulting in a shift in the energy of the scattered light, which has been shown in Fig. 1 [7].

Raman scattering is an inelastic scattering process where photons interact with molecules, transferring energy and changing wavelength [7]. This phenomenon occurs when incident light excites molecular vibrations or rotations, resulting in energy transfer between photons and molecules and subsequently scattering photons with a shifted wavelength, known as Stokes or anti-Stokes shift [6, 8]. It is a key phenomenon used in Raman spectroscopy, which provides information about the vibrational modes of molecules, offering insights into their chemical structure and environment.

## 1.1 *Energy Exchange and Shifts*

Stokes shift occurs if the scattered photon has lower energy than the incident photon, this indicates that the molecule absorbed some energy, shifting the light to a longer wavelength (lower frequency) while Anti-Stokes shift happens when

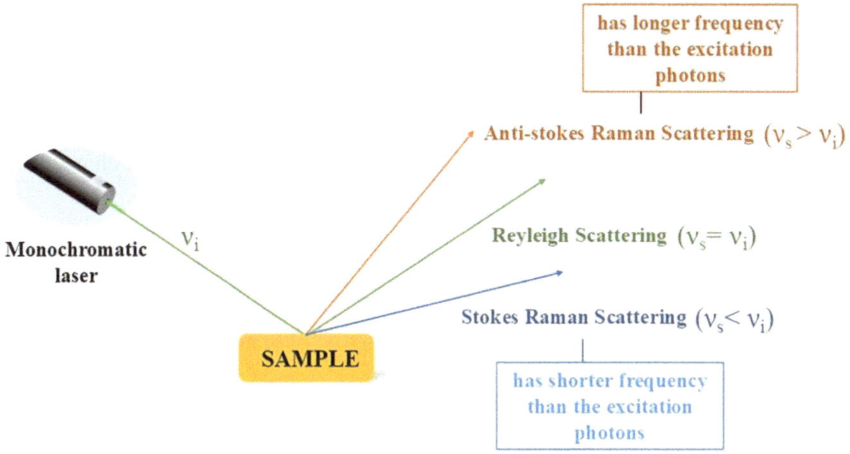


**Fig. 1** Raman scattering

the scattered photon has higher energy than the incident photon, this indicates that the molecule transferred energy to the photon, shifting the light to a shorter wavelength (higher frequency). The incident photon temporarily excites the molecule to a virtual energy state before returning to a different vibrational state, emitting the scattered photon at a different energy which is known as virtual energy state.

Raman spectroscopy is a nondestructive analytical technique that uses Raman scattering to characterize and identify molecular structures and probes the molecular vibrations, rotations, and low-frequency dynamics of a system. Raman spectroscopy has diverse applications across various fields, including biology, materials science, and chemistry [9], where it enables noninvasive, high-resolution chemical analysis and imaging at the microscopic level, without damaging the sample [10]. By utilizing the inelastic scattering of laser light, typically in the visible, near-infrared, or near-ultraviolet spectrum. It stands as a nondestructive and a noninvasive technique for analysis, relying on the inelastic scattering of light to capture the vibrational modes exhibited by molecules [11]. By doing so, it affords a comprehensive insight into the molecular structure, chemical composition, and physical attributes of the sample under examination. Executed as a light scattering approach, this technique requires the sample to be exposed to a potent laser light source. The scattered light consists predominantly of Rayleigh scattering, which has the same wavelength and color as the incident laser light, offering limited valuable insights. A minute proportion of the scattered light, typically around 0.0000001%, constitutes Raman scattering, wherein the light undergoes a shift in wavelengths and colors as a result of interactions with the molecular vibrations of the substance being analyzed. This Raman scattering plays a crucial role in Raman spectroscopy by providing vital insights into the chemical constitution and structure of the sample. Through the examination of the Raman-scattered light, scientists can delve into the molecular signature of the material, facilitating the identification, elucidation, and evaluation of various substances [12]. Hence, it is a powerful scattering technique rooted in the Raman Effect, where a small portion of scattered light exhibits a frequency shift compared to the incident monochromatic radiation. This phenomenon occurs due to the inelastic scattering of incident light as it interacts with vibrating molecules, allowing Raman spectroscopy to probe and characterize molecular vibrations.

A Raman spectrum is a plot of the intensity of Raman scattered light versus its wavelength or wavenumber, featuring a series of peaks that correspond to specific molecular bond vibrations. Each peak in the spectrum represents a unique molecular movement, including individual bond vibrations, such as C–C, C=C, N–O, C–H, and group vibrations which include the breathing mode of benzene ring, Polymer chain vibrations, and Lattice modes (crystalline or molecular lattice vibrations). These peaks provide a molecular fingerprint, allowing researchers to identify and characterize the sample's chemical composition, structure, and properties [9]. The position, shape, and intensity of the peaks can reveal information about the sample's molecular bonds, crystal structure, and functional groups.

The Raman effect is characterized by a fundamental change in the molecular state, where the molecule's final state differs from its initial state due to the interaction with the incident photon. This change results in a shift in the emitted photon's frequency, creating the distinctive Raman spectrum. Notably, Raman scattering is distinct from emission processes such as fluorescence and phosphorescence, where excited molecules release energy and return to their ground state. Instead, Raman scattering involves the inelastic scattering of photons, providing a unique probe into molecular vibrations, rotations, and structure [13].

Raman spectroscopy is a diversified tool employed by biologists and chemists to identify and analyze chemical compounds, functional groups, and complex biomolecules like proteins and DNA [13]. This nondestructive technique rapidly characterizes the chemical composition and nature of diverse samples, including gases, solids, liquids, gels, slurries, and powders [14]. The in vivo measurement capabilities of Raman spectroscopy make it particularly valuable. Raman spectroscopy, alongside infrared absorption spectroscopy, is a leading method for clarifying the structure and characteristics of molecules. By presenting detailed insights into vibrational transitions, these techniques offer comprehensive information about molecular characteristics. Raman scattering has numerous applications in materials science, chemistry, biology, pharmaceutical analysis, and environmental monitoring, enabling molecular identification, structure analysis, bonding analysis, cellular imaging, cancer detection, drug identification, and pollutant detection (Dodo & Fujita, 2022) [14, 15, 16].

## 2 History

Raman scattering generates photons with frequencies that vary based on the light source and the vibrational and rotational characteristics of the molecules being studied. Raman spectroscopy, which operates on the principle of Raman scattering, is widely used by chemists and physicists to analyze materials. Historically, spectra were recorded using mercury lamps and photographic plates, but modern techniques now employ lasers for enhanced precision and efficiency.

In 1930, Sir C.V. Raman received the Nobel Prize in Physics for discovering the Raman effect, also known as Raman scattering. This pioneering work was done in collaboration with his student, K.S. Krishnan, in 1928. Raman spectroscopy is named after the Indian physicist C.V. Raman, who first observed the phenomenon in 1928. He was studying the scattering of light by molecules and discovered that a small portion of the light that is being scattered was shifted in frequency due to interactions with molecular vibrations. This effect, now known as the Raman effect, was a major breakthrough in understanding the behavior of light and molecules [6].

In the early 1930s, Raman and his colleagues developed the first Raman spectrometer, using sunlight and photographic plates to record the spectra. The technique gained popularity in the 1950s and 1960s with the advent of lasers and improved instrumentation.

Over the years, Raman spectroscopy has evolved to include various techniques, such as tip-enhanced Raman spectroscopy (TERS), surface-enhanced Raman spectroscopy (SERS), and coherent anti-Stokes Raman spectroscopy (CARS). Today, Raman spectroscopy is a powerful tool in various fields, including chemistry, biology, materials science, and pharmaceuticals.

## 3 Timeline

In 1928, C.V. Raman discovered the Raman effect, which marked a significant breakthrough in the field of spectroscopy.

During the 1930s, the first Raman spectrometer was developed, paving the way for further advancements in the technique.

In the 1950s and 1960s, the advent of lasers and improved instrumentation greatly increased the popularity and application of Raman spectroscopy.

During 1970s–1980s, surface-enhanced Raman spectroscopy (SERS) and tip-enhanced Raman spectroscopy (TERS) techniques emerged, allowing for enhanced sensitivity and spatial resolution.

Lastly, 1990s–present, coherent anti-Stokes Raman spectroscopy (CARS), stimulated Raman spectroscopy (SRS), and other advanced techniques have been developed, further expanding the capabilities and applications of Raman spectroscopy.

# 4 Types of Raman Spectroscopy

There are a number of advanced types of Raman spectroscopy-

1. Surface-enhanced Raman spectroscopy (SERS): SERS exploits the exceptional optical properties of silver and gold to amplify the Raman signal by up to $10^{11}$ times. This phenomenon relies on the excitation of surface plasmons on a substrate or colloid containing these metals, typically achieved using a laser. The resulting intense electric field surrounding the metal enhances the Raman signal, enabling the detection of molecules at extremely low concentrations. First observed by Martin Fleischmann, this effect was later elucidated by Van Duyne in 1977, who attributed the enhancement to the increased electric field generated by the surface plasmons.
2. Resonance Raman spectroscopy (RRS): RRS enhances specific vibrational modes by aligning the excitation wavelength with the transition of an electron of a molecule or crystal. This selective approach amplifies vibrations linked to the excited electronic state, simplifies complex spectra, and aids in assigning normal modes to observed frequency shifts, making it particularly suitable for studying large molecules like polypeptides. RRS provides valuable insights into molecular structure and dynamics, particularly for biological macromolecules and complex systems [17].
3. Hyper-Raman: Hyper-Raman Spectroscopy is a highly sensitive, nonlinear technique that detects vibrational modes interacting with the second harmonic of the excitation beam. This advanced method requires extremely high-power excitation, enabling the revelation of "silent" vibrational modes that remain inaccessible through conventional Raman spectroscopy. By leveraging Surface-Enhanced Raman Spectroscopy (SERS)-type enhancement, Hyper-Raman Spectroscopy achieves a significant sensitivity boost, allowing for the detection of Weak vibrational signals, low-concentration molecules, and subsurface and interface species [18]
4. Tip-enhanced Raman spectroscopy (TERS): TERS harnesses the power of a metallic-coated atomic force microscopy (AFM) or scanning tunneling microscopy (STM) tip to amplify Raman signals from adjacent molecules. With an unparalleled spatial resolution of approximately 20–30 nm, defined by the tip apex size, TERS has achieved remarkable sensitivity down to the single-molecule level. It offers unparalleled capabilities for bioanalysis and nanoscale chemical imaging, enabling researchers to probe the composition and molecular structure with exceptional precision and accuracy. By merging the strengths of spectroscopy and microscopy, TERS is set to revolutionize our understanding of nanoscale phenomena across biology, chemistry, and materials science [19].
5. Stimulated Raman spectroscopy: A novel spectroscopic technique utilizes two-color pulses to transfer population to a rovibrationally excited state via Raman transition, without electronic resonance. Subsequent two-photon UV ionization enables collection of intra- and intermolecular Raman spectra of clusters, gases, and specific conformations. This technique provides valuable insights into

molecular structure and vibrations, understanding of molecular interactions and dynamics, and has various applications in gas-phase spectroscopy and cluster chemistry, offering a powerful tool for studying molecular properties and behavior. This innovative method advances our understanding of molecular systems [20].

6. Transmission Raman: After being initially overlooked, a groundbreaking technique was rediscovered in 2006, revolutionizing the field of pharmaceutical and medical diagnostics. This revived technique makes it feasible to quickly assay the pharmaceutical dose forms and provides valuable knowledge into turbid materials such as capsules and powders, facilitating noninvasive testing and enhanced diagnostic capabilities. Its rediscovery has paved the way for significant advancements in pharmaceutical science, biomedical research, and materials analysis, transforming the way researchers characterize complex materials and diagnose diseases [21].
7. Spatially offset Raman spectroscopy (SORS): It is a powerful technique that enables noninvasive analysis of turbid materials, such as powders, capsules, and living tissue, by probing beneath their surface. Despite initial investigations by Schrader and Bergmann in 1967, SORS was largely overlooked until its rediscovery in 2006 as a rapid assay method for pharmaceutical dosage forms. Today, SORS holds promise for medical diagnostic applications, offering deep penetration into opaque materials, nondestructive analysis, and rapid sampling capabilities, making it an attractive tool for researchers and clinicians seeking to probe complex biological systems and materials [22]
8. Surface plasmon polariton enhanced Raman scattering (SPPERS): Apertureless metallic conical tips excite molecules in the near-field, suppressing background interference. A laser-generated TM0 mode propagates along the tip without radiation, separating the focal and excitation planes. This design enables precise control and highly localized molecular interactions, ideal for sensitive detection and analysis [23].

## 5 How Does Raman Spectroscopy Work?

The main function of Raman spectroscopy is to analyze variations in a chemical bond's polarizability. Polarizability change describes the distortion of the electron cloud of a molecule caused by light interaction. Variations in polarizability caused by unique energy transitions in chemical bonds result in the generation of Raman active modes. When photons interact with molecules that contain bonds between homonuclear atoms, such as those between carbon, sulfur, and nitrogen atoms, the polarizability of those molecules is changed. These bonding instances result in Raman active spectral bands that are easily visible using Raman spectroscopy.

In general, a Raman spectrometer consists of three major components [24]. These are as follows:

1. Raman laser source
2. Sampling interface
3. Suitable spectrometer

Each of the components is discussed below:

1. Raman laser source: Typically, a source emitting solely at 785 nm is used to record Raman spectra. Because of their high intensity, lasers are almost invariably the sources used in modern Raman spectrometry, allowing for the measurement of Raman scattering with a suitable signal-to-noise ratio. Sources of argon and krypton ions that emit in the blue and green regions of the spectrum have an advantage over other sources because the intensity of Raman scattering varies as the fourth power of the frequency [24].

   Some common laser sources, their wavelength and frequency for Raman spectroscopy are listed in Table 1 (Adapted from [24]).
2. Sampling interface: Sampling interface is the second component of the Raman Spectrometer. A variety of Raman spectrometers with fiber-optic probes are helpful as highly adaptable, flexible sampling interfaces that can be used in conjunction with other relevant sampling chambers, gas, as well as liquid flow cells, along with other related sampling chambers.

   *Liquid samples*: Water is a strong absorber of infrared radiation but a weak Raman scatterer, which makes sample handling in Raman spectroscopy a significant benefit over infrared. This advantage is particularly significant for studies addressing issues with water pollution as well as for inorganic and biological systems.

   *Solid samples*: Solid samples are typically ground to a fine powder and then filled into a tiny cavity to obtain their Raman spectra. Polymers typically do not require sample preprocessing in order to be examined directly.

   *Gas samples*: Glass tubes that are 1–2 cm in diameter and approximately 1 mm thick are typically used to store gases. Additionally, gases can be sealed in tiny capillary tubes [24].
3. Wavelength selector (filter or spectrometer): Usually band-pass filters are used to isolate a single laser beam. A high-quality grating monochromator and notch filter are usually used together when employing dispersive instruments. Additionally, the comparatively weak Raman lines are separated from powerful scattered radiations using holographic notch or edge filters, super notch filters,

**Table 1** Some common laser sources for Raman spectroscopy

| Laser type | Wavelength (nm) | Frequency × $10^7$ ($s^{-1}$) |
|---|---|---|
| Argon ion | 488.0 or 514.5 | 6147 |
| Krypton ion | 530.9 or 647.1 | 5651 |
| Helium/neon | 632.8 | 4740 |
| Diode | 782 or 830 | 3836 |
| Nd/YAG | 1064 | 2819 |

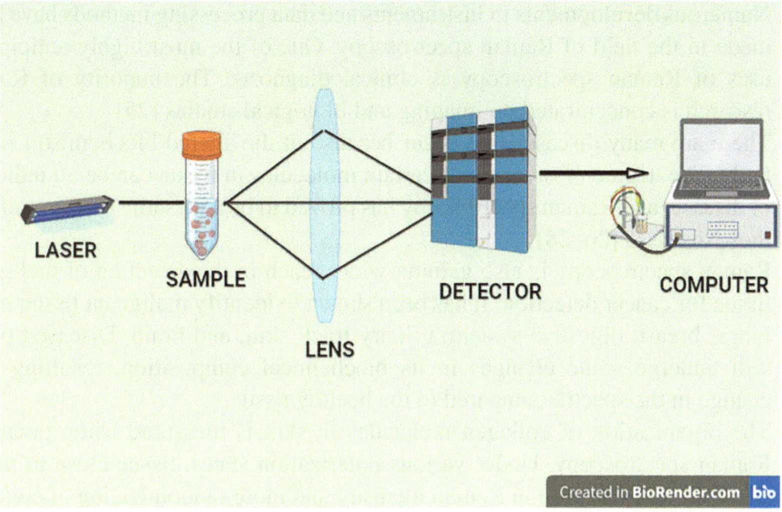


**Fig. 2** Instrumentation of Raman Spectroscopy

rejection filters, double or even triple grating monochromators, and holographic filters.

4. Detector (photodiode array, CCD or PMT): Thermoelectrically cooled photomultiplier tubes and photodiode array detectors were used in early models of dispersive Raman spectrophotometers. More sensitive charge transfer devices (CTDs), such as charge-coupled devices (CCDs) and charge-injection devices (CIDs), have replaced these detectors due to advancements in instrumentation and technology. The detector aids in the detection of the scattered light signal. Modern Raman spectrometers frequently use liquid-crystal display (LCD) array detectors. They are designed to detect extremely weak signals as well as signals at different wavelengths.
5. Computer: A computer equipped with the appropriate software facilitates the production of the final Raman spectroscopy spectrum or graph. The schematic representation of the Instrumentation of Raman Spectroscopy is shown in Fig. 2.

## 6 Applications of Raman Spectroscopy

A wide range of fields, including biology and medicine for analyzing low-frequency DNAs and proteins, chemistry for analyzing the structure and bonding of molecules, and nanotechnology for studying the structure of nanowires, use Raman spectroscopy, which is based on the Raman effect.

1. Numerous developments in instruments and data processing methods have been made in the field of Raman spectroscopy. One of the most highly anticipated uses of Raman spectroscopy is clinical diagnosis. The majority of Raman research is concentrated on imaging and biological studies [25]
2. There are many diseases that occur because of the altered biochemistry in the body. The degree of ordering of certain molecules in tissue can be an indicator of disease, and Raman spectroscopy has proved to be a versatile tool in studying those diseases [26–28]
3. Raman spectroscopy is also gaining wider reach in the detection of malignant tissue for cancer detection. It has been shown to identify malignant tissue of the lungs, breast, digestive system, urinary tract, skin, and brain. Diseased tissue will undergo some changes in its biochemical composition, resulting in a change in the spectra compared to the healthy tissue.
4. The organization of collagen molecules in skin is measured using polarized Raman spectroscopy. Under various polarization states, tissue close to tumor sites had less variation in Raman intensity and more randomization of collagen orientation. This effect was used to classify tumor and healthy skin tissue [26, 28]. For example, Lung cancer results in a distinctive loss of lipids that can be visible in particular Raman bands in the high wavenumber Raman spectrum at around 2850 $cm^{-1}$. In addition to spectral characteristics suggesting alterations in lipids and fatty acids, malignant breast tissue exhibits deviations in Raman bands at 1150 $cm^{-1}$ and 1520 $cm^{-1}$, indicating the presence of carotenoids.
5. Raman spectroscopy can provide in vivo diagnostic results more quickly than conventional diagnostic methods, which may include biopsies and laboratory sample processing. Therefore, the use of Raman scattering techniques is anticipated to significantly improve patient comfort.
6. Since lipids and calcium constitute the majority of plaques in coronary arteries, and Raman signals are highly sensitive to these substances, Raman spectroscopy has also been used to successfully study inflammation and cardiovascular disease. Research provided additional information for the risk management of patients by correlating the biochemical signature of plaques with their severity and stability.
7. Routine endoscopic operations and colonoscopies are two examples of how readily Raman spectroscopy may be integrated into the surgical workflow. It can also be used to evaluate the effectiveness of cancer surgery.
8. Most of the molecules in biological samples are oriented randomly, so the polarizations of excitation and scattered light play a very minimal role in Raman spectroscopy. Whereas, some molecules are highly structured and therefore, under illumination from differently polarized light sources, they scatter differently. In this case, the scattering tensor corresponding to the excitation and detected polarization states will determine the intensity of Raman scattering for each vibrational mode [28, 29].
9. Silk fibres from spiders and silkworms are also studied using Raman spectroscopy. Polarization-dependent differences in the Amide I vibration bands of polarized Raman spectra were caused by the highly organized molecular

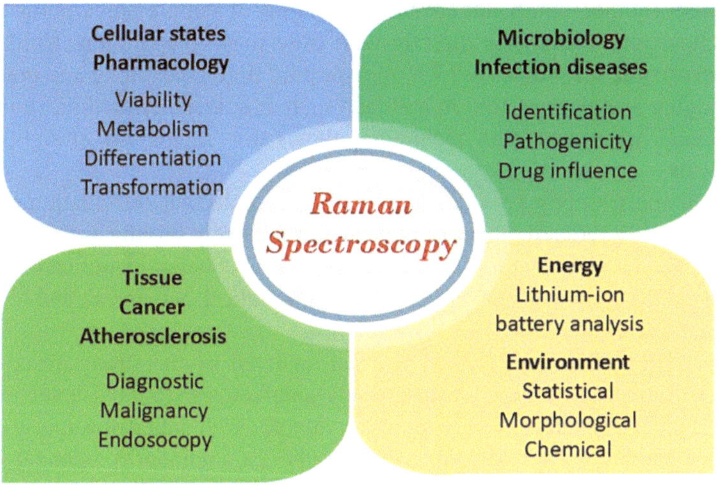


**Fig. 3** Applications of Raman Spectroscopy (adapted from [30])

structure in these fibres. This allows to determine the orientations of β-sheets within the fibres [27, 28].

10. Raman spectroscopy makes use of optical phenomena that precisely show the chemical composition and molecular structures of a material, allowing it to see the chemical responses of molecules in live tissues and cells (Fig. 3). Raman tags have been used to observe small molecules in living cells since they are smaller than bulky fluorophores and can be differentiated from endogenous molecules by molecular vibrations [16]

## 7 Advantages of Raman Spectroscopy

Raman spectroscopy offers numerous advantages, making it a versatile and powerful analytical technique.

1. Raman spectroscopy offers a distinct advantage in material analysis by preserving the sample's integrity, making it an ideal technique for applications where sample preservation is crucial. This nondestructive method helps to analyze delicate materials without damaging or altering them, ensuring the sample remains in its original state. This is particularly valuable in fields such as forensic analysis, where evidence must remain uncontaminated and biological research, where preserving cellular and tissue structures is essential. By maintaining the sample's integrity, Raman spectroscopy facilitates accurate analysis, repeated measurements, and longitudinal studies, making it an indispensable tool in various scientific disciplines

2. It offers another significant advantage in terms of sample preparation, as most samples can be analyzed directly with little to no preparation. Unlike other spectroscopic techniques, which often require time-consuming and potentially damaging sample preparation methods such as solvent extraction, staining, or chemical treatment, this technique allows for rapid and straightforward analysis.
3. Raman spectroscopy provides highly specific molecular information based on the unique vibrational modes of molecules, effectively serving as a "fingerprint" for molecular identification. It is therefore used in material identification, quality control, and chemical analysis across pharmaceuticals, chemical, and biological substances [16].
4. Unlike IR spectroscopy, which is often hindered by the strong absorption of water, Raman scattering remains largely unaffected by the existence of water. This makes Raman spectroscopy exceptionally well-suited for analyzing biological and biomedical samples, where water is a dominant component, such as cell cultures, tissues, or biofluids. As a result, it can be helpful in accurately analyzing the complex biological systems without the need for dehydration or elaborate sample preparation, preserving the sample's natural state and ensuring reliable, high-quality data. This unique benefit has made Raman spectroscopy a valuable tool in biomedical research, enabling insights into biological processes, disease diagnosis, and therapeutic development [9].
5. Raman spectroscopy boasts unparalleled versatility, accommodating a vast array of sample types and forms. Unlike many analytical techniques, Raman spectroscopy can be seamlessly applied to liquids, solids, and gases, making it an indispensable tool for diverse fields. Whether analyzing crystalline solids, aqueous solutions, biological tissues, or atmospheric gases, Raman spectroscopy adapts effortlessly, providing detailed molecular information [6].
6. The resonance Raman effect significantly enhances the sensitivity of Raman spectroscopy by tuning the excitation laser wavelength to match the electronic transitions of a sample. This resonance phenomenon amplifies the Raman scattering signal, allowing for the detection of samples with extremely low concentrations or specific chromophores, such as pigments, dyes, or biomolecules like haemoglobin [16].
7. Raman spectroscopy boasts an expansive spectral range, spanning from low to high wavenumbers, enabling the comprehensive study of various rotational, vibrational, and low-frequency modes. This versatility allows to delve into the molecular structure and bonding of an extensive array of compounds, encompassing both light and heavy molecules.
8. It employs low-energy photons, typically from visible or near-infrared lasers, minimizing the damaging risks of biological samples and enabling quantitative analysis by measuring the intensity of specific peaks in the Raman spectrum, which can be directly correlated with the molecule concentration in the sample [16].

## 8 Challenges

1. The low repeatability of Raman spectroscopy is a significant disadvantage that makes it very difficult to acquire precise quantitative results [31].
2. Insufficient evidence supports the use of Raman spectroscopy to quantify variations in plant composition during the abiotic stress response [32]
3. Raman spectroscopy cannot be used for metals or alloys. Metals are traditionally thought not to produce strong Raman signals. The Raman effect is weak, so the detection requires a sensitive instrument, and Raman signals can also be unreliable due to the rough surfaces of metals, which restricts their usage in practical applications.
4. Assessing low concentrations of a drug is challenging due to the relatively weak Raman effect and low sensitivity.
5. Because biomolecule concentrations in biofluids are very low when compared to tissues and cells, Raman spectroscopy also reveals a significant constraint in the use of biofluids [25].
6. The next challenge is in the case of materials that are hidden beyond a few tens of millimetres. For instance, in the case of strongly scattering media, such as biological tissues, retrieval of Raman signal from a tissue buried under overlying layers noninvasively can be a daunting experimental task [25]
7. Raman scattering is generated using lasers. Since the Raman signal is related to the laser's power, a higher power typically results in a greater signal. High laser strengths can harm samples because every sample has a laser power density threshold that can be exceeded, beyond which structural or chemical changes may occur.
8. Due to extended measurement time and low scattering efficiency, video rate imaging is nearly impossible [33].

## References


1. Settle, F. A. (Ed.). (1997). *Handbook of instrumental techniques for analytical chemistry* (pp. 1–976). Prentice Hall.
2. Lindon, J. C., Tranter, G. E., & Koppenaal, D. W. (Eds.). (2017). *Encyclopedia of spectroscopy and spectrometry* (3rd ed., pp. 469–474). Academic Press.
3. Skoog, D. A., Holler, F. J., & Crouch, S. R. (2019). *Principles of Instrumental Analysis* (7th ed.). Cengage Learning.
4. Raman, C. V., & Krishnan, K. S. (1928). *A new type of secondary radiation. Nature*, 121, 501–502.
5. Willard, H. H., Merritt, L. L., Dean, J. A., & Settle, F. A. (1988). *Instrumental Methods of Analysis* (7th ed.). Wadsworth Publishing Company.
6. Gauglitz, G., & Moore, D. S. (Eds.). (2014). *Handbook of spectroscopy* (Vol. 1). Wiley-VCH.
7. Smith, E., & Dent, G. (2019). Modern Raman Spectroscopy: *A Practical Approach* (2nd ed.). Wiley.

8. McCreery, R. L. (2005). *Raman spectroscopy for chemical analysis* (Vol. 225). Wiley-Interscience.
9. Gala, U., & Chauhan, H. (2015). Principles and applications of Raman spectroscopy in pharmaceutical drug discovery and development. *Expert Opinion on Drug Discovery, 10*(2), 187–206.
10. Gerrard, D. L., & Bowley, H. J. (1988). Raman spectroscopy. *Analytical Chemistry, 60*(12), 368R–377R.
11. Mulvaney, S. P., & Keating, C. D. (2000). Raman spectroscopy. *Analytical Chemistry, 72*(12), 145R–157R.
12. Qu, C., Li, Y., Du, S., Geng, Y., Su, M., & Liu, H. (2022). Raman spectroscopy for rapid fingerprint analysis of meat quality and security: Principles, progress, and prospects. *Food Research International, 161*, 111805.
13. Butler, H. J., Ashton, L., Bird, B., Cinque, G., Curtis, K., Dorney, K., Esmonde-White, N. J., Fullwood, B., Gardner, P. L., Martin-Hirsch, M. J., Walsh, M. R., McAinsh, N. S., & Martin, F. L. (2016). Using Raman spectroscopy to characterize biological materials. *Journal of the American Chemical Society, 138*(4), 664–687.
14. Rebrosova, K., Samek, O., Kizovsky, M., Bernatova, S., Hola, V., & Ruzicka, F. (2022). Raman spectroscopy—A novel method for identification and characterization of microbes on a single-cell level in clinical settings. *Frontiers in Cellular and Infection Microbiology, 12*, 866463.
15. Matuszczyk, J. C., Zijlstra, G., Ede, D., Ghaffari, N., Yuh, J., & Brivio, V. (2023). Raman spectroscopy provides valuable process insights for cell-derived and cellular products. *Current Opinion in Biotechnology, 81*, 102937.
16. Dodo, K., Fujita, K., & Sodeoka, M. (2022). Raman spectroscopy for chemical biology research. *Journal of the American Chemical Society, 144*(43), 19651–19667.
17. Han, X. X., Rodriguez, R. S., & Haynes, C. L. (2021). Surface-enhanced Raman spectroscopy. *Nature Reviews Methods Primers, 1*, 87.
18. Hehlen, B., Amouri, A., Al-Zein, A., & Khemakhem, H. (2014). Hyper-Raman and Raman scattering from the polar modes of PbMg1/3Nb2/3O3. *Journal of Physics: Condensed Matter, 26*(1), 015401.
19. Schultz, J. F. (2023). Tip-enhanced Raman scattering for atomic-scale spectroscopy and imaging. *Nature Reviews Physics, 5*, 1.
20. Xu, Z., Oguchi, K., Taguchi, Y., Sano, Y., Miyawaki, Y., Cheon, D., Katoh, K., & Ozeki, Y. (2022). Stimulated Raman scattering spectroscopy with quantum-enhanced balanced detection. *Optics Express, 30*(11), 18589–18598.
21. Sparén, O., & Svensson, O. (2017). *Transmission Raman: Methods and applications*. Medicine, Materials Science, Chemistry.
22. Mosca, S., Dey, P., Salimi, M., Gardner, B., Palombo, N., Stone, N., & Matousek, P. (2021). Spatially offset Raman spectroscopy—How deep? *Analytical Chemistry, 93*(17), 6755–6762.
23. Mueller, N. S., Pfitzner, E., Okamura, Y., Gordeev, G., Kusch, P., Lange, H., Heberle, J., Schulz, F., & Reich, S. (2021). Surface-enhanced Raman scattering and surface-enhanced infrared absorption by plasmon polaritons in three-dimensional nanoparticle supercrystals. *ACS Nano, 15*(3), 5523–5533.
24. Orlando, A., Franceschini, F., Muscas, C., Pidkova, S., Bartoli, M., Rovere, M., & Tagliaferro, A. (2021). A comprehensive review on Raman spectroscopy applications. *Chemosensors, 9*(9), 262.
25. Kuhar, N., Sil, S., Verma, T., & Umapathy, S. (2018). Challenges in application of Raman spectroscopy to biology and materials. *RSC Advances, 8*(46), 25888–25908.
26. Ly, E., Durlach, A., Antonicelli, F., Bernard, P., Manfait, M., & Piot, O. (2010). Probing tumor and peritumoral tissues in superficial and nodular basal cell carcinoma using polarized Raman microspectroscopy. *Experimental Dermatology, 19*(1), 68–73.
27. Rousseau, M. E., Lefevre, T., Beaulieu, L., Asakura, T., & Pézolet, M. (2004). Study of protein conformation and orientation in silkworm and spider silk fibers using Raman microspectroscopy. *Biomacromolecules, 5*(6), 2247–2257.

28. Shipp, D. W., Sinjab, F., & Notingher, I. (2017). Raman spectroscopy: Techniques and applications in the life sciences. *Advances in Optics and Photonics, 9*(2), 315–428.
29. Turrell, G., & Corset, J. (1996). *Raman microscopy: Developments and applications.* Academic Press.
30. Eberhardt, K., Stiebing, C., Matthäus, C., Schmitt, M., & Popp, J. (2015). Advantages and limitations of Raman spectroscopy for molecular diagnostics: An update. *Expert Review of Molecular Diagnostics*, 15(6), 773–787.
31. Baranska, M., Schütze, W., & Schulz, H. (2006). Determination of lycopene and beta-carotene content in tomato fruits and related products: Comparison of FT-Raman, ATR-IR, and NIR spectroscopy. *Analytical Chemistry, 78*(19), 8456–8461.
32. Don, D., & Zhao, C. (2017). Limitations and challenges of using Raman spectroscopy to detect the abiotic plant stress response. *Proceedings of the National Academy of Sciences of the United States of America, 114*(28), E5486–E5487.
33. Katharina, E., Clara, S., Christian, M., Michael, S., & Popp, J. (2015). Advantages and limitations of Raman spectroscopy for molecular diagnostics: An update. *Expert Review of Molecular Diagnostics, 15*(1), 1–15.

# Basics of Photoacoustic Techniques for Breast Cancer Diagnosis

**Anutee Dolley, Kabyashruti J. Kashyap, and Nima D. Namsa**

**Abstract** Breast cancer continues to be the foremost threat to women's health globally. According to the 2025 cancer statistics, breast cancer is the leading cause of cancer-related mortality among women, highlighting the urgent need for more effective diagnostic and therapeutic strategies. While early detection and advanced treatment options have significantly improved patient survival rates, the complexity and heterogeneity of breast cancer still present significant challenges in personalizing treatment. This underscores the necessity for innovative approaches in precision diagnosis. Currently, X-ray mammogram is the only modality for mass screening of asymptomatic women. However, it has decreased sensitivity in radiographically dense breasts, which is also associated with a higher risk for breast cancer. Photoacoustic or photoacoustic imaging (OAI) is an emerging field that enables deep tissue imaging of optical contrast at ultrasonically defined spatial resolution, which is much higher than that can be achieved in purely optical imaging modalities. By integrating ultrasound guidance, this technique enables precise localization and characterization of tumors. Because of high optical absorption from hemoglobin molecules, photoacoustic imaging can map out hemo-distribution and dynamics in breast tissue and identify malignant lesions based on tumor-associated angiogenesis and hypoxia. Photoacoustic imaging employs nonionizing, laser light to create thermoelastic expansion in tissues and detect the resulting ultrasonic emission. By combining high optical contrast capabilities with the high spatial resolution and anatomic detail of gray-scale ultrasound, OAI offers unique opportunities for visualizing the biological function of tissues in vivo. Over the past decade, human breast applications of OAI, including benign/malignant mass differentiation, distinguishing cancer molecular subtype, and predicting metastatic potential, have significantly increased. In this chapter, we discuss the current state

A. Dolley · K. J. Kashyap · N. D. Namsa (✉)
Department of Molecular Biology and Biotechnology, Tezpur University, Tezpur, Assam, India
e-mail: namsa@tezu.ernet.in



N. Mazumder et al. (eds.), *Advanced Biophysical Techniques in Biosciences*, Lecture Notes in Nanoscale Science and Technology 7,
https://doi.org/10.1007/978-3-031-94551-9_15

of optoacoustic breast imaging, as well as future opportunities and clinical application trends for management and treatment of breast cancer.

**Keywords** Photoacoustic imaging · Breast cancer · Mammogram · Thermal imaging · Tumor identification

## 1 Introduction

Thorough characterization of cancer, including its structural, functional, and metabolic attributes, is crucial for developing effective therapeutic approaches and reducing the risk of recurrence. To achieve this, there is a need for advanced imaging modalities that can provide timely identification of tumors and comprehensive insights into their characteristics [1–3]. Currently, medical practitioners utilize various imaging techniques such as ultrasound imaging (USI), magnetic resonance imaging (MRI), X-ray computed tomography (CT), positron emission tomography (PET), and single photon emission computed tomography (SPECT), photoacoustic imaging (PAI) for tumor detection [4–6] (Table 1). These modalities can be enhanced by targeted contrast agents, enabling the acquisition of molecular data related to malignant tumors (Rowe and Pomper, 2022). Microscopic optical imaging techniques offer superior resolution compared to other modalities, allowing for the detection of smaller quantities of cancer cells within each imaging voxel [7, 8]. However, conventional diffusive optical imaging methods like diffuse optical tomography (DOT) have limited resolution (~5 mm) [9, 10]. Addressing the need for high-resolution imaging with microscale resolution and sufficient penetration depth, photoacoustic imaging (PAI) has emerged as a promising modality [11]. PAI provides high optical contrast images with the capability to visualize tumor locations within deep tissue. It also offers insights into tumor vasculature and facilitates monitoring of angiogenesis. PAI can provide high-resolution and high-contrast information on hemoglobin oxygen saturation without the use of exogenous contrast agents, which is advantageous compared to alternative techniques like blood oxygen level dependent-MRI and PET for assessing tumor hypoxia [12]. Another benefit of PAI is its compatibility with commonly used USI techniques. Integrating PAI with USI allows for the simultaneous acquisition of anatomical and functional information in tumors [13]. Empirical investigations on human breast tissue have demonstrated that ultrasonic imaging effectively illustrates anatomical features of ductal carcinoma, while PAI reveals the dispersed pattern of vascularization associated with the condition [14, 15]. The development of targeted contrast agents, such as gold nanoparticles (AuNPs), has opened up opportunities for in vivo molecular PAI [16, 17]. This technique enables accurate and precise tumor identification with high sensitivity and specificity. Furthermore, combining PAI with other imaging modalities shows potential in guiding cancer treatment. In summary, the advancement of imaging modalities that can provide comprehensive information about tumors is crucial for effective cancer management. PAI has emerged as a promising

**Table 1** Comparative Analysis of Cancer Imaging Technologies: Features, Patient Experience, and Clinical Impact

| Feature | PAI | USI | MRI | CT | PET | SPECT |
|---|---|---|---|---|---|---|
| Imaging principle | Combines optical imaging and ultrasound | High-frequency sound waves | Magnetic fields and radio waves | Ionizing radiation | Radioactive tracers and gamma rays | Radioactive tracers and gamma rays |
| Resolution | High (~50 μm) | Moderate (~100–300 μm) | Very high (~10–100 μm) | High (~50–100 μm) | Moderate (~1–3 mm) | Moderate (~1–3 mm) |
| Depth of penetration | Moderate (~5 cm) | High (~15–20 cm) | Moderate to high (~10–20 cm) | Moderate to high (~5–20 cm) | Whole body scanning possible | Whole body scanning possible |
| Sensitivity to molecular | High (optical absorption contrast) | Low | High (excellent for soft tissue contrast) | Moderate | Very high (metabolic activity) | High (functional activity) |
| Contrast agents | Endogenous or exogenous | Typically none | Gadolinium-based contrast agents | Iodine-based contrast agents | Radioactive tracers (e.g., FDG) | Radioactive tracers (e.g., technetium 99 m) |
| 3D imaging capability | Yes, with tomographic techniques | Limited (requires multiple views) | Yes | Yes | Yes | Yes |
| Functional imaging | Yes (e.g., blood oxygenation) | No | Yes (e.g., fMRI for brain activity) | No | Yes (e.g., glucose metabolism) | Yes (e.g., blood flow, tissue function) |
| Radiation exposure | None | None | None | Yes (moderate to high) | Yes (low to moderate) | Yes (low to moderate) |
| Portability | Potentially portable | Highly portable | Not portable | Not portable | Not portable | Not portable |
| Cost | Moderate to high (emerging technology) | Low | High | Moderate to high | Very high | High |

(continued)

**Table 1** (continued)

| Feature | PAI | USI | MRI | CT | PET | SPECT |
|---|---|---|---|---|---|---|
| Time required for scan | Short (minutes) | Very short (seconds) | Moderate (30–60 min) | Short (minutes) | Long (60–90 min) | Long (60–90 min) |
| Availability | Limited (research-focused) | Widely available | Widely available | Widely available | Limited (mostly in major hospitals) | Limited (mostly in major hospitals) |
| Patient experience | Noninvasive, no discomfort | Noninvasive, minimal discomfort | Noninvasive, enclosed space may cause discomfort | Noninvasive, contrast agent may cause allergy | Invasive (injection of radioactive tracer) | Invasive (injection of radioactive tracer) |
| Clinical applications | Early-stage cancer detection, tumor angiogenesis monitoring | Soft tissue, fetal imaging | Soft tissue contrast, brain imaging | Detailed imaging of structures | Cancer detection, brain disorders, cardiac studies | Cancer detection, cardiac studies |
| Advantages over others | High resolution with molecular contrast, no ionizing radiation | Real-time imaging, widely available, no radiation | Superior soft tissue contrast, no radiation | Detailed structural imaging | Functional imaging at molecular level | Functional imaging with longer half-life tracers |
| Disadvantages | Limited penetration depth, emerging technology | Lower resolution, limited tissue contrast | Expensive, claustrophobic experience, contraindicated with metal implants | Ionizing radiation, potential allergic reactions to contrast agents | Expensive, involves radiation exposure | Expensive, involves radiation exposure |

technique, offering high-resolution and high-contrast imaging with the ability to visualize tumors in deep tissue [18]. PAI also provides insights into tumor vasculature and oxygen saturation without the need for exogenous contrast agents [19]. Integration with USI allows for the simultaneous acquisition of anatomical and functional information [20, 21]. The development of targeted contrast agents further enhances the potential of PAI for molecular imaging. By combining PAI with other modalities, there are promising prospects for improving cancer treatment guidance. The first significant attempt to generate sound from light was described by Alexander Graham Bell [22], stimulating various significant events in the development of the technology, as presented in the timeline (Fig. 1). In vivo studies of optoacoustic imaging (OAI) were made possible by advances in methodology and hardware in the 1970s. In OAI, near-infrared, nonionizing nanosecond laser light pulses induce thermoelastic expansion in tissues generating ultrasonic waves, which are captured by ultrasound transducers [23]. Multispectral excitation and unmixing can provide parametric maps of endogenous molecules such as oxyhemoglobin ($HbO_2$), deoxyhemoglobin (dHb), melanin, and lipids that absorb light and generate acoustic waves, which can then be fused with anatomic images to allow scalable, relative real-time molecular assessment in tissues [24–26]. This chapter aims to provide a concise overview of the extensive research conducted in the field of PAI, specifically focusing on its utility in breast cancer detection and therapy.

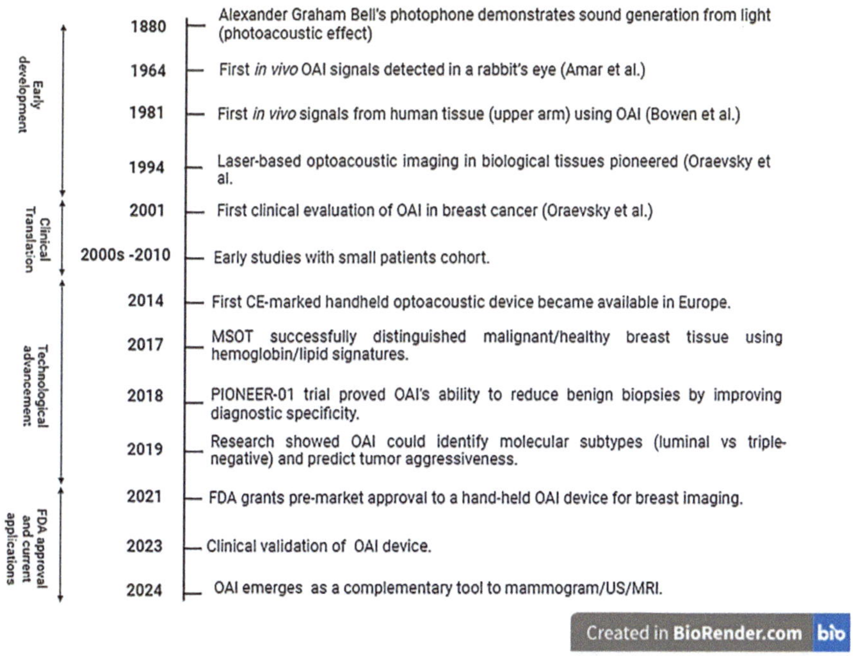


**Fig. 1** A history of optoacoustic imaging. *OAI* Optoacoustic imaging, *CE* Conformité Européenne

## 2 Components of Photoacoustic-Based Techniques

A photoacoustic system uses the principle that when a sample absorbs the incident photons, the electrons get excited and release energy as heat (Fig. 2). The released heat causes thermal expansion, which eventually leads to localized pressure waves that can be measured as acoustic waves [27]. A photoacoustic system consists of the following components.

- A light source—the sample is generally irradiated with a pulsed or modulated laser with a wavelength adjusted according to the tissue sample studied [28].
- Delivery system—the light beam is guided toward the sample through an optic fibre, such that the beam reaches the surface of the sample as spot of size approximately 1 cm [29].
- The sample absorbs the incident light through the presence of chromophores such as hemoglobin and melanin. The tissue then undergoes thermal expansion, which leads to the generation of acoustic waves [30].

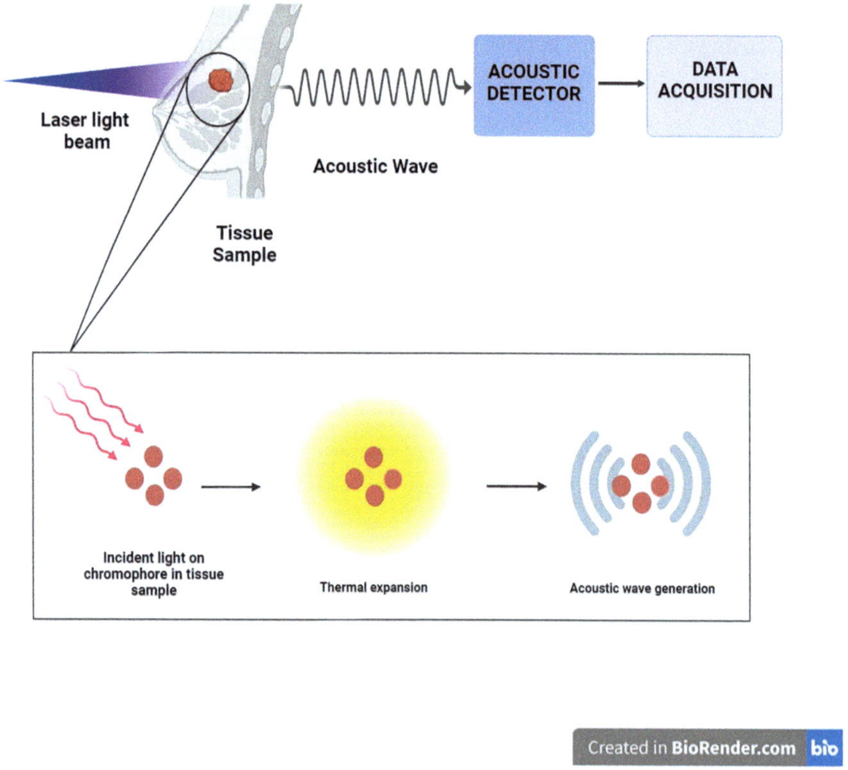


Fig. 2 Schematic diagram of photoacoustic imaging system

- A photoacoustic cell where the photoacoustic effect occurs; this cell is important for wave propagation and resonance. This cell is filled with water, which serves as an acoustic coupling medium. Water permits the laser light beam to pass through with minimal energy loss and allows optimal acoustic wave generation. It acts as a passive transmission medium without hampering thermal expansion [30]. In some systems, the cell is mounted on a vibration isolation stand to minimize or eliminate vibration from the surrounding environment that will reduce the noise [31].
- In traditional photoacoustic systems (PAS), microphones and pressure sensors are used for detecting the sound waves, but ultrasound transducers can also be integrated with PAS, called PAI, where sound waves are converted to create detailed images in breast cancer detection using PAI, positioning the ultrasound transducers near or around the breast tissue. Ideally, there are three positions—spherical, cylindrical, and planar—but ideal geometries are difficult to achieve during diagnosis; hence, modifications are made, like hemispherical or curved for spherical, ring-shaped for cylindrical, and linear or small planar for planar [32].
- The piezoelectric elements collect photoacoustic volumetric data. The detection elements are optimized for high sensitivity and signal-to-noise ratio to ensure accurate detection [30].
- Data acquisition and analysis of the detected sound waves. The acquisition system transmits the signals to a computer that generated images of the sample using advanced algorithms like Fourier transformation, back-projection, etc. [31].

# 3 PAI Contrast Agents

There exist two prevalent categories of contrast agents, namely endogenous contrast agents and exogenous agents, which provide consistent photoacoustic signals for the purpose of imaging. The different types of contrast agents are provided in Table 2.

## *3.1 Endogenous Contrast*

PAI utilizes the inherent presence of diverse endogenous chromophores in bodily tissues, which possess distinct absorption spectra. These chromophores include melanin, water, fat, oxyhemoglobin ($Hb0_2$), and deoxyhemoglobin (Hb). Numerous diseases are associated with alterations in tissue composition, shown by the emergence of neovasculature during cancer progression and the accumulation of vascular fat within atherosclerotic plaques [33, 34]. PAI possesses the capability to visually represent and measure these alterations in contrast to the adjacent healthy tissue. The utilization of endogenous contrast as the sole basis for PAI has been

**Table 2** A detailed information about the different categories and subcategories of PAI contrast agents, including their size range, specific applications, advantages, and disadvantages

| Category | Subcategory | Properties | Examples | Key Points | Size range | Applications | Advantages | Disadvantages |
|---|---|---|---|---|---|---|---|---|
| Endogenous contrast | Melanin | Naturally present in tissues, distinct absorption spectra | – | Utilizes naturally occurring chromophores for imaging | – | Skin and eye imaging, melanoma detection | Noninvasive, no need for external agents | Limited to tissues with high melanin concentration |
| | Water | Absorbs in specific wavelengths, useful for visualizing tissue composition | – | Helps in detecting changes in tissue composition | – | Tissue hydration analysis, edema detection | Abundant in tissues, provides hydration information | Low contrast, limited spectral range |
| | Fat | Different absorption spectra helps in detecting tissue abnormalities | – | Useful for disease diagnosis through tissue composition analysis | – | Atherosclerosis, obesity research | Indicates fat accumulation and distribution | Limited to fatty tissues |
| | Oxyhemoglobin (HbO2) | Reflects oxygen levels in blood, useful in disease diagnosis | – | Provides information on oxygenated blood levels | – | Cancer detection, cardiovascular diseases | Noninvasive, indicates blood oxygenation | Requires accurate calibration |
| | Deoxyhemoglobin (Hb) | Provides information on deoxygenated blood levels | – | Indicates levels of deoxygenated blood | – | Tumor hypoxia, brain imaging | Noninvasive, provides metabolic information | Requires accurate calibration |

| | | | | | | | | |
|---|---|---|---|---|---|---|---|---|
| Exogenous contrast | Nontargeted agents | Tend to extravasate from the bloodstream due to enhanced permeability | Indocyanine green (ICG), Evans blue (EB), methylene blue (MB) | Enhances PAI by using external agents that extravasate from the bloodstream | <10 nm | Tumor imaging, vascular studies | Easily available, approved for clinical use | Nonspecific distribution |
| | Targeted agents | Specifically interact with molecular processes, precise insights into cellular activities | Trastuzumab-linked. dyes, caspase-binder-linked dyes | Offers specific insights into molecular and cellular processes | <100 nm | Molecular imaging, targeted therapy | High specificity, targeted delivery | Potential toxicity, complex synthesis |
| Small Molecule Dyes | | Absorption within the NIR range, high conversion of absorbed energy into PAI signal | Activatable probes with NIR chromophores, MMPSense 680 | Suitable for NIR wavelength PAI, often used in optical fluorescence imaging. | <1 nm | Cancer imaging, apoptosis detection | High resolution, biocompatible | Photobleaching, short circulation time |
| | Biocompatible, efficient elimination from the body susceptible to photobleaching | PEBBLE nanoparticles embedded with ICG | Biocompatible and efficiently eliminated from the body. | <100 nm | Tumor imaging, molecular targeting | Enhanced stability, prolonged circulation | Complex manufacturing, potential toxicity | |

(continued)

**Table 2** (continued)

| Category | Subcategory | Properties | Examples | Key Points | Size range | Applications | Advantages | Disadvantages |
|---|---|---|---|---|---|---|---|---|
| Nanoparticles | Plasmonic nanoparticles | Composed of noble metals, utilize SPR, high optical absorption | Gold nanoshells, nanorods, nanocages | Utilizes surface plasmon resonance for high optical absorption. | 10–200 nm | Cancer imaging, drug delivery | High absorption, tunable properties | Potential long-term toxicity, deformation under irradiation |
| | Nonplasmonic nanoparticles | High light absorption without SPR, often aggregated dye molecules | Single-walled carbon nanotubes (SWNTs), quantum dots, copper sulfide nanoparticles | High light absorption capabilities, often used in molecular | 2–10 nm | Molecular imaging, theranostics | Wide absorption spectrum, prolonged circulation | Potential toxicity, photobleaching |
| Multimodality imaging agents | | Detected by multiple imaging modalities, enhancing diagnostic capabilities | Nanoparticles with plasmonic and superparamagnetic properties, gold nanorods with 125-I | Enhances contrast for PAI while being detectable by other imaging modalities | 10–200 nm | Combined MRI, PAI, and other modalities | Combined MRI, PAI, and other modalities | Complex synthesis, potential regulatory hurdles |

extensively investigated in the context of diverse medical conditions, such as atherosclerosis, breast cancer, and prostate cancer [35]. Nevertheless, the knowledge of absorption spectra for many tissues is incomplete. The utilization of multi-wavelength PAI enables the simultaneous visualization of several tissue components within a single image. However, it is important to note that the incorporation of a tunable pulsed laser source in the system leads to a substantial rise in expenses. The utilization of exogenous molecular imaging agents is occasionally necessary to enhance contrast in the NIR range due to the presence of tissue components with closely resembling absorption spectra. In addition, the utilization of exogenous imaging agents has the potential to enhance the imaging of tissues located at greater depths [36, 37].

## *3.2 Exogenous Contrast*

The utilization of exogenous imaging agents has the potential to significantly augment the contrast of PAI. Nontargeted imaging agents, which exhibit a tendency to extravasate from the bloodstream owing to augmented vascular permeability and the enhanced permeability and retention (EPR) effect in tumorous tissue, can be employed. Alternatively, targeted imaging agents, designed to specifically interact with a particular molecular process such as a receptor, protein, or enzyme, can also be utilized. One of the key benefits associated with targeted imaging agents lies in their ability to offer precise and specific insights into molecular or cellular processes occurring within the tissue [38–40]. A wide range of imaging agents exists that can be employed for the purpose of PAI. The choice of an imaging agent is contingent upon its intended application, necessitating a comprehensive evaluation of several key attributes, including but not limited to size, shape, toxicity, stability, surface chemistry, and targeting moieties. Apart from a limited number of imaging dyes that have obtained approval for human usage, namely Indocyanine Green (ICG), Evans blue (EB), and methylene blue (MB), the remaining imaging agents have not yet received approval and have solely been assessed in phantoms or animal models. Currently, there is limited available information regarding the required mass range for the effective utilization of various imaging agents in the context of PAI research. The matter at hand is of significant importance, as it necessitates the evaluation of the toxicity levels associated with the imaging agents in relation to the quantity required for injection in order to conduct a specific study. It is probable that PAI studies will require higher masses in comparison to highly sensitive techniques like PET and SPECT. The determination of mass ranges must be evaluated for each individual chemical and its respective applications [41, 42].

# 4 PAI for Breast Cancer Imaging and Diagnosis

The precise therapy of cancer is predicated upon the establishment of an accurate diagnosis of malignancies. The clinical treatment of early tumors still presents a significant challenge in terms of enhancing noninvasive imaging and localization. PAI has the capability to identify tumors in their early stages, exhibiting a significant depth of penetration. Furthermore, it can effectively monitor the angiogenesis of early tumor micron levels with remarkable resolution, while also rapidly outlining the boundaries of the tumor [43, 44]. In recent years, advancements in the field of cancer diagnosis have led to improvements in the early detection of various types of cancer. These improvements have been achieved through the optimization of PAI systems, the development of innovative contrast agents, and the integration of multimodal pictures with other imaging methods. This part is dedicated to discussing the cancer models that are considered to be the most representative, with the aim of introducing the application of PAI in their diagnostic processes.

Breast cancer is among the most prevalent and consequential forms of cancer affecting women globally, resulting in significant rates of both mortality and morbidity. In the year 2020, a cumulative number of 2.3 million women received a diagnosis of breast cancer, accounting for approximately 11.7% of the total incidence of new cancer cases. Breast cancer is characterized by its superficial localization and the presence of dense vascular networks [45]. Currently, X-ray mammography is widely regarded as the primary modality for the clinical identification of breast cancer. Nevertheless, the use of X-ray mammography may lead to a reduction in sensitivity for detecting breast cancer due to factors such as the superficial placement of tumors and the presence of thick blood arteries, as well as concerns related to the use of ionizing radiation [46].

In order to enhance the detection sensitivity in women with dense breasts, ultrasonography has been utilized as an adjunctive tool in addressing this issue. Notwithstanding the widespread use of this screening and diagnostic method, the issue of specificity remains a formidable obstacle in enhancing the early detection of breast cancer. The issue of false positive imaging in breast cancer has garnered attention and generated controversy within the realm of clinical practice. One specific instance of concern involves the occurrence of a false positive 18F-FDG PET/CT following administration of the COVID-19 vaccine [47, 48]. Multiple lines of evidence substantiate the assertion that PAI possesses the capability to mitigate false positive results through the discrimination of benign and malignant masses [49]. PAI has the potential to address certain challenges associated with morphologic overlap hence facilitating the differentiation between benign and malignant tumors. Butler et al. [49] conducted a study which demonstrates that the utilization of PAI has the potential to decrease the breast imaging reporting and data system category in the detection of benign masses. Furthermore, the study indicates that PAI exhibits enhanced capabilities in detecting malignant masses when compared to gray scale ultrasound. These findings provide further robust evidence supporting the aforementioned aspect. The enhancement of detection accuracy in angiogenesis can be successfully achieved in the diagnosis of breast cancer by directing attention toward the alteration in microvessel density resulting from the emergence and proliferation of new blood capillaries from preexisting ones [50]. The authors, Li et al. [51], have developed a MSOM that is well-suited for imaging breast cancer. The initial findings of the study showcased the in vivo visualization of variations in optical contrast, achieving a remarkable level of resolution below 50 μm throughout the entirety of the tumor region. The application of multispectral spectral unmixing enabled the researchers to ascertain the hitherto unidentified oxygenation state, vascular morphology, and perfusion characteristics across various types of breast cancer.

The mammography technique utilizing PAI is sometimes referred to as photoacoustic mammography (PAMMG) [52]. A more extensive and advanced analysis of morphological data can be achieved by utilizing the PAMMG technique, which leverages an improved vision field provided by a hemispherical-shaped detector array (HDA) during the breast cancer screening procedure [53]. The Precision Analysis of PAMMG possesses the capability to acquire three-dimensional reconstruction pictures of nearly isotropic blood vessels and conduct in-depth analysis of the morphology and function of blood vessels associated with breast cancer (Kruger et al., 2013). In recent years, numerous researchers have exhaustively evaluated breast cancer as a highly promising advancement in the field of PAI for clinical translation [54]. Toi et al. [55] conducted a study whereby they implemented a PAMMG system equipped with HDA. This system was utilized to visualize the intricate vasculature, microvascular branching patterns, and distinct morphological features of blood vessels in the breast tissue of 30 patients diagnosed with either a benign (fibroadenoma) or malignant tumor [55]. The researchers identified morphological abnormalities in the peritumoral blood vessels, such as the presence of centripetal OAI signal and disruptions or constriction of the vessel signal. Additionally, tumor signals were detected within the breast cancer tissue. Unexpectedly, the OAI system demonstrated the capability to assess alterations in vascular morphology and physiological characteristics caused by a chemotherapeutic agent. Yamaga et al. [56] employed 3D PAI technology in conjunction with HDA to capture visual representations of the anatomical structures within the vascular network of the breasts in a cohort of 22 patients. The PAMMG was utilized to identify vessel branching points (VBPs) and assess disparities between healthy and pathological breasts. It was observed that the presence of a tumor in the breast corresponds to a greater quantity of VBPs compared to the normal breast. Consequently, VBPs have emerged as a promising biomarker for primary breast cancer. The PAMMG system demonstrates a significant advancement by effectively addressing the limited view issue commonly seen in 2D systems. Moreover, it has the capability to capture high-resolution images of the vascular network with a range extending up to 7 mm. The researchers, Lin et al. [57, 58], developed a label-free photoacoustic computed tomography (PACT) system known as single-breath-hold PACT (SBH-PACT). This technique offers a comparable level of contrast to enhanced MRI, but with the added benefit of a greater temporal-spatial resolution. The SBH-PACT technique demonstrated the ability to provide comprehensive visualization of vascular structures over the entire breast in just 15 s. Notably, this technique successfully visualized vessels as small as 258 μm in diameter. The study involved the implementation of SBH-PACT on three patients diagnosed with breast cancer at various stages of their neoadjuvant chemotherapy (NAC) treatment: prior to NAC, during NAC, and following NAC. SBH-PACT imaging has the potential to validate the response of NAC by enabling visualization of vascular density and distribution, tumor size, and vascular heterogeneity. This imaging technique can provide histopathological registration data. Furthermore, the aforementioned findings illustrate the growing potential of PAI as a precise methodology for assessing the advancement of breast cancer and the impact of therapy by identifying systemic pathophysiological signals linked to breast cancer [57, 58]. The surveillance and control of breast cancer metastases pose significant obstacles in the realm of clinical practice. The involvement of lymph nodes is a significant mechanism in the process of metastasis. The sentinel lymph node (SLN) is typically the initial site of breast cancer metastases. Therefore, the identification of SLNs is of utmost importance in the therapeutic approach to breast cancer [59]. During surgical procedures, the exenteration of the SLN may not be required if there is no apparent evidence of metastasis, as determined using precise diagnostic methods. However, the ability to distinguish metastatic SLNs from normal or inflamed ones remains a significant obstacle in the field of clinical surgery. In their study, Dai et al. presented the utilization of a nanoparticle named 5 K-HA-HPPS as a contrast agent for PAI. This nanoparticle was loaded with the fluorescent dye DiR-BOA and employed a dual-targeting approach using hyaluronic acid for imaging sentinel lymph nodes in breast cancer [60]. The nanoparticle paired with PAM enabled the acquisition of distinct photoacoustic signals from metastatic, normal, and inflammatory SLNs. The findings of this study provide evidence that the utilization of a combination of contrast agent and PAM can be a successful approach in accurately detecting malignant SLNs and subsequently guiding their surgical excision. In their study, Garcia Uribe et al. devised a novel imaging system that combined photoacoustic and USI modalities. This integrated system was utilized for the purpose of noninvasively detecting SLNs with the aid of MB dye [61]. Their research could potentially offer valuable assistance in the guidance of percutaneous needle biopsies and axillary staging procedures for breast cancer.

## 5 Spectroscopy Photoacoustic Imaging

Spectroscopy photoacoustic imaging combines spectroscopic analysis along with photoacoustic imaging. The optical absorption spectra of chromophores in tissue is utilized by spectroscopy photoacoustic imaging to provide spatially resolved information of tissue concentration [27]. It leverages the absorption spectra of tissue chromophores (e.g., oxyhemoglobin ($HbO_2$), deoxyhemoglobin (Hb), lipids, and exogenous agents like nanoparticles) within the near infrared (NIR) window (typically 650–900 nm) [62]. The multiple wavelengths are obtained by a specific tuning system which is based on fast mechanical rotation of the optical parametric oscillator crystal. This system allows changing the wavelength to any value between 650 and 900 nm on a per pulse basis. This way, it only takes milliseconds to acquire the total volumetric multispectral data at different wavelengths [27]. The spectral absorption profile of the tissue is a linear combination of oxyhemoglobin and deoxyhemoglobin and other chromophores with concentration extracted via linear least squares method (LLS) [62]. This method can be useful in measuring tumor hypoxia by measuring the oxygen saturation ($sO_2$). The levels of $sO_2$ tell us about the severity ($sO_2 < 10\%$—severe hypoxia) and tumor aggressiveness. This information backs up the treatment procedure for cancer [25, 26].

## 6 Photoacoustic-Based Techniques for Breast Cancer Diagnosis

Early detection of any disease is important in reducing mortality, particularly in the case of life-threatening diseases such as cancer. The gold standard diagnostic techniques X-ray mammogram used for screening breast cancer is often limited by decrease in sensitivity with chances of missed detection [63]. Especially, women with dense breast tissue poses a challenge for detection of tumor or other abnormalities leading to reduced sensitivity and increased risk. Other screening test such as Magnetic resonance imaging (MRI) provides diagnosis with good sensitivity and specificity; however, it is not feasible due to its high cost and the use of contrast agents such as Gadolinium which can cause allergic reaction in some patients [64–66]. Photoacoustic or Optoacoustic imaging is a noninvasive imaging technique which uses a combination of light and acoustic waves to generate image inside biological tissues that are representative of optical absorption. The photoacoustic imaging method involves generating a gentle beam of pulsed laser light into the tissue sample, after which the light is absorbed, which further leads to an increase in temperature, resulting in a thermoelastic expansion. This process generates photoacoustic waves, and sensors can process these signals into high-resolution images [67].

Hemoglobin (Hb) is an important component of the red blood cell that plays a vital role in transporting oxygen throughout the body. Its involvement in cancer progression acts as a key component for detecting breast cancer using PAI [68]. Cancer cell requires a high metabolic demand, and progressive tumor requires a constant supply of oxygen and essential nutrient to grow and multiply. This is accomplished by stimulating the production of new blood vessels via a process known as angiogenesis [69]. Hb, the iron-containing protein responsible for oxygen transport in red blood cells, may contribute to the tumor microenvironment by influencing the availability of iron [70]. Malignant breast tissue contains higher level of Hb compared to normal tissue in the breast [71]. With PAI, different biological chromophores could be detected based on their different absorbance peak and differentiated using different wavelengths of light. The absorbance of Hb compared to other chromophores, like water and lipid, is found to be slightly stronger, which would further assist in detecting tumor cells from normal cells. PAI produces results that are considerably better in terms of clarity and detail and can easily outperform those obtained through traditional CT scans or ultrasound methods. It aids in providing a clearer image by differentiating a tumor cell from a normal tissue and could lead to a more accurate diagnosis of breast cancer.

## 7 Photoacoustic-Based Thermal Imaging

Thermal imaging like thermography has been long used to diagnose and treat diseases [72]. Temperature measurement of tissue through thermal imaging can provide information in a noninvasive and cost-effective way as compared to other commonly used medical imaging techniques such as X-ray mammography and ultrasonic imaging. Photoacoustic-based thermal imaging (PAT) is an approach used in the field of biomedical sciences wherein a combination of photoacoustic and thermal imaging is applied for monitoring the thermal distribution of a biological sample [73]. Photoacoustic effect on a biological sample leads to dissipation of heat, which can be captured by Infrared cameras that detect the emitted thermal radiation. These images help in identifying specific regions such as abnormal blood flow or tumor in the case of breast cancer screening. The radiation causes a thermal distribution of heat or how heat is distributed in the tissue sample. Tumors and inflammatory areas, which are often high in metabolic activity, produce a distinct heat pattern from other healthy tissue that can be detected with the help of PAT imaging [74].

The thermal distribution of skin temperature in breast tissue can be measured using a thermogram to detect any aberrant patterns or changes. Since tumor tissue tends to have higher metabolic activity and increased vascularity, which can create a hot spot on the skin surface [75]. The hot spot region could then be used to study internal tumors by measuring temperature in a specific region of interest (ROI) on the tissue surface with the help of PAI. These ROI areas could be used by physicians to extract tissue for biopsies as well as for targeted radiation therapy. Also, temperature pattern such as localized warmth or diffused increase and anomalies in the heat distribution between the left and right breast can be used for thermal diagnosis of breast cancer. The advantage of using PAI for its high-contrast, noninvasive, and functional imaging could be revolutionary in the field of breast cancer diagnosis for early detection and treatment. With additional advances, it may become a common diagnostic tool alongside mammography and ultrasound.

# 8 Nanoparticles-Based Photothermal Therapy

Cancers are clinically defined into different categories based on their subtypes (such as [ER + |PR+]HER2-, [ER + |PR+]HER2+) and the decision-making in treatment and prognosis of the diseases is determined based on these subtypes [76]. Conventional treatment of breast cancer include treatment such as chemotherapy, surgeries, and radiation. However, their use and applicability are hampered by certain limitations such as microscopic residual disease after the surgery which can lead to local recurrence and side effects on healthy cells from radiotherapy [77]. Photothermal therapy (PTT) is a minimally invasive method that uses photothermal agents such as microwave, ultrasound, and NIR to kill cancer cells. This approach helps in eradicating cancerous cells while preserving healthy cells since cancer cells are more sensitive to heat [78]. The process involves, delivery of biocompatible nanoparticle, such as Gold and carbon nanomaterials, into the tumor site, the nanoparticles further accumulate through enhanced permeability and retention (EPR) effect at the targeted tumor site. Cancerous cells are more prone to heat as compared to healthy cells. An external energy source is generated by a process known as hyperthermia, where a photosensitizer is irradiated by photothermal agent example NIR, as a result of which heat is generated [79]. The nanoparticles accumulated in the tumor site absorbs the NIR light (42 °C–50 °C) and generate high amount of heat that causes the tumor cell apoptosis or irreversible cell damage (necrosis).

Gold nanoparticles (GNP) also known as Gold nanorods (GNRs), Nanoshells, and Nanostars combined with PTTs are efficiently being explored for the treatment of breast cancer. They offer strong NIR absorption and provide excellent biocompatibility [80]. The use of GNPs-aided PTT is known to prevents nonspecific damage of adjacent healthy tissue, such as incorporating light illuminated via NIR, offers wavelength at the visible range of 450–600 nm, which is minimally absorbed by healthy cells compared to the cancerous cells [81]. GNPs can be either injected directly into the tumor site or delivered into the bloodstream. Thermal ablation, which has a high success rate in treating tumors, has side effects following the procedure, including pain, bleeding, infection, and skin burns. GNPs-aided NIR could be used to overcome this problem by targeting specific areas that needs to be treated reducing the risk of side effects on surrounding tissues. In vitro studies have also shown successful results when breast cancer cell lines were treated with GNPs along with thermal therapy with irradiated NIR laser [82]. During one of the experiments, SK-BR-3 cell lines were treated with PEG-conjugated AuNS using an 820 nm NIR laser, which resulted in less surviving cells than cells without AuNS [83]. Also, AuNS conjugated antibodies such as the human epidermal growth factor result in enhanced uptake of the nanoparticles into SK-BR-3 cell lines, promoting increased contrast for microscopic examination and greater cell cytotoxicity upon NIR application [84].

In order to increase its efficiency in delivery and application, GNPs are also modified by incorporating therapeutic agents such as tumor necrosis factor, antisense DNA, small interfering RNA (siRNA), paclitaxel, and docetaxel [84]. The effectiveness of PTT would be achieved by developing nanomaterials that integrate at specific targeted sites, causing high heat deposition in tumor cells and less harm to healthy cells. Nanoparticles possess certain desirable features that serve their applicability for use in PTT such as high biocompatibility, ability to absorb light in the NIR region, and high photothermal conversion efficiency [85].

## 9 Advantages of Optoacoustic Imaging

Optical imaging of the breast suffers from low spatial resolution due to photon scattering, making image interpretation difficult. OAI overcomes this problem by detecting the stimulated US waves, which are combined in a tomographic scheme to generate optical absorption images. Image resolution is determined by the bandwidth of the ultrasound transducer and is limited only by acoustic diffraction. OAI technology enables laser-based imaging of deeper tissue layers. With low US frequencies (2.25–8 MHz), OAI can produce resolutions of 0.5–5 mm at a depth of 3–7 cm [54, 86, 87]. Higher US frequency can increase the spatial resolution, but with lower depth penetration due to the increased attenuation of US waves with increasing frequency. This unique scalability contributes to forming images ranging from the microscopic to the macroscopic scale, which is very challenging using other imaging modalities. A strength of OAI is the ability to achieve rudimentary anatomical images and provide co-registered functional and molecular level tissue information without using ionizing radiation or exogenous contrast agents. Furthermore, OAI can be used safely to image patients with metal implants, claustrophobia, or high BMI that limit access to MRI, without the need to use contrast. Two other modalities that can complement US are ultrasound elastography and microbubble contrast-enhanced ultrasound. Fused ultrasound and OAI with decision support tool assistance showed a lower increase in specificity than has been previously reported for ultrasound elastography, including strain and shear-wave elastography as shown in recent meta-analysis, which ranged from 81% to 88% [88–91]. Of note, there are differences in study design and population, given that studies of elastography included higher numbers of Breast-Imaging Reporting and Data System (BI-RADS) category 2 masses and cysts, whereas OAI studies only included BI-RADS 3, 4, and 5 lesions, where decision-making is more challenging for the radiologist. The previously shown sensitivity range between 80 and 95% as a supplement to US limits its widespread use to down-classify masses [92–94]. Elastography is likely to be used clinically for up-classification or targeted applications, such as differentiating a complicated cyst from a solid mass. Additionally, elastography shows a marked reduction in sensitivity (possibly 2 cm) [91]. This sensitivity for small masses could be particularly important in institutions with supplemental MRI or automated breast ultrasound screening programs, leading to frequent detection of small masses and subsequent need for adjunctive diagnosis and needle biopsies. Similarly, contrast-enhanced US (CEUS) is a potential diagnostic tool for malignant breast tumors due to their generally high vascularization. This method provides real-time information on intralesional vascularization and blood flow by injecting a contrast medium consisting of microbubbles intravenously. CEUS has been shown to improve the diagnostic performance of B-mode US. However, its moderate ability to differentiate benign lesions from malignant ones and the lack of consensus on acquisition and interpretation techniques have prevented its routine use [95, 96]. A meta-analysis by Hu et al. revealed that the pooled weighted estimates of sensitivity and specificity for CEUS in the diagnosis of breast lesions were 0.86 and 0.79, respectively, insufficient to obviate needle biopsy in the majority of breast masses [97]. Finally, the current absence of CPT code for OAI hinders predictions concerning the financial impact of this new technology on patients and the overall healthcare system. However, a recent study by Ozcan et al. assessed the cost-effectiveness of supplemental use of OAI compared to US alone in distinguishing between benign and malignant breast masses [98]. This evaluation was conducted from the perspective of the US healthcare system in a diagnostic context, utilized diagnostic test performance parameters from the PIONEER-01 (NCT01943916) clinical trial, and cost parameters (USD) from the Truven Health Market Scan Databases [24, 98]. The findings indicated that OAI can cut costs and enhance patient quality of life, mainly by reducing false positives and subsequent benign biopsies [98].

## 10 Disadvantages of Optoacoustic Imaging

As a relatively new technique, OAI is still under development to refine and optimize procedures, protocols, and interpretation. Interpretation of OAI can be challenging due to a lack of technical experience in image acquisition and interpretation, together with the need for comprehensive guidelines. This lack of a standardized interpretation system and lexicon makes reporting and comparing the results with other breast imaging techniques challenging. Since there is no consensus on the interpretation schema, it remains subjective and operator-dependent. Additionally, converting the optoacoustic (OA) score to a probability of malignancy and BI-RADS category depends on the reader's judgment, adding additional subjectivity to interpretation. Intra- and inter-observer agreement on OAI scores has demonstrated high reliability and a recent study showed increased accuracy with a machine learning-based decision support tool to address the variability in score interpretation to estimate the probability of malignancy [24, 91]. Progress in source and detector development, such as faster image acquisition at more wavelengths, greater depth penetration, and image reconstruction algorithms, remains area of active research. There are two major physical artifacts that affect photoacoustic images: fluence and acoustic heterogeneity, respectively. Light fluence heterogeneity is caused by superficial layers of an imaging target absorbing photons, lowering the intensity of light that reaches deeper layers. Therefore, the available signal decreases as a function of depth. This nonlinear function, being dependent on the absorption coefficients of different pixels, also varies as a function of the wavelength and is often referred to as spectral coloring. Acoustic heterogeneity induces problems related to resolution as well as to signal intensity; acoustic waves scatter as they pass through interfaces, and they attenuate according to their frequency. Propagation of acoustic waves may be roughly determined by assuming that the wavefronts propagate along geometrical rays. There are effectively two components to acoustic heterogeneity that affect photoacoustic images: Spatially varying speed of sound and frequency-dependent attenuation. Another limitation of OAI systems is the current lack of data format uniformity, which impedes data exchange and comparison. This inconsistency hinders overall technological progress in the field and creates obstacles to effective research collaboration among various institutions and researchers. In 2018, the International Photoacoustic Standardization Consortium (IPASC) was formed to standardize test methods and data handling [99]. IPASC's goal is to ensure that data generated by various OAI systems can be easily shared, compared, and analyzed across different platforms, institutions, and research studies in order to improve the reliability and reproducibility of OAI results and foster advancements in the field.

# 11 Potential Future Applications

## *11.1 Assessing Tumor Hypoxia*

Hypoxia, a hallmark of cancer, is associated with tumor aggressiveness and plays a strong role in genetic stability, stimulating angiogenesis and metastasis. It is also pertinent to therapy resistance, notably radiation. While the importance of hypoxia has long been appreciated, there is still no routine clinical procedure for assessing levels of oxygenation in patients' tumors for radiation therapy planning and monitoring the response [100]. Indeed, there remains the mantra "hypoxia: adored and ignored," which may be attributed to the historic lack of suitable methods for assessing tumor hypoxia [101]. Genetic or proteomic signatures of hypoxia have been proposed for the breast and other disease sites based on biopsy specimens [100], but these are necessarily invasive and not practical for assessment of dynamic changes. OAI is a much newer technology, and applications are open to exploration. OAI allows direct assessment of both oxyhemoglobin ($HbO_2$) and deoxyhemoglobin (dHb), thereby revealing vascular oxygen saturation $sO_2$ [$HbO_2$/($HbO_2$ + dHb)] and total hemoglobin ($HbO_2$ + dHb). This is comparable to BOLD MRI, but has the advantage that changes in dHb resulting from vasodilation are readily discriminated from the interconversion of $HbO_2$ and dHb [102]. Changes in vascular oxygenation monitored by OAI have also been reported with respect to experimental treatments such as vascular disrupting agents, which caused acute hypoxiation [103, 104], and heme targeting agents, which generated improved vascular oxygenation attributable to vascular paring [105]. OAI could be used to provide an early indication of response or resistance to antiangiogenic therapy. Testing the MDA-MB-231 in mice, most tumors were found to be resistant to the treatment, but a subset demonstrated a clear survival benefit. At the endpoint, the PAT data from the responding subset showed significantly lower oxygenation and higher hemoglobin content compared with both resistant and control tumors. Moreover, longitudinal analysis revealed that tumor oxygenation diverged significantly in the responding subset, identifying early treatment response and the evolution of different vascular phenotypes between the subsets. The group also found distinct differences in MCF-7 and MDA-MB-231 tumors by OAI, which coincided with histological phenotypes [106].

## *11.2 Exploring Exogenous OAI Contrast Agents for Targeted Functional and Molecular Imaging*

Beyond endogenous contrast, specific OAI reporter agents have been demonstrated [107]. The strongest contrast may come from metal nanoparticles, such as gold, which can be targeted to specific receptors such as HER-2. Melanin nanoparticles (MNPs) conjugated with cyclic Arg-Gly-Asp (cRGD) have been used to target integrin $\alpha v\beta 3$ in preclinical breast tumor models (MDA-MB-231) and PyMT to reveal tumors and guide surgical removal of tissue identified via the melanin-induced photoacoustic contrast [36, 37]. In other cases, melanin-doped paclitaxel-loaded albumin nanoparticles have been used to visualize uptake and accumulation in breast tumor xenografts in mice [108]. OAI has also been applied to visualize the location of brachytherapy seeds to verify the location of effective radiation distribution in tumors [109]. In terms of radiation, high z-metals such as gold nanoparticles offer a unique opportunity to enhance local radiation dose based on Auger electrons stimulated by tumor irradiation, thereby potentially reducing the required external beam radiation dose, and sparing nearby organs, such as heart, during breast irradiation, as demonstrated in preclinical studies of human tumor xenografts in mice ([110, 111]. Such gold nanoparticles are also highly amenable to OAI observation, providing a distinct opportunity for imaging, providing potential theranostics [112]. Organic chromophores (e.g., 800-CW dye) may also be exploited for tracking agents and indocyanine green (ICG) can provide strong contrast as a perfusion contrast agent [113]. A recent report presented a chromophore, which is directly sensitive to $pO_2$, whereby the optical lifetime of the so-called nanosphere depends on pO2 and therefore generates differential ultrasound contrast. Jo et al. reported a correlation between photoacoustic lifetime (PALTI) and radiation damage assessed by γH2AX in a PDX breast tumor model xenograft in a mouse [114]. There is also an opportunity for instrumentation development. Typically, scans require several seconds for an image acquisition, particularly if specific molecules are to be discriminated (10–100 milliseconds per wavelength, and two to five wavelengths). Multiple acquisitions may be acquired to enhance signal-to-noise ratio; however, motion between individual scans can lead to misregistration, corrupting the ability to assess chromophores, such as oxy- or deoxyhemoglobin. Direct acquisition of 3D scans may partially overcome this issue, as well as acquisition of single excitations, allowing co-registration prior to multispectral unmixing [115]. Another potential innovation is incorporation of photoacoustic elastography, which was shown to provide additional ability to identify tumors in some patients [116].

### *11.3 Noninvasive Assessment of Response to Neoadjuvant Systemic Therapy*

Imaging is crucial to evaluating the response to neoadjuvant therapy as a complement to conventional tumor measurements via physical examination. The degree of response can be used as a prognostic tool since pathologic complete response (pCR) is linked to improved disease-free survival [117, 118]. However, no clear clinical practice guidelines exist for assessing tumor response to neoadjuvant therapy. Generally, patients get mammography, US, and a physical exam. In some cases, they also get a DCE-MRI. An OAI study by Lin et al. aimed to evaluate the response to neoadjuvant therapy in three breast cancer patients, with a particular focus on imaging angiographic structures [57, 58]. They utilized single-breath-hold photoacoustic computed tomography to quantify various parameters, including cancer size, blood vessel density, entropy, and anisotropy. Notably, distinctive angiographic characteristics were observed in the affected breast when compared to the healthy breast at all three predetermined time points: before, during, and after the treatment. These findings were consistent with the dimension information obtained through conventional imaging techniques as well as the final pathology. Considering that successful treatment causes cell death well before the tumor cells lyse and are phagocytized by macrophages, it is likely that functional changes within and around the tumor may precede size changes assessed by conventional imaging. Studies with diffuse optical spectroscopy, which uses near-infrared light to provide spectral information about tissue microstructure and functional parameters including oxygenated and deoxygenated hemoglobin, have demonstrated that changes in oxygenation occur as early as 24 h following the beginning of treatment in responders [119–121]. A clinical trial (ClinicalTrials.gov Identifier: NCT05337280) is currently underway to investigate the feasibility of OAI to detect pathologic complete response. This study aims to evaluate whether OAI can detect changes in the amount of hemoglobin and relative oxygenation/deoxygenation of hemoglobin before changes in size are seen on grayscale (B mode) ultrasound and whether these achieve better sensitivity and/or specificity in predicting pathologic complete response (ClinicalTrials.gov Identifier: NCT05337280).

## 12 Conclusion

OAI is a new imaging technique using nonionizing laser illumination to fuse functional and anatomic imaging in conjunction with US, with promising applications to the human breast. Parametric maps allow scalable, relative real-time molecular assessment of tissue hypoxia and can be used to gain information about breast cancer aggressiveness and tumor biology without the risk of ionizing radiation exposure, intravenous contrast, or radionuclide injection. The main limitations of OAI are the depth limit (~3–7 cm), scatter artifacts, and the lack of an objective interpretation schema. Furthermore, the relatively higher cost of devices compared to the US alone, and the lack of familiarity of the radiologist with OAI imaging features may limit its use in the clinic. OAI is a safe, noninvasive imaging technique with thriving research and high demonstrated clinical impact. Larger multicenter cohort studies and clinical trials are needed to establish the observed trends and accelerate their clinical translation.

## References


1. Huang, R., Pu, Y., Huang, S., Yang, C., Yang, F., Pu, Y., Li, J., Chen, L., & Huang, Y. (2022). FAPI-PET/CT in cancer imaging: A potential novel molecule of the century. *Frontiers Oncology, 12*, 854658.
2. Ju, F., Atyah, M. M., Horstmann, N., Gul, S., Vago, R., Bruns, C. J., Zhao, Y., Dong, Q.-Z., & Ren, N. (2022). Characteristics of the cancer stem cell niche and therapeutic strategies. *Stem Cell Research & Therapy, 13*, 1–17.
3. Pranzini, E., Raugei, G., & Taddei, M. L. (2022). Metabolic features of tumor dormancy: Possible therapeutic strategies. *Cancers, 14*, 547.
4. Sarmadi, A., Razavi, Z. S., van Wijnen, A. J., & Soltani, M. (2024). Comparative analysis of vision transformers and convolutional neural networks in osteoporosis detection from X-ray images. *Scientific Reports, 14*, 18007.
5. Hu, J., Xu, T., Shen, H., Song, Y., Yang, J., Zhang, A., Ding, H., Xing, N., Li, Z., & Qiu, L. (2023). Accuracy of gallium-68 pentixafor positron emission tomography-computed tomography for subtyping diagnosis of primary aldosteronism. *JAMA Network Open, 6*, 2255609–e2255609.
6. Chen, L., Jiang, Z., Yang, L., Fang, Y., Lu, S., Akakuru, O. U., Huang, S., Li, J., Ma, S., & Wu, A. (2023). HPDA/Zn as a CREB inhibitor for ultrasound imaging and stabilization of atherosclerosis plaque. *Chinese Journal of Chemistry, 41*, 199–206.
7. Andreou, C., Weissleder, R., & Kircher, M. F. (2022). Multiplexed imaging in oncology. *Nature Biomedical Engineering, 6*, 527–540.
8. Peddie, C. J., Genoud, C., Kreshuk, A., Meechan, K., Micheva, K. D., Narayan, K., Pape, C., Parton, R. G., Schieber, N. L., & Schwab, Y. (2022). Volume electron microscopy. *Nature Reviews Methods Primers, 2*, 51.
9. Mudeng, V., Nisa, W., & Suprapto, S. S. (2022). Computational image reconstruction for multi-frequency diffuse optical tomography. *Journal of King Saud University-Computer and Information Sciences, 34*, 3527–3538.
10. Okawa, S., & Hoshi, Y. (2023). A review of image reconstruction algorithms for diffuse optical tomography. *Applied Sciences, 13*, 5016.
11. Barbosa, R. C. S., & Mendes, P. M. (2022). A comprehensive review on photoacoustic-based devices for biomedical applications. *Sensors, 22*, 9541.
12. Luo, Q., Shao, N., Zhang, A.-C., Chen, C.-F., Wang, D., Luo, L.-P., & Xiao, Z.-Y. (2023). Smart biomimetic nanozymes for precise molecular imaging: Application and challenges. *Pharmaceuticals, 16*, 249.
13. Zhu, J., Chen, J., Amjadian, M., Liang, S., Qu, Z., Wang, Y., Zhang, Y., & Wang, L. (2023). Simultaneous dual-modal photoacoustic and harmonic ultrasound microscopy with an optimized acoustic combiner. *Biomedical Optics Express, 14*, 1626–1635.

14. Xu, X., Luo, Q., Wang, J., Song, Y., Ye, H., Zhang, X., He, Y., Sun, M., Zhang, R., & Shi, G. (2024). *Large-field objective lens for multi-wavelength microscopy at mesoscale and sub-micron resolution* (Vol. 7, p. 230212-230211-230212-230216). Opto-Electronic Advances.
15. Sangha, G. S., Hu, B., Li, G., Fox, S. E., Sholl, A. B., Brown, J. Q., & Goergen, C. J. (2022). Assessment of photoacoustic tomography contrast for breast tissue imaging using 3D correlative virtual histology. *Scientific Reports, 12*, 2532.
16. Oliveira, B. B., Ferreira, D., Fernandes, A. R., & Baptista, P. V. (2023). Engineering gold nanoparticles for molecular diagnostics and biosensing. *Wiley Interdisciplinary Reviews. Nanomedicine and Nanobiotechnology, 15*, e1836.
17. Kashkooli, F. M., Soltani, M., Souri, M., Meaney, C., & Kohandel, M. (2021). Nexus between in silico and in vivo models to enhance clinical translation of nanomedicine. *Nano Today, 36*, 101057.
18. Hui, X., Malik, M. O., & Pramanik, M. (2022). Looking deep inside tissue with photoacoustic molecular probes: A review. *Journal of Biomedical Optics, 27*, 070901–070901.
19. Lefebvre, T. L., Brown, E., Hacker, L., Else, T., Oraiopoulou, M.-E., Tomaszewski, M. R., Jena, R., & Bohndiek, S. E. (2022). The potential of photoacoustic imaging in radiation oncology. *Frontiers in Oncology, 12*, 803777.
20. Linger, C., Atlas, Y., Winter, R., Vandebrouck, M., Faure, M., Lucas, T., Bridal, S. L., & Gateau, J. (2023). Volumetric and simultaneous photoacoustic and ultrasound imaging with a conventional linear array in a multiview scanning scheme. *IEEE Transactions on Ultrasonics, Ferroelectrics, and Frequency Control, 70*, 1607.
21. Moradi Kashkooli, F., Souri, M., Tavakkoli, J., & Kolios, M. C. (2023). A spatiotemporal computational model of focused ultrasound heat-induced nano-sized drug delivery system in solid tumors. *Drug Delivery, 30*, 2219871.
22. Alexander, G. B., & On the Production and Reproduction of Sound by Light: [Read before the American Association for the Advancement of Science, in Boston, August 27, 1880.]. (1880). Researches of sumner Tainter and Alexander Graham Bell. photophonic transmitters. Experiments to ascertain the nature of the rays that affect selenium. Non-electric photophonic receivers. *American Journal of Science, 20*(118), 305.
23. Ntziachristos, V., Ripoll, J., Wang, L. V., & Weissleder, R. (2005). Looking and listening to light: The evolution of whole-body photonic imaging. *Nature Biotechnology, 23*(3), 313–320. [PubMed: 15765087].
24. Neuschler, E. I., Butler, R., Young, C. A., et al. (2018). A pivotal study of optoacoustic imaging to diagnose benign and malignant breast masses: A new evaluation tool for radiologists. *Radiology, 287*(2), 398–412. [PubMed: 29178816].
25. Dogan, B. E., Menezes, G. L. G., Butler, R. S., et al. (2019a). Optoacoustic imaging and gray-scale US features of breast cancers: Correlation with molecular subtypes. *Radiology, 292*(3), 564–572. [PubMed: 31287388].
26. Dogan, B. E., Menezes, G. L. G., Butler, R. S., Neuschler, E. I., Aitchison, R., Lavin, P. T., & Stavros, A. T. (2019b). Optoacoustic imaging and gray-scale US features of breast cancers: Correlation with molecular subtypes. *Radiology, 182*(071), 564. https://doi.org/10.1148/radiol.2019182071
27. Wang, L. (Ed.). (2009). *Photoacoustic imaging and spectroscopy* (1st ed.). CRC Press. https://doi.org/10.1201/9781420059922
28. Lin, L., & Wang, L. V. (2022). The emerging role of photoacoustic imaging in clinical oncology. *Nature Reviews Clinical Oncology, 19*, 365–384. https://doi.org/10.1038/s41571-022-00615-3
29. Deán-Ben, X. L., & Razansky, D. (2014). Adding fifth dimension to optoacoustic imaging: Volumetric time-resolved spectrally enriched tomography. *Light: Science & Applications, 3*(1), e137. https://doi.org/10.1038/lsa.2014.18
30. Nyayapathi, N., & Xia, J. (2019). Photoacoustic imaging of breast cancer: A mini-review of system design and image features. *Journal of Biomedical Optics, 24*(12), 1–13. https://doi.org/10.1117/1.JBO.24.12.121911

31. Sawada, T. (1982). *Hikari onkyou bunkouhou to sono ouyou—PAS [applications of photoacoustic spectroscopy—PAS]* (p. 10). Japan Scientific Societies Press.
32. Rao, A. P., Bokde, N., & Sinha, S. (2020). Photoacoustic imaging for management of breast cancer: A literature review and future perspectives. *Applied Sciences, 10*(3), 767. https://doi.org/10.3390/app10030767
33. Fakouri, A., Razavi, Z.-S., Mohammed, A. T., Hussein, A. H. A., Afkhami, H., & Hooshiar, M. H. (2024). Applications of mesenchymal stem cell-exosome components in wound infection healing: New insights. *Burns & Trauma, 12*, tkae021.
34. Mahmoudvand, G., Karimi Rouzbahani, A., Razavi, Z. S., Mahjoor, M., & Afkhami, H. (2023). Mesenchymal stem cell therapy for non-healing diabetic foot ulcer infection: New insight. *Frontiers in Bioengineering and Biotechnology, 11*, 1158484.
35. Kumon, R. E., Deng, C. X., & Wang, X. (2011). Frequency-domain analysis of photoacoustic imaging data from prostate adenocarcinoma tumors in a murine model. *Ultrasound in Medicine & Biology, 37*, 834–839.
36. Liu, K., Jiang, Z., Lalancette, R. A., Tang, X., & Jakle, F. (2022a). Near-infrared-absorbing B-N Lewis pair-functionalized anthracenes: Electronic structure tuning, conformational isomerism, and applications in photothermal cancer therapy. *Journal of the American Chemical Society, 144*, 18908–18917.
37. Liu, J. J., Wang, Z., & Nie, L. M. (2022b). RGD-functionalised melanin nanoparticles for intraoperative photoacoustic imaging-guided breast cancer surgery. *European Journal of Nuclear Medicine and Molecular Imaging, 49*(3), 847–860. [PubMed: 34505945].
38. Farooq, A., Sabah, S., Dhou, S., Alsawaftah, N., & Husseini, G. (2022). Exogenous contrast agents in photoacoustic imaging: an in vivo review for tumor imaging. *Nanomaterials, 12*, 393.
39. Taheripak, G., Sabeti, N., Najar, N., Razavi, Z., Saharkhiz, S., & Alipourfard, I. (2024). SIRT1 activation attenuates palmitate induced apoptosis in C2C12 muscle cells. *Molecular Biology Reports, 51*, 354.
40. Farokhi, S., Razavi, Z.-S., Mavaei, M., Shadab, A., Afkhami, H., & Sardarabadi, H. (2024). New perspectives on arteriosclerosis treatment using nanoparticles and mesenchymal stem cells. *Discover Applied Science, 6*, 1–40.
41. Lee, H., Han, S., Kye, H., Kim, T.-K., Choi, W., & Kim, J. (2023). A review on the roles of photoacoustic imaging for conventional and novel clinical diagnostic applications. *Photonics, 10*, 904.
42. Manwar, R., Zafar, M., & Xu, Q. (2020). Signal and image processing in biomedical photoacoustic imaging: A review. *Optics, 2*, 1–24.
43. Xie, W., Zhu, S., Yang, B., Chen, C., Chen, S., Liu, Y., Nie, X., Hao, L., Wang, Z., & Sun, J. (2019). The destruction of laser-induced phase-transition nanoparticles triggered by low-intensity ultrasound: An innovative modality to enhance the immunological, treatment of ovarian cancer cells. *International Journal of Nanomedicine, 14*, 9377–9393.
44. Lv, J., Li, S., Zhang, J., Duan, F., Wu, Z., Chen, R., Chen, M., Huang, S., Ma, H., & Nie, L. (2020). In vivo photoacoustic imaging dynamically monitors the structural and functional changes of ischemic stroke at a very early stage. *Theranostics, 10*, 816.
45. Souri, M., Elahi, S., & Soltani, M. (2024). Intratumoral implantable drug delivery system for targeted localized chemotherapy in breast cancer. *Journal of Drug Delivery Science and Technology, 94*, 105519.
46. Heywang-Kobrunner, S. H., Hacker, A., & Sedlacek, S. (2011). Advantages and disadvantages of mammography screening. *Breast Care, 6*, 199–207.
47. Schapiro, R., Moncayo, V. M., & Meisel, J. L. (2021). Case report of lymph node activation mimicking cancer progression: A false positive F18 FDG PET CT after COVID-19 vaccination. *Current Problems in Cancer: Case Reports, 4*, 100092.
48. Souri, M., Chiani, M., Farhangi, A., Mehrabi, M. R., Nourouzian, D., Raahemifar, K., & Soltani, M. (2022). Anti-COVID-19 nanomaterials: Directions to improve prevention, diagnosis, and treatment. *Nanomaterials, 12*, 783.

49. Butler, R., Lavin, P. T., Tucker, F. L., Barke, L. D., Bohm-Velez, M., Destounis, S., Grobmyer, S. R., Katzen, J., Kist, K. A., & Makariou, E. V. (2018). Optoacoustic breast imaging: Imaging-pathology correlation of optoacoustic features in benign and malignant breast masses. *American Journal of Roentgenology, 211*, 1155–1170.
50. Heijblom, M., Klaase, J., Van Den Engh, F., Van Leeuwen, T., Steenbergen, W., & Manohar, S. (2011). Imaging tumor vascularization for detection and diagnosis of breast cancer. *Technology in Cancer Research & Treatment, 10*, 607–623.
51. Li, J., Chekkoury, A., Prakash, J., Glasl, S., Vetschera, P., Koberstein-Schwarz, B., Olefir, I., Gujrati, V., Omar, M., & Ntziachristos, V. (2020). Spatial heterogeneity of oxygenation and haemodynamics in breast cancer resolved in vivo by conical multispectral optoacoustic mesoscopy. *Light: Science & Applications, 9*, 57.
52. Kitai, T., Torii, M., Sugie, T., Kanao, S., Mikami, Y., Shiina, T., & Toi, M. (2014). Photoacoustic mammography: Initial clinical results. *Breast Cancer, 21*, 146–153.
53. Shiina, T., Toi, M., & Yagi, T. (2018). Development and clinical translation of photoacoustic mammography. *Biomedical Engineering Letters, 8*, 157–165.
54. Manohar, S., & Dantuma, M. (2019). Current and future trends in photoacoustic breast imaging. *Photoacoustics, 16*, 100134.
55. Toi, M., Asao, Y., Matsumoto, Y., Sekiguchi, H., Yoshikawa, A., Takada, M., Kataoka, M., Endo, T., Kawaguchi-Sakita, N., & Kawashima, M. (2017). Visualization of tumor—related blood vessels in human breast by photoacoustic imaging system with a hemispherical detector array. *Scientific Reports, 7*, 41970.
56. Yamaga, I., Kawaguchi-Sakita, N., Asao, Y., Matsumoto, Y., Yoshikawa, A., Fukui, T., Takada, M., Kataoka, M., Kawashima, M., & Fakhrejahani, E. (2018). Vascular branching point counts using photoacoustic imaging in the superficial layer of the breast: A potential biomarker for breast cancer. *Photoacoustics, 11*, 6–13.
57. Lin, L., Tong, X., Hu, P., Invernizzi, M., Lai, L., & Wang, L. V. (2021a). Photoacoustic computed tomography of breast cancer in response to neoadjuvant chemotherapy. *Advancement of Science, 8*, 2003396.
58. Lin, L., Tong, X., Hu, P., Invernizzi, M., Lai, L., & Wang, L. V. (2021b). Photoacoustic computed tomography of breast cancer in response to neoadjuvant chemotherapy. *Advance Science (Weinh), 8*(7), 2003396. [PubMed: 33854889].
59. Cai, X., Liu, X., Liao, L. D., Bandla, A., Ling, J. M., Liu, Y. H., Thakor, N., Bazan, G. C., & Liu, B. (2016). Encapsulated conjugated oligomer nanoparticles for real-time photoacoustic sentinel lymph node imaging and targeted photothermal therapy. *Small, 12*, 4873–4880.
60. Dai, Y., Yu, X., Wei, J., Zeng, F., Li, Y., Yang, X., Luo, Q., & Zhang, Z. (2020). Metastatic status of sentinel lymph nodes in breast cancer determined with photoacoustic microscopy via dual-targeting nanoparticles. *Light: Science & Applications, 9*, 164.
61. Garcia-Uribe, A., Erpelding, T. N., Krumholz, A., Ke, H., Maslov, K., Appleton, C., Margenthaler, J. A., & Wang, L. V. (2015). Dual-modality photoacoustic and ultrasound imaging system for noninvasive sentinel lymph node detection in patients with breast cancer. *Scientific Reports, 5*, 15748.
62. Kim, S. (2011). Ultrasound and photoacoustic imaging for cancer detection and therapy guidance (Doctoral dissertation, The University of Texas at Austin). ProQuest Dissertations & Theses Global.
63. Pinsky, R. W., & Helvie, M. A. (2010). Mammographic breast density: Effect on imaging and breast cancer risk. *Journal of the National Comprehensive Cancer Network, 8*(10), 1157–1165 quiz 1165.
64. Ibrahim, et al. (2006). Heavy metal poisoning: clinical presentations and pathophysiology. *Clinics in Laboratory Medice, 26*(1), 67–97 viii.
65. Murphy, K. J., Brunberg, J. A., & Cohan, R. H. (1996). Adverse reactions to gadolinium contrast media: A review of 36 cases. *AJR. American Journal of Roentgenology, 167*(4), 847–849.

66. Perazella, M. A. (2008). Gadolinium-contrast toxicity in patients with kidney disease: Nephrotoxicity and nephrogenic systemic fibrosis. *Current Drug Safety, 3*(1), 67–75.
67. Xia, J., Yao, J., & Wang, L. V. (2014). Photoacoustic tomography: Principles and advances. *Electromagn Waves, 147*, 1–22.
68. Chen, X., Zhou, H., & Lv, J. (2024). The importance of hypoxia-related to hemoglobin concentration in breast cancer. *Cell Biochemistry and Biophysics, 82*(3), 1893–1906. https://doi.org/10.1007/s12013-024-01386-7. Epub 2024 Jul 3.
69. Yadav, L., Puri, N., Rastogi, V., Satpute, P., & Sharma, V. (2015). Tumour angiogenesis and angiogenic inhibitors: A review. *Journal of Clinical and Diagnostic Research, 9*(6), XE01–XE05. https://doi.org/10.7860/JCDR/2015/12016.6135. Epub 2015 Jun 1. PMID: 26266204; PMCID: PMC4525594.
70. Pfeiffer, R. M., Park, Y., Kreimer, A. R., Lacey, J. V., Jr., Pee, D., Greenlee, R. T., Buys, S. S., Hollenbeck, A., Rosner, B., Gail, M. H., & Hartge, P. (2013). Risk prediction for breast, endometrial, and ovarian cancer in white women aged 50 y or older: Derivation and validation from population-based cohort studies. *PLoS Medicine, 10*(7), e1001492. https://doi.org/10.1371/journal.pmed.1001492. Epub 2013 Jul 30. PMID: 23935463; PMCID: PMC3728034.
71. Tromberg, B. J., Pogue, B. W., Paulsen, K. D., Yodh, A. G., Boas, D. A., & Cerussi, A. E. (2008). Assessing the future of diffuse optical imaging technologies for breast cancer management. *Medical Physics, 35*(6), 2443–2451. https://doi.org/10.1118/1.2919078. PMID: 18649477; PMCID: PMC2809725.
72. Welch, A. J., & Van Gemert, M. J. (Eds.). (2011). *Optical-thermal response of laser-irradiated tissue* (Vol. 2). Springer.
73. Kesztyüs, D., Brucher, S., Wilson, C., & Kesztyüs, T. (2023). Use of infrared thermography in medical diagnosis, screening, and disease monitoring: A scoping review. *Medicina (Kaunas, Lithuania), 59*(12), 2139. https://doi.org/10.3390/medicina59122139. PMID: 38138242; PMCID: PMC10744680.
74. Han, F., Shi, G., Liang, C., Wang, L., & Li, K. (2015). A simple and efficient method for breast cancer diagnosis based on infrared thermal imaging. *Cell Biochemistry and Biophysics, 71*, 491–498.
75. Han, F., Shi, G., Liang, C., Wang, L., & Li, K. (2014). A simple and efficient method for breast cancer diagnosis based on infrared thermal imaging. *Cell Biochemistry and Biophysics, 71*(1), 491–498. https://doi.org/10.1007/s12013-014-0229-5
76. Dai, X., Xiang, L., Li, T., & Bai, Z. (2016). Cancer hallmarks, biomarkers and breast cancer molecular subtypes. *Journal of Cancer, 7*(10), 1281–1294. https://doi.org/10.7150/jca.13141. https://www.jcancer.org/v07p1281.htm
77. Prieto-Vila, M., Takahashi, R.-u., Usuba, W., Kohama, I., & Ochiya, T. (2017). Drug resistance driven by cancer stem cells and their niche. *International Journal of Molecular Sciences, 18*(12), 2574. https://doi.org/10.3390/ijms18122574
78. Riley, R. S., & Day, E. S. (2017). Gold nanoparticle-mediated photothermal therapy: Applications and opportunities for multimodal cancer treatment. *Wiley Interdisciplinary Reviews: Nanomedicine and Nanobiotechnology, 9*(4), e1449.
79. Eskiizmir, G., Ermertcan, A. T., & Yapici, K. (2016). Nanomaterials: Promising structures for the management of oral cancer. In *Nanostructures for Oral Medicine* (pp. 511–544). Elsevier. https://doi.org/10.1016/B978-0-323-47720-8.00018-3
80. Lee, J., Chatterjee, D. K., Lee, M. H., & Krishnan, S. (2014). Gold nanoparticles in breast cancer treatment: Promise and potential pitfalls. *Cancer Letters, 347*(1), 46. https://doi.org/10.1016/j.canlet.2014.02.006
81. Day, E. S., Morton, J. G., & West, J. L. (2009). Nanoparticles for thermal cancer therapy. *Journal of Biomechanical Engineering, 131*(7), 074001. https://doi.org/10.1115/1.3156800
82. Hirsch, L. R., Stafford, R. J., Bankson, J. A., Sershen, S. R., Rivera, B., Price, R. E., Hazle, J. D., Halas, N. J., & West, J. L. (2003). Nanoshell-mediated near-infrared thermal therapy of

tumors under magnetic resonance guidance. *Proceedings of the National Academy of Sciences of the United States America, 100*, 13549–13554. https://doi.org/10.1073/pnas.2232479100
83. Loo, C., Lowery, A., Halas, N., West, J., & Drezek, R. (2005). Immunotargeted nanoshells for integrated cancer imaging and therapy. *Nano Letters, 5*, 709–711. https://doi.org/10.1021/nl050127s
84. Paciotti, G. F., Myer, L., Weinreich, D., Goia, D., Pavel, N., McLaughlin, R. E., & Tamarkin, L. (2004). Colloidal gold: A novel nanoparticle vector for tumor directed drug delivery. *Drug Delivery, 11*, 169–183. https://doi.org/10.1080/10717540490433895
85. Jaque, D., Martínez Maestro, L., del Rosal, B., Haro-Gonzalez, P., Benayas, A., Plaza, J. L., & García Solé, J. (2014). Nanoparticles for photothermal therapies. *Nanoscale, 6*(16), 9494–9530. https://doi.org/10.1039/c4nr00708e
86. Kim, C., Erpelding, T. N., Jankovic, L., Pashley, M. D., & Wang, L. V. (2010). Deeply penetrating in vivo photoacoustic imaging using a clinical ultrasound array system. *Biomedice Optics Express, 1*(1), 284. [PubMed: 21258465].
87. Kruger, R. A., Lam, R. B., Reinecke, D. R., Del Rio, S. P., & Doyle, R. P. (2010). Photoacoustic angiography of the breast. *Medical Physics, 37*(11), 6096–6100. [PubMed: 21158321].
88. Barr, R. G., De Silvestri, A., Scotti, V., Manzoni, F., Rebuffi, C., Capittini, C., & Tinelli, C. (2019). Diagnostic performance and accuracy of the 3 interpreting methods of breast strain Elastography: A systematic review and meta-analysis. *Journal of Ultrasound in Medicine, 38*(6), 1397–1404. [PubMed: 30411806].
89. Pillai, A., Voruganti, T., Barr, R., & Langdon, J. (2022). Diagnostic accuracy of shear-wave elastography for breast lesion characterization in women: A systematic review and meta-analysis. *Journal of the American College of Radiology, 19*(5), 625–634.e0. [PubMed: 35358483].
90. Barr, R. G., Engel, A., Kim, S., Tran, P., & De Silvestri, A. (2023). Improved breast 2D SWE algorithm to eliminate false-negative cases. *Investigative Radiology, 58*(10), 703–709. [PubMed: 36939607].
91. Seiler, S. J., Neuschler, E. I., Butler, R. S., Lavin, P. T., & Dogan, B. E. (2023). Optoacoustic imaging with decision support for differentiation of benign and malignant breast masses: A 15-reader retrospective study. *AJR. American Journal of Roentgenology, 220*(5), 646–658. [PubMed: 36475811].
92. Pu, H., Zhang, X. L., Xiang, L. H., et al. (2019). The efficacy of added shear wave elastography (SWE) in breast screening for women with inconsistent mammography and conventional ultrasounds (US). *Clinical Hemorheology and Microcirculation, 71*(1), 83–94. [PubMed: 29843228].
93. Sadigh, G., Carlos, R. C., Neal, C. H., Wojcinski, S., & Dwamena, B. A. (2013). Impact of breast mass size on accuracy of ultrasound elastography vs. conventional B-mode ultrasound: A meta-analysis of individual participants. *European Radiology, 23*(4), 1006–1014. [PubMed: 23085865].
94. Suvannarerg, V., Chitchumnong, P., Apiwat, W., et al. (2019). Diagnostic performance of qualitative and quantitative shear wave elastography in differentiating malignant from benign breast masses, and association with the histological prognostic factors. *Quantitative Imaging in Medicine and Surgery, 9*(3), 386–398. [PubMed: 31032186].
95. Du, J., Wang, L., Wan, C. F., et al. (2012). Differentiating benign from malignant solid breast lesions: Combined utility of conventional ultrasound and contrast-enhanced ultrasound in comparison with magnetic resonance imaging. *European Journal of Radiology, 81*(12), 3890–3899. [PubMed: 23062280].
96. Liu, H., Jiang, Y. X., Liu, J. B., Zhu, Q. L., & Sun, Q. (2008). Evaluation of breast lesions with contrast enhanced ultrasound using the microvascular imaging technique: Initial observations. *Breast, 17*(5), 532–539. [PubMed: 18534851].
97. Hu, Q., Wang, X. Y., Zhu, S. Y., Kang, L. K., Xiao, Y. J., & Zheng, H. Y. (2015). Meta-analysis of contrast enhanced ultrasound for the differentiation of benign and malignant breast lesions. *Acta Radiologica, 56*(1), 25–33. [PubMed: 24436445].

98. Ozcan, B. B., Xi, Y., & Dogan, B. E. (2023). Supplemental optoacoustic imaging of breast masses: A cost-effectiveness analysis. *Academic Radiology, 31*, 121. https://doi.org/10.1016/j.acra.2023.08.042
99. Gröhl, J., Hacker, L., & Betal, C. (2022). The IPASC data format: A consensus data format for photoacoustic imaging. *Photoacoustics, 26*, 100339. [PubMed: 35287304].
100. Buffa, F. M., Harris, A. L., West, C. M., & Miller, C. J. (2010). Large meta-analysis of multiple cancers reveals a common, compact and highly prognostic hypoxia metagene. *British Journal of Cancer, 102*(2), 428–435. [PubMed: 20087356].
101. Vaupel, P., Schlenger, K., Knoop, C., & Höckel, M. (1991). Oxygenation of human tumors: Evaluation of tissue oxygen distribution in breast cancers by computerized $O_2$ tension measurements. *Cancer Research, 51*(12), 3316–3322. [PubMed: 2040005].
102. Rich, L. J., & Seshadri, M. (2015). Photoacoustic imaging of vascular hemodynamics: Validation with blood oxygenation level-dependent MR imaging. *Radiology, 275*(1), 110–118. [PubMed: 25423146].
103. Liu, L., O'Kelly, D., & Schuetze, R. (2021). Non-invasive evaluation of acute effects of tubulin binding agents: A review of imaging vascular disruption in tumors. *Molecules, 26*(9), 2551.
104. Guo, Y., Wang, H., Gerberich, J. L., et al. (2021). Imaging-guided evaluation of the novel small molecule benzosuberene tubulin-binding agent KGP265 as a potential therapeutic agent for cancer treatment. *Cancers (Basel), 13*(19), 4769.
105. Ghosh, P., Guo, Y., Ashrafi, A., et al. (2020). Oxygen-enhanced optoacoustic tomography reveals the effectiveness of targeting Heme and oxidative phosphorylation at normalizing tumor vascular oxygenation. *Cancer Research, 80*(17), 3542–3555. [PubMed: 32546631] Author Manuscript Author Manuscript Author Manuscript.
106. Quiros-Gonzalez, I., Tomaszewski, M. R., et al. (2022). Photoacoustic tomography detects response and resistance to bevacizumab in breast cancer mouse models. *Cancer Research, 82*(8), 1658–1668. [PubMed: 35404400].
107. MacCuaig, W. M., Jones, M. A., Abeyakoon, O., & McNally, L. R. (2020). Development of multispectral optoacoustic tomography as a clinically translatable modality for cancer imaging. *Radiology: Imaging Cancer, 2*(6), e200066.
108. Sim, C., Kim, H., Moon, H., Lee, H., Chang, J. H., & Kim, H. (2015). Photoacoustic-based nanomedicine for cancer diagnosis and therapy. *Journal of Controlled Release, 203*, 118–125. [PubMed: 25701310].
109. Harrison, T., & Zemp, R. J. (2011). Coregistered photoacoustic-ultrasound imaging applied to brachytherapy. *Journal of Biomedical Optics, 16*(8), 080502. [PubMed: 21895302].
110. Wu, Z., Stangl, S., Hernandez-Schnelzer, A., et al. (2023). Functionalized hybrid iron oxide-gold nanoparticles targeting membrane Hsp70 Radiosensitize triple-negative breast cancer cells by ROS-mediated apoptosis. *Cancers (Basel), 15*(4), 1167.
111. Nosrati, H., Salehiabar, M., Charmi, J., et al. (2023). Enhanced in vivo radiotherapy of breast cancer using gadolinium oxide and gold hybrid nanoparticles. *ACS Applied Bio Materials, 6*(2), 784–792.
112. Schuemann, J., Berbeco, R., Chithrani, D. B., et al. (2016). Roadmap to clinical use of gold nanoparticles for radiation sensitization. *International Journal of Radiation Oncology, Biology, Physics, 94*(1), 189–205. [PubMed: 26700713].
113. Tomaszewski, M. R., Gehrung, M., Joseph, J., Quiros-Gonzalez, I., Disselhorst, J. A., Bohndiek, S. E., et al. (2018). Oxygen-enhanced and dynamic contrast-enhanced optoacoustic tomography provide surrogate biomarkers of tumor vascular function, hypoxia, and necrosis. *Cancer Research, 78*(20), 5980–5991. [PubMed: 30115696].
114. Jo, J., Folz, J., Gonzalez, M. E., et al. (2023). Personalized oncology by in vivo chemical imaging: Photoacoustic mapping of tumor oxygen predicts radiotherapy efficacy. *ACS Nano, 17*(5), 4396–4403. [PubMed: 36847392].
115. O'Kelly, D., Guo, Y., & Mason, R. P. (2020). Evaluating online filtering algorithms to enhance dynamic multispectral optoacoustic tomography. *Photoacoustics, 19*, 100184. [PubMed: 32509522].

116. Biswas, D., Vasudevan, S., Chen, G. C. K., Bhagat, P., Sharma, N., & Phatak, S. (2017). Time–frequency based photoacoustic spectral response technique for differentiating human breast masses. *Biomedical Physics & Engineering Express, 3*(3), 035002.
117. Von Minckwitz, G., Untch, M., & Blohmer, J. U. (2012). Definition and impact of pathologic complete response on prognosis after neoadjuvant chemotherapy in various intrinsic breast cancer subtypes. *Journal of Clinical Oncology, 30*(15), 1796–1804. [PubMed: 22508812].
118. Cortazar, P., Zhang, L., Untch, M., et al. (2014). Pathological complete response and long-term clinical benefit in breast cancer: The CTNeoBC pooled analysis. *Lancet, 384*(9938), 164–172. [PubMed: 24529560].
119. Schaafsma, B. E., van de Giessen, M., Charehbili, A., et al. (2015). Optical mammography using diffuse optical spectroscopy for monitoring tumor response to neoadjuvant chemotherapy in women with locally advanced breast cancer. *Clinical Cancer Research, 21*(3), 577–584. [PubMed: 25473002].
120. Roblyer, D., Ueda, S., Cerussi, A., et al. (2011). Optical imaging of breast cancer oxyhemoglobin flare correlates with neoadjuvant chemotherapy response one day after starting treatment. *Proceedings of the National Academy of Sciences of the United States of America, 108*(35), 14626–14631. [PubMed: 21852577].
121. Ntziachristos, V., & Chance, B. (2001). Probing physiology and molecular function using optical imaging: Applications to breast cancer. *Breast Cancer Research, 3*(1), 41–46. [PubMed: 11250744].

# Current Updates on Photoacoustic-Based Techniques for Breast Cancer Diagnosis

**Pronama Biswas, Merla Sudha, Khushi Agarwal, Madhavi Bhatt, Bhoomika Sridhar, and Belaguppa Manjunath Ashwin Desai**

**Abstract** Breast cancer is the most diagnosed cancer worldwide, with rising incidence rates globally. Early detection is essential for improving survival rates, yet conventional imaging methods have limitations. Mammography, the gold standard, struggles with reduced sensitivity in dense breast tissue. Ultrasound (US) aids in distinguishing lesions but is highly operator-dependent, while MRI offers detailed imaging but remains costly and less accessible. These challenges underscore the need for improved imaging technologies. Photoacoustic imaging (PAI) is an emerging, noninvasive modality that integrates optical and ultrasound imaging to provide high-resolution, real-time visualization of breast tissue. Unlike conventional methods, PAI enables the assessment of tumor vascularization, blood oxygenation, and molecular characteristics, enhancing early cancer detection and treatment monitoring. Various PAI techniques, including photoacoustic tomography, photoacoustic microscopy, multispectral PAI, and photoacoustic endoscopy, offer unique advantages in imaging tumor structure and function. Recent advancements, such as molecular contrast agents and multimodal integration with ultrasound and MRI, further enhance PAI's diagnostic accuracy. This chapter explores the principles, techniques, and applications of PAI in breast cancer detection, tumor characterization, and therapy assessment. We also address challenges in clinical translation and future prospects. With continued development, PAI has the potential to revolutionize breast cancer diagnostics and improve patient outcomes.

**Keywords** Photoacoustic imaging · Breast cancer · Mammography · Photoacoustic tomography · Tumour characterization

P. Biswas (✉) · M. Sudha · K. Agarwal · M. Bhatt · B. Sridhar
School of Basic and Applied Sciences, Dayananda Sagar University, Bengaluru, India
e-mail: pronama-sbas@dsu.edu.in

B. M. A. Desai
School of Engineering, Dayananda Sagar University, Bengaluru, India



N. Mazumder et al. (eds.), *Advanced Biophysical Techniques in Biosciences*, Lecture Notes in Nanoscale Science and Technology 7,
https://doi.org/10.1007/978-3-031-94551-9_16

# 1 Introduction

Breast cancer remains a major global health concern, and it has emerged as the most prevalent cancer among women worldwide, with a significant rise in incidence rates over recent decades. According to the Global Cancer Statistics 2020, breast cancer accounted for approximately 2.26 million new cases, representing 24.5% of incidence and 15.5% of breast cancer-related deaths globally in women [1, 2]. Breast cancer is a complex cancer due to its heterogeneous behavior, leading to multiple biological and molecular subtypes, with each exhibiting distinct clinical behaviors and responses to treatment [3]. Improvements in cancer biology and genomics have advanced our understanding of the disease, facilitating the identification of the molecular subtypes and their associated risk factors [3]. The risk factors for breast cancer are diverse as it also includes genetic predispositions such as BRCA1 and BRCA2 mutations, hormonal influences, and environmental exposures [3]. Age plays a crucial role in onset of disease, with the likelihood of developing breast cancer increasing over time due to the accumulation of genetic and epigenetic alterations [4]. Breast cancer incidence and patient outcomes vary significantly, with pronounced disparities in low- and middle-income countries (LMICs), where limited access to screening and treatment remains a major challenge [5]. Additionally, the emergence of novel therapeutic strategies has significantly improved breast cancer management. Targeted therapies have been developed for specific molecular subtypes, such as anti-HER2 treatments for HER2-positive breast cancers. Currently, research continues to explore the role of HER2-low expression in early-stage disease [6].

The diagnosis of breast cancer is a complex process that involves a combination of imaging techniques, histopathological evaluation, and clinical assessment. The primary methods used for initial detection include mammography, ultrasound (US), and magnetic resonance imaging (MRI), all of which help identify suspicious lesions that may require further examination through biopsy [7]. Among these, mammography remains the most preferred method for screening of breast cancer, as it can diagnose asymptomatic women [8]. Ultrasound is another method often employed for individuals with dense breast tissue. Despite their widespread use, conventional diagnostic approaches, such as mammography, ultrasound, and clinical examination, possess certain limitations that can hinder early detection and precise diagnosis. Mammography's sensitivity is significantly reduced in women with dense breast tissue, prompting legislative efforts in certain regions to ensure patients are informed about these limitations [8]. While ultrasound is advantageous due to its noninvasive nature and the absence of ionizing radiation, its diagnostic accuracy is often influenced by operator expertise, leading to variability in results [9]. Although automated breast ultrasound systems have been introduced to mitigate some of these challenges, ultrasound remains insufficient as a standalone screening tool [9]. Another challenge in breast cancer diagnosis is the reliance on histopathological evaluation for confirmation. These limitations necessitate advancements in breast cancer diagnostic technologies. Advancements in imaging technologies and the incorporation of molecular diagnostics have the potential to enhance detection accuracy and reliability. These improvements may lead to early detection, better patient care, and more effective treatment outcomes.

Photoacoustic imaging (PAI) is a promising technique for breast cancer diagnosis and management, combining the strengths of both optical and ultrasound imaging. This hybrid approach relies on the photoacoustic effect, where absorbed light energy is transformed into ultrasound waves, enabling high-resolution imaging at depths relevant to breast tissue [10]. A significant advantage of PAI is its capability to visualize vascular structures and monitor tumor-associated angiogenesis, a key characteristic of cancer progression [11]. By detecting variations in blood oxygenation and hemodynamics, PAI provides valuable insights into the tumor microenvironment, potentially facilitating early cancer detection [12]. Recent developments in PAI technology have aimed at improving imaging specificity and sensitivity. One notable advancement is the use of molecularly targeted contrast agents, such as indocyanine green (ICG) and various nanoparticles, which enhance tumor detection by producing signals linked to specific tumor markers [13, 14]. Additionally, integrating PAI with established imaging modalities, such as ultrasound and MRI, offers the potential to improve diagnostic accuracy and provide a more comprehensive evaluation of breast lesions. This chapter discusses the principle of PAI and its significance, focussing on the need for this technology and its advantages over other imaging techniques. Different types of PAI methods available for breast cancer detection and screening are explored in Table 1. Additionally, important applications of PAI in early detection and screening, tumor characterization, and treatment response assessment are examined. Recent advancements in PAI, such as innovations in contrast agents, dual-modality imaging, and multimodal approaches, are highlighted. Regulatory and ethical considerations, challenges, and future perspectives are also discussed. This chapter aims to provide a comprehensive understanding of the role of PAI in breast cancer detection and management.

## 2 Principles of Photoacoustic Imaging

### *2.1 The Photoacoustic Effect*

Photoacoustic effect refers to the generation of acoustic pressure waves resulting from the interaction between light and target materials. This effect depends on the thermodynamic and optical properties of the materials as well as the characteristics of the incident light [21]. Photoacoustic technology measures the ultrasonic (acoustic) signals produced by target molecules when they absorb light energy. The principle underlying this effect is based on the conversion of absorbed optical energy into acoustic waves through thermal expansion in a material. This effect is fundamental to advanced imaging techniques, particularly in the biomedical field, where it enables high-resolution visualization of functional and molecular information in

**Table 1** Summary of the photoacoustic methods

| Photoacoustic methods | Principle | Key feature | Clinical relevance in breast cancer detection | References |
|---|---|---|---|---|
| Photoacoustic tomography (PAT) | Use light absorption in tissue to generate ultrasound waves for 3D imaging based on optical properties | High resolution, noninvasive, deep penetration | Used for early breast cancer detection and vascular imaging | [15] |
| Photoacoustic microscopy (PAM) | Based on light absorption, it creates images by detecting sound waves | High resolution at different depths, subcellular imaging | Prognostic tool for local recurrence risk | [16] |
| Optical-resolution PAM (OR-PAM) | Provides high-resolution imaging by tightly focusing the excitation laser, with resolution governed by optical focus. | Optical focus, high resolution, shallow imaging | Detects breast tumor microvasculature and CTCs with high resolution but is limited to superficial tissues | [17] |
| Acoustic-resolution PAM (AR-PAM) | Enables deeper imaging by using a loosely focused laser, with resolution determined by acoustic focus | Acoustic focus, moderate resolution, deep imaging | Provides deep imaging of breast tumors by assessing vascular abnormalities and oxygenation | [17] |
| Multispectral photoacoustic imaging (MS-PAI) | Based on multiple wavelengths, it generates acoustic waves to identify and quantify chromophores through their unique absorption spectra | Noninvasive, improve tissue contrast | Identify tumor based on their blood supply and metabolic activity | [18] |
| Photoacoustic flow cytometry (PAFC) | Detect circulating tumor cells (CTCs) in blood using photoacoustics | Noninvasive, high contrast, good penetration depth | The structural and compositional differences between healthy breast tissue and cancerous tumors can be detected | [15] |
| Thermoacoustic imaging (TAI) | Uses energy sources such as microwaves or radiofrequency pulses to generate ultrasound waves in tissue | High penetration depth, hybrid imaging technique | Deep tissue imaging and distinguish cancerous tissues | [19] |
| Photoacoustic computed tomography (PACT) | Based on optical absorption, it generates ultrasonic waves to form 3D breast tissue images that reveal regions with high blood vessel density | Deep tissue imaging, highly sensitive to blood vessels | Provides detailed tumor mapping for precision treatment planning | [20] |

biological tissues [22]. When biological tissues are illuminated with pulsed laser light, specific chromophores such as hemoglobin, melanin, or other pigments, absorb the light energy. This absorption causes a rapid temperature rise, leading to localized thermal expansion. As the tissue expands, it generates stress waves that propagate as acoustic signals. These ultrasonic waves are detected by ultrasound transducers, which convert them into electrical signals (Fig. 1). Using advanced reconstruction algorithms, these signals are processed to produce detailed images that reveal structural and functional information [10]. Various methods, including Radon transforms and finite element methods, are employed to reconstruct images from the detected acoustic signals [23–25]. A standard PAI system consists of two key components: a laser source: typically a tunable, nanosecond-pulse laser, and an ultrasound (US) system to detect the generated ultrasound waves. Many advanced image processing techniques used in conventional ultrasound imaging, such as delay and sum beamforming and computed tomography (CT)-based reconstruction algorithms, can be directly applied to PAI. These methods enhance imaging precision and enable three-dimensional visualization of tissue structures [26].

The origins of the photoacoustic effect trace back to 1880, when Alexander Graham Bell first observed that modulated light could generate sound waves in solid materials [21]. Although this discovery was significant, the photoacoustic effect remained underexplored until the 1960s, when the invention of laser technology provided the high peak power and spectral purity required for its practical application. The development of lasers marked a turning point, allowing researchers to systematically study and utilize the photoacoustic effect for scientific and medical

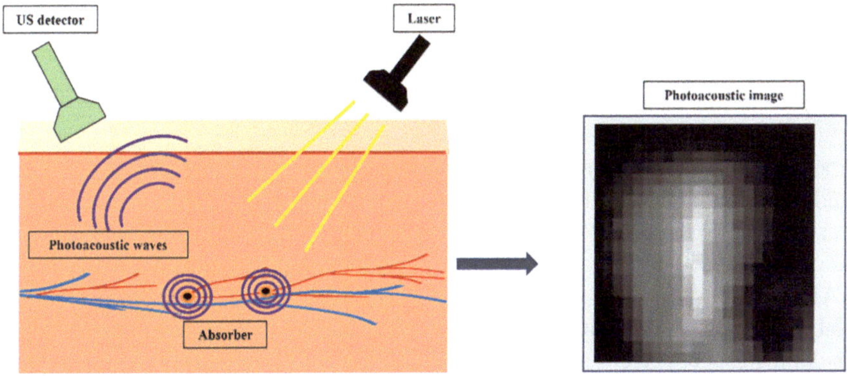


**Fig. 1** Schematic representation of the photoacoustic effect in biological tissues. This illustration depicts the fundamental mechanism of the photoacoustic effect used in biomedical imaging. A pulsed laser beam irradiates biological tissue, where endogenous chromophores like hemoglobin in blood vessels absorb the optical energy. This absorption leads to localized heating, causing rapid thermoelastic expansion and the generation of acoustic ultrasound (US) waves. These waves propagate through the tissue and are detected by an US transducer. The collected signals are processed to reconstruct images with high resolution, which are based on optical absorption contrast. This figure was made using Microsoft® PowerPoint® 2021 MSO (Version 2501 Build 16.0.18429.20132) 64-bit

applications. In the 1970s, researchers Rosencwaig and Gersho advanced the theoretical understanding of the photoacoustic effect. They introduced mathematical models that described how absorbed light energy leads to acoustic wave generation through thermal expansion. These models laid the foundation for the development of modern PAI technologies [27]. Subsequent research in the 2000s expanded these models, refining our understanding of how light–material interactions generate acoustic signals in biological tissues [28, 29]. PAI also enables functional imaging by mapping oxygen saturation and metabolic processes, making it ideal for disease diagnostics progression analysis. Moreover, this technique is noninvasive and label-free, as it does not require ionizing radiation, and endogenous chromophores provide natural contrast. Ongoing research continues to enhance PAI by improving laser sources, transducer sensitivity, and computational algorithms. The contrast in PA images is primarily determined by the wavelength-dependent optical absorption coefficient (μa), which varies according to the tissue's molecular composition. Different biological chromophores, such as oxyhemoglobin ($HbO_2$), deoxyhemoglobin (Hb), melanin, water, and lipid, each possess distinct optical absorption spectra that can be utilized to produce high contrast images. Multi-wavelength and spectroscopic imaging techniques leverage the unique optical properties of tissue components to enhance imaging contrast and differentiation. $HbO_2$ and Hb exhibit distinct absorption patterns at various wavelengths, enabling PAI to effectively distinguish between veins and arteries. Moreover, at approximately 805 nm, both $HbO_2$ and Hb share the same optical absorption, making this wavelength particularly valuable for assessing total hemoglobin concentration (CHb) and blood oxygenation.

## 2.2 *Types of Photoacoustic Methods for Cancer Detection*

### 2.2.1 Photoacoustic Tomography

Photoacoustic tomography (PAT) is a hybrid imaging technique combining optical and ultrasound imaging [10]. It has the ability to achieve high-resolution imaging at varying depths, giving PAT a distinct advantage over conventional imaging techniques, which often struggle to balance resolution and penetration depth effectively. PAT can achieve imaging depths of several centimetres while maintaining high spatial resolution, making it suitable for a wide range of biomedical applications, including cancer detection and monitoring [25]. A major limitation of optical imaging is light diffusion, which restricts deep-tissue resolution. PAT overcomes this by employing the photoacoustic effect, which allows detailed imaging in both ballistic and diffusive optical regimes [10, 30]. Jin et al. highlighted that PAT could sample optical phenomena within tissues to depths of several centimetres, making it particularly useful for imaging structures that are not accessible through standard optical methods [31]. This depth advantage is crucial for applications in oncology, where the detection of tumors often requires imaging through layers of surrounding tissue [32]. This capability is further enhanced using nanoparticles and nanotubes as contrast agents, which can improve the sensitivity and specificity of the imaging process [33]. The integration of these nanoparticles into PAT systems has been shown to significantly enhance the contrast of the images obtained, facilitating better detection of pathological changes in tissues [33]. Similarly, the incorporation of carbon nanotubes enhances the imaging capabilities of PAT by improving the contrast and specificity of the images obtained [17]. The noninvasive nature of PAT, coupled with its use of nonionizing radiation, positions it as a safer alternative to other imaging modalities such as X-ray, CT, and MRI [34]. This aspect is particularly important in clinical settings, where minimizing patient exposure to ionizing radiation is a priority. The ability of PAT to provide real-time imaging also adds to its clinical utility, allowing for dynamic monitoring of physiological processes [35].

The multiscale imaging potential of PAT was demonstrated through studies on lacZ, a widely used molecular imaging reporter gene [36]. lacZ encodes β-galactosidase, an enzyme that breaks down lactose into glucose and galactose, producing a blue compound that strongly absorbs light between 605 and 665 nm, making it suitable for deep PA imaging. Cai et al. demonstrated that PAI combined with the lacZ/X-gal reporter system enables precise tumor identification. As shown in Fig. 2a, a tumor-bearing nude mouse had 9 L gliosarcoma tumors, implanted on its right and left flanks, with only the right-side tumor expressing the lacZ transgene. Before imaging, both tumors were injected with an X-gal (5-bromo-4-chloro-3-indolyl-β-D-galactopyranoside) solution, which is a chromogenic substrate used to detect β-galactosidase activity. When β-galactosidase, encoded by the lacZ gene, cleaves X-gal, it produces an insoluble blue product. This color change allows researchers to confirm the presence of lacZ expression visually. After sacrificing the mouse and removing the skin over the tumors, only the lacZ+ tumor exhibited a blue stain. This confirmed the specific detection of lacZ expression using the X-gal probe (Fig. 2b). The tumor region was then examined in vivo using PAT and ultrasound imaging. The imaging produced dual-modality images that included volumetric, cross-sectional, and projection views (Fig. 2c–e). In these images, photoacoustic signals appeared green, while ultrasound signals were displayed in grayscale. The lacZ+ tumor was clearly identified in three-dimensional space (Fig. 2c). The tumor appeared as a dark (hypoechoic) area in US images. However, this signal alone was not a specific tumor marker (Fig. 2d). The lacZ/X-gal system, visualized through photoacoustic imaging, provided high specificity in detecting the lacZ-expressing tumor (Fig 2e). Furthermore, chicken breast tissue was placed over the tumor to assess deep-tissue detection. The lacZ expression remained visible up to 5.0 cm deep in biological tissue. The contrast ratio was 3.04. These results demonstrate the capability of photoacoustic imaging (PAI) for deep-tissue tumor detection [17, 37].

A multiscale PAT experiment on a mouse tumor assessed imaging depth by overlaying tissue layers and capturing images using a linear ultrasound transducer array [17]. Another Advantage of PAT includes super-resolution imaging, which allows visualization of cellular and subcellular structures beyond the optical diffraction limit, significantly benefiting biological research [37]. Conventional super-resolution techniques mainly focus on fluorescence imaging, requiring complex laser setups and fluorophore manipulation. In contrast, PAT has the potential to

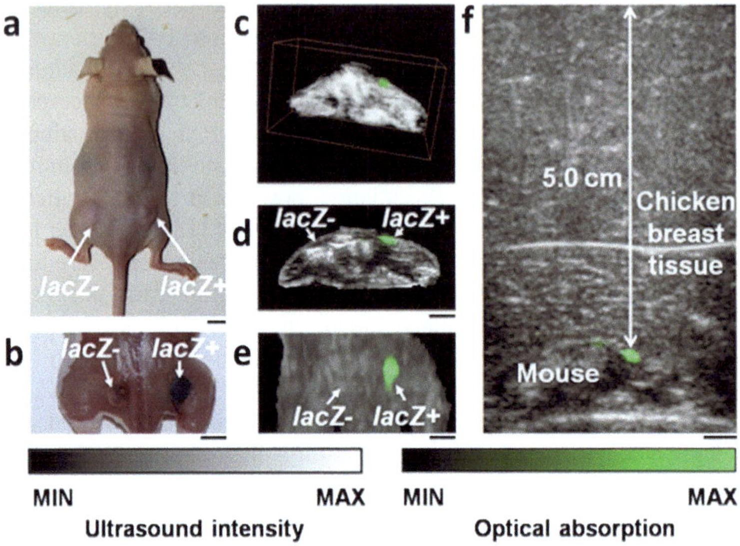


**Fig. 2** PAT case study 1. Visualization of 9 L tumors with and without the *lacZ* reporter gene using photoacoustic and ultrasonic imaging. (**a**) Image of a tumor-bearing mouse before scanning, (**b**) Post-euthanasia image showing *lacZ*$^-$ and *lacZ*$^+$ tumors after removal of skin, (**c–e**) Merged photoacoustic and ultrasonic images, (**c**) 3D visualization, (**d**) B-scan, and **e** maximum amplitude projection, (**f**) B-scan of the *lacZ*-positive tumor at a depth of 5 cm, with an overlay of chicken breast tissue on the mouse. Photoacoustic images are displayed in green, while ultrasonic images appear in gray. Scale bars: 5 mm. Reprinted with permission from [17]

image both fluorescent and nonfluorescent molecules at specific wavelengths. Recent advancements have enabled PAT to break past traditional diffraction limits, making it a promising tool for high-resolution biomedical imaging. PAT signals provide detailed anatomical, functional, and metabolic insights into tissue microenvironments. Since single-parameter imaging may not fully capture physiological and pathological conditions, multiparameter PAT enhances diagnostic accuracy. Key measurable parameters include total CHb and oxygen saturation ($sO_2$), which are crucial for evaluating blood oxygenation and perfusion. Increased CHb due to angiogenesis and reduced $sO_2$ due to hypoxia are associated with late-stage cancers, while hyperoxia is linked to early-stage malignancies [38]. Fluence-compensated PA measurements at hemoglobin's isosbestic wavelengths allow for the accurate assessment of total CHb without being affected by variations in oxygenation levels. Fluence compensation refers to the correction of light intensity variations within tissue, ensuring that the PA signal accurately reflects the chromophore concentration rather than changes in light distribution. Whereas Hemoglobin's isosbestic wavelengths are specific wavelengths where the absorption of $HbO_2$ and Hb is equal, meaning that PA signals at these wavelengths remain independent of blood oxygenation levels. By utilizing these wavelengths, researchers can measure CHb blood without interference from oxygenation fluctuations, enabling more reliable and quantitative PA imaging in medical applications [39]. Integrating PAT with diffuse reflectance spectroscopy [40] or diffuse optical tomography (DOT) [41] to assess tissue optical properties can facilitate better calculation of $sO_2$ quantification of CHb and oxygenation, improving the accuracy and reliability of functional imaging in biomedical applications. Alternatively, analyzing PA signal frequency spectra at multiple wavelengths can determine absolute $HbO_2$ and HbR concentrations without requiring fluence compensation [42]. Leveraging hemoglobin's strong contrast between intra- and extravascular spaces, PAT is also applied for blood flow analysis [43, 44]. Various PA-based methods assess blood flow speed, including photoacoustic Doppler flowmetry, which measures axial flow velocity using Doppler frequency shifts [45, 46].

Additionally, the development of multispectral PAT allows for the differentiation of various chromophores within tissues, providing further insights into the biochemical composition of the imaged areas [47]. Another extension to PAT has also been employed for PA molecular imaging, which integrates molecular contrast agents to improve sensitivity and specificity for visualizing biochemical and cellular processes. Nyayapathi et al. investigated the application of PAI for breast cancer detection by examining tumor vascular structures and molecular features. Their research demonstrated that PAI could accurately differentiate tumors by analyzing hemoglobin oxygenation levels, aiding in the identification of hypoxic regions, which are often associated with aggressive cancer development. The results indicated that using multi-wavelength PAI improves tumor visibility and enhances detection precision [48]. Balasundaram et al. investigated the potential of molecular PAI for precise breast cancer detection by utilizing conjugated polymer nanoparticles. Their research aimed to enhance tumor imaging contrast through the application of targeted contrast agents that specifically bind to breast cancer cells. The study demonstrated that these nanoparticles significantly increase PAI sensitivity in identifying malignant tissues compared to conventional imaging methods. The findings highlighted the benefits of molecular PAI in delivering high-resolution, real-time imaging of breast tumors. However, the researchers emphasized the necessity of further clinical studies to validate the safety and effectiveness of these targeted imaging agents for routine medical use [49]. PAT can further be enhanced with the use of a pulsed laser system. For instance, Upputuri and Pramanik discussed a low-cost, high-speed pulsed laser diode photoacoustic tomography (PLD-PAT) system, which addressed the limitations of traditional Q-switched Nd:YAG/OPO lasers. Their work illustrates the integration of a compact pulsed laser diode to facilitate real-time imaging, to overcome the challenges faced due to the bulkiness and high cost of conventional systems. Together, these studies underscore the diversity of pulsed laser systems available, ranging from sophisticated fiber lasers to more accessible diode-based solutions, each contributing uniquely to the field of PAI [50]. The image formation in PAT can be categorized into two primary systems, which are photoacoustic computed tomography (PACT) and Photoacoustic Microscopy (PAM). PACT utilizes reconstruction algorithms to generate images from the collected photoacoustic signals, while PAM employs direct point-by-point detection methods to render images through faster scanning [51]. This distinction is essential as it influences the resolution and the type of information that can be extracted from the images, such as blood vessel parameters and tissue morphology [52].

### 2.2.2 Photoacoustic Microscopy and Photoacoustic Computed Tomography

PAM is a specialized modality of PAT that provides optical absorption-based imaging with high spatial resolution. Unlike conventional PAT, which operates in a wide-field imaging mode, PAM employs a focused optical beam for precise imaging at microscopic scales. Depending on the optical and acoustic resolution, PAM is classified into two primary types, Optical-Resolution PAM (OR-PAM) and acoustic-resolution PAM (AR-PAM), each distinguished by their resolution and imaging depth characteristics. The fundamental principle of PAM is based on the PA effect. PAM provides intrinsic optical absorption contrast, allowing for the visualization of structures such as blood vessels and tumors with high sensitivity and resolution [16, 53]. OR-PAM utilizes tightly focused laser light to achieve high lateral resolution, typically in the range of 1–5 μm. This high resolution is primarily due to the diffraction-limited optical focusing, which enables detailed imaging of cellular structures and fine anatomical features [54, 55]. However, the penetration depth of OR-PAM is limited, generally reaching up to 1.2 mm, due to the scattering of light in biological tissues. This limitation makes OR-PAM particularly suitable for imaging superficial tissues or thin samples where high resolution is critical [56, 57]. But the ability of OR-PAM to provide high-resolution images makes it valuable for applications such as studying microvascular structures, cellular dynamics, and other fine details within tissues. It is often employed in research settings where detailed morphological information is required, such as in cancer research or developmental biology [58, 59]. A recent study demonstrated the capability of OR-PAM in high-resolution cellular imaging by visualizing 9 L/lacZ cells stained with 1 mg/ml X-gal solution at 37 °C for 12 h. PA images (Fig. 3a) revealed strong absorbers in the cytoplasm surrounding low-absorbing nuclei, indicating that β-galactosidase primarily localizes in the cytoplasm. A corresponding micrograph (Fig. 3b) confirmed the 20–40 μm diameter of the cells, validating OR-PAM's precision in subcellular imaging. This study underscores the potential of OR-PAM in disease diagnostics, cellular imaging, and biomedical research, offering noninvasive, high-resolution visualization of intracellular structures [17].

In contrast, AR-PAM is designed to achieve greater imaging depths while sacrificing some lateral resolution. AR-PAM employs weak optical illumination combined with tight acoustic focusing, allowing it to penetrate deeper into tissues, often exceeding 3 mm [60, 61]. The lateral resolution in AR-PAM is typically around 40–45 μm, which is sufficient for imaging larger structures such as blood vessels and tissue layers [62, 63]. This technique is particularly advantageous for in vivo imaging applications where deeper tissue visualization is necessary, such as in monitoring vascular changes or assessing tumor characteristics in live

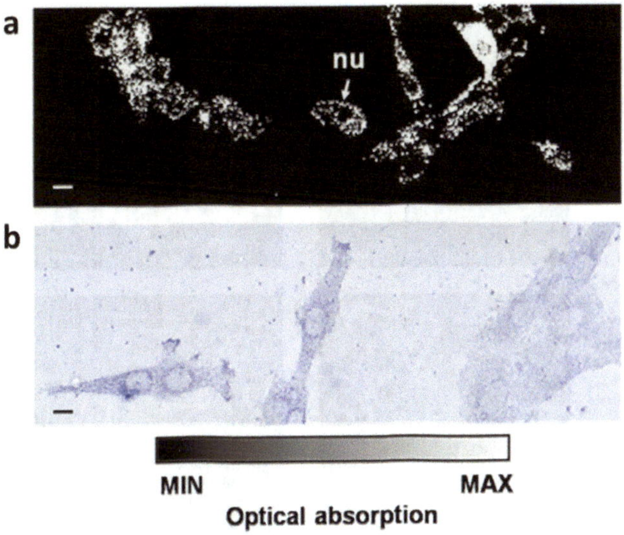


**Fig. 3** OR-PAM imaging of lacZ-stained 9 L cells. (**a**) MAP image of lacZ-expressing cells stained with X-gal, (**b**) micrograph showing cell structure, with "nu" indicating the nucleus. Scale bars: 10 μm. Reprinted with permission from [17]

animal models [64, 65]. A notable application of AR-PAM is in the simultaneous imaging of tumor microenvironments and surrounding blood vessels. In a study employing dual-wavelength PAI, researchers successfully mapped lacZ-marked tumors and their associated vasculature using two specific wavelengths (584 and 635 nm) to distinguish between hemoglobin and a blue absorption product [17]. At 635 nm, AR-PAM effectively visualized the tumor region, as the selected wavelength maximized contrast from the lacZ marker (Fig. 4a). Meanwhile, at 584 nm, the imaging system highlighted blood microvasculature, providing a relative measure of total CHb (Fig. 4b). A combined overlay image demonstrated the spatial correlation between the tumor and its surrounding vascular network, offering insights into tumor angiogenesis and metabolic interactions (Fig 4c). This case study illustrated the unique advantage of AR-PAM in functional imaging, where distinct chromophores can be mapped with high precision by selecting appropriate optical wavelengths. The ability to separately visualize tumors and vascular structures is particularly useful in oncology, aiding in tumor characterization, progression monitoring, and treatment evaluation [17]. AR-PAM's ability to image at greater depths makes it suitable for applications that require visualization of structures located beneath layers of tissue, where OR-PAM would be ineffective due to its limited penetration depth. This capability is crucial for understanding physiological processes and disease progression in more complex biological systems [53, 55].

The choice between OR-PAM and AR-PAM largely depends on the specific imaging requirements of a given study. OR-PAM is favored when high spatial

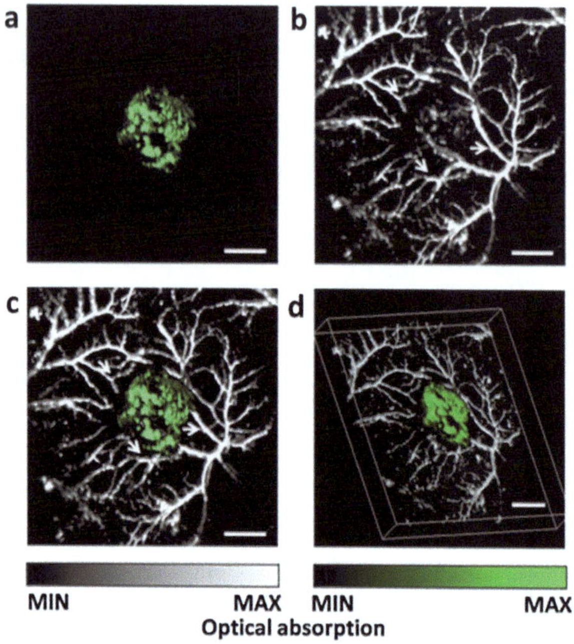


**Fig. 4** Dual-wavelength AR-PAM imaging was used to examine a *lacZ*-expressing tumor and its surrounding microvasculature in vivo. (**a**) MAP image visualised at 635 nm reveals the tumor morphology, (**b**) MAP image taken at 584 nm, highlights the microvascular network, (**c**) A combined image displays the spatial arrangement between the tumor and blood vessels, (**d**) 3D composite PA image illustrates the tumor in green along with the blood vessels supplying it (arrows). Scale bars: 2 mm. Reprinted with permission from [17]

resolution is paramount, particularly for applications involving detailed cellular imaging. On the other hand, AR-PAM is preferred for studies requiring deeper tissue penetration, where the resolution is less critical than the ability to visualize structures at greater depths [56, 57]. Recent advancements in PAM technology have led to hybrid systems that can switch between OR-PAM and AR-PAM capabilities, allowing researchers to tailor their imaging approach based on the specific needs of their experiments [64, 66]. Such systems enhance the versatility of PAM, enabling comprehensive studies that require both high resolution and deep tissue imaging. Additionally, the implementation of automated laser-fibre coupling systems has significantly improved imaging efficiency and accuracy, streamlining workflows and enhancing usability in clinical applications [67]. Pulsed laser systems are pivotal in advancing PAI technologies, with various configurations customized to specific applications. Aytac-Kipergil et al. highlighted the development of a fiber laser designed for optical resolution photoacoustic microscopy, emphasizing its high repetition frequency and the ability to independently adjust its properties. This flexibility allows for enhanced imaging capabilities, making fiber lasers a preferred choice in modern photoacoustic systems [68]. Beyond hardware improvements, noise

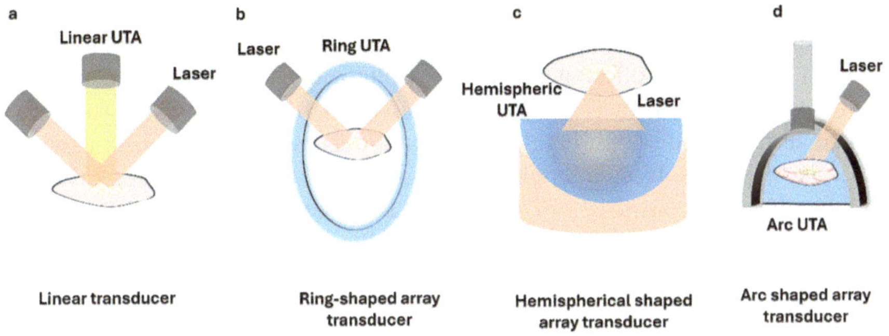


**Fig. 5** Various applications of PACT systems, (**a**) PACT system with linear transducer, (**b**) PACT system with ring-shaped array transducer. (**c**) PACT system with hemispherical array transducer, (**d**) arc-shaped array transducer. Figure was made using Microsoft® PowerPoint® for Microsoft 365 MSO (Version 2501 Build 16.0.18429.20132) 64-bit

reduction has been a key focus in PAI research to ensure sharper and more reliable imaging. The application of score-based diffusion generative models has demonstrated substantial advancements in denoising optical-resolution PAI, leading to clearer image outputs with reduced artifacts [69].

PACT is another technique, which works on the principle of photoacoustic effect, where the choice of laser wavelength plays an important role in imaging. For breast cancer detection, near-infrared (NIR) light at 1064 nm is often used, as it penetrates deep into tissue while maintaining strong contrast between healthy and malignant tissues [70]. Depending on the imaging configuration, the PACT system can further be designed with various transducer arrangements to optimize imaging based on the target application (Fig. 5). The linear transducer detects ultrasound waves generated by tissues after absorbing pulsed laser light, enabling high-resolution imaging (Fig. 5a). Ring-shaped arrays provide a 360-degree field of view, allowing detailed cross-sectional imaging of structures like the breast (Fig. 5b). Hemispherical arrays enable volumetric imaging, capturing 3D tissue structures with improved depth sensitivity (Fig. 5c). Arc-shaped arrays offer a balanced approach, optimizing both resolution and depth while allowing flexible positioning around the target area (Fig. 5d). These transducer configurations are essential for breast tissue imaging, as they help achieve comprehensive coverage, improved image quality, and precise tumor localization**.** These configurations determine imaging depth, spatial resolution, and field of view, making PACT adaptable for various clinical applications, including whole-breast imaging, tumor margin assessment, and treatment monitoring [71].

One of the most significant advancements in this field is the development of single-breath-hold PACT (SBH-PACT) by Lin et al., which allows for detailed imaging of the entire breast within a single 15 s breath hold. This system enables high resolution visualization of breast vasculature and tumors, allowing for accurate tumor localization, differentiation between benign and malignant lesions, and tracking of tumor regression during neoadjuvant chemotherapy (NAC). As shown in

Fig. 6, the study on breast cancer patients undergoing NAC at multiple time points demonstrated that SBH-PACT could effectively monitor changes in tumor vasculature and oxygenation, offering a noninvasive alternative to contrast-enhanced MRI for treatment assessment [72]. Figure 6 presents SBH-PACT breast images of a patient at three time points: baseline (T1), mid-treatment (T2), and post-treatment (T3). The depth-resolved, color-encoded angiograms of the healthy breast (Fig. 6a) and the affected breast (Fig. 6c) are shown. The contralateral images of the unaffected breast serve as controls, showing no significant angiographic changes due to NAC. In the SBH-PACT images, the cancerous region appears as an area with increased blood vessel density. To enhance visualization of these regions, vessel skeletons were extracted, and vessel density maps were generated for each image (Fig. 6b, d). The highest vessel density regions correspond to the tumor location (Fig. 6d). The images illustrate lesion shrinkage following NAC, though the patient did not achieve a complete response. Areas with elevated vessel density outside the

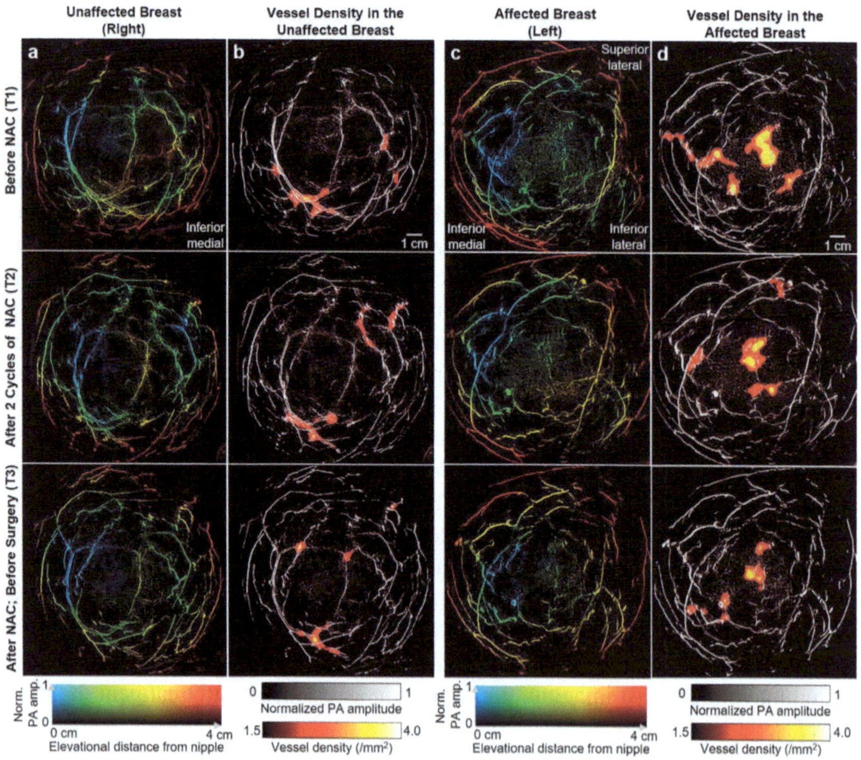


**Fig. 6** Imaging analysis of a patient receiving NAC treatment. (**a**) Depth-encoded angiograms of the unaffected (right) breast were obtained using SBH-PACT at three different time intervals. (**b**) Vessel density maps were superimposed on maximum amplitude projections (MAPs) of the breast images from (**a**). (**c**) Depth-encoded angiograms captured from the affected (left) breast. (**d**) Vessel density maps superimposed on the MAPs of the images in (**c**), marking areas with higher vascular density. Reprinted with permission from [72]

pathologically confirmed cancer site may represent false positives; however, they could also indicate early-stage lesions that have not yet been clinically diagnosed.

Another important application of PACT in breast cancer detection is its ability to differentiate between benign and malignant tumors based on optical absorption spectra. Youwei et al. investigated multi wavelength PACT for breast tumor classification, distinguishing between benign fibroadenomas, ductal carcinoma in situ (DCIS), and invasive breast cancer (IBC). By utilizing multiple laser wavelengths, they successfully identified spectral differences between tumor types, allowing for molecular characterization of breast lesions without the need for a biopsy [73]. Additionally, PACT has been used to evaluate chemotherapy response by measuring tumor oxygenation levels. Zhang et al. utilized PACT to assess tumor hypoxia by analyzing $HbO_2$ and Hb ratios, revealing that chemotherapy-sensitive tumors exhibited increased oxygenation levels over time, while chemotherapy-resistant tumors remained hypoxic [74]. Similarly, PACT has emerged as a transformative imaging modality, and its ability to track response in real time offers a significant advantage in personalized oncology. With continuous advancements in imaging technology, PACT is poised to become a key tool in improving diagnostic precision and patient outcomes.

### 2.2.3 Multispectral Photoacoustic Imaging

Multispectral photoacoustic imaging (MS-PAI) represents a significant progression in biomedical imaging, which combines the principles of PA effects with multispectral imaging techniques. The integration of multispectral imaging allows for the simultaneous acquisition of data in various wavelengths, causing their optical absorption characterization [75]. One of the primary benefits of MS-PAI is the ability to provide quantitative measurements of tissue oxidation levels. Godefroy et al. revealed that by accounting for electronic noise in multispectral PA fluctuation imaging, a fair absorption spectrum can be obtained, which is important for accurate oxygenation assessments [76]. The specificity of this method for hemoglobin allows for straightforward spectral inversion, minimizing the complexities arising from spectral unmixing required in conventional imaging techniques [76]. This internal ability of MS-PAI to focus on hemoglobin makes it particularly valuable in clinical settings, where accurate evaluation of blood oxidation is important to diagnose various conditions, including heart diseases. The versatility of MS-PAI extends to its application in cancer diagnostics. For instance, Liu et al. explored the use of broadband copper sulfide (CuS) nanoparticles in MS-PAI, which can cover both near-infrared I and II biological windows [77]. This approach allows for the simultaneous reconstruction of endogenous and exogenous chromophore concentrations, providing a comprehensive view of tumor microenvironments. The ability to visualize both blood oxygenation and the presence of nanoparticles enhances the diagnostic capabilities of MS-PAI, making it a powerful tool for cancer detection and monitoring.

Technological progress in MS-PAI is complemented by the development of better data processing techniques. Dolet et al. highlighted the importance of spatial and spectral regularization methods to increase tissue discrimination in MS-PAI [78]. By employing advanced signal processing techniques, researchers can improve the accuracy of tissue characterization, which is important for effective diagnosis and treatment planning. In the analysis of multispectral photoacoustic data, the integration of artificial intelligence (AI) and machine learning (ML) techniques in MS-PAI is paving the way for automatic image analysis and interpretation. Jnawali et al. demonstrated the ability to detect automatic cancer tissue using multispectral photoacoustic imaging, which can largely streamline the clinical process and reduce the burden on pathologists [79]. The ability to remove optical signatures and volumetric distribution of biomarkers such as $HbO_2$ and Hb may increase initial cancer detection, which is important for improving the results of treatment [79]. MS-PAI can also facilitate in the differentiation between healthy and cancerous tissues during surgery or biopsy. Zhang et al. developed a method to analyze tissue types based on absorption of light at different wavelengths. Their approach involved calculating photoacoustic contrast for various wavelengths and then using principal component analysis (PCA) to create representative spectral patterns for healthy and cancerous tissues. To classify test samples, they compared each pixel's spectrum to these reference patterns using cosine similarity. In this study, each pixel represented a small area of tissue that was captured in the imaging process, with its color or intensity corresponding to how the tissue absorbs light. The results showed that healthy tissue had a high similarity of 0.967 to the healthy reference spectrum and a lower similarity of 0.801 to the cancer reference spectrum. In contrast, cancerous tissue had a strong similarity of 0.954 to the cancer reference and a lower similarity of 0.762 to the healthy spectrum. The method was further validated by comparing it with traditional histopathology images stained with H&E. Their findings showed that healthy tissue was mainly associated with the optical absorption of deoxyhemoglobin (a component of blood), while invasive ductal carcinoma breast cancer tissue was linked to lipid absorption. This label-free imaging technique could provide real-time tumor margin assessment during surgeries and biopsies, potentially improving accuracy in cancer treatment [80].

### 2.2.4 Thermoacoustic Imaging

Thermoacoustic Imaging (TAI) is an imaging technique that exploits the photoacoustic effect, using far-infrared light or microwaves as excitation sources. TAI operates on the principle of the thermoacoustic effect, where pulsed microwaves are used to excite biological tissues, causing transient thermal expansion. This rapid expansion results in the generation of acoustic waves, known as thermoacoustic signals (TAS), which travel through the tissue and are detected by ultrasound transducers. The received signals are then processed to reconstruct detailed images of internal structures. Unlike PAI, which relies on light excitation, TAI uses microwaves as its energy source, allowing for deeper tissue penetration while maintaining high contrast and spatial resolution. The unique advantage of TAI lies in its ability to combine the strong absorption differentiation of microwave imaging with the spatial resolution of ultrasound imaging [81]. Since microwave absorption depends on the dielectric properties of biological matter, TAI can effectively map physiological and pathological differences, such as the presence of polar molecules, ion distributions, and abnormal cellular structures [82]. A key application of TAI is microwave induced thermoacoustic tomography (TAT), which is a hybrid imaging modality which integrates microwave and ultrasound imaging. TAT utilizes the high contrast of microwave absorption and the precise spatial resolution of ultrasound imaging to visualize the distribution of microwave absorption within tissues, which directly correlates with their physiological state. This technique is particularly promising for cancer diagnostics, especially in detecting breast tumors, as malignant tissues often exhibit distinct dielectric properties compared to normal tissues [83]. One key development in TAI research was the introduction of quantitative thermoacoustic imaging (qTAI) by Huang et al., where they successfully recovered an absolute conductivity map to improve tissue characterization and disease detection. Unlike conventional TAI, which offers qualitative contrast, qTAI utilizes mathematical models to generate absolute conductivity maps, improves tissue differentiation and disease detection. This technique is particularly useful for identifying abnormalities in soft tissues, where conductivity variations may signal pathological changes [84]. Later, Wu et al. developed circular scanning techniques, miniaturized systems, a handheld dipole antenna, and a low-cost microwave generator to enhance accessibility and imaging resolution. Experimental studies demonstrated the effectiveness of TAI in detecting breast, vascular, and liver diseases by distinguishing healthy and diseased tissues. Validation was performed using tissue phantoms, excised breast tumors, and in vivo imaging on animal models (mice and rabbits) as well as human subjects. These findings highlighted the potential of TAI as a reliable diagnostic tool for tissue differentiation in various disease conditions. The integration of dual modality imaging TAI and ultrasound further improved diagnostic accuracy by providing both structural and functional tissue insights. Furthermore, researchers have explored terahertz-induced thermoacoustic signals (Thz-TAS) for their strong interaction with biological macromolecules, highlighting their potential in TAI-assisted surgical procedures. Their findings indicate that THz-TAS could enhance tumor resection accuracy by providing precise tissue differentiation, making it a promising tool for future clinical applications [85].

Another study explored by Song et al. aimed to evaluate the effectiveness of Microwave-Induced thermoacoustic tomography (MITAT) in detecting and classifying breast cancer. To test the capability of MITAT, the researchers conducted experiments using real breast tissue samples, including both normal and malignant specimens. They further experimented T1 (early-stage) and T2 (advanced-stage) tumors using millimeter-sized tissue samples, simulating early breast tumors. The study demonstrated that MITAT could successfully differentiate between normal and cancerous tissues and even distinguish tumors at different clinical stages based on their microwave absorption properties [86]. Despite its widespread application, TAI possesses certain limitations, including limited penetration depth, complex image reconstruction, and thermal safety concerns, which must be overcome to facilitate clinical adoption. Continued research, technological innovations, and regulatory approvals will be essential in refining TAI, ensuring its accuracy, safety, and effectiveness. Another study improved microwave-induced TAI for breast tumor detection by developing a new screening method. The researchers created an Integrated Thermoacoustic Breast Screening System (ITBCS), which includes a flexible detector that adapts to breast tissue for better imaging and a multiaxis traction device that allows scanning from different angles. This system makes tumor detection more accurate, easier to use, and faster, improving early diagnosis and making it suitable for large-scale screenings [87].

### 2.2.5 Photoacoustic Flow Cytometry

Photoacoustic flow cytometry (PAFC) has emerged as a powerful technique for the detection and analysis of circulating tumor cells (CTCs). CTCs are released from the primary tumor to the bloodstream, and they can serve as biomarkers for the metastasis, prognosis, and treatment efficacy [88, 89]. Traditional methods to detect CTC, such as cell search systems, often suffer from limitations in sensitivity and specificity, especially in the presence of platelets that can mask CTCs, blocking tumor cell surfaces [90]. The PAFC addresses these challenges using the photoacoustic effect, which generates ultrasound signals in response to laser-inspired light absorption by CTC, allowing them to be detected without the need for labelling or wider sample preparation [91, 92]. The data collected can be processed to analyze various parameters, including the size and acoustic properties of the detected cells. This information can be used to separate CTCs from other circulating cells, such as leukocytes and platelets, depending on their unique photoacoustic abilities [93, 94]. This innovative approach connects the principles of PAI with flow cytometry, which enables the noninvasive enumeration and characterization of CTC in vivo. Figure 7 illustrates the PAFC system that uses a 1064 nm laser as the primary excitation source, which is directed and focused onto a blood vessel through a two-lens assembly. The laser wavelength is chosen depending on its ability to penetrate biological tissues with minimal scattering while being selectively absorbed by melanin-rich CTCs (e.g., in melanoma) or CTCs labeled with contrast agents such as gold nanoparticles. A pilot laser aids in the optical alignment of the system, ensuring precise focusing on the bloodstream. Mirrors and a series of spherical and cylindrical lenses further refine the beam to enhance energy delivery to the target cells. When the laser irradiates the circulating blood, normal red blood cells exhibit minimal absorption at 1064 nm. In the case of CTCs, due to their intrinsic or extrinsically enhanced optical absorption properties, absorb significantly higher energy. This differential absorption leads to rapid thermoelastic expansion of the CTCs, generating photoacoustic waves that propagate through the surrounding tissue. The generated acoustic waves are then detected by an ultrasonic transducer, which converts the signals into electrical impulses for further analysis. The strength of the photoacoustic signal correlates with the optical absorption properties of the cells, allowing selective identification of CTCs amid normal blood components. The system incorporates a trigger photodiode and an energy meter to monitor laser pulse

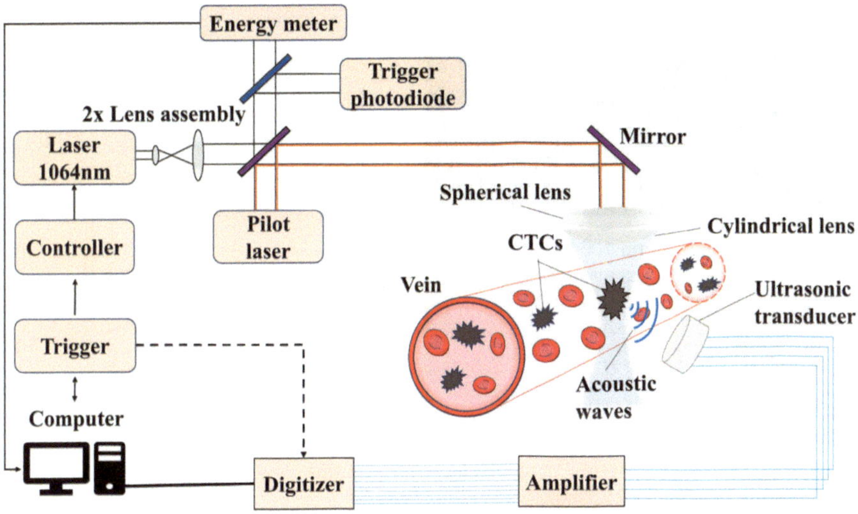


**Fig. 7** Schematics of photoacoustic flow cytometry (PAFC) device with the optical and electronic recording system used in measurement of cell responses, where the circulating tumor cells (CTCs) are irradiated using a pulsed laser system inducing photoacoustic (PA) signals. The ultrasound transducer detects the generated acoustic waves. This figure was made using Microsoft® PowerPoint® 2021 MSO (Version 2501 Build 16.0.18429.20132) 64-bit

energy and synchronize signal acquisition, ensuring reproducibility. The detected acoustic signals are amplified, hence enhancing sensitivity, followed by digitization and computational analysis through a digitizer and computer interface [91, 95]. Advanced signal processing algorithms further differentiate CTCs based on unique acoustic signatures, minimizing false positives from blood components.

He et al. illustrated the real-time detection and targeted destruction of CTC using the dual-purpose 1064 nm laser system (Fig. 7) [96]. Figure 8a represents, a low-energy 1064 nm imaging laser was employed to identify the CTC, generating a PA signal that made it visible. The dashed cyan line represents the region analyzed in the signal profile. Following detection, Fig. 8b shows a high-energy 1064 nm therapy laser pulse that was applied with a 50-μm focal diameter, ensuring that the CTC was irradiated at a lethal fluence level. The orange circle marks the therapy laser spot, while the CTC location is indicated by an “X.” The surrounding blood vessels were visualized using a 532 nm imaging laser for anatomical reference. In Fig. 8c, the PA signal profile along the dashed cyan line from Fig. 8a is shown. The red solid line corresponds to the PA signal from the imaging laser, which detected the CTC with a contrast-to-noise ratio of ~25, as highlighted in the inset. The orange dashed line represents the stronger PA signal, which is approximately 14 times higher than the imaging signal. Thus, confirming that the therapy laser accurately targeted and destroyed the CTC, with the signal peak appearing within 10 μm of the detected CTC’s position. Using this technique, He et al. successfully demonstrated a highly precise approach for real-time CTC detection and photothermal therapy, utilizing a

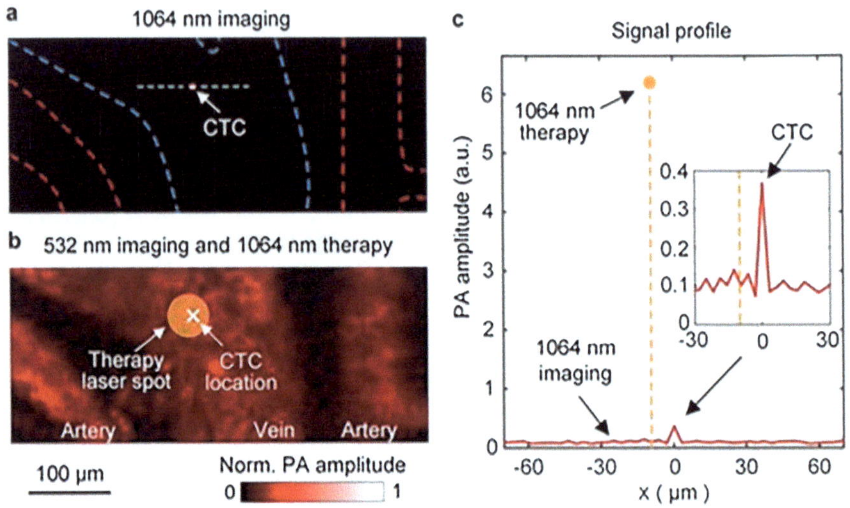


**Fig. 8** (**a**) The circulating tumor cell (CTC) was initially identified using 1064 nm laser-induced flow cytography. (**b**) It was then exposed to a therapy laser pulse at the same wavelength, with a focal diameter of 50 μm for targeted irradiation. The therapy laser pulse generated a significantly stronger photoacoustic (PA) signal, represented by a filled yellow circle. (**c**) The PA signal profile across the CTC corresponds to the dashed cyan line in (**a**). The x-axis, which follows the scanning direction of the imaging laser (moving from right to left), was centered on the CTC's position. The red solid line represents the PA signals induced by the 1064 nm imaging laser, with the CTC signal (shown in the inset) achieving a contrast-to-noise ratio of about 25. The peak PA signal from the therapy laser (dashed orange line) was observed approximately 10 μm from the detected CTC position, confirming that the CTC (marked by a white cross in (**b**)) was located within the therapy laser's focal region (circle in (**b**)) [96]

single 1064 nm laser source with adaptable energy levels to selectively eliminate CTCs while minimizing damage to surrounding tissues.

Edgar et al. indicated that PAFC can effectively be applied to CTC in patients with melanoma [91, 92]. Additionally, PAFC's ability to detect CTCs in real time and in vivo enhances its clinical utility. For instance, studies have shown that PAFC can effectively identify CTC in the mouse model of melanoma, a contrasting agent [95, 97] as an internal optical property of melanin can be used. This ability not only facilitates the calculation of CTCs but also the evaluation of their functional states, which is important to understand the dynamics and therapeutic reactions of the tumor. Figure 9 demonstrates the effectiveness of the laser-based system in targeting and eliminating CTCs to prevent metastasis. B16F10 melanoma cells, a commonly used murine (mouse) skin cancer cell line, were mixed with blood and pumped through a flow system to simulate CTCs in circulation. Figure 9a outlines the experimental setup when therapy group is represented at the top, where the system detects and destroys CTCs using a laser, while in the control group, the laser was turned off (bottom), allowing CTCs to remain intact. Once the ultrasound waves are detected, a second high-energy laser pulse is applied to destroy the CTCs,

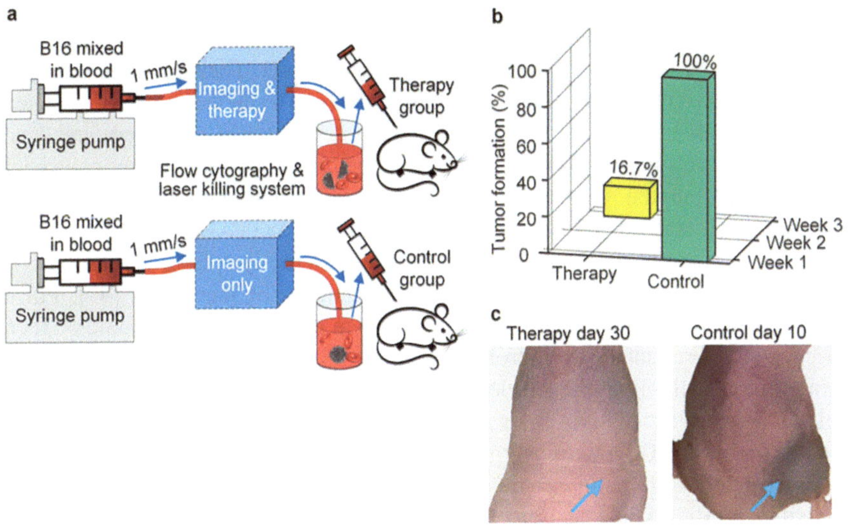


**Fig. 9** In the pseudo-therapy study, (**a**) the control experiment (bottom) was performed after the therapy experiment (top) by simply turning off the therapy laser. The flow rate was maintained at approximately 80 μL/h, with around 20 circulating tumor cells (CTCs) passing through the system per second. (**b**) Out of six therapy experiments, only one resulted in tumor formation by the third week, whereas in the control group, tumors developed in all cases within the first week. (**c**) Representative images of mice post experiment show no tumor formation at the inoculation site (blue arrow) within 30 days following a therapy experiment, while a raised tumor (blue arrow) was observed 10 days after a control experiment. Reprinted with permission from [96]

generating a stronger PA signal to confirm destruction of the cell. Figure 9b presents the results of tumor, where only 1 out of 6 therapy experiments resulted in a tumor by week 3 (16.7%), compared to 100% tumor formation within a week in the control group. Figure 9c visually confirms these findings, with the left image (therapy group, day 30) shows no visible tumor at the inoculation site, whereas the right image (control group, day 10) reveals a raised tumor, indicating rapid cancer growth. These findings highlight the potential of laser-based CTC destruction, combined with PA imaging, as an effective strategy to reduce metastasis risk in melanoma and other cancers.

PAFC has been adapted to detect a variety of CTCs, including those who can undergo epithelial-to-mesenchymal transformation (EMT). PAFC has been demonstrated across various cancer types, including breast cancer. In 2013, López et al. highlighted the importance of PAFC in providing insights into CTCs and its advantages over conventional techniques. PAFC can capture multiparameter images of CTCs, allowing for a deeper understanding of their role in cancer progression [98]. Bhattacharyya et al. demonstrated that PAFC effectively detects and isolates breast cancer cells, emphasizing its potential as a prognostic tool for evaluating CTC concentrations, which are associated with disease progression. Moreover, PAFC can detect CTCs even at extremely low concentrations (0–1000 cells/mL of blood), making it superior to conventional detection methods. Recently, Galanzha et al. studied the role of lymphatic circulating tumor cells (L-CTCs) in metastasis using PAFC in mouse models of breast cancer and melanoma. They employed a highly sensitive in vivo detection technique to monitor L-CTCs at early stages of cancer. Their observations revealed that L-CTCs were detectable even in very small primary tumors, around 0.0004 $mm^3$. From this, they concluded that tumor cells begin spreading through the lymphatic system even before metastasis is observed in clinical tests. Their study demonstrated the feasibility of using PAFC for in vivo L-CTC detection and emphasized the importance of integrating lymphatic and blood CTC assessment for a more comprehensive understanding of metastasis [99]. Pulsed laser systems have found extensive applications in various fields of oncology. It can spectrally differentiate signals from melanoma CTCs and red blood cells (RBCs) using multiple illumination wavelengths enhancing detection sensitivity and specificity [100]. Advanced algorithms and ML techniques can be employed to increase the accuracy of CTC identification and perception, which facilitates the development of individual treatment strategies [101]. For instance, studies have demonstrated their efficacy in detection of melanoma CTC, which provides insight into the dynamics of tumor progression and reaction to therapy [92, 102]. Additionally, PAFC has been used to monitor CTCs in breast cancer patients. Integration of other forms of imaging techniques with PAFC, such as MRI and CT, can provide a comprehensive approach to tumor behavior and metastasis, enhancing overall clinical abilities [91, 103]. The ability to monitor the real-time CTCs in vivo opens new avenues to understand tumor biology and develop targeted treatments. Despite its ability, many challenges in widely adopting PAFC for clinical applications are present. Various laboratories and clinical settings [104, 105] are required to standardize protocols and carry out verification studies to ensure the reliability of the results. Additionally, the development of more sophisticated contrast agents that can selectively target CTC by reducing the noise of the background are required to improve detection sensitivity. Future research should also focus on increasing PAFC's spatial resolution, which can enable its interaction with CTC clusters [92, 102].

# 3 Applications of Photoacoustic Imaging in Breast Cancer

## *3.1 Early Detection and Screening*

Breast cancer diagnosis involves imaging techniques such as MRI, ultrasound, and mammography. Mammography may be limited, particularly in women with dense breast tissue, where both dense tissue and tumors appear white on X-ray images leading to potentially masking of malignancies. PAI holds promise for the early detection and screening of breast cancer by employing the PA effect. In the case of breast cancer, early detection plays a crucial role, as it significantly improves survival rates at early stages [106]. PAI has a major advantage in screening as it can visualize tumor-associated vascular changes. The Twente Photoacoustic Mammoscope has demonstrated the capability to generate 3D images of vascular networks in both normal and malignant breast tissue, highlighting abnormalities indicative of cancer [107]. The robustness of photoacoustic imaging (PAI) is well-established, as studies have confirmed its reliability through comparative analyses with other screening methods. Figure 10 displays the average intensity projections (AIPs) in the cranio-caudal (CC) direction for MR (left) and PA (right) reconstructed data from three cases, demonstrating a progression in histopathological lesion sizes. In Fig. 10a, b, representing case P38, and Fig. 10c, d, representing case P55, the lesions measure up to 35 mm in conventional X-ray mammography (XRM). These figures show strong co-localization of lesions in MR and PA images, with a remarkable similarity in appearance and shape. In contrast, Fig. 10e, f, corresponding to case P52, depict a lesion larger than 40 mm. While the lesion's position aligns well between MR and PA images, the similarity in shape and appearance is less pronounced. This observation suggests that for larger lesions, such as those in P52, PA imaging may struggle to reconstruct the complete malignancy due to limitations in field of view (FOV). However, despite this limitation, PA imaging demonstrates high potential in detecting and localizing breast lesions, particularly for small- to medium-sized tumors, offering a promising, noninvasive imaging technique that can complement MR for improved diagnostic accuracy [7].

In recent years, advancements in PAI have facilitated the development of whole-breast imaging systems designed for effective detection and screening. Like X-ray

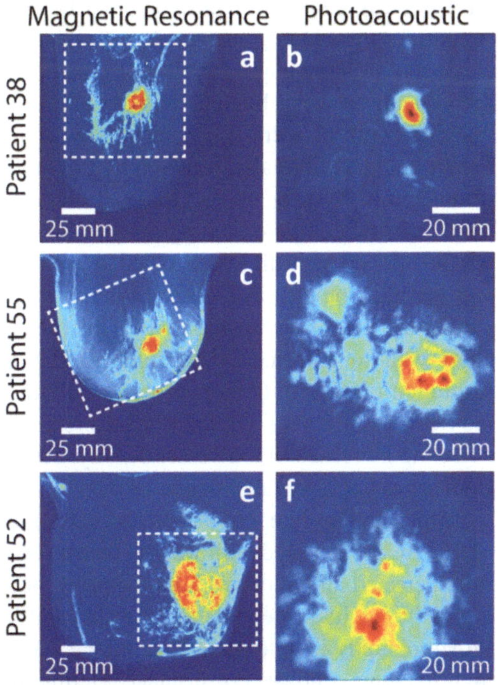


**Fig. 10** Comparison of photoacoustic imaging (PAI) and magnetic resonance imaging (MRI), MR (left) and PA (right) cranio-caudal (CC) average intensity projections (AIP) of the breast for three patients, (**a**, **b**) patient P38 with a 19 mm grade 3 invasive ductal carcinoma (IDC), (**c**, **d**) patient P55 with a 34 mm grade 2 IDC, (**e**, **f**) patient P52 with an IDC exceeding 50 mm (grade unknown). The box in the MR image highlights the region where the PA image was captured. Reprinted with permission from [7]

mammography, these systems involve breast compression with a water tank, followed by scanning the breast's surface with an ultrasonic linear array probe, which captures photoacoustic signals that are generated by high-energy laser pulses that are transmitted through a transparent support plate. In this setup, the patient stands in front of the system, gently placing one breast onto a transparent glass plate. The breast is lightly compressed against the bottom of a water tank, which helps reduce its thickness for better light penetration while also minimizing movement caused by breathing. Inside the tank, a 1D unfocused linear ultrasound (US) array moves smoothly along a track at a speed of 3 mm/s, capturing detailed images. The tank's bottom is made of low-density polyethylene (LDPE), a material carefully chosen for its acoustic properties, which closely matches those of breast tissue, ensuring high-quality imaging (Fig. 11a) [108]. Another study had demonstrated the diagnostic potential of PAI by analyzing the regional variations in the photoacoustic signals. This study evaluated 14 benign and 26 malignant breast nodules using a handheld dual-modal PA/US imaging system. Both quantitative and semiquantitative analyses revealed notable differences in PA signal distributions between benign and malignant lesions. The findings highlight the strong diagnostic potential of PAI for clinical use, supported by its ability to assess regional variations in photoacoustic signals [110]. Many advancements are being explored in PACT to revolutionize breast cancer detection using PAI. For instance, Li et al. overcame the challenges posed by PACT by developing a 3D PACT system (Fig. 11b), which facilitated depth in large images, high-speed imaging, and a scalable field of view. It was used as a multipurpose tool to reveal information in biological samples. The setup featured an imaging platform placed above the ultrasonic array housing, offering stable support for the imaging subject. A disposable plastic wrap holding cup was securely clamped onto the imaging aperture to maintain separation between the subject and the system, ensuring hygiene and reducing interference. In this study, an in vivo imaging depth of 4 cm was achieved by scanning the breast within a single 10-s breath-hold. Additionally, research by Lin et al. highlighted the use of photoacoustic imaging (PAI) for breast cancer detection by developing a single-breath-hold photoacoustic computed tomography (SBH-PACT) system (Fig. 11c). This system was positioned beneath a patient bed, minimizing the gap between the bed's surface and the ultrasonic transducer array's top scanning position. During imaging, the patient lay in a prone position while the breast was gently compressed against the chest wall to enhance image quality. The SBH-PACT system delivered high-resolution (255 μm), deep-penetration (4 cm in vivo) imaging and captured volumetric breast images within a 15-s breath-hold, reducing motion artifacts. By detecting increased blood vessel density, SBH-PACT effectively identified tumors, even in radiographically dense breasts. Its rapid imaging speed also supported dynamic studies like photoacoustic elastography for assessing tumor stiffness. SBH-PACT was tested with various breast sizes and skin tones, the system successfully detected all tumors without the need for ionizing radiation or contrast agents, highlighting its potential as a safe and noninvasive diagnostic tool [70]. PAMmography is an effective technique for visualizing breast malignancies with high imaging contrast. For instance, Heijblom et al. conducted a study where PAI has been explored for detecting breast malignancies by analyzing tumor vascularity.

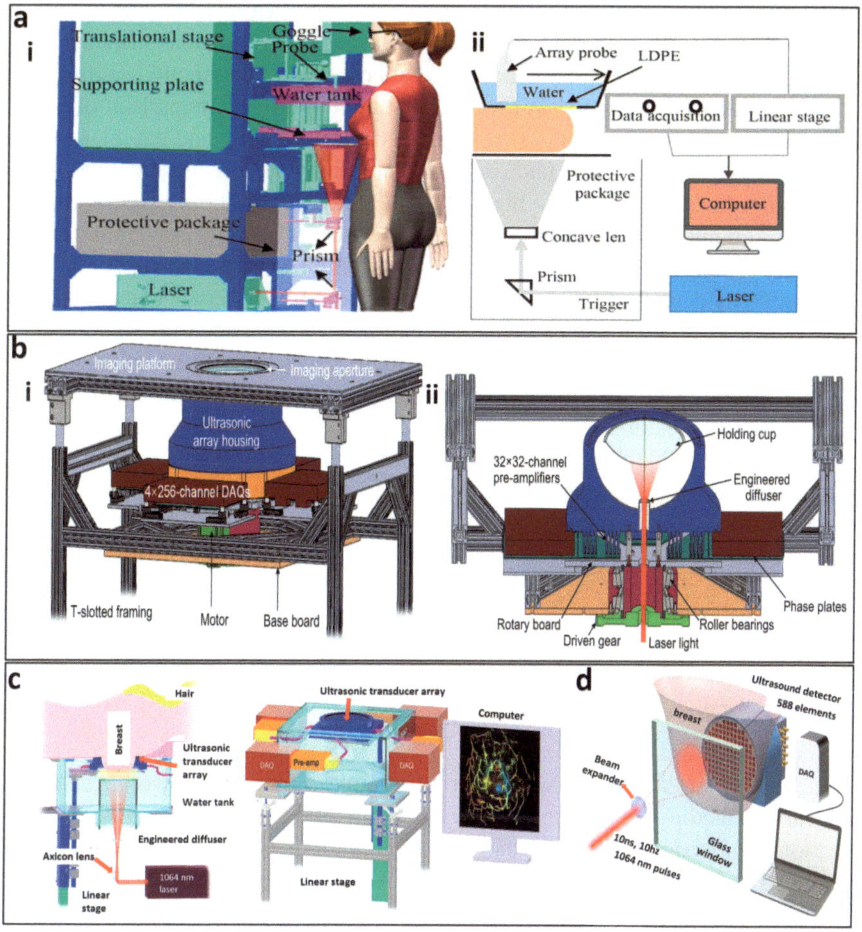


**Fig. 11** Different types of instruments used to detect breast cancer using photoacoustic imaging (PAI) imaging technique: (**a**) PAI system for full-breast imaging. (i) A 3D representation of the internal design of the PAI system. (ii) A schematic representation depicting the system's architecture [108]. (**b**) Structure and components of 3D-PACT system. (i) Perspective view of the system. (ii) A cross-sectional view of the system with the imaging platform removed [109]. (**c**) Schematic illustration of the SBH-PACT system. (i) A perspective cut-away view reveals the system without the data acquisition (DAQ) components. (ii) The system is depicted in two different views, a perspective cut-away view reveals the internal setup without the DAQ components, while another perspective view displays the system with the patient bed and optical components removed. Key elements, including the DAQ system and pre-amplifier circuits (pre-amp), are highlighted for better understanding [70]. (**d**) Schematic of the twente photoacoustic mammoscope [7]. Reprinted with permission from [7, 70, 108, 109]

Using the upgraded Twente Photoacoustic Mammoscope (Fig. 11d), researchers identified lesions in 28 of 29 cases, comparing results with x-ray, ultrasound, MR imaging, and histopathology. They observed three distinct photoacoustic patterns, which include mass-like structures that appear as solid regions, non-mass patterns that are more diffuse, and ring-shaped appearances surrounding a central area.

These patterns resemble contrast enhancement types seen in MR imaging, where contrast agents help differentiate tumor types based on vascular characteristics. Similarly, the variations in photoacoustic patterns likely correspond to differences in blood vessel structure and oxygenation within tumors. A comparison of photoacoustic image features with contrast-enhanced MR imaging and vascular-stained histopathology suggests that photoacoustic intensities are directly linked to tumor vascularity. Overall, this study highlights the potential of photoacoustic imaging (PAI) as a complementary approach for breast cancer screening, providing valuable insights into tumor visualization and characterization based on vascular features [7]. Another study demonstrated the effectiveness of PAMmography in detecting breast malignancies with high imaging contrast. The Twente Photoacoustic Mammoscope (PAM) was utilized in planar mode with a single wavelength for imaging, successfully identifying all detected lesions in alignment with conventional imaging methods. Tumor size estimation showed an average deviation of 28% from the gold standard, comparable to conventional imaging techniques. From a patient comfort perspective, most participants strongly favored PAMmography over X-ray mammography due to its less invasive nature and improved comfort during the procedure [111].

## 3.2 *Tumor Characterization and Monitoring Treatment Response*

Tumor characterization in breast cancer is a complex process that involves considering various factors such as tumor microenvironment, genetic, and epigenetic modifications. Understanding these factors is essential for developing effective treatment strategies and enhancing patient outcomes. Thakur et al. examined the role of external environmental factors and the internal cellular microenvironment in driving epigenetic modifications that lead to tumor development. Their findings emphasized the significance of a multidisciplinary approach in understanding the impact of the interactions between genes and environment on the progression of breast cancer [112]. The tumor microenvironment also has a significant role in tumor behavior and patient prognosis. Westreich et al. introduced the concept of tumor-associated collagen signatures, which describe the structural characteristics of the stroma surrounding tumors. Their findings revealed that specific collagen fibre arrangements are associated with tumor progression and patient survival, demonstrating that stromal properties can serve as prognostic markers [113]. Tumor characterization requires a detailed assessment of biological, morphological, and functional aspects of tumors. This process is crucial for making accurate diagnoses, planning treatments, and monitoring cancer progression or response to therapy. Advancements in imaging technologies, such as PAI, have significantly enhanced the ability to evaluate tumors without invasive procedures. In breast cancer, tissue characterization involves analyzing structural, functional, and molecular properties to distinguish normal, benign, and malignant conditions. Studies have demonstrated that PAI distinguishes malignant from benign lesions by analyzing CHband oxygen saturation, offering valuable insights into tumor metabolism and aggressiveness [12]. Since malignant tumors display abnormal vascularization and hypoxia, PAI enables a detailed assessment of these characteristics and aids in tumor differentiation. Research has demonstrated that mapping CHband oxygen saturation levels, which are key indicators of tumor aggressiveness, allows PAI to effectively differentiate between benign and malignant breast lesions [48]. Traditional imaging techniques such as mammography, ultrasound, and MRI provide essential structural and functional information but have certain limitations. Mammography is less effective in dense breast tissue, which can lead to false positives and false negatives [8]. Ultrasound helps in detecting non-palpable lesions but lacks specificity [9]. MRI provides high soft tissue contrast and enables the detection of angiogenesis, but it is expensive and requires contrast agents [114]. PAI has advantage over these techniques due to its ability to visualize tumor-associated angiogenesis and hypoxia, which are critical indicators of malignancy. By quantifying endogenous photo absorbers such as hemoglobin and lipids, this technique allows for the noninvasive characterization of focal breast lesions [115]. Identifying these biomarkers not only improves diagnostic accuracy but also provides important insights into tumor aggressiveness and potential treatment responses [72]. Additionally, PAI captures both intratumoral and peritumoral features, enhancing the ability of radiomic models to differentiate malignant from benign lesions.

A simple optoacoustic imaging technique was developed by Abeyakoon et al. to distinguish between benign and malignant breast lesions. In this study, 94 female patients with different types of breast lesions, such as benign, indeterminate, and suspicious ones, were imaged with optoacoustic ultrasound (OA-US). Initially, an OA-US feature set based on a single wavelength was designed from 38 patients, among whom 14 had malignant and 8 had benign solid lesions. To test the accuracy of the feature set, two independent radiologists blindly evaluated OA-US images from another 56 patients (31 malignant and 13 benign solid lesions). They observed 84.6% specificity and 96.8% sensitivity. From these results, the researchers concluded that OA-US imaging is effective for cancer detection based on its high sensitivity and specificity in distinguishing malignant from benign lesions. With a sensitivity of 96.8%, the method accurately identified nearly all malignant cases, minimizing false negatives. Additionally, its specificity of 84.6% indicated a strong ability to correctly classify benign cases, reducing false positives. The evaluation by two independent radiologists further confirmed the reliability and reproducibility of the technique. Since OA-US imaging detects vascular patterns, which are closely associated with malignancy due to increased angiogenesis in tumors, the study suggests that this imaging approach can serve as a valuable tool for identifying cancerous lesions in breast tissue. Figure 12 depicts how OA imaging at 800 nm detects neoangiogenesis (new blood vessel formation), which is a key feature in differentiating malignant from benign lesions. This supports the study's main finding that OA-US can reveal vascular patterns linked to malignancy, reinforcing its potential as a diagnostic tool [116].

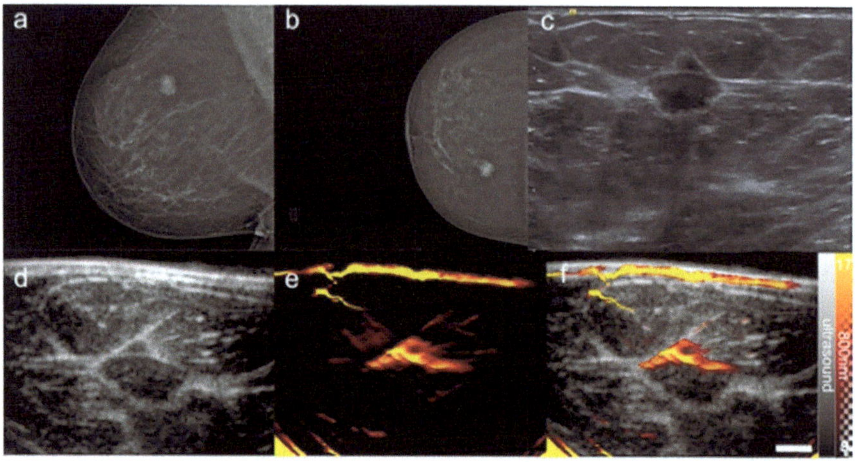


**Fig. 12** An example of a grade 2 invasive ductal carcinoma ($ER^+$ $PR^+$ $Her2^-$) is presented. Mammography images in the Mediolateral Oblique (MLO) (**a**) and Craniocaudal (CC) (**b**) views display a lobulated mass in the upper inner quadrant of the left breast, characterized by a slightly spiculated outline, (**c**) Standard ultrasound imaging reveals an irregular hypoechoic mass, while (**d**) the OA-US scan also detects an irregular hypoechoic mass, though with lower resolution compared to clinical ultrasound. (**e**) The OA image at 800 nm highlights an irregular cap of signal at the anterior margin of the tumor. (**f**) The overlaid OA-US image demonstrates the correlation between OA-detected neoangiogenesis and the tumor's position in US imaging. Strong OA signals observed at the breast surface are attributed to hemoglobin absorption in the skin's microvasculature and melanin pigmentation. Scale bar: 1 cm. Reprinted with permission from [116]

One of the key applications of PAI in assessing breast cancer treatment response is its ability to track changes in vascularity and oxygenation. Recent studies have demonstrated that this technique can be employed for quantitative analysis of the angiographic profile of tumors, offering valuable information on their response to treatments like neoadjuvant chemotherapy [109]. For instance, research was conducted using high-speed three-dimensional PACT to monitor changes in blood vessel density and morphology during neoadjuvant chemotherapy, serving as a predictive tool for relapse-free survival. Additionally, PAI has shown a strong association with tumor progression and metastasis by enabling real-time visualization of blood flow and oxygen dynamics [109]. This capability is especially valuable for assessing targeted therapy effectiveness since changes in vascularity can serve as early indicators of treatment response. Hinshaw et al. demonstrated that PAT could identify early treatment responses and resistance to bevacizumab therapy. By analyzing tumor oxygenation, the differentiation of control responsive and resistant tumors was identified within weeks of treatment initiation [117]. Moreover, the ability to monitor tumor oxygenation using PAI has been linked to predicting radiation therapy outcomes, allowing for more personalized treatment approaches. Beyond its functional imaging capabilities, PAI enhances conventional techniques like mammography and ultrasound by providing additional insights into tumor biology and treatment response [118].

# 4 Modern Innovations Transforming Photoacoustic Imaging

## *4.1 Integration of Artificial Intelligence and Deep Learning Models*

Advancements in image processing, innovative contrast agents, and transducer technology are continuously enhancing PAI, improving its effectiveness in disease diagnosis and monitoring. The introduction of deep learning (DL) techniques has played a crucial role in enhancing PAI by improving image reconstruction and artifact removal. One of the major challenges in PAI is the presence of artifacts, which are unwanted distortions or noise that can obscure minor details that are crucial for disease diagnosis. The integration of deep learning techniques in medical imaging has significantly improved image reconstruction and artifact removal. He et al. explored various strategies for eliminating artifacts from photoacoustic images of brain tumors by comparing conventional filtering techniques with advanced DL models. The dataset consisted of MRI-based brain tumor images that have been processed to generate simulated photoacoustic images with artifacts. To enhance image quality, multiple methods were applied and evaluated using two key metrics. The peak signal-to-noise ratio measures image clarity, while the structural similarity index assesses the resemblance between the processed and original images. The study compared traditional approaches such as Time Reversal (TR) and Gaussian filtering with DL models, including CycleGAN, YOLOv8, and Pix2Pix (Fig. 13). Additionally, combinations of these DL models were tested to achieve optimal artifact removal. The results indicated that traditional techniques like Time Reversal and Gaussian filtering struggle with effective artifact removal (Fig. 13b, c), often resulting in blurry images with substantial information loss. Among DL approaches, CycleGAN improves internal image clarity but does not fully eliminate external artifacts (Fig. 13d). YOLOv8, known for object detection, is more effective at removing external artifacts but still leaves internal distortions (Fig. 13e). The Pix2Pix model enhances internal image quality but does not completely address external artifacts (Fig. 13f). The YOLOv8 algorithm combined with CycleGAN improves internal and external artifact removal but remains limited in addressing artifacts at the brain boundary (Fig. 13g). However, the combined approach of YOLOv8 and Pix2Pix delivers the best performance, achieving the highest peak signal-to-noise ratio and structural similarity index values (Fig. 13h). This combination effectively removes both internal and external artifacts while preserving the tumor's structural integrity, leading to clearer images and more accurate tumor segmentation [119].

Recent advancements in computational techniques have significantly improved the performance of PAT. For instance, DL algorithms have been applied to enhance image reconstruction and artifact removal, thereby increasing the accuracy and reliability of the imaging results [120, 121]. These AI-driven algorithms generate clearer, more precise images, ultimately enhancing clinical decision-making [66, 119]. DL has also been applied to accelerate multiscale super-resolution

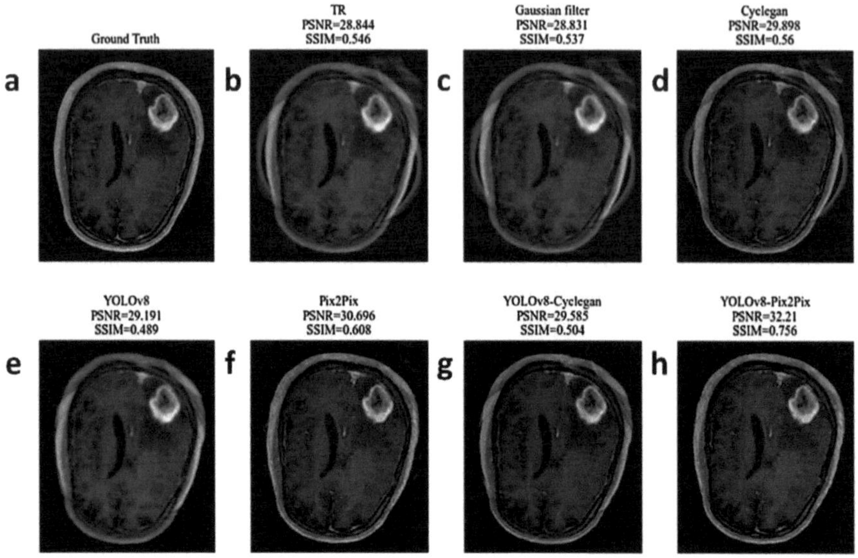


**Fig. 13** Removal of artifacts in photoacoustic images using different methods. Reprinted with permission from [119]

localization, enabling detailed imaging from cellular structures to entire organs [122]. Such precision is invaluable in personalized medicine, where high-resolution imaging plays a key role in directing targeted therapies with greater accuracy. PAI poses certain limitations due to heterogeneity in cells, leading to acoustic distortions including amplitude reduction, signal broadening, and mode transformation, which ultimately amplify low-frequency vibrations while affecting imaging of nanostructures and sharp edges. This leads to significant distortions, artifacts and PA image is compromised. DL approaches improved breast cancer detection through PAI by enhancing image quality and diagnostic precision. Various DL models, including multilayer perceptrons and stacked denoising autoencoders, contributed to these advancements. Multilayer perceptrons process complex patterns to differentiate between healthy and cancerous tissues, while stacked denoising autoencoders specialize in noise reduction, refining high-dimensional medical images for more accurate analysis. DL is also used in cancer research for gene connection prediction, which identifies relationships between genes based on expression patterns. Understanding these connections provides valuable insights into cancer progression and potential therapeutic targets. Convolutional neural networks (CNN) have been employed in this field to detect gene expression patterns, improving breast cancer classification and in breast cancer imaging, where large pathology slides are segmented into smaller sections for more precise tumor detection. A major challenge in medical imaging with DL is managing large datasets, which require substantial computational resources. To enhance the performance of the model, researchers have incorporated optimization techniques such as particle swarm optimization and artificial bee colony algorithms. Particle swarm optimization is employed to improve classification accuracy, while artificial bee colony optimization is used for optimization of solution exploration across extensive datasets. These techniques enhance DL models' accuracy and adaptability in breast cancer detection. Deep CNN have demonstrated remarkable effectiveness in PAI for breast cancer detection by improving image quality, identifying microcalcifications, and enhancing mass classification [123].

For instance, Li et al. focused on improving the preoperative assessment of whether breast cancer has spread to the axillary lymph nodes (ALNs), which are small immune system glands in the armpit area. The researchers employed PA imaging and deep learning (DL); a type of artificial intelligence (AI) that helps computers recognize patterns in images. A total of 324 patients with early-stage breast cancer (2022–2024) were analyzed, with their data split into two set, which are training group (259 patients) to develop the model and a test group (65 patients) to evaluate the accuracy of the developed model. The study compared two approaches: a traditional clinical model based on patient information, which achieved an accuracy score (AUC) of 0.775 in the training group and 0.783 in the test group, and a new AI-based model (nomogram) that combined PA imaging and deep learning, which performed better, with an AUC of 0.906 in the training group and 0.868 in the test group. The results suggest that this AI-assisted approach provides a more precise method for predicting ALN status before surgery. By improving accuracy, it could help doctors make better treatment decisions leading to better therapeutic outcomes by avoiding the need of unnecessary surgeries. Similarly, Li et al. developed a DL approach using PAI to improve the accuracy of breast cancer diagnosis. They collected data from patients with breast lesions and used a ResNet50-based DL model with attention mechanisms (PAUS-ResAM50) to analyze photoacoustic-ultrasound (PA-US) images. An attention mechanism in DL helps a model focus on the most important regions of an image while minimizing irrelevant details. In PAUS-ResAM50, it enhances the model's ability to detect key features in photoacoustic-ultrasound (PA-US) images, improving breast cancer diagnosis accuracy. Their model was tested against other similar models and showed superior performance in distinguishing cancerous from noncancerous lesions. Additionally, it enhanced radiologists ability to diagnose breast cancer more accurately without compromising sensitivity [124].

Another study by John et al. suggested that integrating DL with PA imaging could serve as a powerful tool for early breast cancer detection and clinical decision-making. They focused on improving PAT by overcoming the challenge of limited-view problem, where imaging systems cannot capture a full 360-degree view, resulting in blurry images. To address this, they developed a DL-based approach that enhances image reconstruction quality in a reduced amount of time. They employed mathematical techniques Stochastic Block Total Variation (SBTV) and Relaxed Bregman Proximal Alternating Direction Method of Multipliers (rBP-ADMM) to clean up images and integrated a U-Net denoiser, a DL model that removes noise for sharper visuals. They further tested different ways of placing the sensors in PAI to find the best arrangement to obtain a good quality image. A semicircular sensor arrangement, where sensors are placed in a half-circle manner around the target, was observed to produce images that were almost as good as those from a full-view setup, where sensors surround the target completely. Their approach outperformed other traditional methods and accelerated image reconstruction, making PAT more practical for early breast cancer detection [125]. Another study used a preprocessing algorithm to improve the clarity and consistency of breast cancer images, followed by transfer learning, where a pretrained AI model was adapted to improve classification accuracy. They compared different ML models, such as support vector machines (SVM), AlexNet, and GoogLeNet, to evaluate their effectiveness in distinguishing cancerous from noncancerous images. Their findings suggesed that integrating DL and PAT can improve clinical diagnostics. Additionally, they categorized breast cancer images into six grades based on the Breast Imaging Reporting and Data System (BI-RADS), which is a classification system used by radiologists. To make this process easier, they developed a segmentation software capable of identifying these six grades, which was tested using a mammography image database (LAPIMO EESC/USP) and showed promising results. This study focused on highlighting the potential of AI-based PAT to improve image quality [126]. By improving both efficiency and accuracy, DL brings PA imaging closer to better detection and monitoring of treatment outcomes in a clinical setup.

## 4.2 Dual-Modality and Multimodal-With US and MRI

In recent times, the integration of US and MRI with PAI in dual-modality and multimodal imaging systems has attracted considerable attention. These hybrid imaging platforms employ the strengths of individual modalities to enhance diagnostic precision, improve spatial resolution, and provide a comprehensive understanding of biological tissues. MRI is well-known for its superior soft tissue contrast and ability to visualize three-dimensional structures with high spatial resolution. While PAI provides functional insights by detecting optical absorption properties of tissues, allowing for real-time assessments of blood oxygenation and hemodynamic key factors in cancer diagnostics and treatment monitoring [127, 128]. The introduction of gadolinium-doped nanoparticles has further enhanced photoacoustic signals, making them highly effective as dual-modality contrast agents for MRI and PAI applications [129]. The integration of PAI with US and MRI has been significantly facilitated by innovations in imaging technology. One such development is the design of transparent ultrasound transducers, which enable the simultaneous use of PAI and endoscopic ultrasound (EUS), providing high-resolution tissue imaging while preserving the benefits of both modalities. This technology enhances spatial resolution and extends the imaging range, making it especially useful in applications such as gastrointestinal and vascular imaging. AI-driven algorithms simplify the analysis and interpretation of images, boosting both diagnostic accuracy and efficiency [50, 123]. Another key development is the integration of multimodal imaging in theranostics, combining therapy and diagnostics. By integrating PAI with MRI, real-time feedback during therapeutic interventions has become possible, allowing clinicians to monitor treatment responses and adjust strategies [130]. For example, PEGylated polypyrrole nanoparticles conjugated with gadolinium chelates are used in dual-modal MRI/PAI imaging to guide photothermal therapy for cancer [131].

Goh et al. explored the use of a handheld US-PA imaging probe to analyze excised breast specimens from female patients who had undergone breast-conserving surgery (BCS). Their research focused on establishing the potential of US-PA to detect biochemical markers such as lipids, collagen, and haemoglobin and compare the results with conventional imaging techniques and histopathology across different breast densities. This study involved 12 excised breast samples from patients with a mean age 59.7 ± 9.5 years. The sample tissues were illuminated with a US-PA probe at a wavelength of 700–1100 nm, followed by assessment for biochemical distribution. Their findings indicated that collagen intensity rose with increasing breast density, whereas lipid intensity exhibited a decreasing trend. Moreover, their findings were validated with imaging and histopathology results. The study revealed a significant difference in collagen intensity between cancerous and healthy tissue ($P < 0.001$), indicating its potential as a biomarker for risk assessment [132]. Figure 14 depicts the characteristics of breast tissue categorized as category B (scattered fibroglandular tissue) using different imaging techniques. These include an in vivo mammogram (Fig. 14a), ultrasound imaging of the specimen (Fig. 14b), histological analysis (Fig. 14c), PA imaging showing HbT distribution (Fig. 14d), and a combined visualization of lipid and collagen signals (Fig. 14e). In this category, normal breast tissue displayed a high presence of lipid signals, occasionally interrupted by linear or curvilinear collagen strands (Fig. 14e). These collagen structures are closely aligned with the normal fibrous stroma of the breast, as confirmed through histopathology (Fig. 14c). Moreover, the collagen distribution within the fibrous stroma closely resembled its pattern in the mammogram [132].

These technological advancements are essential for making multimodal imaging more practical and accessible in clinical settings. As the role of AI in multimodal imaging expands, ML models are increasingly applied to combine and analyze data from different imaging techniques. AI-assisted image fusion methods improve diagnostic precision by automatically aligning and registering images from various modalities, minimizing human error and enhancing clinical workflow [133]. Additionally, AI-powered pattern recognition tools can identify subtle disease markers that traditional imaging might miss, further amplifying the diagnostic capabilities of multimodal imaging systems. Multimodal imaging continues to expand its capabilities with advancements like PAT, which combines the high contrast of optical imaging with the deep penetration of ultrasound, offering detailed visualization of biological samples at significant depths [134, 135]. Jiang et al. highlighted the significance of integrating US and MRI. The potential of multimodal imaging was explored to understand the advantages of multimodal ultrasound imaging technology in breast cancer. Multiple parameter evaluations were considered for the characterization of breast tumors and to monitor treatment progression. In this

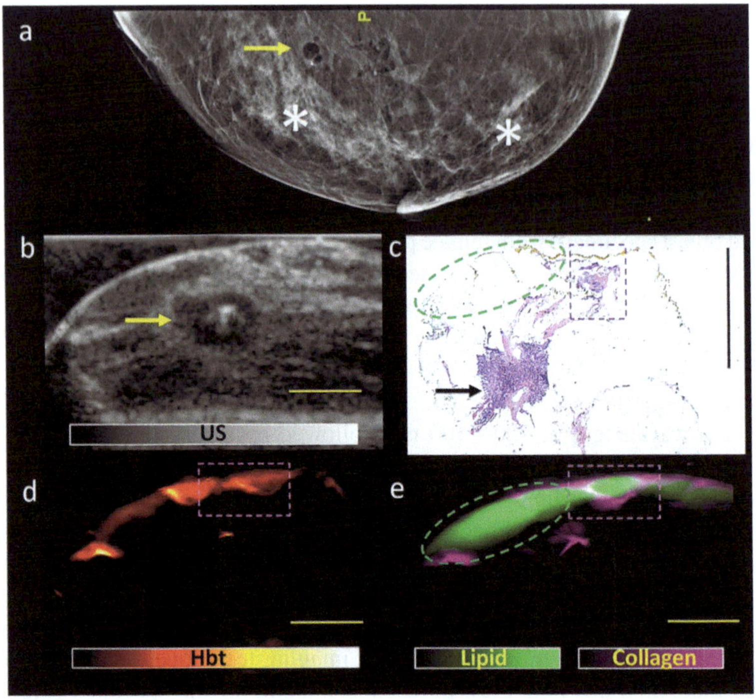


**Fig. 14** Multimodal imaging of a grade 1 invasive carcinoma (category B) in a 51-year-old woman. (**a**) A mammogram reveals the biopsy cavity with gas locules in the outer right breast, indicating a confirmed malignancy (arrow). The surrounding breast parenchyma comprises scattered fibroglandular tissue (*), classified as BIRADS category B. The patient underwent a hookwire-guided wide local excision. (**b**) Ex vivo ultrasound, coregistered with PA images (**d**, **e**) verifies the removal of the primary breast tumor, which appears as a highly hypoechoic mass with slightly irregular borders (arrow). A central echogenic focus within the mass represents the hookwire. (**c**) A hematoxylin and eosin (H&E) stained microscopic image of the tumor shows non-neoplastic breast tissue located between the tumor (black arrow) and the scanned margin. This tissue mainly consists of adipose tissue (green dotted circle), interspersed with a small amount of fibrous tissue containing breast lobules and ductules (purple dotted box), (**d**) Photoacoustic (PA) imaging of the total hemoglobin (HbT) distribution map, corresponding to (**c**), reveals a thin layer of blood on the specimen's surface along with several small traversing blood vessels. (**e**) A composite PA image displays the distribution of lipid (green) and collagen (magenta). The collagen and lipid regions, highlighted by the purple dotted box and green dotted circle, respectively, correspond to the same structures identified in (**c**). Scale bars: (**c**) 10 mm; (**b**, **d**, **e**) 5 mm. Figure reprinted with permission from [132]

chapter, a novel detection algorithm was designed by integrating deep CNN with ultrasound imaging to identify benign and malignant lumps. The algorithm's performance was observed to be better compared to the traditional CNN models, indicating that the integration of multimodal imaging can improve detection and diagnosis. Moreover, breast density grading from mammography was incorporated to show the efficiency of both these modalities in diagnosis [136]. Another study elucidated the effect of incorporating multiple imaging modalities and clinical data on the performance of AI-based diagnostic algorithms in breast cancer detection. This study compared AI models trained based on unimodal ultrasound data and dual-modality imaging on ultrasound and MRI. The results of unimodal data were rendered to be inconsistent and lack contextual integration. Apart from dual-modality imaging, including clinical and demographic data can further improve the AI-powered diagnostics. In a clinical setup, the radiologists evaluate the patient by considering the age, patient history, and risk factors, in addition to the images. The limitations of AI-driven algorithms is due to the lack of these contextual factors [137]. The integration of dual-modality and multimodal imaging techniques, particularly the combination of ultrasound and MRI with PAI, represents a major leap forward in biomedical imaging. These hybrid systems provide detailed anatomical and functional information, improving diagnostic accuracy and enabling more personalized treatment strategies. As research continues to progress, these imaging modalities will play an increasingly critical role in shaping the future of medical diagnostics and therapeutic interventions.

## 4.3 Innovations in Contrast Agents

Advancements in contrast agents have played a pivotal role in enhancing medical imaging, significantly improving the sensitivity, specificity, and overall effectiveness of modalities such as MRI, CT, and US. In addition to endogenous chromophores, PAI contrast can be greatly enhanced using exogenous contrast agents, such as nanoparticles and dyes, which can be engineered to absorb light at specific wavelengths within the near infrared (NIR) optical window (600–1100 nm). This spectral range is advantageous for deep tissue imaging, as it minimizes absorption by natural biological components such as water and hemoglobin, facilitating greater penetration depth while maintaining strong signal contrast. Furthermore, the integration of targeted molecular probes in PAI supports not only structural imaging but also functional and molecular imaging, enhancing the ability to detect and monitor cancer progression more effectively [138]. The emergence of novel contrast agents, particularly those leveraging nanotechnology, has revolutionized the field by addressing limitations of traditional agents and enhancing diagnostic precision. A significant breakthrough in multimodal imaging has been the development of nanoparticle-based contrast agents, which offer enhanced imaging capabilities due to their unique physical and chemical properties [139]. These agents improve signal strength, extend circulation time in the body, and can be functionalized for targeted imaging of specific tissues. For instance, gold nanoparticles (AuNPs) are known for their excellent X-ray attenuation, making them ideal for CT imaging [140], while metal oxide nanoparticles enhance MRI contrast by addressing the sensitivity limitations of traditional agents [141]. Another major advancement is the development of activatable contrast agents that respond to specific environmental changes, such as pH or enzyme activity, allowing for more precise imaging of pathological conditions [142]. For instance, ROS-degradable gold nanoparticle-polyphosphazene nanoprobes have shown promise in detecting inflammation and cancer [142].

The use of contrast agent is also well explored in breast cancer diagnosis. Nitric oxide (NO) plays a crucial role in cellular signaling, oxidative stress regulation, and tumorigenesis. It exhibits a dual role as a pro-oncogenic and tumor-suppressive agent thus making it a key factor in cancer biology. NO is involved in metabolic pathways, epigenetic modifications, and posttranslational protein modifications, influencing tumor growth. Its interaction with reactive oxygen species (ROS) and reactive nitrogen species (RNS) regulates apoptosis and autophagy, both of which are significant in cancer pathophysiology. Given its biological importance, NO can serve as a functional biomarker for tumor microenvironments in PA imaging, leveraging its optical absorption properties and interactions with cellular redox states [143]. Reinhardt et al. developed a series of acoustogenic probe(s) for nitric oxide (APNO) for analyte-specific detection of NO in vivo [144]. APNO-5 was observed to be the best probe developed, where it reacted with NO rapidly to form N-nitroso product, which led to a 91 nm hypsochromic shift, where the absorption or emission peak moved to a shorter wavelength (higher energy) in a spectroscopic measurement. This enabled ratiometric detection of NO, as the probe absorbs light at different wavelengths before and after reacting with NO. This shift is crucial for accurate detection, as it provides a clear distinction between the unreacted probe and its NO-bound form. The probes were developed by incorporating o-phenylenediamine (OPD) onto the aza-BODIPY dye platform. OPD selectively reacts with NO, leading to the hypsochromic shift, and the aza-BODIPY core ensures strong NIR absorption, photostability, and deep-tissue penetration, enhancing sensitivity and selectivity of PA imaging. Another study was conducted by the same group of researchers, where they optimized the APNO probe by planarizing the aza-BODIPY dye platform through steric relaxation (SR), leading to the development of the novel contrast agent (SR-APNO-3) [145]. In this study, the distribution of NO within tumors was examined by locally administering SR-APNO-3 to a murine breast cancer model. Following administration, a 1.22-fold increase in the PA signal was observed between normal tissue and the tumor, confirming that the probe underwent nitrosation by NO (Fig. 15a, b). To further confirm NO detection specificity, NG-monomethyl-L-arginine (L-NMMA), a NO synthase inhibitor, was co-administered with SR-APNO-3, resulting in an 81% reduction in the PA response, thereby verifying the selectivity of NO imaging (Fig. 15c). Control experiments involving L-NMMA injection without SR-APNO-3 showed no alterations in the PA response, eliminating the possibility of interference from endogenous absorbers like hemoglobin. Figure 15d presents the quantified relative ratiometric PA signal at 3 and 6 h, with statistical analysis conducted using two-way ANOVA ($\alpha = 0.05$) to assess differences among the flank, tumor, and L-NMMA-treated tumor. Sidak's multiple comparison test ($\alpha = 0.05$) was employed to compare relative signals against the tumor at each time point, revealing significant differences where $*p < 0.05$ and $**p < 0.01$. These advancements in molecular probe development underscore the growing potential of NO-targeted contrast agents in enhancing breast cancer diagnosis and therapeutic monitoring through PAI.

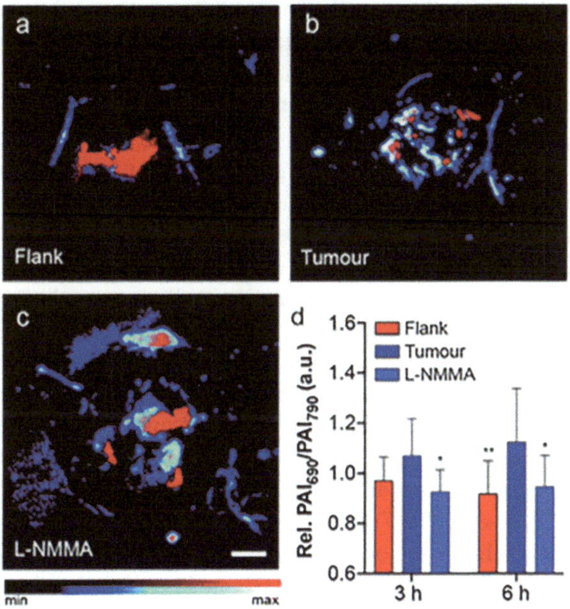


**Fig. 15** SR-APNO-3 was employed for photoacoustic (PA) imaging to detect nitric oxide (NO) derived from cancer cells. (**a–c**) Representative image overlays illustrate the distribution of the N-nitrosated product (690 nm, blue) and SR-APNO-3 (790 nm, red) 6 h following subcutaneous or intratumoral administration (25 μL, 50 mM, 15% DMF in saline v/v), (**d**) Quantification of the relative ratiometric PA signal was conducted at 3 and 6 h, with statistical comparisons made between the flank, tumor, and L-NMMA-inhibited tumor using two-way ANOVA ($\alpha = 0.05$). Sidak's multiple comparison test ($\alpha = 0.05$) was used to evaluate relative signal differences at each time point, where $*p < 0.05$ and $**p < 0.01$. Data is presented as mean ± standard deviation (flank: $n = 11$, tumor: $n = 18$, L-NMMA: $n = 9$). The scale bar represents 2.0 mm. Figure reprinted with permission from [145]

Another study by Li et al. focused on developing a noninvasive imaging approach to address intratumor heterogeneity (ITH) in breast cancer. They developed and employed multitargeted, multiplexed patchy polymeric photoacoustic contrast agents (MTMPPPCAs), designed to selectively bind to breast tumors that express estrogen and progesterone receptors. These contrast agents were then integrated with single-impulse panoramic photoacoustic computed tomography (SIP-PACT), a cutting-edge imaging method capable of capturing high-resolution images while providing real-time visualization of tumor features. Followed by this, they further validated their results by performing fluorescence and photoacoustic measurements, which confirmed the ability of MTMPPPCAs to differentiate between tumor subtypes. Additionally, they performed tissue pathology analysis to further validate the molecular specificity of their imaging system. Their findings showed that this hybrid imaging technology can effectively identify ITH in nonmetastatic tumors, which could significantly improve breast cancer diagnosis and treatment strategies by enabling more personalized therapeutic approaches [146]. Yang et al. developed a novel contrast agent called indocyanine green (ICG)-conjugated superparamagnetic iron oxide nanoworms (NW–ICG), designed to enhance noninvasive breast cancer imaging. When tested in a spontaneous mouse breast cancer model, NW–ICG demonstrated high tumor-to-normal tissue contrast across multiple imaging modalities, including magnetic particle imaging, near-infrared fluorescence imaging, and photoacoustic imaging, thereby improving the accuracy of breast cancer detection [147]. Similarly, Xiao et al. explored the dynamics of a tumor-targeted contrast agent, HypoxiSense 680 (HS680), in breast xenograft mouse models using Dynamic Contrast-Enhanced Multispectral Optoacoustic Tomographic (DCE-MSOT) imaging. They employed an arterial input function (AIF) approach, which tracks the concentration of the contrast agent in blood plasma over time, serving as a reference for kinetic modeling of tissue uptake. By applying multispectral unmixing, the distinct spectral signature of HS680, which peaks at 680 nm, was isolated from background signals, enabling accurate localization and intensity differentiation within the tumor. Cross-sectional images of the breast tumor region were acquired, with multiple visualization layers aiding in interpretation (Fig. 16). The first row in the Fig. 16 displays grayscale MSOT images, providing a general anatomical reference. The second row highlights the distribution of $HbO_2$, which helps to define internal structures. The third row depicts the HS680 signal intensity, showing its biodistribution within the mouse. Additionally, an overlay image combined the HS680 signal gradient with a single-wavelength background (900 nm), enhancing visualization of the contrast agent within the tumor environment. The $HbO_2$ signal played a crucial

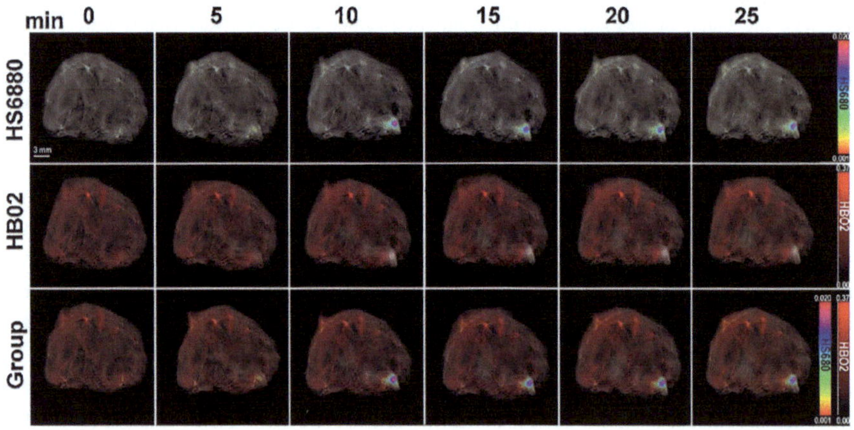


**Fig. 16** The uptake of HypoxiSense 680 (HS680) in breast tumors was assessed using multispectral optoacoustic tomographic (MSOT) imaging. Female mice implanted with breast tumors were fitted with tail-vein catheters, equilibrated for 5 min, and imaged every 10 s before HS680 injection. Over time at 10, 15, 20, and 25 min, HS680 accumulated in the tumor microenvironment, shown in a rainbow color bar. The first row displays grayscale background MSOT images, the second row shows oxygenated hemoglobin ($HbO_2$) distribution in red, the third row represents HS680 biodistribution, and the fourth row presents an overlay of HS680 and $HbO_2$ on a 900 nm background image. Figure reprinted with permission from [148]

role in defining anatomical structures, ensuring accurate localization of HS680. This optoacoustic imaging approach enabled detailed tracking of the contrast agent's dynamics in vivo, offering valuable insights into its accumulation and interaction within the tumor microenvironment [148].

## 5 Regulatory and Ethical Considerations

PAI is emerging as a valuable technique in breast cancer diagnostics due to its ability for the early diagnosis and treatment. However, its progression from experimental research to widespread clinical implementation is impeded by various regulatory and ethical challenges that need to be addressed to ensure both patient safety and the effective use of the technology in healthcare [149]. A primary regulatory challenge in PAI is the absence of standardized imaging protocols across different systems. Variations in hardware, such as laser sources and ultrasound transducers, contribute to inconsistencies in imaging practices, making it difficult to establish universal clinical standards. The development of standardized guidelines for image acquisition, processing, and the interpretation is essential to improve reliability and comparability. Organizations such as the International Photoacoustic Standardization Consortium (IPASC) aim to develop unified standards to facilitate large scale clinical studies and advance the integration of PAI into routine medical practice [150, 151]. Ethical considerations also play a crucial role in the clinical deployment of PAI. Although it is a non-invasive imaging method, the use of laser-based imaging raises concerns about potential tissue damage, especially when contrast agents are employed to enhance imaging quality [152]. Patients must be fully informed about the risks and benefits of the procedure before undergoing imaging. This is particularly for individuals with dense breast tissue, where traditional imaging modalities may be less effective, and PAI could offer enhance diagnostic accuracy [153]. Another major ethical concern is data privacy. The high-resolution nature of PAI images, when combined with other medical records could enhance the risk of patient identity exposure. Developing stringent data protection protocols and ensuring compliance with existing privacy regulations, such as the health insurance Portability and Accountability Act (HIPPA) in the United States and the General Data protection Regulation (GDPR) in Europe, will be essential steps in the responsible clinical implementation of PAI [154]. The use of contrast agents in PAI further introduces regulatory and ethical concerns. While these agents significantly enhance image contrast and diagnostic accuracy, their long-term effects on human health remain uncertain [10]. Regulatory bodies require extensive toxicological assessments and clinical trials to confirm their safety before approving them for widespread medical use [10]. Furthermore, the capability of PAI to assess tumor microenvironments offers both benefits and ethical challenges. This technique allows detailed visualization of vascular networks and oxygenation levels, aiding in the development of personalized treatment strategies. However, detection of incidental findings can cause concerns about potential distress and the ethical obligation of healthcare providers to disclose such results responsibly [10]. In conclusion, although PAI holds great promise for improving breast cancer diagnostics, its successful clinical implementation depends on overcoming regulatory and ethical challenges. Addressing these issues through standardized imaging protocols, patient safety measures, and data security policies will be critical for ensuring the safe and effective adoption of PAI in healthcare [155].

## 6 Challenges and Future Perspectives

PAI is emerging as a promising technique for breast cancer diagnostics, providing high resolution, noninvasive functional imaging. However, before it can be widely adopted as a standard clinical tool, several challenges must be addressed. A key obstacle is the necessity for extensive clinical validation. Unlike well-established imaging methods such as mammography and MRI, PAI has yet to undergo large scale clinical trials to confirm its accuracy and reliability in detecting breast cancer. Due to the lack of substantial clinical evidence, regulatory authorities remain cautious about granting approval for its widespread medical application. Overcoming this barrier will require coordinated efforts to conduct rigorous studies and accumulate sufficient data to support diagnostic capabilities of PAI [151]. Integrating PAI into existing diagnostic workflows remains challenging, as mammography, ultrasound, and MRI are the primary imaging tools for breast cancer detection. For PAI to be widely adopted, it needs to establish its effectiveness in early tumor detection and prove its cost efficiency, as healthcare facilities carefully evaluate the technologies based on their clinical and financial impact [156]. Contrast agents in PAI offer both advantages and challenges. While they enhance imaging contrast and specificity, their long-term safety and biocompatibility are still being studied. Regulatory agencies mandate rigorous toxicological assessments and clinical trials to ensure their safety before approving widespread medical use. Advancements in nanoparticle-based contrast agents with biodegradable properties may help address these concerns and enhance the clinical applicability of PAI [157]. Advancements in AI and ML have the potential to significantly enhance PAI by improving image analysis and streamlining diagnostic workflows. AI-powered algorithms can improve tumor detection and classification accuracy, thereby reducing the workload of radiologists while enhancing diagnostic efficiency. However, integrating AI into PAI also introduces regulatory and ethical challenges, including concerns about algorithm transparency, biases in training data, and the necessity for clinical validation to ensure accuracy in medical decision-making [155]. Future progress in PAI should prioritize the development of compact and cost-effective imaging devices, improving accessibility for healthcare centres, especially in resource limited areas. Additionally, continued research into innovative contrast agents targeting specific breast cancer biomarkers may enhance accuracy of PAI in early detection. Furthermore, PAI holds great potential for monitoring treatment response over time, allowing real time assessment of therapeutic effectiveness and facilitating personalized treatment modifications [156]. In conclusion, while PAI presents a transformative opportunity for breast cancer diagnostic, addressing its current limitations and driving future advancements are essential for its widespread adoption in clinical practice. Efforts to standardize imaging protocols, conduct extensive clinical validation, and advance technological innovations will be key to establishing PAI as a reliable diagnostic tool. Addressing these challenges will enable PAI to play a crucial role in early breast cancer detection and personalized treatment approaches, ultimately enhancing patient outcomes.

## 7 Conclusion

Breast cancer continues to be a significant global health challenge, where early diagnosis plays a vital role in enhancing survival rates and treatment effectiveness. Traditional imaging methods like mammography, ultrasound, and MRI face challenges in detecting tumors in dense breast tissue and distinguishing malignant from benign lesions. This has driven the search for more advanced, accurate, and noninvasive diagnostic tools, with PAI emerging as a promising alternative. PAI combines optical and ultrasound techniques, providing high-resolution, real-time imaging of tumor vascularization, oxygenation, and tissue structure. Unlike conventional methods, it offers functional insights into tumor biology and does not rely on ionizing radiation, making it safer for frequent screenings. Recent advancements, including contrast agents like nanoparticles and indocyanine green, have enhanced PAI's sensitivity and specificity. Additionally, combining PAI with MRI and ultrasound improves breast lesion assessment and diagnostic accuracy. Despite its potential, challenges remain before PAI can be widely adopted in clinical practice. Standardizing imaging protocols, obtaining regulatory approvals, and conducting large-scale clinical studies are essential. Cost-effectiveness and accessibility, particularly in low-income regions, also need consideration. Future advancements in AI and ML could enhance image analysis, while portable and cost-effective PAI devices may improve accessibility. In conclusion, PAI is a groundbreaking approach to breast cancer diagnostics, offering noninvasive, high-resolution, and functionally informative tumor detection. With continued research and technological advancements, PAI could revolutionize breast cancer screening, improve early detection, and reduce mortality worldwide.

## References

1. Sung, H., Ferlay, J., Siegel, R. L., Laversanne, M., Soerjomataram, I., Jemal, A., et al. (2021). Global cancer statistics 2020: GLOBOCAN estimates of incidence and mortality worldwide for 36 cancers in 185 countries. *CA: A Cancer Journal for Clinicians, 71*, 209–249. https://doi.org/10.3322/caac.21660

2. Wilkinson, L., & Gathani, T. (2022). Understanding breast cancer as a global health concern. *The British Journal of Radiology, 95*, 20211033. https://doi.org/10.1259/bjr.20211033
3. Feng, Y., Spezia, M., Huang, S., Yuan, C., Zeng, Z., Zhang, L., et al. (2018). Breast cancer development and progression: Risk factors, cancer stem cells, signaling pathways, genomics, and molecular pathogenesis. *Genes and Diseases, 5*, 77–106. https://doi.org/10.1016/j.gendis.2018.05.001
4. Takeshima, H., & Ushijima, T. (2019). Accumulation of genetic and epigenetic alterations in normal cells and cancer risk. *npj Precision Oncology, 3*, 7. https://doi.org/10.1038/s41698-019-0079-0
5. Pace, L. E., & Shulman, L. N. (2016). Breast cancer in sub-Saharan Africa: Challenges and opportunities to reduce mortality. *The Oncologist, 21*, 739–744. https://doi.org/10.1634/theoncologist.2015-0429
6. Theivendren, P., Kunjiappan, S., Mariappa Hegde, Y., Ravi, K., Vellaichamy, S., Gopal, M., et al. (2022). The future perspectives of drug repurposing and treatment for the drug resistant breast cancer: A review. In *Drug repurposing - Molecular aspects and therapeutic applications*. IntechOpen. https://doi.org/10.5772/intechopen.100143
7. Heijblom, M., Piras, D., Brinkhuis, M., van Hespen, J. C. G., van den Engh, F. M., van der Schaaf, M., et al. (2015). Photoacoustic image patterns of breast carcinoma and comparisons with magnetic resonance imaging and vascular stained histopathology. *Scientific Reports, 5*, 11778. https://doi.org/10.1038/srep11778
8. Pisano, E. D., Gatsonis, C., Hendrick, E., Yaffe, M., Baum, J. K., Acharyya, S., et al. (2005). Diagnostic performance of digital versus film mammography for breast-cancer screening. *New England Journal of Medicine, 353*, 1773–1783. https://doi.org/10.1056/NEJMoa052911
9. Bachawal, S. V., Jensen, K. C., Wilson, K. E., Tian, L., Lutz, A. M., & Willmann, J. K. (2015). Breast cancer detection by B7-H3–targeted ultrasound molecular imaging. *Cancer Research, 75*, 2501–2509. https://doi.org/10.1158/0008-5472.CAN-14-3361
10. Wang, L. V., & Hu, S. (2012). Photoacoustic tomography: In vivo imaging from organelles to organs. *Science, 335*, 1458–1462. https://doi.org/10.1126/science.1216210
11. Hu, J., Yu, M., Ye, F., & Xing, D. (2011). In vivo photoacoustic imaging of osteosarcoma in a rat model. *Journal of Biomedical Optics, 16*, 020503. https://doi.org/10.1117/1.3544502
12. Ripping, T. M., Hubbard, R. A., Otten, J. D. M., den Heeten, G. J., Verbeek, A. L. M., & Broeders, M. J. M. (2016). Towards personalized screening: Cumulative risk of breast cancer screening outcomes in women with and without a first-degree relative with a history of breast cancer. *International Journal of Cancer, 138*, 1619–1625. https://doi.org/10.1002/ijc.29912
13. Wilson, K. E., Bachawal, S. V., Abou-Elkacem, L., Jensen, K., Machtaler, S., Tian, L., et al. (2017). Spectroscopic photoacoustic molecular imaging of breast cancer using a B7-H3-targeted ICG contrast agent. *Theranostics, 7*, 1463–1476. https://doi.org/10.7150/thno.18217
14. Sun, I.-C., Ahn, C.-H., Kim, K., & Emelianov, S. (2019). Photoacoustic imaging of cancer cells with glycol-chitosan-coated gold nanoparticles as contrast agents. *Journal of Biomedical Optics, 24*, 1. https://doi.org/10.1117/1.JBO.24.12.121903
15. Sangha, G. S., Hu, B., Li, G., Fox, S. E., Sholl, A. B., Brown, J. Q., et al. (2022). Assessment of photoacoustic tomography contrast for breast tissue imaging using 3D correlative virtual histology. *Scientific Reports, 12*, 2532. https://doi.org/10.1038/s41598-022-06501-3
16. Hosseinaee, Z., Le, M., Bell, K., & Reza, P. H. (2020). Towards non-contact photoacoustic imaging. *Photoacoustics, 20*, 100207. https://doi.org/10.1016/j.pacs.2020.100207
17. Cai, X., Paratala, B. S., Hu, S., Sitharaman, B., & Wang, L. V. (2012). Multiscale photoacoustic microscopy of single-walled carbon nanotube-incorporated tissue engineering scaffolds. *Tissue Engineering. Part C, Methods, 18*, 310–317. https://doi.org/10.1089/ten.tec.2011.0519
18. Becker, A., Masthoff, M., Claussen, J., Ford, S. J., Roll, W., Burg, M., et al. (2018). Multispectral optoacoustic tomography of the human breast: Characterisation of healthy tissue and malignant lesions using a hybrid ultrasound-optoacoustic approach. *European Radiology, 28*, 602–609. https://doi.org/10.1007/s00330-017-5002-x

19. Yang, Y., Xia, Z., Xia, H., Li, Y., Liu, G., & Xu, J. (2020). Applied current thermoacoustic imaging for biological tissues. *Technology and Health Care, 28*, 57–65. https://doi.org/10.3233/THC-209007
20. Huang, Z., Mo, S., Wu, H., Kong, Y., Luo, H., Li, G., et al. (2024). Optimizing breast cancer diagnosis with photoacoustic imaging: An analysis of intratumoral and peritumoral radiomics. *Photoacoustics, 38*, 100606. https://doi.org/10.1016/j.pacs.2024.100606
21. Bell, A. G. (1880). On the production and reproduction of sound by light. *American Journal of Science, 20*, 305–324. https://doi.org/10.2475/ajs.s3-20.118.305
22. Huang, C., Wang, K., Nie, L., Wang, L. V., & Anastasio, M. A. (2013). Full-wave iterative image reconstruction in photoacoustic tomography with acoustically inhomogeneous media. *IEEE Transactions on Medical Imaging, 32*, 1097–1110. https://doi.org/10.1109/TMI.2013.2254496
23. Sun, Y., & Jiang, H. (2022). Enhancing finite element-based photoacoustic tomography by localized reconstruction method. *Photonics, 9*, 337. https://doi.org/10.3390/photonics9050337
24. Moon, S. (2016). A radon-type transform arising in photoacoustic tomography with circular detectors. *Journal of Inverse and Ill-Posed Problems, 24*, 233–244. https://doi.org/10.1515/jiip-2014-0008
25. Yao, J., & Wang, L. V. (2011). Photoacoustic tomography: Fundamentals, advances and prospects. *Contrast Media & Molecular Imaging, 6*, 332–345. https://doi.org/10.1002/cmmi.443
26. Gharieb, R. (2020). Photoacoustic imaging for cancer diagnosis: A breast tumor example. In *Photoacoustic imaging - Principles, advances and applications*. IntechOpen. https://doi.org/10.5772/intechopen.92084
27. Rosencwaig, A., & Gersho, A. (1976). Theory of the photoacoustic effect with solids. *Journal of Applied Physics, 47*, 64–69. https://doi.org/10.1063/1.322296
28. Weng, Y.-X., Xiao, H., Chan, K.-C., & Che, C.-M. (2000). Direct observation of mass transfer at solid–liquid interface by laser flash photolysis of the interface probe molecules. *The Journal of Physical Chemistry. B, 104*, 7713–7724. https://doi.org/10.1021/jp001071n
29. Knyazev, G., Ignatyeva, D., Sopko, I., Belotelov, V., & Romanov, O. (2020). Amplification of the electrostriction mechanism of photoacoustic conversion in layered media. *Journal of Physics D: Applied Physics, 53*, 475101. https://doi.org/10.1088/1361-6463/aba872
30. Wang, L. V. (2009). Multiscale photoacoustic microscopy and computed tomography. *Nature Photonics, 3*, 503–509. https://doi.org/10.1038/nphoton.2009.157
31. Jin, Y., Jia, C., Huang, S.-W., O'Donnell, M., & Gao, X. (2010). Multifunctional nanoparticles as coupled contrast agents. *Nature Communications, 1*, 41. https://doi.org/10.1038/ncomms1042
32. Bal, G., & Kirkeby, A. (2019). Photo-acoustic tomography in the rotating setting. *Inverse Problems, 35*, 105001. https://doi.org/10.1088/1361-6420/ab315c
33. Qu, M., Mehrmohammadi, M., Truby, R., Graf, I., Homan, K., & Emelianov, S. (2014). Contrast-enhanced magneto-photo-acoustic imaging in vivo using dual-contrast nanoparticles. *Photoacoustics, 2*, 55–62. https://doi.org/10.1016/j.pacs.2013.12.003
34. Slump, C. H., Willemink, R. G. H., Manohar, S., & Hoenders, B. J. (2011). On the role of passive elements in photoacoustic reconstruction. *Proceedings of SPIE, 12*, 78993. https://doi.org/10.1117/12.875327
35. Neprokin, A., Broadway, C., Myllylä, T., Bykov, A., & Meglinski, I. (2022). Photoacoustic imaging in biomedicine and life sciences. *Life, 12*, 588. https://doi.org/10.3390/life12040588
36. Li, L., Zemp, R. J., Lungu, G., Stoica, G., & Wang, L. V. (2007). Photoacoustic imaging of lacZ gene expression in vivo. *Journal of Biomedical Optics, 12*, 020504. https://doi.org/10.1117/1.2717531
37. Huang, B., Bates, M., & Zhuang, X. (2009). Super-resolution fluorescence microscopy. *Annual Review of Biochemistry, 78*, 993–1016. https://doi.org/10.1146/annurev.biochem.77.061906.092014

38. Hanahan, D., & Weinberg, R. A. (2011). Hallmarks of cancer: The next generation. *Cell, 144*, 646–674. https://doi.org/10.1016/j.cell.2011.02.013
39. Yao, J., Maslov, K. I., Zhang, Y., Xia, Y., & Wang, L. V. (2011). Label-free oxygen-metabolic photoacoustic microscopy in vivo. *Journal of Biomedical Optics, 16*, 076003. https://doi.org/10.1117/1.3594786
40. Ranasinghesagara, J. C., & Zemp, R. J. (2010). Combined photoacoustic and oblique-incidence diffuse reflectance system for quantitative photoacoustic imaging in turbid media. *Journal of Biomedical Optics, 15*, 046016. https://doi.org/10.1117/1.3470336
41. Bauer, A. Q., Nothdurft, R. E., Erpelding, T. N., Wang, L. V., & Culver, J. P. (2011). Quantitative photoacoustic imaging: Correcting for heterogeneous light fluence distributions using diffuse optical tomography. *Journal of Biomedical Optics, 16*, 096016. https://doi.org/10.1117/1.3626212
42. Guo, Z. (2012). Quantitative photoacoustic microscopy of optical absorption coefficients from acoustic spectra in the optical diffusive regime. *Journal of Biomedical Optics, 17*, 066011. https://doi.org/10.1117/1.JBO.17.6.066011
43. Guo, Z., Hu, S., & Wang, L. V. (2010). Calibration-free absolute quantification of optical absorption coefficients using acoustic spectra in 3D photoacoustic microscopy of biological tissue. *Optics Letters, 35*, 2067. https://doi.org/10.1364/OL.35.002067
44. Wang, R. K., & Hurst, S. (2007). Mapping of cerebro-vascular blood perfusion in mice with skin and skull intact by optical micro-angiography at 13μm wavelength. *Optics Express, 15*, 11402. https://doi.org/10.1364/OE.15.011402
45. Sheinfeld, A., Gilead, S., & Eyal, A. (2010). Time-resolved photoacoustic Doppler characterization of flow using pulsed excitation. *Proceedings of SPIE, 2010*, 75643. https://doi.org/10.1117/12.858340
46. Fang, H., Maslov, K., & Wang, L. V. (2007). Photoacoustic Doppler flow measurement in optically scattering media. *Applied Physics Letters, 91*, 264103. https://doi.org/10.1063/1.2825569
47. Prakash, R., Benavides, J., & Avanaki, K. (2024). Evaluation of multispectral unmixing algorithm for HbO2 and HbR. In A. A. Oraevsky & L. V. Wang (Eds.), *Photons plus ultrasound: Imaging and sensing* (p. 153). SPIE. https://doi.org/10.1117/12.3003563
48. Nyayapathi, N., & Xia, J. (2019). Photoacoustic imaging of breast cancer: A mini review of system design and image features. *Journal of Biomedical Optics, 24*, 1. https://doi.org/10.1117/1.JBO.24.12.121911
49. Balasundaram, G., Ho, C. J. H., Li, K., Driessen, W., Dinish, U., Wong, C. L., et al. (2015). Molecular photoacoustic imaging of breast cancer using an actively targeted conjugated polymer. *International Journal of Nanomedicine, 10*, 387. https://doi.org/10.2147/IJN.S73558
50. Upputuri, P. K., & Pramanik, M. (2015). Performance characterization of low-cost, high-speed, portable pulsed laser diode photoacoustic tomography (PLD-PAT) system. *Biomedical Optics Express, 6*, 4118. https://doi.org/10.1364/BOE.6.004118
51. Jayapal, N., & Shetty, G. (2013). Photoacoustic imaging: An orchestra of light and sound. *Journal of Evolution of Medical and Dental Sciences, 2*, 8713–8723. https://doi.org/10.14260/jemds/1512
52. Zhang, L., & Wang, X. (2022). Optical resolution photoacoustic microscopy imaging in the detection of early oral cancer under image reconstruction algorithm. *Computational and Mathematical Methods in Medicine, 2022*, 1–11. https://doi.org/10.1155/2022/6077748
53. Matthews, T. P., Zhang, C., Yao, D.-K., Maslov, K. I., & Wang, L. V. (2014). Label-free photoacoustic microscopy of peripheral nerves. *Journal of Biomedical Optics, 19*, 1. https://doi.org/10.1117/1.JBO.19.1.016004
54. Avti, P. K., Hu, S., Favazza, C., Mikos, A. G., Jansen, J. A., Shroyer, K. R., et al. (2012). Detection, mapping, and quantification of single walled carbon nanotubes in histological specimens with photoacoustic microscopy. *PLoS One, 7*, e35064. https://doi.org/10.1371/journal.pone.0035064

55. Chen, S.-L., Burnett, J., Sun, D., Wei, X., Xie, Z., & Wang, X. (2013). Photoacoustic microscopy: A potential new tool for evaluation of angiogenesis inhibitor. *Biomedical Optics Express, 4*, 2657. https://doi.org/10.1364/BOE.4.002657
56. Moothanchery, M., Bi, R., Kim, J. Y., Jeon, S., Kim, C., & Olivo, M. (2018). Optical resolution photoacoustic microscopy based on multimode fibers. *Biomedical Optics Express, 9*, 1190. https://doi.org/10.1364/BOE.9.001190
57. Attia, M. F., Anton, N., Wallyn, J., Omran, Z., & Vandamme, T. F. (2019). An overview of active and passive targeting strategies to improve the nanocarriers efficiency to tumour sites. *Journal of Pharmacy and Pharmacology, 71*, 1185–1198. https://doi.org/10.1111/jphp.13098
58. Yao, J., & Wang, L. V. (2013). Photoacoustic microscopy. *Laser & Photonics Reviews, 7*, 758–778. https://doi.org/10.1002/lpor.201200060
59. Wang, P.-H., & Li, M.-L. (2012). DVD pickup head based optical resolution photoacoustic microscopy. *Proceedings of the SPIE, 2012*, 82232V. https://doi.org/10.1117/12.907680
60. Li, X., Huang, B., Yeung, K., Tsang, V. T. C., Lo, C. T. K., & Wong, T. T. (2023). Dual-wavelength low-cost high-resolution photoacoustic microscopy system for blood oxygenation imaging. In A. A. Oraevsky & L. V. Wang (Eds.), *Photons plus ultrasound: Imaging and sensing* (p. 143). SPIE. https://doi.org/10.1117/12.2647870
61. Zhou, Y., Liang, S., Li, M., Liu, C., Lai, P., & Wang, L. (2020). Optical-resolution photoacoustic microscopy with ultrafast dual-wavelength excitation. *Journal of Biophotonics, 13*, e201960229. https://doi.org/10.1002/jbio.201960229
62. Nteroli, G., Dasa, M. K., Messa, G., Koutsikou, S., Bondu, M., Moselund, P. M., et al. (2022). Two octaves spanning photoacoustic microscopy. *Scientific Reports, 12*, 10590. https://doi.org/10.1038/s41598-022-14869-5
63. Rao, B., Leng, X., Zeng, Y., Lin, Y., Chen, R., Zhou, Q., et al. (2019). Optical resolution photoacoustic microscopy of ovary and fallopian tube. *Scientific Reports, 9*, 14306. https://doi.org/10.1038/s41598-019-50743-7
64. Yao, J., & Wang, L. V. (2014). Sensitivity of photoacoustic microscopy. *Photoacoustics, 2*, 87–101. https://doi.org/10.1016/j.pacs.2014.04.002
65. Xie, Z., Chen, S.-L., Ling, T., Guo, L. J., Carson, P. L., & Wang, X. (2011). Pure optical photoacoustic microscopy. *Optics Express, 19*, 9027. https://doi.org/10.1364/OE.19.009027
66. Hu, G., Hu, Y., Wang, R., Yu, X., & Song, X. (2023). Enhancement of the depth of field of photoacoustic microscopy based on deep learning. In V. V. Tuchin, Q. Luo, & L. V. Wang (Eds.), *Sixteenth international conference on photonics and imaging in biology and medicine (PIBM 2023)* (p. 37). SPIE. https://doi.org/10.1117/12.2683129
67. Han, S., Kye, H., Kim, C.-S., Kim, T.-K., Yoo, J., & Kim, J. (2023). Automated laser-fiber coupling module for optical-resolution photoacoustic microscopy. *Sensors, 23*, 6643. https://doi.org/10.3390/s23146643
68. Aytac-Kipergil, E., Demirkiran, A., Uluc, N., Yavas, S., Kayikcioglu, T., Salman, S., et al. (2016). Development of a fiber laser with independently adjustable properties for optical resolution photoacoustic microscopy. *Scientific Reports, 6*, 38674. https://doi.org/10.1038/srep38674
69. Gong, W., Cao, Y., Wang, Y., & Song, X. (2024). Denoising for optical-resolution photoacoustic microscopy via a score-based diffusion generative model. In Y. Yue & G. Mendes Pacheco (Eds.), *International academic conference on optics and photonics (IACOP 2023)* (p. 17). SPIE. https://doi.org/10.1117/12.3022561
70. Lin, L., Hu, P., Shi, J., Appleton, C. M., Maslov, K., Li, L., et al. (2018). Single-breath-hold photoacoustic computed tomography of the breast. *Nature Communications, 9*, 2352. https://doi.org/10.1038/s41467-018-04576-z
71. Kim, J., Lee, J., Choi, S., Lee, H., Yang, J., Jeon, H., et al. (2024). 3D multiparametric photoacoustic computed tomography of primary and metastatic tumors in living mice. *ACS Nano, 18*, 18176–18190. https://doi.org/10.1021/acsnano.3c12551
72. Lin, L., Tong, X., Hu, P., Invernizzi, M., Lai, L., & Wang, L. V. (2021). Photoacoustic computed tomography of breast cancer in response to neoadjuvant chemotherapy. *Advanced Science, 8*, 2003396. https://doi.org/10.1002/advs.202003396

73. Bao, Y., Deng, H., Wang, X., Zuo, H., & Ma, C. (2021). Development of a digital breast phantom for photoacoustic computed tomography. *Biomedical Optics Express, 12*, 1391. https://doi.org/10.1364/BOE.416406
74. Zhang, L., Li, M., Zhou, Q., Dang, M., Tang, Y., Wang, S., et al. (2020). Computed tomography and photoacoustic imaging guided photodynamic therapy against breast cancer based on mesoporous platinum with insitu oxygen generation ability. *Acta Pharmaceutica Sinica B, 10*, 1719–1729. https://doi.org/10.1016/j.apsb.2020.05.003
75. Spreinat, A., Selvaggio, G., Erpenbeck, L., & Kruss, S. (2020). Multispectral near infrared absorption imaging for histology of skin cancer. *Journal of Biophotonics, 13*, 201960080. https://doi.org/10.1002/jbio.201960080
76. Godefroy, G., Arnal, B., & Bossy, E. (2023). Full-visibility 3D imaging of oxygenation and blood flow by simultaneous multispectral photoacoustic fluctuation imaging (MS-PAFI) and ultrasound Doppler. *Scientific Reports, 13*, 2961. https://doi.org/10.1038/s41598-023-29177-9
77. Liu, Y., Gao, D., Xu, M., & Yuan, Z. (2019). Multispectral photoacoustic imaging of cancer with broadband CuS nanoparticles covering both near-infrared I and II biological windows. *Journal of Biophotonics, 12*, 201800237. https://doi.org/10.1002/jbio.201800237
78. Dolet, A., Varray, F., Mure, S., Grenier, T., Liu, Y., Yuan, Z., et al. (2018). Spatial and spectral regularization to discriminate tissues using multispectral photoacoustic imaging. *EURASIP Journal on Advances in Signal Processing, 2018*, 39. https://doi.org/10.1186/s13634-018-0554-8
79. Jnawali, K., Chinni, B., Dogra, V., & Rao, N. (2020). Automatic cancer tissue detection using multispectral photoacoustic imaging. *International Journal of Computer Assisted Radiology and Surgery, 15*, 309–320. https://doi.org/10.1007/s11548-019-02101-1
80. Zhang, J., Arroyo, J., & Lediju Bell, M. A. (2025). Multispectral photoacoustic imaging of breast cancer tissue with histopathology validation. *Biomedical Optics Express, 16*, 995. https://doi.org/10.1364/BOE.547262
81. Wang, F., Yang, Z., Peng, W., Song, L., Luo, Y., Zhao, Z., et al. (2024). RPCA-based thermoacoustic imaging for microwave ablation monitoring. *Photoacoustics, 38*, 100622. https://doi.org/10.1016/j.pacs.2024.100622
82. Wu, D., Huang, L., Jiang, M., & Jiang, H. (2014). Contrast agents for photoacoustic and thermoacoustic imaging: A review. *International Journal of Molecular Sciences, 15*, 23616–23639. https://doi.org/10.3390/ijms151223616
83. Ji, Z., Fu, Y., & Yang, S. (2013). Microwave-induced thermoacoustic imaging for early breast cancer detection. *Journal of Innovative Optical Health Sciences, 6*, 1350001. https://doi.org/10.1142/S1793545813500016
84. Huang, L., Yao, L., Liu, L., Rong, J., & Jiang, H. (2012). Quantitative thermoacoustic tomography: Recovery of conductivity maps of heterogeneous media. *Applied Physics Letters, 101*, 24. https://doi.org/10.1063/1.4772484
85. Huang, Y.-J., Feng, Q.-K.-Y., & Huang, L. (2023). Thermoacoustic imaging: From single-element scanning to portable and array-based imaging. Journal of electronic. *Science and Technology, 21*, 100209. https://doi.org/10.1016/j.jnlest.2023.100209
86. Song, J., Zhao, Z., Wu, J., Zhu, X., Nie, Z., & Liu, Q.-H. (2013). Thermo-acoustic imaging for different breast tissues in microwave induced thermo-acoustic tomography system. In *2013 IEEE antennas and propagation society international symposium (APSURSI)* (pp. 2032–2033). IEEE. https://doi.org/10.1109/APS.2013.6711674
87. Zhao, Y., Ji, Z., Qin, B., & Xing, D. (2018). A thermoacoustic imaging system with variable curvature and multi-dimensional detection adapted to breast tumor screening. *Journal of Applied Physics, 124*, 144902. https://doi.org/10.1063/1.5042121
88. Lipson, E. J., Velculescu, V. E., Pritchard, T. S., Sausen, M., Pardoll, D. M., Topalian, S. L., et al. (2014). Circulating tumor DNA analysis as a real-time method for monitoring tumor burden in melanoma patients undergoing treatment with immune checkpoint blockade. *Journal for Immunotherapy of Cancer, 2*, 42. https://doi.org/10.1186/s40425-014-0042-0

89. George, T. J., Ogunwobi, O. O., Sheng, W., Fan, Z. H., & Liu, C. (2014). "Tissue is the issue": Circulating tumor cells in pancreatic cancer. *Journal of Gastrointestinal Cancer, 45*, 222–225. https://doi.org/10.1007/s12029-014-9638-3
90. Schott, D. S., Pizon, M., Schill, E., & Pachmann, K. (2021). How circulating cancer cells disguise: The role of platelets. *Journal of Clinical Oncology, 40*, e15011. https://doi.org/10.21203/rs.3.rs-892511/v1
91. Menyaev, Y. A., Nedosekin, D. A., Sarimollaoglu, M., Juratli, M. A., Galanzha, E. I., Tuchin, V. V., et al. (2013). Optical clearing in photoacoustic flow cytometry. *Biomedical Optics Express, 4*, 3030. https://doi.org/10.1364/BOE.4.003030
92. Edgar, R. H., Tarhini, A., Sander, C., Sanders, M. E., Cook, J. L., & Viator, J. A. (2021). Predicting metastasis in melanoma by enumerating circulating tumor cells using photoacoustic flow cytometry. *Lasers in Surgery and Medicine, 53*, 578–586. https://doi.org/10.1002/lsm.23286
93. Vora, N., Shekhar, P., Esmail, M., Patra, A., & Georgakoudi, I. (2022). Label-free flow cytometry of rare circulating tumor cell clusters in whole blood. *Scientific Reports, 12*, 10721. https://doi.org/10.1038/s41598-022-14003-5
94. Nolan, J., Sarimollaoglu, M., Nedosekin, D. A., Jamshidi-Parsian, A., Galanzha, E. I., Kore, R. A., et al. (2016). In vivo flow cytometry of circulating tumor-associated exosomes. *Analytical Cellular Pathology, 2016*, 1–12. https://doi.org/10.1155/2016/1628057
95. Galanzha, E. I., & Zharov, V. P. (2012). Photoacoustic flow cytometry. *Methods, 57*, 280–296. https://doi.org/10.1016/j.ymeth.2012.06.009
96. He, Y., Wang, L., Shi, J., Yao, J., Li, L., Zhang, R., et al. (2016). In vivo label-free photoacoustic flow cytography and on-the-spot laser killing of single circulating melanoma cells. *Scientific Reports, 6*, 39616. https://doi.org/10.1038/srep39616
97. Nedosekin, D. A., Sarimollaoglu, M., Ye, J., Galanzha, E. I., & Zharov, V. P. (2011). In vivo ultra-fast photoacoustic flow cytometry of circulating human melanoma cells using near-infrared high-pulse rate lasers. *Cytometry Part A, 79A*, 825–833. https://doi.org/10.1002/cyto.a.21102
98. López-Riquelme, N., Minguela, A., Villar-Permuy, F., Ciprian, D., Castillejo, A., Álvarez-López, M., et al. (2013). Imaging cytometry for counting circulating tumor cells: Comparative analysis of the cellsearch vs imagestream systems. *APMIS, 121*, 1139–1143. https://doi.org/10.1111/apm.12061
99. Han, M., Watts, J. A., Jamshidi-Parsian, A., Nadeem, U., Sarimollaoglu, M., Siegel, E. R., et al. (2020). In vivo lymphatic circulating tumor cells and progression of metastatic disease. *Cancers, 12*, 2866. https://doi.org/10.3390/cancers12102866
100. Hai, P., Qu, Y., Li, Y., Zhu, L., Shmuylovich, L., Cornelius, L. A., et al. (2020). Label-free high-throughput photoacoustic tomography of suspected circulating melanoma tumor cells in patients in vivo. *Journal of Biomedical Optics, 25*, 1. https://doi.org/10.1117/1.JBO.25.3.036002
101. Suo, Y., Liu, T., Xie, C., Wei, D., Tan, X., Wu, L., et al. (2015). Near infrared in vivo flow cytometry for tracking fluorescent circulating cells. *Cytometry Part A, 87*, 878–884. https://doi.org/10.1002/cyto.a.22711
102. Hai, P., Zhou, Y., Zhang, R., Ma, J., Li, Y., & Wang, L. V. (2017). Linear-array-based photoacoustic tomography for label-free high-throughput detection and quantification of circulating melanoma tumor cell clusters. In *Photons plus ultrasound: Imaging and sensing* (p. 100641). SPIE. https://doi.org/10.1117/12.2252819
103. Sindeeva, O. A., Verkhovskii, R. A., Sarimollaoglu, M., Afanaseva, G. A., Fedonnikov, A. S., Osintsev, E. Y., et al. (2019). New frontiers in diagnosis and therapy of circulating tumor markers in cerebrospinal fluid in vitro and in vivo. *Cells, 8*, 1195. https://doi.org/10.3390/cells8101195
104. Sun, R. W., Tuchin, V. V., Zharov, V. P., Galanzha, E. I., & Richter, G. T. (2018). Current status, pitfalls and future directions in the diagnosis and therapy of lymphatic malformation. *Journal of Biophotonics, 11*, e201700124. https://doi.org/10.1002/jbio.201700124

105. Gnyawali, V., Strohm, E. M., Wang, J.-Z., Tsai, S. S. H., & Kolios, M. C. (2019). Simultaneous acoustic and photoacoustic microfluidic flow cytometry for label-free analysis. *Scientific Reports, 9*, 1585. https://doi.org/10.1038/s41598-018-37771-5
106. Arnold, M., Morgan, E., Rumgay, H., Mafra, A., Singh, D., Laversanne, M., et al. (2022). Current and future burden of breast cancer: Global statistics for 2020 and 2040. *The Breast, 66*, 15–23. https://doi.org/10.1016/j.breast.2022.08.010
107. Schoustra, S. M., De Santi, B., Opt Root, T. J., Klazen, C. A., van der Schaaf, M., Veltman, J., et al. (2023). Imaging breast malignancies with the twente photoacoustic mammoscope 2. *PLoS One, 18*, e0281434. https://doi.org/10.1371/journal.pone.0281434
108. Zhang, G., Li, W., Yang, M., & Li, C. (2020). Developing a photoacoustic whole-breast imaging system based on the synthetic matrix array. *Frontiers of Physics, 8*, 600589. https://doi.org/10.3389/fphy.2020.600589
109. Lin, L., Hu, P., Tong, X., Na, S., Cao, R., Yuan, X., et al. (2021). High-speed three-dimensional photoacoustic computed tomography for preclinical research and clinical translation. *Nature Communications, 12*, 882. https://doi.org/10.1038/s41467-021-21232-1
110. Zhang, R., Zhao, L., Zhao, C., Wang, M., Liu, S., Li, J., et al. (2021). Exploring the diagnostic value of photoacoustic imaging for breast cancer: The identification of regional photoacoustic signal differences of breast tumors. *Biomedical Optics Express, 12*, 1407. https://doi.org/10.1364/BOE.417056
111. Heijblom, M., Piras, D., van den Engh, F. M., van der Schaaf, M., Klaase, J. M., Steenbergen, W., et al. (2016). The state of the art in breast imaging using the twente photoacoustic mammoscope: Results from 31 measurements on malignancies. *European Radiology, 26*, 3874–3887. https://doi.org/10.1007/s00330-016-4240-7
112. Thakur, C., Qiu, Y., Fu, Y., Bi, Z., Zhang, W., Ji, H., et al. (2022). Epigenetics and environment in breast cancer: New paradigms for anti-cancer therapies. *Frontiers in Oncology, 12*, 971288. https://doi.org/10.3389/fonc.2022.971288
113. Westreich, J., Khorasani, M., Jones, B., Demidov, V., Nofech-Mozes, S., & Vitkin, A. (2019). Novel methodology to image stromal tissue and assess its morphological features with polarized light: Towards a tumour microenvironment prognostic signature. *Biomedical Optics Express, 10*, 3963. https://doi.org/10.1364/BOE.10.003963
114. Choi, E. J., Choi, H., Choi, S. A., & Youk, J. H. (2016). Dynamic contrast-enhanced breast magnetic resonance imaging for the prediction of early and late recurrences in breast cancer. *Medicine, 95*, e5330. https://doi.org/10.1097/MD.0000000000005330
115. Wilson, K. E., Bachawal, S. V., Tian, L., & Willmann, J. K. (2014). Multiparametric spectroscopic photoacoustic imaging of breast cancer development in a transgenic mouse model. *Theranostics, 4*, 1062–1071. https://doi.org/10.7150/thno.9922
116. Abeyakoon, O., Woitek, R., Wallis, M. G., Moyle, P. L., Morscher, S., Dahlhaus, N., et al. (2022). An optoacoustic imaging feature set to characterise blood vessels surrounding benign and malignant breast lesions. *Photoacoustics, 27*, 100383. https://doi.org/10.1016/j.pacs.2022.100383
117. Hinshaw, D. C., & Shevde, L. A. (2019). The tumor microenvironment innately modulates cancer progression. *Cancer Research, 79*(18), 4557–4566.
118. Schoustra, S. M., Huijink, R., Alink, L., Opt Root, T. J. P. M., Sprünken, D., Piras, D., et al. (2019). The twente photoacoustic mammoscope 2: 3D vascular network visualization. In A. A. Oraevsky & L. V. Wang (Eds.), *Photons plus ultrasound: Imaging and sensing* (p. 37). SPIE. https://doi.org/10.1117/12.2507448
119. He, R., Chen, Y., Jiang, Y., Lei, Y., Yan, S., Zhang, J., et al. (2024). Deep learning realizes photoacoustic imaging artifact removal. *Applied Sciences, 14*, 5161. https://doi.org/10.3390/app14125161
120. Schwab, J., Antholzer, S., Nuster, R., & Haltmeier, M. (2018). *Real-time photoacoustic projection imaging using deep learning*.
121. Allman, D., Reiter, A., & Bell, M. A. L. (2018). Photoacoustic source detection and reflection artifact removal enabled by deep learning. *IEEE Transactions on Medical Imaging, 37*, 1464–1477. https://doi.org/10.1109/TMI.2018.2829662

122. Kim, J., Kim, G., Li, L., Zhang, P., Kim, J. Y., Kim, Y., et al. (2022). Deep learning acceleration of multiscale superresolution localization photoacoustic imaging. *Light: Science & Applications, 11*, 131. https://doi.org/10.1038/s41377-022-00820-w
123. Sobhanan Warrier, G., Amirthalakshmi, T. M., Nimala, K., Thaj Mary Delsy, T., Stella Rose Malar, P., Ramkumar, G., et al. (2022). Automated recognition of cancer tissues through deep learning framework from the photoacoustic specimen. *Contrast Media & Molecular Imaging, 2022*, 4356744. https://doi.org/10.1155/2022/4356744
124. Li, G., Huang, Z., Tian, H., Wu, H., Zheng, J., Wang, M., et al. (2024). Deep learning combined with attention mechanisms to assist radiologists in enhancing breast cancer diagnosis: A study on photoacoustic imaging. *Biomedical Optics Express, 15*, 4689. https://doi.org/10.1364/BOE.530249
125. John, M., & Barhumi, I. (2025). Unrolled deep learning for breast cancer detection using limited-view photoacoustic tomography data. *Medical & Biological Engineering & Computing*. https://doi.org/10.1007/s11517-025-03302-4
126. Zhang, J., Chen, B., Zhou, M., Lan, H., & Gao, F. (2019). Photoacoustic image classification and segmentation of breast cancer: A feasibility study. *IEEE Access, 7*, 5457–5466. https://doi.org/10.1109/ACCESS.2018.2888910
127. Basal, L. A., Yan, Y., Shen, Y., Haacke, E. M., Mehrmohammadi, M., & Allen, M. J. (2017). Oxidation-responsive, $Eu^{II/III}$-based, multimodal contrast agent for magnetic resonance and photoacoustic imaging. *ACS Omega, 2*, 800–805. https://doi.org/10.1021/acsomega.6b00514
128. Wang, Z., Yang, F., Ma, H., Cheng, Z., Zhang, W., Xiong, K., et al. (2022). Bifocal 532/1064 nm alternately illuminated photoacoustic microscopy for capturing deep vascular morphology in human skin. *Journal of the European Academy of Dermatology and Venereology, 36*, 51–59. https://doi.org/10.1111/jdv.17677
129. Kim, J., Heo, D., Park, J., Cho, S., Ahn, J., Kim, J., et al. (2023). High-resolution photoacoustic and ultrasound endoscope based on the transparent ultrasound transducer. In A. A. Oraevsky & L. V. Wang (Eds.), *Photons plus ultrasound: Imaging and sensing* (p. 15). SPIE. https://doi.org/10.1117/12.2650804
130. Song, X., Wang, X., Yu, S., Cao, J., Li, S., Li, J., et al. (2015). $Co_9Se_8$ nanoplates as a new theranostic platform for photoacoustic/magnetic resonance dual-modal-imaging-guided chemo-photothermal combination therapy. *Advanced Materials, 27*, 3285–3291. https://doi.org/10.1002/adma.201405634
131. Liang, X., Li, Y., Li, X., Jing, L., Deng, Z., Yue, X., et al. (2015). PEGylated polypyrrole nanoparticles conjugating gadolinium chelates for dual-modal MRI/photoacoustic imaging guided photothermal therapy of cancer. *Advanced Functional Materials, 25*, 1451–1462. https://doi.org/10.1002/adfm.201402338
132. Goh, Y., Balasundaram, G., Tan, H. M., Putti, T. C., Renzhe, B., Hartman, M., et al. (2024). Ultrasound-guided photoacoustic (US-PA) tomography of the breast: Biochemical differentiation using intrinsic tissue markers—Lipids, collagen and hemoglobin with histopathologic correlation. *Scientific Reports, 14*, 18054. https://doi.org/10.1038/s41598-024-65114-0
133. Jiang, D., Shi, Y., Chen, X., Wang, M., & Song, Z. (2017). Fast and robust multimodal image registration using a local derivative pattern. *Medical Physics, 44*, 497–509. https://doi.org/10.1002/mp.12049
134. Pan, C., Hou, S., Wan, Q., Hu, G., Zeng, J., Zhong, W., et al. (2023). Effect of center frequency and bandwidth of ultrasonic transducer on photoacoustic tomography based on virtual PAT. In V. V. Tuchin, Q. Luo, & L. V. Wang (Eds.), *Sixteenth international conference on photonics and imaging in biology and medicine (PIBM 2023)* (p. 39). SPIE. https://doi.org/10.1117/12.2683131
135. Li, M., Tang, Y., & Yao, J. (2018). Photoacoustic tomography of blood oxygenation: A mini review. *Photoacoustics, 10*, 65–73. https://doi.org/10.1016/j.pacs.2018.05.001
136. Jiang, M., Lei, S., Zhang, J., Hou, L., Zhang, M., & Luo, Y. (2022). Multimodal imaging of target detection algorithm under artificial intelligence in the diagnosis of early breast cancer. *Journal of Healthcare Engineering, 2022*, 1–10. https://doi.org/10.1155/2022/9322937

137. Pfob, A., Sidey-Gibbons, C., Barr, R. G., Duda, V., Alwafai, Z., Balleyguier, C., et al. (2022). The importance of multi-modal imaging and clinical information for humans and AI-based algorithms to classify breast masses (INSPiRED 003): An international, multicenter analysis. *European Radiology, 32*, 4101–4115. https://doi.org/10.1007/s00330-021-08519-z
138. Mallidi, S., Luke, G. P., & Emelianov, S. (2011). Photoacoustic imaging in cancer detection, diagnosis, and treatment guidance. *Trends in Biotechnology, 29*, 213–221. https://doi.org/10.1016/j.tibtech.2011.01.006
139. Cormode, D. P., Jarzyna, P. A., Mulder, W. J. M., & Fayad, Z. A. (2010). Modified natural nanoparticles as contrast agents for medical imaging. *Advanced Drug Delivery Reviews, 62*, 329–338. https://doi.org/10.1016/j.addr.2009.11.005
140. Silvestri, A., Zambelli, V., Ferretti, A. M., Salerno, D., Bellani, G., & Polito, L. (2016). Design of functionalized gold nanoparticle probes for computed tomography imaging. *Contrast Media & Molecular Imaging, 11*, 405–414. https://doi.org/10.1002/cmmi.1704
141. Anderson, D., Anderson, T., & Fahmi, F. (2019). Advances in applications of metal oxide nanomaterials as imaging contrast agents. *Physica Status Solidi (a), 216*, 1801008. https://doi.org/10.1002/pssa.201801008
142. Hsu, J. C., Nieves, L. M., Betzer, O., Sadan, T., Noël, P. B., Popovtzer, R., et al. (2020). Nanoparticle contrast agents for X-ray imaging applications. *WIREs Nanomedicine and Nanobiotechnology, 12*, e1642. https://doi.org/10.1002/wnan.1642
143. Biswas, P., Patra, B., & Verma, R. S. (2021). Nitric oxide increases mitomycin-C cytotoxicity by generation of reactive oxygen species and induces cell death by apoptosis in $FancC^{-/-}$ cells. *Annals of Hematology & Oncology, 8*, 1360. https://doi.org/10.26420/annhematoloncol.2021.1360
144. Reinhardt, C. J., Zhou, E. Y., Jorgensen, M. D., Partipilo, G., & Chan, J. (2018). A ratiometric acoustogenic probe for *in vivo* imaging of endogenous nitric oxide. *Journal of the American Chemical Society, 140*, 1011–1018. https://doi.org/10.1021/jacs.7b10783
145. Reinhardt, C. J., Xu, R., & Chan, J. (2020). Nitric oxide imaging in cancer enabled by steric relaxation of a photoacoustic probe platform. *Chemical Science, 11*, 1587–1592. https://doi.org/10.1039/C9SC05600A
146. Li, L., Patil, D., Petruncio, G., Harnden, K. K., Somasekharan, J. V., Paige, M., et al. (2021). Integration of multitargeted polymer-based contrast agents with photoacoustic computed tomography: An imaging technique to visualize breast cancer intratumor heterogeneity. *ACS Nano, 15*, 2413–2427. https://doi.org/10.1021/acsnano.0c05893
147. Yang, C.-W., Liu, K., Yao, C.-Y., Li, B., Juhong, A., Qiu, Z., et al. (2022). Indocyanine green-conjugated superparamagnetic iron oxide nanoworm for multimodality breast cancer imaging. *ACS Applied Nano Materials, 5*, 18912–18920. https://doi.org/10.1021/acsanm.2c04687
148. Xiao, T. G., Weis, J. A., Gayzik, F. S., Thomas, A., Chiba, A., Gurcan, M. N., et al. (2018). Applying dynamic contrast enhanced MSOT imaging to intratumoral pharmacokinetic modeling. *Photoacoustics, 11*, 28–35. https://doi.org/10.1016/j.pacs.2018.07.003
149. Wang, L. V., & Yao, J. (2016). A practical guide to photoacoustic tomography in the life sciences. *Nature Methods, 13*, 627–638. https://doi.org/10.1038/nmeth.3925
150. Palma-Chavez, J., Pfefer, T. J., Agrawal, A., Jokerst, J. V., & Vogt, W. C. (2021). Review of consensus test methods in medical imaging and current practices in photoacoustic image quality assessment. *Journal of Biomedical Optics, 26*, 090901. https://doi.org/10.1117/1.JBO.26.9.090901
151. Attia, A. B. E., Balasundaram, G., Moothanchery, M., Dinish, U. S., Bi, R., Ntziachristos, V., et al. (2019). A review of clinical photoacoustic imaging: Current and future trends. *Photoacoustics, 16*, 100144. https://doi.org/10.1016/j.pacs.2019.100144
152. Kim, C., Favazza, C., & Wang, L. V. (2010). In vivo photoacoustic tomography of chemicals: High-resolution functional and molecular optical imaging at new depths. *Chemical Reviews, 110*, 2756–2782. https://doi.org/10.1021/cr900266s
153. Cox, B., Laufer, J. G., Arridge, S. R., & Beard, P. C. (2012). Quantitative spectroscopic photoacoustic imaging: A review. *Journal of Biomedical Optics, 17*, 061202. https://doi.org/10.1117/1.JBO.17.6.061202

154. Galbusera, F., & Cina, A. (2024). Image annotation and curation in radiology: An overview for machine learning practitioners. *European Radiology Experimental, 8*, 11. https://doi.org/10.1186/s41747-023-00408-y
155. Lee, H., Han, S., Kye, H., Kim, T.-K., Choi, W., & Kim, J. (2023). A review on the roles of photoacoustic imaging for conventional and novel clinical diagnostic applications. *Photonics, 10*, 904. https://doi.org/10.3390/photonics10080904
156. Heijblom, M., Piras, D., Xia, W., van Hespen, J. C. G., Klaase, J. M., van den Engh, F. M., et al. (2012). Visualizing breast cancer using the twente photoacoustic mammoscope: What do we learn from twelve new patient measurements? *Optics Express, 20*, 11582. https://doi.org/10.1364/OE.20.011582
157. Tsai, W.-Y., Breimann, S., Shen, T.-W., & Frishman, D. (2023). Photoacoustic and absorption spectroscopy imaging analysis of human blood. *PLoS One, 18*, e0289704. https://doi.org/10.1371/journal.pone.0289704

# Correction to: Quantum Dot Facilitated Cancer Diagnosis and Treatment

**Debasmita Paul, Brajadulal Chattopadhyay, Kunal Biswas, and Abiral Tamang**

## Correction to: Chapter 4 in: N. Mazumder et al. (eds.), *Advanced Biophysical Techniques in Biosciences*, Lecture Notes in Nanoscale Science and Technology 7, https://doi.org/10.1007/978-3-031-94551-9_4

In chapter 4 "Quantum Dot Facilitated Cancer Diagnosis and Treatment", the affiliation "Centre for Nanoscience and Nanotechnology, International Research Centre, Sathyabama Institute of Science and Technology, Chennai, India" of the authors "Debasmita Paul and Brajadulal Chattopadhyay" has been changed to "Department of Physics, Jadavpur University, Kolkata, India".

---

The updated version of this chapter can be found at https://doi.org/10.1007/978-3-031-94551-9_4



N. Mazumder et al. (eds.), *Advanced Biophysical Techniques in Biosciences*, Lecture Notes in Nanoscale Science and Technology 7,
https://doi.org/10.1007/978-3-031-94551-9_17

# Index




N. Mazumder et al. (eds.), *Advanced Biophysical Techniques in Biosciences*, Lecture Notes in Nanoscale Science and Technology 7,
https://doi.org/10.1007/978-3-031-94551-9

MIX
Papier aus verantwortungsvollen Quellen
Paper from responsible sources
FSC® C105338

If you have any concerns about our products,
you can contact us on
**ProductSafety@springernature.com**

In case Publisher is established outside the EU,
the EU authorized representative is:
**Springer Nature Customer Service Center GmbH**
**Europaplatz 3, 69115 Heidelberg, Germany**

Printed by Libri Plureos GmbH
in Hamburg, Germany